AF333699

# Palaeomagnetism in Fold and Thrust Belts: New Perspectives

The Geological Society of London
**Books Editorial Committee**

**Chief Editor**
RICK LAW (USA)

**Society Books Editors**
JIM GRIFFITHS (UK)
DAVE HODGSON (UK)
PHIL LEAT (UK)
NICK RICHARDSON (UK)
DANIELA SCHMIDT (UK)
RANDELL STEPHENSON (UK)
ROB STRACHAN (UK)
MARK WHITEMAN (UK)

**Society Books Advisors**
GHULAM BHAT (India)
MARIE-FRANÇOISE BRUNET (France)
ANNE-CHRISTINE DA SILVA (Belgium)
JASPER KNIGHT (South Africa)
MARIO PARISE (Italy)
SATISH-KUMAR (Japan)
VIRGINIA TOY (New Zealand)
MARCO VECOLI (Saudi Arabia)

## Geological Society books refereeing procedures

The Society makes every effort to ensure that the scientific and production quality of its books matches that of its journals. Since 1997, all book proposals have been refereed by specialist reviewers as well as by the Society's Books Editorial Committee. If the referees identify weaknesses in the proposal, these must be addressed before the proposal is accepted.

Once the book is accepted, the Society Book Editors ensure that the volume editors follow strict guidelines on refereeing and quality control. We insist that individual papers can only be accepted after satisfactory review by two independent referees. The questions on the review forms are similar to those for *Journal of the Geological Society*. The referees' forms and comments must be available to the Society's Book Editors on request.

Although many of the books result from meetings, the editors are expected to commission papers that were not presented at the meeting to ensure that the book provides a balanced coverage of the subject. Being accepted for presentation at the meeting does not guarantee inclusion in the book.

More information about submitting a proposal and producing a book for the Society can be found on its website: www.geolsoc.org.uk.

It is recommended that reference to all or part of this book should be made in one of the following ways:

PUEYO, E. L., CIFELLI, F., SUSSMAN, A. J. & OLIVA-URCIA, B. (eds) 2016. *Palaeomagnetism in Fold and Thrust Belts: New Perspectives*. Geological Society, London, Special Publications, **425**.

MÁRTON, E., GRABOWSKI, J., TOKARSKI, A. K. & TÚNYI, I. 2016. Palaeomagnetic results from the fold and thrust belt of the Western Carpathians: an overview. *In:* PUEYO, E. L., CIFELLI, F., SUSSMAN, A. J. & OLIVA-URCIA, B. (eds) *Palaeomagnetism in Fold and Thrust Belts: New Perspectives*. Geological Society, London, Special Publications, **425**, 7–36. First published online August 12, 2015, http://dx.doi.org/10.1144/SP425.1

GEOLOGICAL SOCIETY SPECIAL PUBLICATION NO. 425

# Palaeomagnetism in Fold and Thrust Belts: New Perspectives

EDITED BY

**E. L. PUEYO**
IGME, Spain

**F. CIFELLI**
Università degli Studi Roma TRE, Italy

**A. J. SUSSMAN**
Los Alamos National Laboratory, USA

and

**B. OLIVA-URCIA**
Universidad Autónoma de Madrid, Spain

2016
Published by
The Geological Society
London

# THE GEOLOGICAL SOCIETY

The Geological Society of London (GSL) was founded in 1807. It is the oldest national geological society in the world and the largest in Europe. It was incorporated under Royal Charter in 1825 and is Registered Charity 210161.

The Society is the UK national learned and professional society for geology with a worldwide Fellowship (FGS) of over 10 000. The Society has the power to confer Chartered status on suitably qualified Fellows, and about 2000 of the Fellowship carry the title (CGeol). Chartered Geologists may also obtain the equivalent European title, European Geologist (EurGeol). One fifth of the Society's fellowship resides outside the UK. To find out more about the Society, log on to www.geolsoc.org.uk.

**The Geological Society Publishing House** (Bath, UK) produces the Society's international journals and books, and acts as European distributor for selected publications of the American Association of Petroleum Geologists (AAPG), the Indonesian Petroleum Association (IPA), the Geological Society of America (GSA), the Society for Sedimentary Geology (SEPM) and the Geologists' Association (GA). Joint marketing agreements ensure that GSL Fellows may purchase these societies' publications at a discount. The Society's online bookshop (accessible from www.geolsoc.org.uk) offers secure book purchasing with your credit or debit card.

To find out about joining the Society and benefiting from substantial discounts on publications of GSL and other societies worldwide, consult www.geolsoc.org.uk, or contact the Fellowship Department at: The Geological Society, Burlington House, Piccadilly, London W1J 0BG: Tel. +44 (0)20 7434 9944; Fax +44 (0)20 7439 8975; E-mail: enquiries@geolsoc.org.uk.

For information about the Society's meetings, consult *Events* on www.geolsoc.org.uk. To find out more about the Society's Corporate Affiliates Scheme, write to enquiries@geolsoc.org.uk.

Published by The Geological Society from:
The Geological Society Publishing House, Unit 7, Brassmill Enterprise Centre, Brassmill Lane, Bath BA1 3JN, UK

The Lyell Collection: www.lyellcollection.org
Online bookshop: www.geolsoc.org.uk/bookshop
Orders: Tel. +44 (0)1225 445046, Fax +44 (0)1225 442836

The publishers make no representation, express or implied, with regard to the accuracy of the information contained in this book and cannot accept any legal responsibility for any errors or omissions that may be made.



**British Library Cataloguing in Publication Data**

A catalogue record for this book is available from the British Library.
ISBN 978-1-86239-737-8
ISSN 0305-8719

**Distributors**
For details of international agents and distributors see:
www.geolsoc.org.uk/agentsdistributors

Typeset by Nova Techset Private Limited, Bengaluru & Chennai, India
Printed and bound by CPI Group (UK) Ltd, Croydon CR0 4YY

# Contents

# Foreword

Palaeomagnetism provides a unique method to define vertical axis rotations associated with fold-and-thrust belt evolution; for this reason it has been successfully used in most of the ancient and recent fold-and-thrust belts of the world. Palaeomagnetic research in the Mediterranean, Appalachians, Rocky Mountains, Andean and Cantabrian orogens has been fundamental for understanding their tectonic evolution through time and for comprehending the limits and difficulties in applying palaeomagnetic techniques to orogenic belts.

This volume includes contributions that explore the potential of up-to-date palaeomagnetic techniques to define fold-and-thrust belt kinematics, deformation mechanisms and tectonic evolution through time. The range of topics in these 13 papers is broad, and covers most of the important topics of palaeomagnetism applied to the study of orogens at different scales. They include the causes of orogenic curvature, precise definition of the age and magnitude of vertical axis rotations in fold–and-thrust belts and their distribution in space and time; the use of magnetostratigraphy to constrain the timing of deformation of regional tectonic structures; the integration of structural analyses (including AMS) and palaeomagnetism to reconstruct the entire tectonic history of inverted sedimentary basins. Furthermore, several papers are dedicated to some of the main issues in palaeomagnetic research applied to tectonics: magnetic overprinting; complex deformation mechanisms that enable the application of 'simple' bedding corrections; and the necessity to use 3D restoration to reconstruct tectonic deformation. This book should stimulate new perspectives in the field as it demonstrates that some of traditional issues in palaeomagnetic research can be used to better understand the way rocks acquire their magnetization and how orogenic belts deform during their evolution.

MASSIMO MATTEI
April 2016

# Acknowledgements

Aleksandra Abrazhevich (Australian Nacional University, Australia)

Bjarne Almqvist (Uppsala Universitet, Sweden)

Cesar Arriagada (Universidad de Chile, Chile)

Mikhail Bazhenov[†] (Russian Academy of Sciences, Russia)

Kurtis Burmeister (University of the Pacific, California, USA)

Nuria Carrera (Universitat de Barcelona)

Antonio M. Casas (Universidad de Zaragoza, Spain)

Ian Dalziel (The University of Texas at Dallas)

John Geissman (The University of Texas at Dallas)

Ann Hirt (Eidgenössische Technische Hochschule Zürich, Switzerland)

Luca Lanci (Urbino University, Italy)

Juan Cruz Larrasoaña (Inst. Geológico y Minero de España, Spain)

Adolfo Maestro (Inst. Geológico y Minero de España, Spain)

Samantha Nemkin-Wilson (University of Michigan, USA)

Oscar Pueyo Anchuela (Universidad de Zaragoza, Spain)

Robert Scholger (Montanuniversität Leoben, Austria)

Ruth Soto (Inst. Geológico y Minero de España, Spain)

Fabio Speranza (Istituto Nazionale di Geofisica e Vulcanologia, Roma, Italy)

Rob Van der Voo (University of Michigan, USA)

Arlo Weil (Bryn Mawr College, Pennsilvania, USA)

Maodu Yan (Institute of Tibetan Plateau Research, China)

and some anonymous reviewers.

# Introduction: Palaeomagnetism in fold and thrust belts: new perspectives


E. L. PUEYO[1]*, F. CIFELLI[2], A. J. SUSSMAN[3] & B. OLIVA-URCIA[4]

[1]*Instituto Geológico y Minero de España, Unidad de Zaragoza c/Manuel Lasala 44, 9°, 50006 Zaragoza, Spain*

[2]*Dipartimento Scienze, Università degli Studi di Roma TRE, Largo San Leonardo Murialdo 1, 00146 Rome, Italy*

[3]*Earth and Environmental Sciences Division, MS-D452, Los Alamos National Laboratory, Los Alamos, NM 87545, USA*

[4]*Departamento Geología y Geoquímica Facultad de Ciencias, Universidad Autónoma de Madrid, Campus de Cantoblanco C/Francisco Tomás y Valiente, 7, M-06, 6ª planta, 28049 Madrid, Spain*

**Corresponding author (e-mail: unaim@igme.es)*




Palaeomagnetism, that is, the study of the ancient magnetic field recorded in rocks, is the only vectorial indicator in the Earth sciences that is capable of associating geological bodies with their original location (primary vectors) or with intermediate locations (secondary vectors) during their geological history. For this reason, palaeomagnetism has played a key role in supporting continental drift theory.

Beyond tectonic plate-scale applications, palaeomagnetism has become a fundamental tool for assessing the evolution of mountain ranges owing to its unique potential for quantifying vertical axis rotations (VAR). Since the pioneering applications of authors such as Norris & Black (1961) and Tarling (1969), palaeomagnetism has been applied to problems at a variety of scales in many orogenic systems (e.g. Elredge *et al.* 1985; Kissel & Laj 1989; Weil & Sussman 2004; Elmore *et al.* 2012). In particular, palaeomagnetic data have been increasingly used as key quantitative information for determining the timing, distribution and magnitude of vertical axis rotations (Van der Voo & Channell 1980; McCaig & McClelland 1992; Allerton 1998).

Together with structural analyses, palaeomagnetic investigations help to unravel the deformation history of fold and thrust belts, For instance, palaeomagnetism in growth strata has shed much light not only on the kinematics of folding and thrusting but also on the deformation timing such that quantitative geological reconstructions can be accomplished even in four dimensions (Pueyo *et al.* 2004; Arriagada *et al.* 2008; Ramón *et al.* 2012). These approaches reduce the levels of uncertainty of other structural variables, such as shortening and internal deformation.

Overall, this Geological Society Special Publication compiles contributions presenting recent developments in palaeomagnetism, in terms of both methodology and applications of palaeomagnetic studies to fold and thrust belts. The scientific breadth of this volume includes:

- compilations and overviews of fold and thrust belts with abundant palaeomagnetic data;
- case studies showing how palaeomagnetic data are used in understanding the kinematic evolution of geological structures at different scales;
- case studies showing how magnetostratigraphy contributes to the assessment of the magnitude and the timing of vertical axis rotations and deformation rates;
- studies of complex tectonic structures, including the integration of palaeomagnetism with anisotropy of magnetic susceptibility (AMS) and/or structural analyses; and
- analyses pertaining to data reliability, resolution analysis and error sources.

The volume starts with two papers that present large palaeomagnetic datasets (Fig. 1); the Western Carpathians (**Márton *et al.* 2015**) and the Western and Central Mediterranean arcs (**Cifelli *et al.* 2016**). In both cases, synthetic and robust palaeomagnetically deduced VARs are used to constrain the kinematic and geodynamic evolution in these regions.

The second part of the volume presents a suite of papers focused on different structural and tectonic


*From*: Pueyo, E. L., Cifelli, F., Sussman, A. J. & Oliva-Urcia, B. (eds) 2016. *Palaeomagnetism in Fold and Thrust Belts: New Perspectives*. Geological Society, London, Special Publications, **425**, 1–6.
First published online July 11, 2016, http://doi.org/10.1144/SP425.13
© 2016 The Author(s). Published by The Geological Society of London.
Publishing disclaimer: www.geolsoc.org.uk/pub_ethics

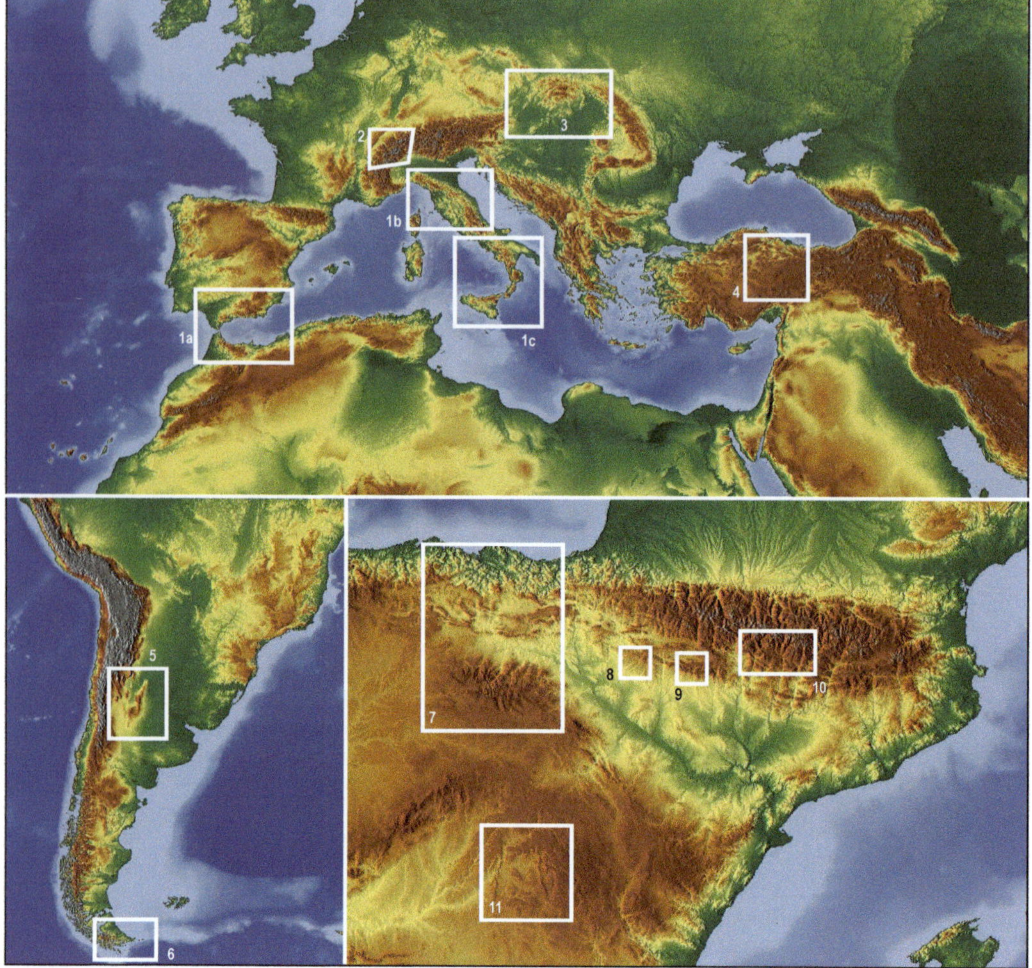

**Fig. 1.** Topographic map with the locations of the study area comprising this volume: **1a**, Betics and Rif arc; **1b**, Apeninnes; **1c**, Calabrian arc (Cifelli *et al.* 2016); **2**, Alps (Cardello *et al.* 2015); **3**, Western Carpathians (Márton *et al.* 2015); **4**, Central Anatolia (Cinku *et al.* 2015); **5**, Pampean Ranges (Japas *et al.* 2015); **6**, Patagonia (Rapalini *et al.* 2015); **7**, Polientes, Cabuerniga and Cameros basins (Villalaín *et al.* 2015); **8**, Western Pyrenean External Sierras (Oliva-Urcia *et al.* 2015); **9**, Pico del Aguila Anticline (Anastasio *et al.* 2015); **10**, Cotiella (Garcés *et al.* 2015); **11**, Altomira Range (Valcárcel *et al.* 2015). Source map: http://www.maps-for-free.com (licensed under the Creative Commons Attribution-Share Alike 2.0 Generic licence.).

problems in the Andes, Pyrenees, the Iberian Range, the Alps and Anatolia. **Rapalini *et al.* (2015)** show new data focused on unravelling the primary or secondary nature of the Patagonian orocline, **Japas *et al.* (2015)** discuss Neogene rotations measured in the Andean Precordillera fold and thrust belt. **Oliva-Urcia *et al.* (2015)** present a new magneto-stratigraphic section from the southwestern portion of the Pyrenean main thrust that allows for both the refinement of the chronostratigraphy of the region and dating of primary folding and thrusting events. The Pico del Aguila anticline, in the central

part of the basal thrust, is the target of the paper by **Anastasio *et al.* (2015)**, which uses AMS and published chronostratigraphy to decipher fold kinematics. The work by **Cardello *et al.* (2015)** in the Helvetic Alps and that by **Valcárcel *et al.* (2015)** in the Southern Iberian Range share some common features in that they represent detailed studies of individual structures and aim to obtain well-constrained kinematic models. These papers both integrate their palaeomagnetic data with structural, tectonic and AMS data, and bracket the limitations of their palaeomagnetic analyses. **Cinku *et al.***

**(2015)** investigate the palaeomagnetic rotations that occurred in the Kırşehir block during the closure of the remnant of the northern branch of the Neotethys Ocean, in the İzmir–Ankara–Erzincan suture zone, from Late Cretaceous to Middle Eocene times.

The third section of this volume is devoted to new approaches for the application of palaeomagnetism to fold and thrust belts. The paper by **Garcés** *et al.* **(2015)** introduces a new concept for the restoration of remagnetizations to the ancient reference system in the Central Pyrenees that takes into account the geometry of unconformities that formed during the extensional-basin inversion instead of the bedding plane. **Villalaín** *et al.* **(2015)** present the reconstruction of highly subsident extensional basins that, in turn, reset the palaeomagnetic signal; these basins were inverted during Tertiary development of a fold and thrust belt. In this case, the geometric analysis of the remagnetized directions has allowed for reconstruction of the geometry of those basins at the time interval of the remagnetization.

While quantitative use of palaeomagnetically deduced rotations is scarce, some researchers have addressed this problem in recent years. Palinspastic restorations of orogens can be much more precise if palaeomagnetic vectors are combined with structural features in a numerical approach. This was first applied in the Bolivian orocline by Arriagada *et al.* (2008). In a similar way, the quantification of errors in shortening estimates deduced from restored and balanced sections in fold and thrust belts has been recently introduced (e.g. Pueyo *et al.* 2004; Sussman *et al.* 2012). Here, **Ramón** *et al.* **(2015a)** propose a new 3-D restoration method that combines the use of palaeomagnetic data and a parametric approach for the definition and processing of the structural surfaces. These authors demonstrate that using palaeomagnetism clearly improves the restoration result of complex geological settings. Finally, **Pueyo** *et al.* **(2016)** present an overview of the potential sources of error in palaeomagnetic studies applied to deformed regions, review field, laboratory and processing techniques and propose an extension of the classic reliability criteria established by Van der Voo (1990).

## Future and challenges of palaeomagnetic studies in fold and thrust belts

The application of palaeomagnetism to fold and thrust belts still has many challenges that need to be met in the future:

- *Data resolution* – resolution studies on palaeomagnetic data can be found in the literature specifically applied to tectonic problems (Demarest 1983; Bazhenov 1988). However, the definition

of palaeomagnetic means at the dip-domain scale (*sensu* Suppe 1985; Groshong 1999) along thrust fronts is still required, perhaps with an approach similar to that of Deenen *et al.* (2011) but more focused on the geometry of the fold and thrust belts.

- *Reliability* – many potential sources of error can affect palaeomagnetic datasets in fold and thrust belts (see **Pueyo** *et al.* **2016**). Apart from other processing and laboratory problems, the effect of internal deformation, non-resolved overlapping of components or errors in the restoration of complex settings may limit the value of the palaeomagnetic vectors as a robust kinematic indicator. In this sense, some advances on demonstrating the quality and reliability of the palaeomagnetic data have been made (MacDonald 1980; Chan 1988; Sellés-Martínez 1988; Weinberger *et al.* 1995; Rodríguez-Pintó *et al.* 2011). However, numerical modelling of errors and the establishment of specific procedures are still to be set.

- *Age constraining palaeomagnetic vectors* – constraining the age of the magnetization in palaeomagnetic studies is a key variable and it is usually done via the fold test (Graham 1949; McElhinny 1964), a powerful tool that is rarely used for other classic structural indicators. The statistical significance of the fold test is mature (e.g. Tauxe & Watson 1994; McFadden 1998; Enkin 2003), but mathematical approaches to the fold test do not consider phenomena such as folding kinematics. While some authors have already considered this problem (Cairanne *et al.* 2002; Delaunay *et al.* 2002; Waldhör & Appel 2006; **Villalaín** *et al.* **2015**), further advances in this regard need to be accomplished.

- *Understanding the architecture of rotated areas* – understanding of the spatial and temporal variability of rotational patterns in fold and thrust belts is still in its infancy. In this sense, integration of classic determination of VARs and magnetostratigraphic studies may shed much more light in future on these systems in 4-D.

- *Quantitative implications* – palaeomagnetic results should play a more active role in the 2-D/3-D shortening estimation (cross-section, map view, 3-D restorations, etc.), since such data account for the out-of-plane movements biasing classic approaches (e.g. McCaig & McClelland 1992; Pueyo *et al.* 2004; Sussman *et al.* 2012). In this sense, palaeomagnetic vectors, together with the bedding plane, are the only reliable 3-D reference frame to be used in 3-D restoration and validation techniques, since they are accurately known in both the deformed and undeformed stages. Although some recent works have demonstrated this (Ramón *et al.* 2012, **2015a**, *b*), many challenges remain. We encourage future investigators to meet such challenges.

We acknowledge the reviewers who have helped to complete this Special Publication as well as the editorial work done by the Geological Society staff. We are particularly thankful to Massimo Mattei for writing the foreword to this volume.

# References


ALLERTON, S. 1998. Geometry and kinematics of vertical-axis rotations in fold and thrust belts. *Tectonophysics*, **299**, 15–30.

ANASTASIO, D., PARÉS, J.M., KODAMA, K.P., TROY, J. & PUEYO, E.L. 2015. Anisotropy of magnetic susceptibility (AMS) records synsedimentary deformation kinematics at Pico del Aguila anticline, Pyrenees, Spain. *In*: PUEYO, E.L., CIFELLI, F., SUSSMAN, A.J. & OLIVA-URCIA, B. (eds) *Palaeomagnetism in Fold and Thrust Belts: New Perspectives*. Geological Society, London, Special Publications, **425**. First published online June 16, 2015, http://doi.org/10.1144/SP425.8

ARRIAGADA, C., ROPERCH, P., MPODOZIS, C. & COBBOLD, P.R. 2008. Paleogene building of the Bolivian Orocline: Tectonic restoration of the central Andes in 2-D map view. *Tectonics*, **27**, 1–14.

BAZHENOV, M.L. 1988. Analysis of the resolution of the paleomagnetic method in solving Tectonic Problems. *Geotectonics*, **22**, 204–212.

CAIRANNE, G., AUBOURG, C. & POZZI, J.P. 2002. Synfolding remagnetization and the significance of the small circle test: examples from the Vocontian trough (SE France). *Physics and Chemistry of the Earth, Parts A/B/C*, **27**, 1151–1159.

CARDELLO, G.L., ALMQVIST, B.S.G., HIRT, A.M. & MANCKTELOW, N.S. 2015. Determining the timing of formation of the Rawil Depression in the Helvetic Alps by palaeomagnetic and structural methods. *In*: PUEYO, E.L., CIFELLI, F., SUSSMAN, A.J. & OLIVA-URCIA, B. (eds) *Palaeomagnetism in Fold and Thrust Belts: New Perspectives*. Geological Society, London, Special Publications, **425**. First published online July 22, 2015, updated version published online August 18, 2015, http://doi.org/10.1144/SP425.4

CHAN, L.S. 1988. Apparent tectonic rotations, declination anomaly equations and declination anomaly charts. *Journal of Geophysical Research*, **93**, 12 151– 12 158.

CIFELLI, F., CARICCHI, C. & MATTEI, M. 2016. Formation of arc-shaped orogenic belts in the Western and Central Mediterranean: a paleomagnetic review. *In*: PUEYO, E.L., CIFELLI, F., SUSSMAN, A.J. & OLIVA-URCIA, B. (eds) *Palaeomagnetism in Fold and Thrust Belts: New Perspectives*. Geological Society, London, Special Publications, **425**. First published online February 17, 2016, http://doi.org/10.1144/SP425.12

CINKU, M.C., HISARLI, M. *ET AL.* 2015. Evidence of Late Cretaceous oroclinal bending in north-central Anatolia: palaeomagnetic results from Mesozoic and Cenozoic rocks along the İzmir–Ankara–Erzincan Suture Zone. *In*: PUEYO, E.L., CIFELLI, F., SUSSMAN, A.J. & OLIVA-URCIA, B. (eds) *Palaeomagnetism in Fold and Thrust Belts: New Perspectives*. Geological

Society, London, Special Publications, **425**. First published online August 3, 2015, updated version published online August 14, 2015, http://doi.org/10.1144/SP425.2

DEENEN, M.H.L., LANGEREIS, C.G., VAN HINSBERGEN, D.J.J. & BIGGIN, A.J. 2011. Geomagnetic secular variation and the statistics of palaeomagnetic directions. *Geophysical Journal International*, **186**, 509–520.

DELAUNAY, S., SMITH, B. & AUBOURG, C. 2002. Asymmetrical fold test in the case of overfolding: two examples from the Makran accretionary prism (Southern Iran). *Physics and Chemistry of the Earth, Parts A/B/C*, **27**, 1195–1203.

DEMAREST, H.H. 1983. Error analysis for the determination of tectonic rotation from paleomagnetic data. *Journal of Geophysical Research: Solid Earth (1978–2012)*, **88**, 4321–4328.

ENKIN, R.J. 2003. The direction–correction tilt test: an all-purpose tilt/fold test for paleomagnetic studies. *Earth and Planetary Science Letters*, **212**, 151–166.

ELMORE, R.D., MUXWORTHY, A.R. & ALDANA, M. 2012. Remagnetization and chemical alteration of sedimentary rocks. *In*: ELMORE, R.D., MUXWORTHY, A.R., ALDANA, M.M. & MENA, M. (eds) *Remagnetization and Chemical Alteration of Sedimentary Rocks*. Geological Society, London, Special Publications, **371**, 1–21, http://doi.org/10.1144/SP371.15

ELREDGE, S.; BACHTADSE, V. & VAN DER VOO, R. 1985. Paleomagnetism and the orocline hypothesis. *Tectonophysics*, **119**, 153–179.

GARCÉS, M., GARCÍA-SENZ, J., MUÑOZ, J.A., LÓPEZ-MIR, B. & BEAMUD, E. 2015. Timing of magnetization and vertical-axis rotations of the Cotiella massif (Late Cretaceous, South Central Pyrenees) *In*: PUEYO, E.L., CIFELLI, F., SUSSMAN, A.J. & OLIVA-URCIA, B. (eds) *Palaeomagnetism in Fold and Thrust Belts: New Perspectives*. Geological Society, London, Special Publications, **425**. First published online October 23, 2015, http://doi.org/10.1144/SP425.11

GRAHAM, J.W. 1949. The stability and significance of magnetism in sedimentary rocks. *Journal of Geophysical Research*, **54**, 131–167.

GROSHONG, R.H., JR, 1999. *3-D Structural Geology*. Springer-Verlag, Heidelberg.

JAPAS, M.S., RE, G.H., ORIOLO, S. & VILAS, J.F. 2015. Palaeomagnetic data from the Precordillera fold and thrust belt constraining Neogene foreland evolution of the Pampean flat-slab segment (Central Andes, Argentina). *In*: PUEYO, E.L., CIFELLI, F., SUSSMAN, A.J. & OLIVA-URCIA, B. (eds) *Palaeomagnetism in Fold and Thrust Belts: New Perspectives*. Geological Society, London, Special Publications, **425**. First published online August 18, 2015, http://doi.org/10.1144/SP425.9

KISSEL, C. & LAJ, C. 1989. Paleomagnetic rotations and continental deformation. *NATO ASI Science Series C*, **254**, Kluwer Academic Publishers, Boston.

MACDONALD, W.D. 1980. Net tectonic rotation, apparent tectonic rotation and the structural tilt correction in paleomagnetism studies. *Journal of Geophysical Research*, **85**, 3659–3669.

MÁRTON, E., GRABOWSKI, J., TOKARSKI, A.K. & TÚNYI, I. 2015. Palaeomagnetic results from the fold and thrust belt of the Western Carpathians: an overview.

*In*: PUEYO, E.L., CIFELLI, F., SUSSMAN, A.J. & OLIVA-URCIA, B. (eds) *Palaeomagnetism in Fold and Thrust Belts: New Perspectives*. Geological Society, London, Special Publications, **425**. First published online August 12, 2015, http://doi.org/10.1144/SP425.1

MCCAIG, A.M. & MCCLELLAND, E. 1992. Palaeomagnetic techniques applied to thrust belts. *In*: MCCLAY, K.R. (ed.) *Thrust Tectonics*. Chapman & Hall, London, 209–216.

MCELHINNY, M.W. 1964. Statistical significance of the fold test in palaeomagnetism. *Geophysical Journal International*, **8**, 338–340.

MCFADDEN, P.L. 1998. The fold test as an analytical tool. *Geophysical Journal International*, **135**, 329–338.

NORRIS, D.K. & BLACK, R.F. 1961. Application of palaeomagnetism to thrust mechanics. *Nature (London)*, **192**, 933–935.

OLIVA-URCIA, B., BEAMUD, E., GARCÉS, M., ARENAS, C., SOTO, R., PUEYO, E.L. & PARDO, G. 2015. New magnetostratigraphic dating of the Palaeogene syntectonic sediments of the west-central Pyrenees: tectonostratigraphic implications. *In*: PUEYO, E.L., CIFELLI, F., SUSSMAN, A.J. & OLIVA-URCIA, B. (eds) *Palaeomagnetism in Fold and Thrust Belts: New Perspectives*. Geological Society, London, Special Publications, **425**. First published online August 3, 2015, updated version published online August 21, 2015, http://doi.org/10.1144/SP425.5

PUEYO, E.L., POCOVÍ, A., MILLÁN, H. & SUSSMAN, A. 2004. Map-view models for correcting and calculating shortening estimates in rotated thrust fronts using paleomagnetic data. *In*: WEIL, A. & SUSSMAN, A. (eds) *Special Publication on Orogenic Curvature: Integrating Paleomagnetic and Structural Analyses*. Geological Society of America, Boulder, CO, **383**, 57–71.

PUEYO, E.L., SUSSMAN, A.J. & OLIVA-URCIA, B. & CIFELLI, F. 2016. Palaeomagnetism in fold and thrust belts: use with caution. *In*: PUEYO, E.L., CIFELLI, F., SUSSMAN, A.J. & OLIVA-URCIA, B. (eds) *Palaeomagnetism in Fold and Thrust Belts: New Perspectives*. Geological Society, London, Special Publications, **425**. First published online July 7, 2016, http://doi.org/10.1144/SP425.14

RAMÓN, M.J., PUEYO, E.L., BRIZ, J.L., POCOVÍ, A. & CIRIA, J.C. 2012. Flexural unfolding in 3D using paleomagnetic vectors. *Journal of Structural Geology*, **35**, 28–39, http://doi.org/10.1016/j.jsg.2011.11.015

RAMÓN, M.J., PUEYO, E.L., CAUMON, G. & BRIZ, J.L. 2015*a*. Parametric unfolding of flexural folds using palaeomagnetic vectors. *In*: PUEYO, E.L., CIFELLI, F., SUSSMAN, A.J. & OLIVA-URCIA, B. (eds) *Palaeomagnetism in Fold and Thrust Belts: New Perspectives*. Geological Society, London, Special Publications, **425**. First published online August 12, 2015, http://doi.org/10.1144/SP425.6

RAMÓN, M.J., BRIZ, J.L., PUEYO, E.L. & FERNÁNDEZ, O. 2015*b*. Horizon restoration by best fitting of finite elements and rotation constraints: sensitivity to the mesh geometry and pin-element location. *Mathematical Geosciences*, 1–19.

RAPALINI, A.E., PERONI, J. *ET AL.* 2015. Palaeomagnetism of Mesozoic magmatic bodies of the Fuegian Cordillera: implications for the formation of the Patagonian Orocline *In*: PUEYO, E.L., CIFELLI, F., SUSSMAN, A.J. & OLIVA-URCIA, B. (eds) *Palaeomagnetism in Fold and Thrust Belts: New Perspectives*. Geological Society, London, Special Publications, **425**. First published online July 22, 2015, http://doi.org/10.1144/SP425.3

RODRÍGUEZ-PINTÓ, A., RAMÓN, M.J., OLIVA-URCIA, B., PUEYO, E.L. & POCOVÍ, A. 2011. Errors in paleomagnetism: structural control on overlapped vectors, mathematical models. *Physics of the Earth and Planetary Interiors*, **186**, 11–22.

SELLÉS-MARTÍNEZ, J. 1988. Las correcciones estructural y tectónica en el tratamiento de los datos magnéticos. *Geofísica Internacional*, **27–3**, 379–393.

SUPPE, J. 1985. *Principles of Structural Geology*. Prentice Hall, NJ.

SUSSMAN, A.J., PUEYO, E.L., CHASE, C.G., MITRA, G. & WEIL, A.J. 2012. The impact of vertical-axis rotations on shortening estimates. *Lithosphere*, **4**, 383–394, http://doi.org/10.1130/L177.1

TARLING, D.H. 1969. The palaeomagnetic evidence of displacements within continents. *In*: KENT, P.E., SATTERTHWAITE, E. & SPENCER, A.M. (eds) *Time and Place in Orogeny*. Geological Society, London, Special Publications, **3**, 95–113, http://doi.org/10.1144/GSL.SP.1969.003.01.06

TAUXE, L. & WATSON, G.S. 1994. The fold test: an Eigen analysis approach. *Earth and Planetary Science Letters*, **122**, 331–341.

VALCÁRCEL, M., SOTO, R., BEAMUD, E., OLIVA-URCIA, B., MUÑOZ, J.A. & BIETE, C. 2015. Integration of palaeomagnetic data, basement-cover relationships and theoretical calculations to characterize the obliquity of the Altomira–Loranca structures (central Spain). *In*: PUEYO, E.L., CIFELLI, F., SUSSMAN, A.J. & OLIVA-URCIA, B. (eds) *Palaeomagnetism in Fold and Thrust Belts: New Perspectives*. Geological Society, London, Special Publications, **425**. First published online August 21, updated version published online September 3, 2015, http://doi.org/10.1144/SP425.7

VAN DER VOO, R. 1990. The reliability of paleomagnetic data. *Tectonophysics*, **184**, 1–9.

VAN DER VOO, R. & CHANNELL, J.E.T. 1980. Paleomagnetism in orogenic belts. *Reviews of Geophysics and Space Physics*, **18**, 455–481.

VILLALAÍN, J.J., CASAS-SAINZ, A.M. & SOTO, R. 2015. Reconstruction of inverted sedimentary basins from syn-tectonic remagnetizations. A methodological proposal. *In*: PUEYO, E.L., CIFELLI, F., SUSSMAN, A.J. & OLIVA-URCIA, B. (eds) *Palaeomagnetism in Fold and Thrust Belts: New Perspectives*. Geological Society, London, Special Publications, **425**. First published online October 2, 2015, http://doi.org/10.1144/SP425.10

WALDHÖR, M. & APPEL, E. 2006. Intersections of remanence small circles: new tools to improve data processing and interpretation in palaeomagnetism. *Geophysical Journal International*, **166**, 33–45.

WEIL, A.B. & SUSSMAN, A. 2004. Classification of curved ororgens based on the timing relationships between

structural development and vertical-axis rotations. *In:*
WEIL, A.B. & SUSSMAN, A. (eds) *Paleomagnetic and
Structural Analysis of Orogenic Curvature.* Geological
Society of America, Special Papers, **383**, 1–17.

WEINBERGER, R., AGNON, A., RON, H. & GARFUNKEL, Z.
1995. Rotation about an inclined axis: three dimensional
matrices for reconstructing paleomagnetic and structural data. *Journal of Structural Geology*, **17**, 777–782.

# Palaeomagnetic results from the fold and thrust belt of the Western Carpathians: an overview

EMŐ MÁRTON[1]*, JACEK GRABOWSKI[2], ANTEK K. TOKARSKI[3] & IGOR TÚNYI[4†]

[1]*Geological and Geophysical Institute of Hungary, Palaeomagnetic Laboratory, Columbus 17-23, H-1145 Budapest, Hungary*

[2]*Polish Geological Institute – National Research Institute, Palaeomagnetic Laboratory, Rakowiecka 4, 00-975 Warsaw, Poland*

[3]*Institute of Geological Sciences, Polish Academy of Sciences, Research Centre in Kraków, Senacka 1, 31-002 Kraków, Poland*

[4]*Slovak Academy of Sciences, Geophysical Institute, Palaeomagnetic Laboratory, Dúbravská cesta 9, 845 28 Bratislava, Slovakia*

**Corresponding author (e-mail: paleo@mfgi.hu)*


**Abstract:** The Western Carpathians are separated into an Outer and Inner Carpathians (both comprising several nappe systems) by the extremely narrow and highly deformed Pieniny Klippen Belt. The main phase of deformation and thrusting took place during the Late Cretaceous in the Inner Carpathians, at the end of Cretaceous–Paleocene in the Pieniny Klippen Belt and in the Miocene in the Outer Carpathians. In this paper a large amount of palaeomagnetic results of different quality available from several nappe stacks and from overstep sequences were reviewed and interpreted in terms of tectonics. The data suggest that all three main units participated in two phases of anticlockwise rotation starting at 18.5 Ma, that is, the Outer Carpathian nappes in front of the already consolidated Alpine–Carpathian–Pannonian block became accreted to the block. Late Cretaceous nappe transport, Neogene uplift of 'core mountains' and possibly oroclinal bending of pre-Oligocene age can account for important differences in pre-Cenozoic palaeomagnetic declinations. Most of them exhibit less or no anticlockwise rotation suggested by the overstep sequences, implying pre-Cenozoic clockwise rotations of variable angles.


The Western Carpathians (Fig. 1 inset) are the northernmost segment of the European Alpides. To the north, they are thrusted over the southern margin of the European Platform (Figs 1 & 2). To the south, the Western Carpathians *sensu lato* contact the Tisza unit along the Mid-Hungarian Line (Fig. 2), a Cenozoic strike-slip fault zone (for review see Fodor 2006). The Western Carpathians are usually subdivided into the Inner Carpathians, which formed during the Mesozoic, and the Outer Carpathians, which formed during the Cenozoic. The two units are separated by the Pieniny Klippen Belt (PKB), a narrow arcuate structure which was formed during two deformational stages in the latest Cretaceous–Paleocene and in the Miocene. Some authors (e.g. Plašienka 1995) suggest subdividing the Inner Carpathians into a Central and an Internal Carpathians which are separated by the Meliata Suture (Fig. 2). The Western Carpathians, north of the Pieniny Klippen Belt (Fig. 1) are composed exclusively of non-metamorphosed rocks

showing thin-skinned tectonics, whereas the Inner Carpathians are thick-skin orogens (Picha *et al.* 2005), composed of basement fragments and sedimentary cover.

The deformation in the Western Carpathians took place between the Late Jurassic and the Miocene, becoming younger outwards (Plašienka *et al.* 1997). The main tectonic features of the Carpathians were formed largely due to the convergence of the European and African plates and, more specifically, as a result of successive closures of three branches of the Tethys oceanic realm resulting in three successive subductions (Froitzheim *et al.* 2008). The final stage of thrusting in the Western Carpathians was partly concomitant with a lateral, NE-directed material escape from the Eastern Alps into the Carpathian–Pannonian realm (Nemčok *et al.* 1998).

Products of volcanism related to the youngest Cenozoic subduction cut and cover the Inner Carpathians, the Pieniny Klippen Belt and the inner part of the Outer Carpathians (Fig. 1a). Intramontane

---

[†]Deceased.

*From*: Pueyo, E. L., Cifelli, F., Sussman, A. J. & Oliva-Urcia, B. (eds) 2016. *Palaeomagnetism in Fold and Thrust Belts: New Perspectives*. Geological Society, London, Special Publications, **425**, 7–36.
First published online August 12, 2015, http://doi.org/10.1144/SP425.1

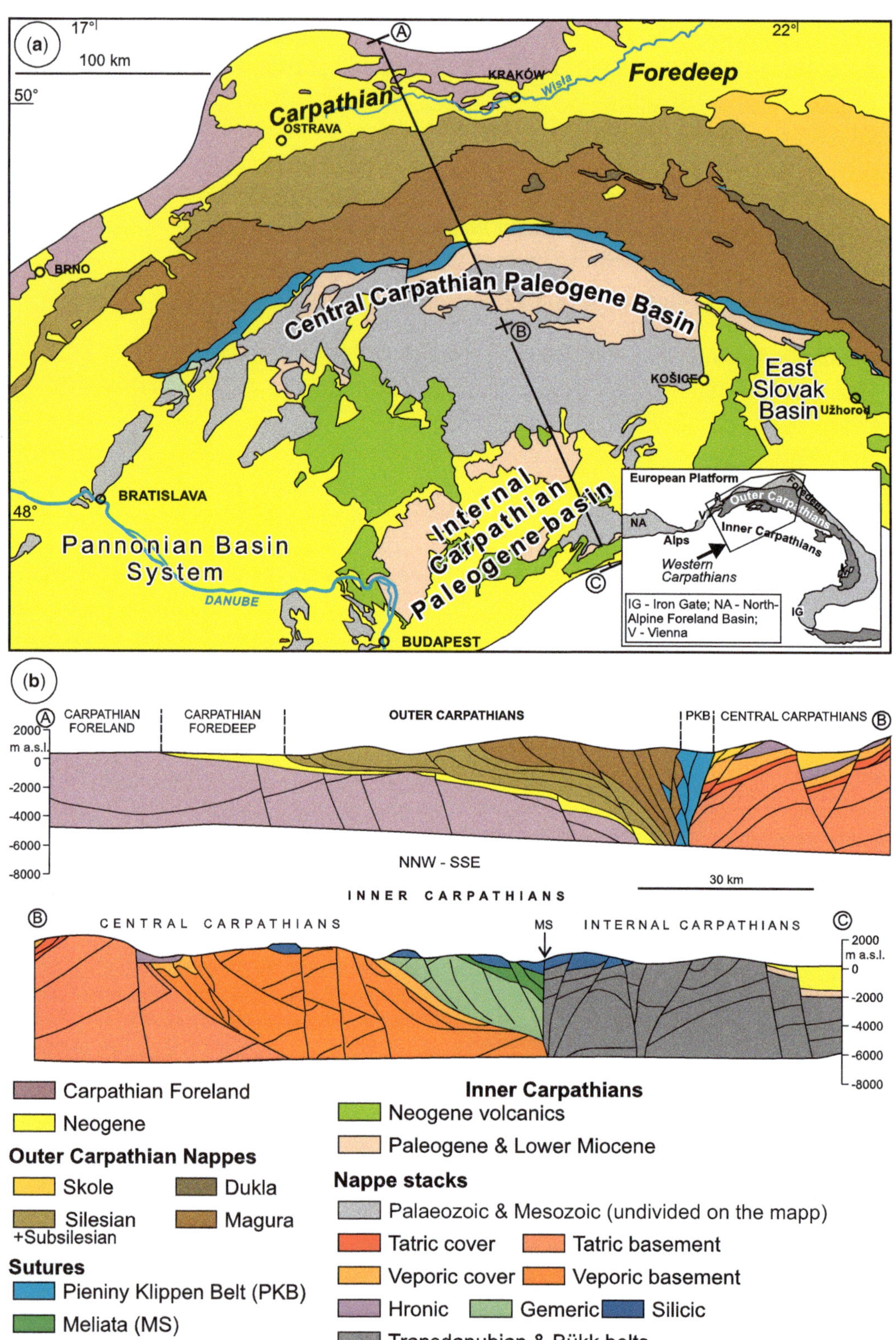
(a)
17°
100 km
50°
22°
KRAKÓW
Wisła
Foredeep
Carpathian
OSTRAVA
BRNO
Central Carpathian Paleogene Basin
KOŠICE
East Slovak Basin
Užhorod
BRATISLAVA
48°
Internal Carpathian Paleogene-basin
Pannonian Basin System
DANUBE
BUDAPEST
European Platform
Foredeep
NA
Outer Carpathians
Alps
Inner Carpathians
Western Carpathians
IG
IG - Iron Gate; NA - North-Alpine Foreland Basin; V - Vienna
(b)
A
CARPATHIAN FORELAND
CARPATHIAN FOREDEEP
OUTER CARPATHIANS
PKB
CENTRAL CARPATHIANS
B
2000
m a.s.l.
0
-2000
-4000
-6000
-8000
NNW - SSE
30 km
INNER CARPATHIANS
B
CENTRAL CARPATHIANS
MS
INTERNAL CARPATHIANS
C
2000
m a.s.l.
0
-2000
-4000
-6000
-8000
Carpathian Foreland
Neogene
Inner Carpathians
Neogene volcanics
Paleogene & Lower Miocene
Outer Carpathian Nappes
Skole
Dukla
Silesian +Subsilesian
Magura
Nappe stacks
Palaeozoic & Mesozoic (undivided on the mapp)
Tatric cover
Tatric basement
Veporic cover
Veporic basement
Sutures
Pieniny Klippen Belt (PKB)
Meliata (MS)
Hronic
Gemeric
Silicic
Transdanubian & Bükk belts

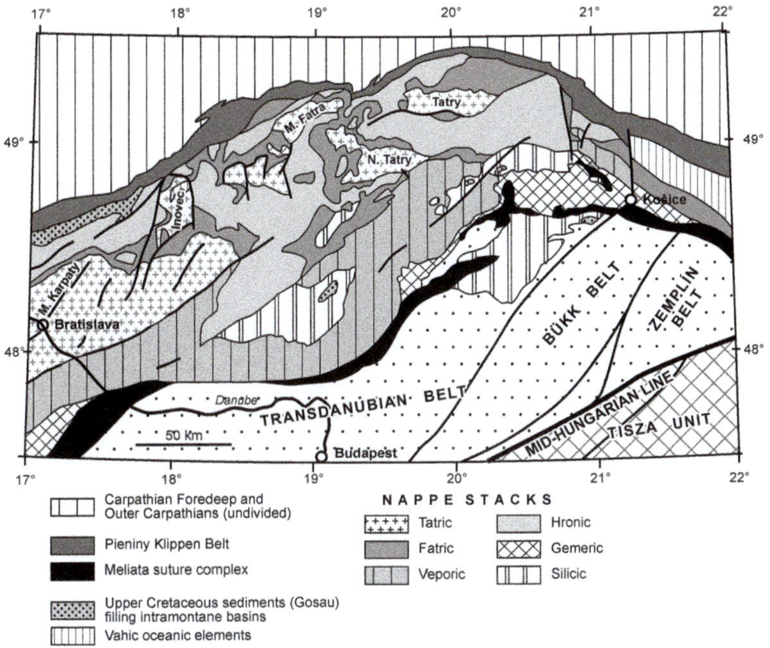


**Fig. 2.** Basement map of the Inner Western Carpathians (simplified after Plašienka *et al.* 1997). The map shows the locations of the most important 'core mountains' which are the High Tatra (Tatry), the Low Tatra (N. Tatry), the Little Fatra (M. Fatra), the Innovec and the Little Carpathians (M. Karpaty).

basins are filled with Neogene sediments in the Outer Carpathians, whereas Upper Cretaceous, Palaeogene and Neogene sediments occur in the Inner Carpathians.

Palaeomagnetic investigations in the Western Carpathians *sensu stricto* started in the 1960s. The first rocks studied were the Lower Permian red shales in the Central Western Carpathians of Slovakia (Kotásek & Krs 1965) and Neogene andesites in Poland (Birkenmajer & Nairn 1968). Pioneering studies of Mesozoic sedimentary rocks (Pieniny Klippen Belt, Upper Cretaceous red marls) were published by Bazhenov *et al.* (1980). In the early phase of palaeomagnetic research, most results were obtained in the former Czechoslovakia (for summary see Krs *et al.* 1982, 1996).

The Mesozoic sequences in the Central Western Carpathians (the Tatra Mountains and Slovakian 'core mountains') were first studied by Kądziałko-Hofmokl & Kruczyk (1987) and Kruczyk *et al.*

(1992). The first publications were followed by several papers by Grabowski (1995, 1997, 2000, 2005; Grabowski *et al.* 2009, 2010) and by a recent article by Szaniawski *et al.* (2012). The Triassic sediments of the Silica Nappe, the innermost unit of the Central Carpathians which also extends to the Internal Western Carpathians, were investigated by Márton *et al.* (1988), Márton *et al.* (1991), Kruczyk *et al.* (1998) and Channell *et al.* (2003). The Variscan granitoids of the 'core mountains' have not yet been studied palaeomagnetically, except that of the High Tatra Mountains (Grabowski & Gawęda 1999).

More recently, increased interest in the palaeomagnetism of the Pieniny Klippen Belt has resulted in several papers. Systematic palaeomagnetic studies were reported from the Neogene andesites and Upper Cretaceous red marls by Márton *et al.* (2004, 2013), while Grabowski *et al.* (2008) and Jeleńska *et al.* (2011) investigated the Jurassic

---

**Fig. 1.** The geological building of the Western Carpathians. (**a**) Simplified geological map of the area (modified after Lexa *et al.* 2000) and location of the cross-section ABC. The inset shows the location of the Western Carpathians in the East Alpine–Carpathian–Pannonian system. (**b**) The NNW–SSE-oriented cross-section shows the positions of most of the tectonostratigraphic units of the Western Carpathians. Note that the sediments of the Central Carpathian Palaeogene basin were preserved in one larger (surrounding the High Tatra Mountains) and in several smaller outcroping areas, and the Palaeogene rocks of the Internal Carpathian Palaeogene Basin are also accessible in outcrops in the areas covered by Neogene sediments and volcanic rocks.

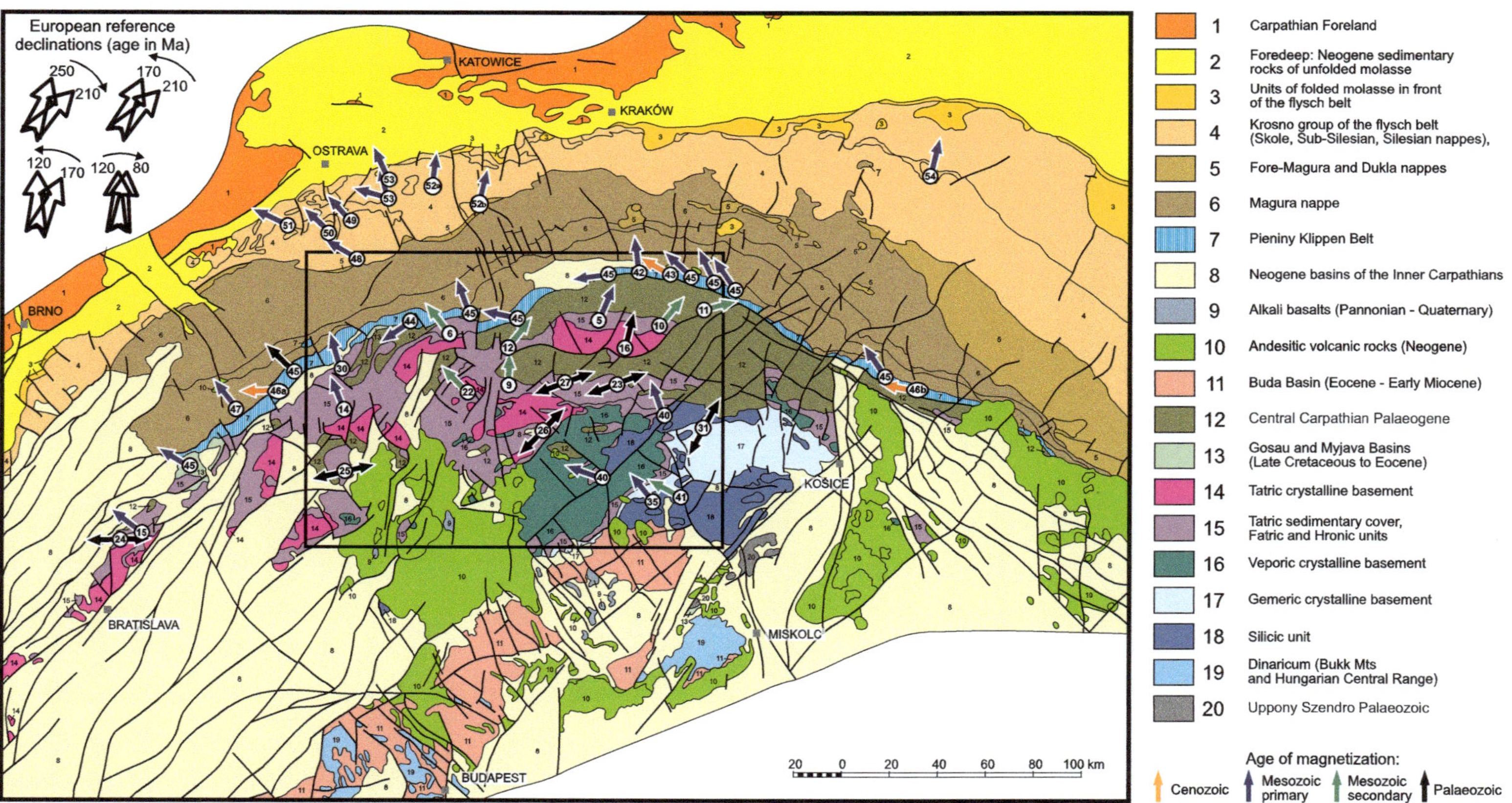



**Fig. 3.** Structural scheme of the Western Carpathians and adjacent areas (after Lexa *et al.* 2000, slightly modified) with distribution of palaeomagnetic results (palaeodeclinations) obtained from the Palaeozoic and Mesozoic rocks, with age of magnetization indicated. Arrow (palaeodeclination) number corresponds to the entries in Table 1. Rectangle indicates area depicted in the Figure 4. Some Palaeozoic declinations show two alternative interpretations of vertical axes rotations (clockwise and anticlockwise).

rocks in the Polish and the Slovak sectors of the Pieniny Klippen Belt.

The systematic studies of Cenozoic basins (over-step sequences) in northern Hungary, Slovakia and Poland were carried out by Márton et al. (e.g. 1992, 1996, 1999, 2000, 2007a, b), Márton & Márton (1996), Túnyi & Márton (1996), Márton & Pécskay (1998), Póka et al. (2004), Túnyi et al. (2004), and Karátson et al. (2007). An extensive database of palaeomagnetic results from the geographically distributed Tertiary sedimentary rocks of the Outer Carpathian nappes (Magura and Silesian) was presented by Márton et al. (2009a). A few Mesozoic results were also reported from the Outer Western Carpathians. They represent geographically limited areas in the Lower Cretaceous teschenitic rocks in the western part of Silesian Nappe (Krs et al. 1982; Grabowski et al. 2006), in sediments from the western part of the Silesian Nappe and from the Dukla Nappe (Krs et al. 1982), and in red pelagic marls from the eastern sector of Subsilesian Nappe (Szaniawski et al. 2013).

## Geology of the Western Carpathians

### Carpathian Foredeep

The foredeep of the Western Carpathians is part of an elongated basin extending from Vienna Forest to the Iron Gates on the Danube (Fig. 1, inset). Near Vienna it joins the North Alpine Foreland Basin. The Carpathian Foredeep is filled by Miocene, mostly clastic sediments, up to 3.5 km thick and contains a single intercalation of evaporate (Oszczypko et al. 2005). The sediments of the foredeep basin discordantly overlie the southern fringe of the European Platform. The contact, where exposed, is sedimentary (Márton et al. 2011). In the innermost part of the foredeep, the Miocene strata are incorporated into the imbricated frontal zone of the Outer Carpathians or form separate tectonic units composed of northwards-thrusted slices (Oszczypko et al. 2005; Picha et al. 2005). In the outer part, the Miocene strata become subhorizontal except close to map-scale faults (Márton et al. 2011). In eastern Poland, numerous NE–SW- and NW–SE-trending normal and strike-slip faults cut the Miocene fill, whereas strike-slip faults parallel to the Carpathian front are common in the innermost part of the Carpathian Foredeep in the western part (Márton et al. 2011).

### Outer Western Carpathians

The Outer Western Carpathians are a north-verging, north-convex fold and thrust belt thrusted over the Miocene sediments of the Carpathian Foredeep in the north and separated from the Inner Carpathians by the Pieniny Klippen Belt in the south. They comprise the Skole, Subsilesian–Zdanice, Silesian, Dukla and Magura rootless nappes (Fig. 1), composed largely of Lower Cretaceous–Lower Miocene flysch, locally over 10 km thick (Ślączka et al. 2005).

The nappe pile started to form during the Eocene (Świerczewska & Tokarski 1998; Nemčok et al. 2006). The deformation later progressed from the hinterland towards the foreland. The regular sequential succession of the foreland-verging structures is locally disturbed by out-of-sequence thrust sheets.

The traditional view was that the age of the termination of the nappe stacking was younger eastwards (Márton et al. 2009a), from about 14.5 Ma (end of Langhian according to Mediterranean, and middle Badenian according to Central Paratethyan stages; for correlation of the Mediterranean and Central Paratethyan stages, see Rögl 1996) at the westernmost part, until about 12 Ma (Serravalian according to Mediterranean, and Sarmatian according to Central Paratethyan stages) at the easternmost part. It was recently discovered, however, that strata younger than 11.5 Ma (Tortonian according to Mediterranean, and Sarmatian–Pannonian according to Central Paratethyan stages) are overthrusted by the Outer Carpathian nappes in the western segment (Wójcik & Jugowiec 1998). This implies that the final episode of thrusting, possibly affecting the whole Outer Carpathians, must have occurred after 11.5 Ma.

The nappe stacking took place in two successive stages during which tectonic transport was directed (in present co-ordinates) towards the NW and then towards the NE (Aleksandrowski 1985). During the second stage, the first stage folds were partly refolded and overprinted by folds of second generation (Aleksandrowski 1985), whereas some of the NE-striking thrusts of the first shortening event were reactivated as sinistral strike-slip faults (Fodor et al. 1995; Decker et al. 1997).

The thrusting in the Outer Carpathians was followed by regional collapse resulting in the formation of intermontane depressions filled by Neogene and Quaternary sediments (Zuchiewicz et al. 2002; Zattin et al. 2011). Some of the faults and thrusts formed during the nappe stacking and subsequent collapse were reactivated during Quaternary time (Tokarski et al. 2007).

In plate tectonic terms, the Outer Carpathians are a Tertiary accretionary complex with the backstop located at the Pieniny Klippen Belt. The accretionary complex was related to the southwards subduction of the oceanic or sub-oceanic crust, intervening between the continental crust of the European plate and the continental crust of the Inner Carpathians and their subsequent collision (e.g. Tomek & Hall 1993). This process resulted in considerable shortening. The minimum amount of the shortening was calculated by traditional methods as 60–100 km (Książkiewicz 1977; Oszczypko & Ślączka 1985) and by restoration of balanced cross-sections as 160 km (Picha *et al.* 2005) to 507 km (Gągała *et al.* 2012).

Most of the palaeogeographic reconstructions show the pre-folding shapes of the Magura and Silesian basins as north convex, similar to the present-day shapes of the Magura and Silesian Nappes (Książkiewicz 1960; Nemčok *et al.* 2000; Oszczypko & Oszczypko-Clowes 2006). However, in some reconstructions (e.g. Picha *et al.* 2005) both basins are depicted as rectilinear until Late Oligocene times. However, no arguments on the shapes of the basins have been presented in any of the above papers. According to Mastella & Konon (2002) the present-day shape of the Silesian Nappe results from tectonic bending; Nemčok *et al.* (2006) suggest that the Magura Nappe has radically changed its shape since the Late Oligocene.

Numerous small-scale andesite intrusions of Miocene age cut the inner part of the Magura Nappe (Birkenmajer & Pécskay 2000; Picha *et al.* 2005). Furthermore, numerous small-scale Lower Cretaceous hypabyssal intrusions and lava flows of alkaline lamprophyres (Teschenite Association Rocks) occur in the western part of the Silesian Nappe (Lucińska-Anczkiewicz *et al.* 2002; Grabowski *et al.* 2003).

*Pieniny Klippen Belt (PKB)*

The PKB (Figs 1–3) is a narrow, steeply dipping zone of extreme shortening and wrenching (Birkenmajer 1986). It mainly involves Jurassic and Cretaceous sediments with extremely variable lithology and intricate internal structures. Numerous lithostratigraphic and tectonic units of distant provenances were recognized in the Pieniny Klippen Belt, suffering excessive shortening and dispersal within this restricted zone. In general, two types of rock units are distinguished in the belt. The 'klippen'are rigid blocks of Jurassic–Lower Cretaceous limestones. They are embedded in an incompetent matrix, the 'klippen mantle', composed of Upper Cretaceous–Palaeogene marlstones, claystones and flysch (Birkenmajer 1986; Plašienka 2012*a*). The structure results from two successive deformation stages: a Late Cretaceous–Palaeogene stage when the nappe stack was formed; and a Miocene deformation stage when it was strongly modified and almost entirely disintegrated by left-lateral strike-slip movements, subparallel to the PKB trend (Birkenmajer 1986; Plašienka 2012*a*). The two deformation stages resulted in the 'block-in-matrix' structure of the PKB. Despite the complicated tectonic history the rocks of the PKB show

very little, if any, macroscopically observable, ductile strain.

*Inner Western Carpathians*

The Inner Western Carpathians are built up of numerous horst blocks (core mountains) separated by intramontane basins and embayments of the Pannonian Basin System. They comprise several stacks of nappes composed of metamorphic and plutonic Palaeozoic basement and of non- or moderately metamorphosed Permian–Cretaceous sedimentary cover (Fig. 2). All nappe stacks are north-verging in the Central Western Carpathians. South of the inferred Meliata suture, in the Internal Western Carpathians, the nappes verge to the south. The Central Carpathians comprise six nappe stacks. These are (from south to north): Silicic, Gemeric, Hronic, Veporic, Fatric and Tatric nappe stack. Thick-skinned stacks which comprise both basement and cover are the Tatric, Veporic and Gemeric stacks. Detached cover nappe stacks are the Fatric, Hronic and Silicic stacks, containing Late Palaeozoic–Mesozoic sedimentary rocks with rare volcanic rocks (Froitzheim *et al.* 2008). The Central Carpathians show a distinct progradation of Mesozoic shortening and collision events from the south towards the north (in present-day co-ordinates) from Late Jurassic to Late Cretaceous times (Plašienka *et al.* 1997; Picha *et al.* 2005; Froitzheim *et al.* 2008) related to successive closures of the Meliata and Vahic oceans (Plašienka *et al.* 1997).

Slivers of oceanic sediments and dismembered ophiolites occur in the Meliata suture. South of this suture, non-metamorphosed or slightly metamorphosed Palaeozoic and Mesozoic sediments are found in the south-verging Silica Nappe as well as in the different nappe units of the Bükk belt. In the Transdanubian belt the thick Permian–Cretaceous cover complex of a quite simple structure forms the basement, which is a huge SW–NE-trending synform (Plašienka *et al.* 1997).

*Overstep sequences*

The basement units are discordantly covered by Upper Cretaceous and younger sediments and volcanic rocks filling intramontane depressions and embayments of the Pannonian basin system (Froitzheim *et al.* 2008). Except for the Upper Cretaceous Gosau facies, the sediments filling the intramontane basins and the Pannonian basin system are unfolded or deformed by open folds.

## Permian and Mesozoic palaeomagnetic results

The data for all Palaeozoic and Mesozoic results is presented in Table 1. Most (but not all)

**Table 1.** *Summary of palaeomagnetic results from Palaeozoic and Mesozoic rocks in the Western Carpathians*

| | Latitude, longitude (°N, °E) | Age of rocks | N/No | n/no | $D$ (°) | $I$ (°) | $k$ | $\alpha_{95}$ | $D_c$ (°) | $I_c$ (°) | $k$ | $\alpha_{95}$ | Rotation (°) | Pole latitude (°N) | Pole longitude (°E) | $\delta p$ (°) | $\delta m$ (°) | Remark | Ref. |
|---|---|---|---|---|---|---|---|---|---|---|---|---|---|---|---|---|---|---|---|
| *Krížna nappe* | | | | | | | | | | | | | | | | | | | |
| 1 Tatra Mountains– Bobrowiec + Havran unit | 49.3, 19.8 | J2–J3 | 7/8 | | – | – | – | – | 22 | 59 | 198 | 4 | +22 ± 8 | 72 | 132 | 5 | 6 | *? | 1 |
| 2 Tatra Mountains– Bobrowiec and Gladkie unit | 49.3, 19.9 | Oxfordian | 4/5 | | 159 | 79 | 17 | 23 | 37 | 53 | 96 | 9 | +37 ± 15 | 59 | 125 | 9 | 13 | f+,* | 2 |
| 3 Tatra Mountains– Suchy Wierch unit | 49.3, 20.0 | T2 | 2/2 | 14/14 | 163 | 55 | 10 | 13 | 62 | 67 | 83 | 4 | +62 ± 10 | – | – | – | – | f+,* | 3 |
| 4 Tatra Mountains– Bobrowiec unit | 49.3, 19.8 | T3–K1 | 7/8 | | 176 | 80 | 46 | 9 | 34 | 60 | 74 | 7 | +34 ± 14 | 65 | 116 | 8 | 11 | f+,* | 3 |
| 5 Tatra Mountains | 49.3, 19.8 | Berriasian | 3/3 | | 346 | 84 | 182 | 9 | 23 | 47 | 499 | 5 | +23 ± 7 | 63 | 152 | 5 | 7 | f+, $R_B$ | 4 |
| 6 Malá Fatra | 49.24, 19.16 | J2–J3 | 21 sites | | 328 | 85 | 66 | 4 | 321 | 44 | 96 | 3 | −39 ± 4 | 58 | 253 | ? | ? | *? | 5 |
| 7 Malá Fatra– Zázrivská Dolina | 49.24, 19.15 | J2–J3 | 1 | 15/? | – | – | – | – | 331 | 42 | 55 | 5 | −29 ± 7 | 56 | 252 | 4 | 6 | ** | 6 |
| 8 Malá Fatra– Zázrivská Dolina | 49.23, 19.16 | Hettangian | 1 | 14/? | – | – | – | – | 358 | 43 | 35 | 7 | −2 ± 10 | 66 | 203 | 5 | 8 | ** | 6 |
| 9 Low Tatra | 49.03, 19.28 | J2 | 3 sites | | 35 | 76 | 243 | 8 | 2 | 56 | 73 | 14 | +2 ± 25 | 78 | 192 | ? | ? | *? | 5 |
| 10 Belianske Tatra | 49.25, 20.20 | J2–J3 | 4 sites | | 190 | 73 | 58 | 12 | 40 | 59 | 118 | 8 | +40 ± 16 | 61 | 113 | ? | ? | *? | 5 |
| 11 Magura Spisska | 49.28, 20.53 | J2 | 3 sites | | 312 | 70 | 28 | 24 | 75 | 46 | 114 | 12 | +75 ± 17 | 30 | 102 | ? | ? | *? | 5 |
| 12 Choč hills | 49.12, 19.23 | J1 | 3 sites | | 116 | 80 | 104 | 12 | 39 | 63 | 104 | 12 | +39 ± 26 | 63 | 105 | ? | ? | *? | 5 |
| 13 Strážov Mountains (component B) | 48.94, 18.47 | Tithonian– Hauterivian | 3/3 | | 137 | 54 | 38 | 20 | 28 | 75 | 317 | 7 | +28 ± 27 | – | – | – | – | *, f+ | 7 |
| 14 Strážov Mountains (component C) | 48.94, 18.47 | Tithonian– Berriasian | 2/3 | 18/64 | 177 | 84 | 12 | 10 | 338 | 49 | 75 | 4 | −22 ± 6 | 65 | 247 | 4 | 5 | f+ | 7 |
| 15 Male Karpaty | 48.51, 17.36 | Tithonian– Berriasian | 1 | 34/47 | 215 | 65 | 15 | 7 | 307 | 49 | 17 | 6 | −53 ± 9 | 46 | 282 | 5 | 8 | ** | 8 |
| *Tatricum, Tatra Mountains* | | | | | | | | | | | | | | | | | | | |
| 16 Tatra Mountains– crystalline core | 49.2, 20.1 | C1 (345– 340 Ma?) | 4/7 | | 193 | 17 | 59 | 12 | 194 | 47 | 59 | 12 | +4 ± 18 | −12 | 7 | 10 | 15 | ** | 9 |
| 17 Tatra Mountains (parautochthon) | 49.2, 19.9 | T1 | 6/8 | | 10 | 63 | 47 | 10 | 8 | 21 | 213 | 5 | +8 ± 5 | 51 | 188 | 3 | 6 | f+, | 10 |
| 18 Tatra Mountains (parautochthon) | 49.2, 19.9 | J2–K1 | 4/8 | | 211 | 76 | 29 | 17 | 23 | 50 | 256 | 6 | +23 ± 9 | 64 | 149 | 5 | 8 | *, f+ | 11 |
| 19 Tatra Mountains (allochthon) | 49.25, 19.87 | J2–J3 | 1 | 18/? | 301 | 50 | 5 | 48 | 312 | 48 | 48 | 5 | −48 ± 7 | 49 | 279 | – | – | *? | 11 |
| 20 Tatra Mountains (parautocht. + allochton) | 49.2, 19.9 | J3–K1 | 2/10 | 28/? | 70 | −68 | – | – | 180 | −69 | – | – | −110 ± ?? | 27 | 339 | – | – | **? | 11 |

(*Continued*)

**Table 1.** *Summary of palaeomagnetic results from Palaeozoic and Mesozoic rocks in the Western Carpathians* (*Continued*)

| | Latitude, longitude (°N, °E) | Age of rocks | N/No | n/no | D (°) | I (°) | k | $\alpha_{95}$ | $D_c$ (°) | $I_c$ (°) | k | $\alpha_{95}$ (°) | Rotation (°) | Pole latitude (°N) | Pole longitude (°E) | $\delta p$ (°) | $\delta m$ (°) | Remark | Ref. |
|---|---|---|---|---|---|---|---|---|---|---|---|---|---|---|---|---|---|---|---|
| *Tatricum, Mala and Velka Fatra* | | | | | | | | | | | | | | | | | | | |
| 21 Lúčivná | 49.21, 19.17 | Berriasian | 1 | 15/? | – | – | – | – | **341** | **54** | **39** | **6** | −19 ± 10 | 69 | 250 | 6 | 9 | ** | 6 |
| 22 Došná Dolinka | 48.97, 19.09 | Callovian–Oxfordian | 1 | 13/? | – | – | – | – | **316** | **48** | **11** | **13** | −44 ± 19 | 51 | 274 | 11 | 17 | ** | 6 |
| *Choč nappe* | | | | | | | | | | | | | | | | | | | |
| 23 Central Slovakia, red shales | 49.0, 20.0 | P1 | 13 | | – | – | – | – | **72** | **20** | **4** | **6** | ? | −20 | 297 | 3 | 7 | ** | 12 |
| 24 Malé Karpaty, melaphyres | 48.47, 17.3 | P | 6 | | 265 | 17 | 13 | 19 | **269** | **−2** | **13** | **19** | ? | 1 | 287 | 9 | 19 | ** | 13 |
| 25 Tříbec Mountains, melaphyres | 48.49, 18.53 | P | 3 | | 258 | −21 | 8 | 10 | **255** | **−18** | **8** | **10** | ? | −17 | 294 | 5 | 10 | ** | 13 |
| 26 Nizke Tatra S melaphyres | 48.85, 19.55 | P | 16 | | 242 | −24 | 4 | 20 | **223** | **−13** | **5** | **18** | ? | −35 | 325 | 9 | 19 | ** | 13 |
| 27 Nizke Tatra N melaphyres | 48.97, 19.70 | P | 32 | | 262 | −21 | 7 | 10 | **250** | **−16** | **6** | **11** | ? | −19 | 300 | 6 | 12 | ** | 13 |
| 28 Tatra Mountains, Reifling limestone | 49.27, 19.82 | T2 | 1 | 5/5 | 144 | 38 | 109 | 7 | **100** | **60** | **89** | **8** | +100 ± 16 | 24 | 75 | – | – | Synf., * | 3 |
| *Manin nappe* | | | | | | | | | | | | | | | | | | | |
| 29 Záskalie | 49.14, 18.52 | Albian | 1 | 14/? | – | – | – | – | **334** | **48** | **53** | **14** | −26 ± 21 | 62 | 252 | 5 | 7 | ** | 6 |
| 30 Butkov | 49.02, 18.36 | Oxfordian | 1 | 26/? | – | – | – | – | **343** | **46** | **14** | **8** | −17 ± 11 | 65 | 237 | 6 | 10 | ** | 6 |
| *Gemericum* | | | | | | | | | | | | | | | | | | | |
| 31 Eastern Slovakia, red shales | 48.83, 20.50 | P1 | 9 | | – | – | – | – | **29** | **17** | **5** | **5** | +29 ± 5 | −43 | 339 | 5 | 3 | ** | 12 |
| *Silica nappe* | | | | | | | | | | | | | | | | | | | |
| 32 Aggtelek Mountains | | T1 | 3 | | 294 | 51 | 7 | 49 | **294** | **24** | **31** | **23** | −66 ± 25 | 25 | 281 | 13 | 25 | f+ | 14 |
| 33 Aggtelek Mountains | | T2 | 4 | | 294 | 42 | 30 | 17 | **272** | **40** | **34** | **16** | −88 ± 44 | 18 | 305 | 12 | 19 | f+ | 14 |
| 34 Aggtelek Mountains | | T3 | 6 | | 259 | 46 | 12 | 20 | **289** | **59** | **38** | **11** | −71 ± 21 | 40 | 309 | 12 | 16 | f+ | 14 |
| 35 Slovak Karst | 48.5, 20.3 | T2–T3 | 5/5 | | 299 | 57 | 22 | 17 | **320** | **42** | **168** | **6** | −40 ± 8 | 50 | 267 | – | – | Synf., * | 15 |
| 36 Slovak Karst normal | 48.5, 20.3 | T2–T3 | 5/6 | | 318 | 50 | 22 | 16 | **322** | **50** | **66** | **9** | −38 ± 14 | 56 | 273 | – | – | Synf., * | 16 |
| 37 Slovak Karst reversed | 48.5, 20.3 | T3 | 1/6 | 69/? | **98** | **−69** | **12** | **5** | 89 | −54 | 12 | 5 | −82 ± 14 | 41 | 328 | – | – | | 16 |
| 38 Slovak karst Silicka Brezova LBT | 48.5, 20.5 | T3 | 8 | | **318** | **61** | **64** | **7** | 316 | 43 | 48 | 8 | −42 ± 14 | 60 | 295 | 8 | 11 | ** | 17 |
| 39 Slovak karst Silicka Brezova HBT | 48.5, 20.5 | T3 | 8 | | 319 | 64 | 46 | 8 | **316** | **46** | **112** | **5** | −44 ± 7 | 50 | 274 | 4 | 7 | f+ | 17 |
| 40 Stratenska Hornatina (North. Gemericum) | 48.8, 20.1 | Carnian | 2/8 | | 308 | 44 | – | – | **314** | **55** | – | – | −46 ± ?? | 54 | 286 | – | – | ** | 15 |

| No. | Locality | Coordinates | Age | N/No | n/no | $D$ | $I$ | $k$ | $\alpha_{95}$ | $D_c$ | $I_c$ | $k$ | $\alpha_{95}$ | Palaeolat. | Plat | Plong | $\delta p$ | $\delta m$ | R | Ref. |
|---|---|---|---|---|---|---|---|---|---|---|---|---|---|---|---|---|---|---|---|---|
| | *Bódva nappe* | | | | | | | | | | | | | | | | | | | |
| 41 | Rudabánya Mountains | | T2 | 5 | | 33 | 72 | 5 | 41 | **298** | **43** | **20** | **17** | −62 ± 23 | 37 | 288 | 13 | 21 | f+ | 14 |
| | *Pieniny Klippen Belt* | | | | | | | | | | | | | | | | | | | |
| 42 | Czorsztyn unit– Primary component | 49.4, 20.0 | J2–J3 | 3/5 | | 354 | 5 | 6 | 55 | **353** | **40** | **29** | **23** | −7 ± 30 | 63 | 215 | 17 | 28 | f+ | 18 |
| 43 | Czorsztyn unit– Secondary component | 49.4, 20.4 | J2–J3 | 2/5 | | **120** | **−67** | **–** | **–** | 166 | −65 | – | – | −60 ± ?? | 52 | 314 | – | – | ** | 18 |
| 44 | Brodno | 49.26, 18.75 | Tithonian– Berriasian | 1 | 104 | – | – | – | – | **236** | **45** | **10** | **6** | −124 ± 14 | 1 | 331 | 4 | 7 | ** | 19 |
| 45 | Red marls | 48.6–49.25, 17.3–21.3 | K2 | 11/14 | | 301 | 13 | 2 | 62 | **311** | **53** | **17** | **11.3** | −49 ± 28 | 51 | 284 | 11 | 15 | f+ | 20 |
| 46a | Red marls, secondary magnetization, W | 49.00, 18.10 | K2 | | 8 | **91** | **−66** | **20** | **13** | 352 | 69 | 6 | 24 | −89 ± 32 | 35 | 324 | 17 | 21 | f−, ** | 20 |
| 46b | Red marls, secondary magnetization, E | 49.01, 21.58 | K2 | | 10/14 | **111** | **−51** | **48** | **7** | 179 | −37 | 37 | 8 | −69 ± 11 | 37 | 296 | 6 | 9 | f−, ** | 20 |
| | *Outer West Carpathians, Magura unit* | | | | | | | | | | | | | | | | | | | |
| 47 | Biele Karpaty | 48.9, 17.8 | K2 | 9 | | 327 | 56 | 12 | 16 | **321** | **44** | **86** | **6** | −39 ± 8 | 52 | 265 | 4 | 7 | f+ | 21 |
| | *Outer West Carpathians, Silesian unit* | | | | | | | | | | | | | | | | | | | |
| 48 | NE Moravia, Silesia, Grey sandstones | 49.48, 18.43 | Coniacian | no data | 94 | – | – | – | – | **300** | **46** | **3** | **10** | −60 ± 14 | 40 | 285 | 8 | 13 | ** | 13 |
| 49 | NE Moravia , Silesia, Ondrejnik Mountains, red claystones | 49.57, 18.30 | Cenomanian– Turonian | no data | | – | – | – | – | **318** | **73** | **11** | **4** | −42 ± 14 | 64 | 323 | 7 | 8 | ** | 13 |
| 50 | NE Moravia, Silesia, Red claystones | 49.52, 18.27 | Cenomanian– Turonian | no data | | – | – | – | – | **313** | **53** | **9** | **3** | −47 ± 5 | 52 | 281 | 3 | 5 | ** | 13 |
| 51 | NE Moravia, Silesia teschenites | 49.57, 18.07 | K1 | no data | | – | – | – | – | **295** | **56** | **6** | **6** | −65 ± 11 | 42 | 297 | 6 | 8 | ** | 22 |
| 52a | Teschenitic rocks Poland, E, Świętoszówka | 49.7, 18.9 | K1 | 2 | | 345 | 55 | – | – | **7** | **59** | **–** | **–** | +7 ± ?? | | | | | f+,**? | 23 |
| 52b | Teschenitic rocks Poland, E, Żywiec | 49.6, 19.2 | K1 | 3 | | 298 | −29 | 1 | 139 | **13** | **69** | **272** | **7** | +13 ± 20 | 81 | 82 | 11 | 13 | f+,**? | 23 |
| 53 | Teschenitic rocks Poland, W | 49.6, 18.6 | K1 | 3 | | 335 | 67 | 1 | 149 | **319** | **66** | **26** | **24** | −41 ± 59 | 63 | 301 | 32 | 40 | **? | 23 |
| | *Outer West Carpathians, Subsilesian unit* | | | | | | | | | | | | | | | | | | | |
| 54 | Węglówka marls | 49.8, 21.8 | K2 | 4/6 | | 191 | 11 | 4 | 50 | **194** | **−18** | **21** | **20** | ? | 48 | 181 | 11 | 21 | ** | 24 |

Key: N/No (n/no:): number of used/collected localities (samples, if only one locality is tabulated); $D$, $I$ ($D_c$, $I_c$): declination, inclination before (after) tectonic/tilt correction; $k$ and $\alpha_{95}$: statistical parameters (Fisher 1953); Values in heavy letters are interpreted in terms of tectonics ; $\delta p$, $\delta m$: statistical parameters. Remarks: R: classification of reversal tests (McFadden & McElhinny 1990), a, b, c: positive, −: negative; i: test with isolated observation; rev.: the population has originally reversed polarity (in the table shown as normal); f: fold test (Enkin 2003) +: positive, −: negative: i: indeterminate. Constraints on the age of the magnetization: *Late Cretaceous remagnetization; **Palaeogene–Neogene remagnetization; Synf*: synfolding remagnetization. References: 1. Kądziałko-Hofmokl & Kruczyk (1987); 2. Grabowski (1995); 3. Grabowski (2000); 4. Grabowski (2005); 5. Kruczyk *et al.* (1992); 6. Pruner *et al.* (1998); 7. Grabowski *et al.* (2009); 8. Grabowski *et al.* (2010); 9. Grabowski & Gawęda (1999); 10. Szaniawski *et al.* (2012); 11. Grabowski (1997); 12. Kotásek & Krs (1965); 13. Krs *et al.* (1982); 14. Márton *et al.* (1988); 15. Márton *et al.* 1991; 16. Kruczyk *et al.* (1998); 17. Channell *et al.* (2003); 18. Grabowski *et al.* (2008); 19. Krs *et al.* (1996); 20. Márton *et al.* (2013); 21. Krs *et al.* (1993); 22. Krs & Šmíd (1979); 23. Grabowski *et al.* (2006); 24. Szaniawski *et al.* (2013).

palaeodeclinations are located on the geological maps in Figures 3 and 4. Some data were omitted in areas where many results are available (especially the area of Tatra Mountains and Silica unit).

## Krížna unit (Fatricum)

The most numerous pre-Cenozoic data come from the Krížna Nappe of the Tatra Mountains in Poland. The first palaeomagnetic results were presented by Kądziałko-Hofmokl & Kruczyk (1987). The characteristic direction, obtained from Bathonian–Kimmeridgian radiolarian and nodular limestones (1 in Table 1) did not differ significantly between particular stages, with mean value of pre-folding component close to the expected Eurasian reference direction for the Middle–Late Jurassic. The shortcomings of the abovementioned paper were that they did not discuss either the problem of remagnetization or the effect of tectonic correction. In the early 1980s the phenomena of remagnetization was not as well known (see McCabe & Elmore 1989) as today. The almost exclusively normal polarity of magnetization (except one site) was in accordance with the concept of a 'Jurassic Quiet Zone', a long normal interval embracing Callovian–Lower Kimmeridgian (Tominaga *et al.* 2008). A discovery of frequent magnetic reversals in the Late Jurassic (see Opdyke & Channell 1996; Gradstein *et al.* 2012 for review) meant that

interpretation of the data of Kądziałko-Hofmokl & Kruczyk (1987) was not straightforward. The abovementioned Oxfordian radiolarian limestones were restudied by Grabowski (1995) (2 in Table 1). It appeared that clustering of characteristic magnetizations, identical to those of Kądziałko-Hofmokl & Kruczyk (1987), is indeed better after tectonic correction, but their primary origin was questioned: it appeared that other components with higher unblocking temperatures and mixed polarity are better candidates for primary magnetizations. The age of remagnetization was interpreted as Late Cretaceous, since it was definitely older than the final emplacement of the Krížna Nappe over the High Tatric (Tatricum) substratum (Grabowski 1995). In fact, Krížna Nappe in the Tatra Mountains forms a complicated structure consisting of several imbricated units (duplexes) which behaved quite independently during the Late Cretaceous thrusting.

It was then discovered that the remagnetization is also ubiquitous in the Middle Triassic–Lower Cretaceous rocks of the Krížna Nappe (3 and 4 in Table 1; Fig. 4). Data collected from many localities situated in several duplex structures proved that the remagnetization was acquired during early phases of thrusting (Grabowski 2000). It must have occurred during the Cretaceous Quiet Zone, before the Coniacian (termination of thrusting in the Central Western Carpathians). Nevertheless, primary magnetization was documented in the

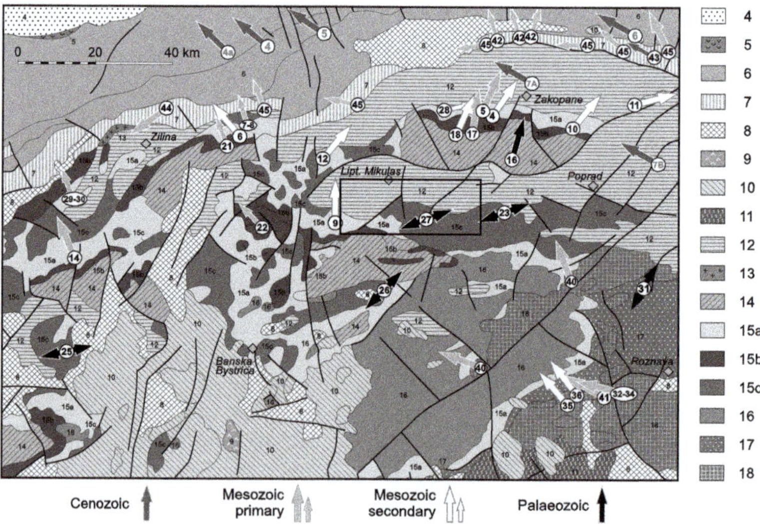



**Fig. 4.** Geological sketch of the middle part of the Inner Carpathians (after Lexa *et al.* 2000, slightly modified) with all available palaeomagnetic data indicated. Arrow (palaeodeclination) number corresponds to the entries in Tables 1 (black numbers) and 2 (grey numbers). Big arrows indicate result based on several sites or detailed magnetostratigraphic study in a single section; small arrows indicate result based on single locality. For lithological key (except items 15a–c) and other explanations, see Figure 3. 15a: Fatric unit, sediments; 15b: Tatricum, sedimentary cover; 15c: Hronic unit, sediments.

calpionellid-bearing upper Tithonian–Berriasian limestones (5 in Table 1; Figs 3 & 4) of the Krížna Nappe by Grabowski (2005) and the rocks yielded valuable magnetostratigraphic results (Grabowski & Pszczółkowski 2006). The calculated Lower–Middle Berriasian palaeopole is indeed located close to the coeval Eurasian data (Galbrun 1985), implying a net post-Berriasian clockwise rotation of 15–20°.

The Jurassic pelagic carbonates of the Krížna Nappe were also studied by Kruczyk *et al.* (1992) in several geographically distributed localities in Slovakia: Malá Fatra, Nizke Tatra Mountains, Belianske Tatra Mountains, Magura Spišska and Choč Mountains (6, 9–12 in Table 1; Figs 3 & 4). The characteristic remanences of exclusively normal polarity were interpreted as being of pre-folding age, with a fan-like pattern of palaeodeclinations coinciding with tectonic vergencies of the studied nappe fragments. The results were interpreted as support of the oroclinal bending model of the Central Western Carpathians (Burtman 1988). However, the age of bending and tectonic rotations was not constrained in the paper, and the age of magnetization was not clearly discussed. It was subsequently suggested (Grabowski & Nemčok 1999) that these data also might represent the Cretaceous remagnetizations acquired during the 'Cretaceous Quiet Zone'. The evidence was: (1) exclusively normal polarity of magnetization from rocks belonging to different parts of the Jurassic; and (2) unblocking temperatures of 350–450 °C which are typical for undoubtedly remagnetized carbonates in the Tatra Mountains (Grabowski 2000). An additional argument for the syntectonic nature of the components is a large scatter of 'pre-folding' inclinations from 44° to 63°. This problem was also pointed out by Kruczyk *et al.* (1992). In fact, an inclination-only test (Enkin & Watson 1996, not performed in the source paper) applied to these data gives negative results. This could easily be explained assuming a syntectonic (synthrusting) age of magnetization. Despite the suspicion of remagnetization, the data of Kruczyk *et al.* (1992) yielded useful constraints for the rotation pattern in the Krížna Nappe. These were confirmed by subsequent studies of Pruner *et al.* (1998) in the area of Malá Fatra (7 and 8 in Table 1; Fig. 4), where slight anticlockwise rotations of the Jurassic rocks were documented.

Further evidence for Mesozoic remagnetizations in the Krížna Nappe was obtained from the high-resolution study of Strážovce section (Tithonian–Neocomian) in the Strážov Mountains (Grabowski *et al.* 2009). The statistically well-defined secondary magnetization (13 in Table 1) of normal polarity must have been acquired during thrusting episodes during the Late Cretaceous. The most plausible tectonic corrections indicate that the rocks must have been magnetized when they dipped opposite to the thrusting direction, forming a hinterland-dipping duplex. The interpretation is concordant with the internal structure of the Krížna Nappe, consisting of imbricated units of duplex-type structure (Prokešova *et al.* 2012). The primary component of double polarity (14 in Table 1; Figs 3 & 4), which was possible to extract from some stratigraphic horizons, indicates a slight anticlockwise rotation of similar magnitude as in the Mala Fatra Mountains (Kruczyk *et al.* 1992).

Successful magnetostratigraphic study of the Tithonian–Berriasian limestones was performed by Grabowski *et al.* (2010) in the Krížna Nappe of the Male Karpaty Mountains, in the SW termination of the Carpathian arc (15 in Table 1; Fig. 3). The palaeodeclinations of 300° account for the largest anticlockwise rotation of Krížna Nappe fragment studied so far. The studied sequence was apparently not affected by the synthrusting remagnetization of Late Cretaceous age.

*Tatricum*

Palaeomagnetic data from the Tatric units are less numerous than from the Krížna unit. Firstly, the Tatric rocks mostly represent a shallow-water development and are not suitable palaeomagnetic material such as, for example, Jurassic–Cretaceous pelagic carbonates of the Krížna unit. Secondly, during the Alpine Orogeny the Tatricum was deeply buried under the overriding Fatric and Hronic units, which resulted in anchimetamorphic conditions in some Central Western Carpathian massifs (e.g. Plašienka 1995). Most of the published palaeomagnetic data were obtained in the High Tatra Mountains: (1) the Lower Carboniferous granitoids of the High Tatra Mountains (Grabowski & Gawęda 1999; 16 in Table 1; Figs 3 & 4); (2) the Lower Triassic sandstones of the autochthonous sedimentary cover of the Tatricum (Szaniawski *et al.* 2012; 17 in Table 1; Fig. 4); and (3) Jurassic–Lower Cretaceous limestones (Grabowski 1997) overlying the autochthonous Triassic (18 in Table 1; Fig. 4); or included in the overthrusted units (19 in Table 1). The granitoids and the Mesozoic autochthonous cover rocks reveal consistent N–NE palaeodeclinations similar to the palaeodeclinations of the Krížna Nappe. According to Szaniawski *et al.* (2012), the Lower Triassic hematite-bearing sandstones preserved the primary magnetization while the Jurassic–Cretaceous limestones were remagnetized in the Late Cretaceous during the Cretaceous Normal Superchron (Grabowski 1997). One Upper Jurassic locality situated in an allochthonous position (19 in Table 1) indicates significant anticlockwise rotation in relation to the autochthon. Two localities (20 in Table 1; Fig. 4) bear evidence of peculiar reversed polarity remagnetizations, residing most probably in pyrrhotite (maximum unblocking temperatures between 300 °C and 350°C) which might be interpreted as Cenozoic, acquired before the Neogene tilting event which is connected to the uplift of the High Tatra Mountains (Grabowski 1997).

Two results are available from the Mesozoic cover of Tatricum outside of the High Tatra Mountains, in the Malá and Velká Fatra (Pruner *et al.* 1998; 21 and 22 in Table 1; Figs 3 & 4). Detailed geological setting and tectonic correction is not discussed in the source paper. Mixed polarity in both localities might point to primary magnetization. Slight anticlockwise-rotated declinations seem to mirror the trend of rotations described from the Krížna Nappe of the Malá Fatra (Table 1, items 6–9) and Strážov Mountains (Table 1, item 14).

### Choč (Hronicum) and Manin units

Several geographically distributed localities were studied from the Permian melaphyres and red sediments of the Choč Nappe in 1960s and 1970s (23–27 in Table 1; Figs 3 & 4) and the results summarized in a review by Krs *et al.* (1982). The palaeomagnetic vectors were obtained using the technique typical for that time, that is, the magnetization exhibiting high stability on thermal or alternating field (AF) demagnetization was isolated and statistically evaluated on locality level. Although the within locality scatter is fairly high, the statistical parameters in most cases satisfy the minimum criteria for acceptable palaeomagnetic direction. Field tests for constraining the age of the magnetizations are lacking, but the shallow inclinations are in accordance with those expected for the Permian. Originally, these results were interpreted in terms of large clockwise rotations.

The only Mesozoic palaeomagnetic data from the Choč Nappe come from a single locality of Reifling limestones (Middle Triassic) from the Western Tatra Mountains in Poland (Grabowski 2000; 28 in Table 1; Fig. 4). The rock was remagnetized during the Late Cretaceous; the palaeodeclination indicates large clockwise rotation (almost 80°) in relation to the Tatric autochthon and the Krížna Nappe. However, as the remagnetization is of synthrusting age and the exact structural position during the acquisition of the secondary magnetization is not known (it might have dipped even 30–40° to the south; see Grabowski 2000) the amount of vertical axis rotation cannot be precisely estimated. The remagnetization might have been synchronous with resetting of the K–Ar ages, indicating 90 Ma, of the Middle Triassic tuffites in the same locality (Środoń *et al.* 2006).

Two results from the Manin unit (Pruner *et al.* 1998; 29–30 in Table 1; Figs 3 & 4) indicate slight anticlockwise rotation, similarly to the Mesozoic palaeodeclinations from Malá Fatra and Strážov Mountains. Mixed polarity might indicate primary magnetization in the Butkov locality (30 in Table 1); however directions before tectonic corrections and fold test results are not presented.

### Gemericum

There is a single Permian palaeomagnetic direction from this unit (Krs *et al.* 1982; 31 in Table 1; Figs 3 & 4). The comments about the Permian palaeomagnetic directions from the Choč Nappe (see section 'Choč (Hronicum) and Manin units') are also applicable to that result.

### Silica unit

Primary magnetization was documented in the Triassic of Aggtelek Mountains (Silica Nappe s.s.) and Rudabánya Mountains (Bodva Nappe) in northern Hungary (Márton *et al.* 1988; 32–34, 41 in Table 1; Fig. 4). The palaeomagnetic record in the Slovakian part of the Silica unit is more complex. Independent studies of Middle–Upper Triassic rocks indicate the presence of remagnetization. Its age is interpreted as syntectonic acquired during the Late Cretaceous (Márton *et al.* 1991; 35 in Table 1; Figs 3 & 4) or post-tectonic acquired during the Oligocene–Miocene (Kruczyk *et al.* 1998; 36, 37 in Table 1; Fig. 4). In their magnetostratigraphic study in the Silická Brezová locality, Channell *et al.* (2003) documented a post-folding secondary component (38 in Table 1). However, they were also able to isolate a primary component (39 in Table 1). Both primary and secondary components in Slovakia account for a *c.* 40–50° anticlockwise rotation. The rotation in the southern part of Silica unit seems slightly larger (65–90°) in relation to present-day north. The palaeomagnetic data from the northern periphery of the Silica unit (north of Gemeric unit) also indicate net 45° anticlockwise rotation (Márton *et al.* 1991; 40 in Table 1; Figs 3 & 4). However, differential rotations occur between the particular tectonic sub-units (Muran and Stratena nappes).

### Pieniny Klippen Belt

The three sites of Callovian–Tithonian limestones from the Czorsztyn unit in the central (Polish segment) of the Pieniny Klippen Belt (Grabowski *et al.* 2008) reveal a slight rotation in the anticlockwise sense (42 in Table 1; Figs 3 & 4). The mixed polarity magnetization was interpreted as primary. The palaeodeclinations measured in the Middle–Upper Jurassic limestones of the western Slovak sector of the Pieniny Klippen Belt (Jeleńska *et al.*

2011) are badly scattered; the authors were not able to present any explanation of this observation and the results are not included in Table 1. The palaeodeclination from the Tithonian–Berriasian of the western Slovak sector of the Pieniny Klippen Belt revealed a very large anticlockwise rotation (almost 120°). As the studied strata are overturned and refolded, the application of additional tectonic correction for the plunge of the fold axis cannot be excluded (Houša *et al.* 1996; 44 in Table 1; Figs 3 & 4).

Recently a systematic palaeomagnetic study of the Upper Cretaceous red marls was carried out along the strike of the Pieniny Klippen Belt from western Slovakia through Poland to eastern Slovakia (Márton *et al.* 2013). Remanences of pre-folding age were documented for 11 localities pointing to a net anticlockwise rotation of *c.* 50° in relation to the present-day north (45 in Table 1; Figs 3 & 4). Differences in palaeodeclinations between particular localities might be interpreted in favour of the secondary bending of the Pieniny Klippen Belt before the Oligocene.

Some palaeomagnetic directions in the Mesozoic rocks of the Pieniny Klippen Belt are Cenozoic overprints. It concerns two Upper Cretaceous localities in the western and eastern Slovak sector (Márton *et al.* 2013), with negative within-locality fold test and *c.* 75° anticlockwise rotated declinations (46a and 46b in Table 1; Fig. 3). They correspond quite well to the Oligocene–Early Miocene declinations, which are characteristic of the Inner Carpathian Palaeogene. Another evidence of Neogene remagnetization was found in Middle–Upper Jurassic rocks in the central segment of the Pieniny Klippen Belt (Grabowski *et al.* 2008; 43 in Table 1; Figs 3 & 4). Although the fold test was not conclusive, their direction before tectonic correction (component C of Grabowski *et al.* 2008) reveals a similar amount of anticlockwise rotation as the above directions of post-folding age (Márton *et al.* 2013). It should be noted that all these overprints are of reversed polarity, as for the primary magnetization of the Neogene andesites in the Pieniny Klippen Belt (Márton *et al.* 2004; 6 in Table 2; Fig. 4).

*Outer Western Carpathians*

Palaeomagnetic results were obtained from the Lower Senonian siliciclastics of the Biele Karpaty (Magura Nappe; Krs *et al.* 1993; 47 in Table 1; Fig. 3). The double polarity of the characteristic direction, which is based on hematite, suggests a primary origin of magnetization. Tectonic correction is not discussed in detail, however. The palaeodeclination suggests a *c.* 40° anticlockwise rotation.

Most Mesozoic data from the Outer Western Carpathians come from the siliciclastic and volcanic rocks of the Silesian Nappe in NE Moravia (Krs *et al.* 1982, 1996; 48–51 in Table 1; Fig. 3). All of them reveal significant anticlockwise-rotated declinations, in accordance with the result from Biele Karpaty. The quality of these results is moderate; since clustering parameter *k* seldom exceeds 10, characteristic magnetizations are only given *in situ* without any tectonic correction and there is no discussion on possible remagnetization. For example, stereograms from the teschenitic rocks (Krs & Šmíd 1979) clearly indicate the presence of rotated and non-rotated palaeodeclinations, which suggests that some intrusions might carry a secondary magnetization of normal polarity. An inclination-only test, performed for the teschenite localities in Poland (Grabowski *et al.* 2006; 52–53 in Table 1; Fig. 3) was not fully positive, indicating that either the rocks were remagnetized during folding and thrusting or magnetization, although primary, is not synchronous. Indeed, radiometric dating of teschenitic rocks gave quite a broad spectra of ages of 149–122 Ma (from the Jurassic-Cretaceous boundary (J/K) up to the Barremian–Aptian boundary). At two localities (52a and b in Table 1; Fig. 3) where positive-contact and within-locality fold tests were performed, almost no rotation is observed. The other localities (53 in Table 1; Fig. 3) reveal an anticlockwise rotation of magnitude comparable to that in the Czech part of the Silesian Nappe.

Recently, a result from the Upper Cretaceous red marls of the eastern part of the Subsilesian Nappe (Szaniawski *et al.* 2013; 54 in Table 1) was published. The authors claim that although significantly affected by inclination shallowing, their result accounts for the lack of significant rotation in relation to the reference Cretaceous palaeomeridian.

## Cenozoic palaeomagnetic results

In the Outer Western and in the Central Carpathians the Cenozoic is represented by flysch of Palaeogene and subordinately Lower Miocene age. In the former area the flysch occurs in rootless nappes, while in the latter it forms the autochthonous cover of the basement units.

From the Silesian Nappe of the Outer Western Carpathians, mostly Krosno beds (and mostly claystone members) of Oligocene (also Lower Miocene in the east) age were collected from 24 localities concentrated in the western, central and eastern segments of the nappe, respectively. Using standard laboratory procedures, tectonically interpretable results were obtained for most localities (Fig. 5, light arrows plotted on the Silesian Nappe). In the

Table 2. *Summary of palaeomagnetic results from Cenozoic rocks in the Western Carpathians*

E. MÁRTON *ET AL.*

| | | Latitude, longitude (°N, °E) | Age of rocks | N/No | n/no | $D$ (°) | $I$ (°) | $k$ | $\alpha_{95}$ | $D_c$ (°) | $I_c$ (°) | $k$ | $\alpha_{95}$ (°) | Rotation (°) | Pole latitude (°N) | Pole longitude (°E) | $\delta p$ (°) | $\delta m$ (°) | Remark | Ref. |
|---|---|---|---|---|---|---|---|---|---|---|---|---|---|---|---|---|---|---|---|---|
| | *Outer Carpathian Nappes* | | | | | | | | | | | | | | | | | | | |
| 1 | Silesian nappe Claystone, siltstone | 49.3–49.9, 17.6–22.4 | Oligocene (1 locality Miocene) | **17/25** | | 328 | 40 | 8 | 13.1 | **307** | **62** | **20** | **8.2** | $-53 \pm 17$ | 54 | 302 | 10 | 13 | Rb, f+ | 1, 0 |
| 1a | Silesian nappe W Claystone, siltstone | 49.3–49.6, 17.6–18.7 | Oligocene | **5/6** | | 304 | 49 | 20 | 17.6 | **279** | **64** | **48** | **11.2** | $-81 \pm 26$ | 38 | 318 | 14 | 18 | Rc, f+ | 1 |
| 1b | Silesian nappe C Claystone, siltstone | 49.7–49.9, 20.7–21.9 | Oligocene | **7/8** | | 331 | 45 | 13 | 17.4 | **317** | **67** | **20** | **13.8** | $-43 \pm 35$ | 63 | 307 | 19 | 23 | Rc, f+ | 1 |
| 1c | Silesian nappe E Claystone, siltstone | 49.3–49.6, 21.6–22.4 | Oligocene (1 locality Miocene) | **5/11** | | 342 | 19 | 8 | 29.8 | **318** | **49** | **46** | **11.4** | $-42 \pm 17$ | 53 | 277 | 10 | 15 | Rc, f+ | 1, 0 |
| 1d | Silesian nappe E Claystone, siltstone | 49.3–49.6, 21.6–22.4 | Oligocene, post-folding | **5/11** | | **315** | **61** | **59** | **10.0** | 279 | 42.8 | 4 | 44.6 | $-45 \pm 16$ | 58 | 296 | 12 | 15 | f– | 1 |
| 2 | Dukla nappe Claystone, siltstone | 49.3–49.5, 21.2–22.1 | Oligocene | **7/10** | | 327 | 48 | 8 | 23.1 | **312** | **52** | **11** | **19.5** | $-48 \pm 32$ | 51 | 285 | 18 | 27 | syntilt | 0 |
| 3 | Dukla nappe Red Claystone | 49.16, 22.19 | Middle–Late Eocene | **5/15** | | | | | | 339 | 40 | | | $-21 \pm ??$ | 59 | 242 | 3 | 4 | | 2 |
| 4 | Magura nappe Claystone, siltstone | 49.5–49.7, 19.1–21.4 | Late Eocene–Oligocene, post-folding | **13/34** | | **310** | **56** | **18** | **10.0** | 284 | 52 | 6 | 18.4 | $-50 \pm 18$ | 52 | 290 | 10 | 14 | f– | 1 |
| 4a | Magura nappe W Claystone, siltstone | 49.5–49.6, 19.1–19.6 | Late Eocene–Oligocene | **5/10** | | **314** | **53** | **26** | **15.2** | 310 | 60 | 14 | 21.5 | $-46 \pm 25$ | 53 | 282 | 15 | 21 | fi | 1 |
| 4b | Magura nappe C Claystone, siltstone | 49.5–49.7, 20.0–20.6 | Late Eocene–Oligocene | **4/17** | | **323** | **53** | **14** | **25.8** | 306 | 48 | 5 | 47.6 | $-37 \pm 43$ | 58 | 274 | 25 | 36 | fi | 1 |
| 4c | Magura nappe E Claystone, siltstone | 49.5–49.6, 21.1–21.4 | Late Eocene-Oligocene | **4/7** | | **290** | **60** | **18** | **22.2** | 248 | 32 | 22 | 20.2 | $-70 \pm 44$ | 42 | 310 | 25 | 34 | fi | 1 |
| 5 | Oravska Magura Sandstone | 49.3–49.5, 19.1–19.4 | Late Paleocene–Middle Eocene | **4/15** | | | | | | 307 | 59 | 41 | 14.4 | $-53 \pm 28$ | 52 | 295 | 16 | 21 | | 3 |
| | *Pieniny Klippen Belt* | | | | | | | | | | | | | | | | | | | |
| 6 | Pieniny intrusions Andesite | 49.4–49.6, 20.2–20.5 | 25?–11 Ma? | **7/13** | | 314 | 56 | 26 | 12.1 | **314** | **56** | **26** | **12.1** | $-46 \pm 21$ | 55 | 287 | 12 | 17 | Rbi | 4 |
| | *Central Carpathian Palaeogene Basin* | | | | | | | | | | | | | | | | | | | |
| 7 | Podhale + Levoca basin Claystone, siltstone | 48.9–49.4, 19.9–21.2 | Eocene–Oligocene | **10/18** | | 299 | 55 | 39 | 7.8 | **301** | **55** | **76** | **5.6** | $-59 \pm 10$ | 46 | 295 | 6 | 8 | Rci, f+ | 5 |
| 7a | Podhale basin Claystone, siltstone | 49.3–49.4, 19.9–20.3 | Oligocene | **6/10** | | 298 | 54 | 37 | 11.1 | **298** | **53** | **121** | 6.1 | $-62 \pm 10$ | 42 | 295 | 6 | 8 | f+ | 6 |
| 7b | Levoca basin Claystone, siltstone | 48.9–49.2, 20.5–21.2 | Eocene–Oligocene | **4/8** | | 302 | 56 | 32 | 16.4 | **308** | **59** | **55** | **12.5** | $-52 \pm 24$ | 52 | 296 | 14 | 19 | Rci, fi | 5 |
| 8 | Sološnica limestone | 48.47, 17.23 | Eocene | | **11/15** | 163 | 62 | 13 | 13 | **286** | **66** | **13** | **13** | $-74 \pm 32$ | 43 | 316 | 17 | 21 | rev. | 7 |
| 9 | Roh Motel siltstone | 48.66, 17.50 | Eocene | | **6/6** | 21 | 6 | 29 | 13 | **250** | **21** | **29** | **13** | $-110 \pm 14$ | −5 | 310 | 7 | 14 | | 7 |
| 10 | Omastiná flysch | 48.78, 18.39 | Eocene | | **10/25** | 265 | 25 | 26 | 10 | **283** | **28** | **24** | **10** | $-77 \pm 11$ | 20 | 289 | 6 | 11 | rev. | 8 |
| 11 | Demjata flysch | 49.11, 21.32 | Eocene | | **9/13** | 295 | 12 | 18 | 13 | **255** | **54** | **9** | **19** | $-105 \pm 32$ | 17 | 325 | 19 | 27 | rev. | 8 |
| 12 | Lada claystone | 49.04, 21.36 | 20–18 Ma | | **20/21** | 289 | 57 | 14 | 9 | **281** | **30** | **14** | **9** | $-79 \pm 10$ | 19 | 294 | 6 | 10 | | 9 |

*Internal Carpathian Palaeogene Basin*

| | | | | | | | | | | | | | | | | | | | |
|---|---|---|---|---|---|---|---|---|---|---|---|---|---|---|---|---|---|---|---|
| 13 | Esztergom Castle hill siltstone | 47.79, 18.73 | Oligocene | | 9/25 | 279 | 43 | 24 | 11 | 293 | 33 | 24 | 11 | −67 ± 13 | 29 | 284 | 7 | 12 | rev. | 7 |
| 14 | Esztergom town marl | 47.80, 18.74 | Oligocene | | 10/10 | 303 | 38 | 44 | 7 | 315 | 49 | 44 | 7 | −45 ± 11 | 51 | 278 | 6 | 9 | | 7 |
| 15 | Lábatlan limestone | 47.72, 18.52 | Eocene | | 23/30 | 312 | 53 | 41 | 11 | 295 | 48 | 75 | 9 | −65 ± 13 | 37 | 292 | 8 | 12 | | 7 |
| 16 | Recsk, Lahóca hill andesite & contact metamorphosed Anisian limestone | 47.9−47.9, 20.1−20.1 | Late Eocene | 8/10 | | 287 | 55 | 62 | 7.1 | 287 | 55 | 62 | 7.1 | −73 ± 12 | 36 | 305 | 7 | 10 | rev. | 10 |
| 17 | Bükk limestone, turbidite, clay | 47.9−48.0, 20.4−20.6 | Late Eocene−Oligocene | 3/3 | | 295 | 39 | 28 | 23.6 | 275 | 55 | 33 | 21.9 | −85 ± 38 | 29 | 313 | 22 | 31 | fi, Roi | 10 |
| 18 | Nógrád + S. Slovak basin bentonite, clay, siltstone, sandstone | 48.0−48.4, 19.4−20.3 | 20.0−18.0 Ma | 8/9 | | 303 | 45 | 37 | 9.3 | 300 | 43 | 43 | 8.5 | −60 ± 12 | 38 | 286 | 7 | 11 | fi, Rc | 11, 14 |
| 19 | Nógrád + S. Slovak basin Rhyolite, tuff, ignimbrite | 48.0−48.2, 19.7−19.9 | 21.0−18.5 Ma | 9/9 | | 267 | 57 | 33 | 9.0 | 267 | 57 | 33 | 9.0 | −93 ± 16 | 25 | 318 | 10 | 13 | rev. | 11, 10 |
| 20 | Bükk Mountains Ignimbrite & tuff | 47.8−48.0, 20.3−20.7 | 21.0−18.5 Ma | 11/11 | | 283 | 48 | 51 | 6.5 | 283 | 48 | 51 | 6.5 | −77 ± 10 | 30 | 302 | 6 | 8 | rev. | 10, 12, 13 |
| 21 | Ipolytarnóc sediment & ignimbrite | 48.2−48.3, 19.6−19.7 | 17.5−17.2 Ma | 15/15 | | 328 | 57 | 70 | 4.6 | 328 | 57 | 70 | 4.6 | −32 ± 8 | 64 | 277 | 5 | 7 | Rb | 14 |
| 22 | Bükk & Cserhát Mountains aleurit, fluviatile silt | 47.8−48.3, 19.7−20.6 | 18.0−13.0 Ma | 4/4 | | 324 | 56 | 96 | 9.4 | 327 | 57 | 76 | 10.6 | −33 ± 19 | 64 | 279 | 11 | 15 | Fi | 10 |
| 23 | Bükk Mountains Ignimbrite & tuff | 47.9−48.1, 20.4−20.8 | 17.5−16.0 Ma | 28/28 | | 335 | 37 | 40 | 4.4 | 335 | 37 | 40 | 4.4 | −25 ± 5 | 56 | 246 | 3 | 5 | Rb | 10, 13 |
| 24 | Bükk & Mátra Mountains Ignimbrite & tuff | 47.8−48.3, 19.7−20.6 | 14.0−13.0 Ma | 5/6 | | 13 | 62 | 42 | 11.9 | 13 | 62 | 42 | 11.9 | +13 ± 26 | 80 | 133 | 14 | 19 | Rci | 10, 13, 15 |
| 24a | Demjén, Nagyeresztvény Rhyolite tuff | 47.84, 20.35 | 14.0 Ma | | 21/22 | 8 | 50 | 92 | 3.3 | 8 | 50 | 92 | 3.3 | +8 ± 5 | 72 | 178 | 3 | 4 | rev. | 10 |
| 24b | Felsőnyárád Dacie tuff | 48.33, 20.60 | 13.7−14.0 Ma | | 5/6 | 18 | 50 | 20 | 18 | 18 | 50 | 20 | 18 | +18 ± 28 | 68 | 156 | 16 | 24 | | 13 |
| 24c | Felnémet, quarry Dacitic ignimbrite | 47.93, 20.38 | 14.0−14.1 Ma | | 10/10 | 25 | 71 | 523 | 2.1 | 25 | 71 | 523 | 2.1 | +25 ± 6 | 73 | 75 | 3 | 4 | rev. | 13 |
| 24d | Felnémet, Bajusz-völgy Dacite-rhyolite tuff | 47.93, 20.39 | 14.0−14.1 Ma | | 8/9 | 353 | 75 | 210 | 3.8 | 353 | 75 | 210 | 3.8 | −7 ± 15 | 76 | 7 | 6 | 7 | rev. | 13 |
| 24e | Tar Dacite tuff | 47.95, 19.76 | 13.0−13.9 Ma | | 12/12 | 15 | 64 | 54 | 6 | 15 | 64 | 54 | 6 | +15 ± 14 | 80 | 116 | 8 | 10 | rev. | 15 |
| 24f | Lénárddaróc Rhyolite tuff | 48.15, 20.36 | 13.5−11.5 Ma | | 8/9 | 354 | 23 | 33 | 10.0 | 354 | 23 | 33 | 10.0 | −6 ± 11 | 53 | 210 | 6 | 11 | | 10 |
| 25 | Cserhát Mountains andesite, ignimbrite | 47.9−48.1, 19.5−19.7 | 16.6−14.7 Ma | 6/7 | | 322 | 56 | 19 | 15.6 | 322 | 56 | 19 | 15.6 | −38 ± 28 | 60 | 281 | 16 | 22 | rev. | 10, 16 |
| 26 | Cserhát Mountains andesite, andesite tuff | 47.9−48.1, 19.5−19.7 | 14.5−12.0 Ma | 7/7 | | 6 | 64 | 81 | 6.7 | 6 | 64 | 81 | 6.7 | +6 ± 15 | 85 | 137 | 8 | 11 | Rc | 16 |
| 27 | Börzsöny Mountains dacite tuff & andesite | 47.8−48.1, 18.8−19.1 | 16.5−14.5 Ma | 40/42 | | 339 | 61 | 31 | 4.1 | 339 | 61 | 31 | 4.1 | −21 ± 8 | 75 | 275 | 5 | 6 | Ra | 17, 18 |
| 28 | Visegrád Mountains dacite tuff & andesite | 47.7−47.8, 18.8−19.1 | 15.5−15.3 Ma | 12/12 | | 337 | 56 | 21 | 9.0 | 337 | 56 | 21 | 9.0 | −23 ± 16 | 70 | 264 | 9 | 13 | rev. | 18, 19, 20 |

(*Continued*)

**Table 2.** *Summary of palaeomagnetic results from Cenozoic rocks in the Western Carpathians* (*Continued*)

| | | Latitude, longitude (°N, °E) | Age of rocks | N/No | n/no | $D$ (°) | $I$ (°) | $k$ | $\alpha_{95}$ | $D_c$ (°) | $I_c$ (°) | $k$ | $\alpha_{95}$ (°) | Rotation (°) | Pole latitude (°N) | Pole longitude (°E) | $\delta p$ (°) | $\delta m$ (°) | Remark | Ref. |
|---|---|---|---|---|---|---|---|---|---|---|---|---|---|---|---|---|---|---|---|---|
| 29 | Börzsöny Mountains dacite tuff & andesite | 47.8–48.1, 18.8–19.1 | younger than 13.7 Ma | **23/24** | | **20** | **61** | **44** | **4.6** | **20** | **61** | **44** | **4.6** | $+20 \pm 9$ | 75 | 124 | 5 | 7 | Ra | 17, 18 |
| 30 | Visegrád Mountains andesite & dacite | 47.7–47.8, 18.8–19.1 | younger than 14.5 Ma | **17/17** | | **12** | **55** | **43** | **5.5** | **12** | **55** | **43** | **5.5** | $+12 \pm 9$ | 75 | 159 | 6 | 8 | rev. | 18, 19 |
| 31 | East Slovak Basin Zeolithized tuff | 48.7–48.9, 21.7–21.8 | 16.0–11.5 Ma | **5/8** | | 351 | 53 | 12 | 23.1 | **326** | **65** | **35** | **13.1** | $-34 \pm 31$ | 67 | 300 | 17 | 21 | f+, Ri– | 9 |
| 32 | Zemplín Mountains & East Slovak Lowland andesite, rhyolite, rhyodacite, tuff, zeolithized tuff | 48.3–48.5, 21.7–22.0 | 16.0–11.5 Ma | **23/23** | | **330** | **51** | **18** | **7.4** | **330** | **51** | **18** | **7.4** | $-30 \pm 12$ | 62 | 266 | 7 | 10 | R– | 21 |
| 33 | Vihorlatské Mountains andesite & rhyodacite | 48.8–49.0, 21.9–22.4 | 12.6–11.9 Ma | **5/5** | | 341 | 46 | 21 | 17.1 | **341** | **46** | **21** | **17.1** | $-19 \pm 25$ | 64 | 243 | 14 | 22 | Ra | 22 |
| 34 | Vihorlatské Mountains andesite & rhyodacite | 48.8–49.0, 21.9–22.4 | 11.9–9.4 Ma | **4/4** | | **0** | **59** | **119** | **8.4** | **0** | **59** | **119** | **8.4** | $0 \pm 16$ | 81 | 202 | 9 | 12 | rev. | 22 |

References: 0. Kiss *et al.* (2015); 1. Márton *et al.* (2009*a*); 2. Korab *et al.* (1981); 3. Krs *et al.* (1991); 4. Márton *et al.* (2004); 5. Márton *et al.* (2009*b*); 6. Márton *et al.* (1999); 7. Márton *et al.* (1992); 8. Túnyi & Márton (1996); 9. Márton *et al.* (2000); 10. Márton & Márton (1996); 11. Márton *et al.* (1996); 12. Márton & Pécskay (1998); 13. Márton *et al.* (2007*b*); 14. Márton *et al.* (2007*a*); 15. Zelenka *et al.* (2005); 16. Póka *et al.* (2004); 17. Karátson *et al.* (2000); 18. Balla & Márton-Szalay (1978); 19. Karátson *et al.* (2007); 20. Lantos pers. comm., 2005; 21. Orlický (1996); 22. Túnyi *et al.* (2004).

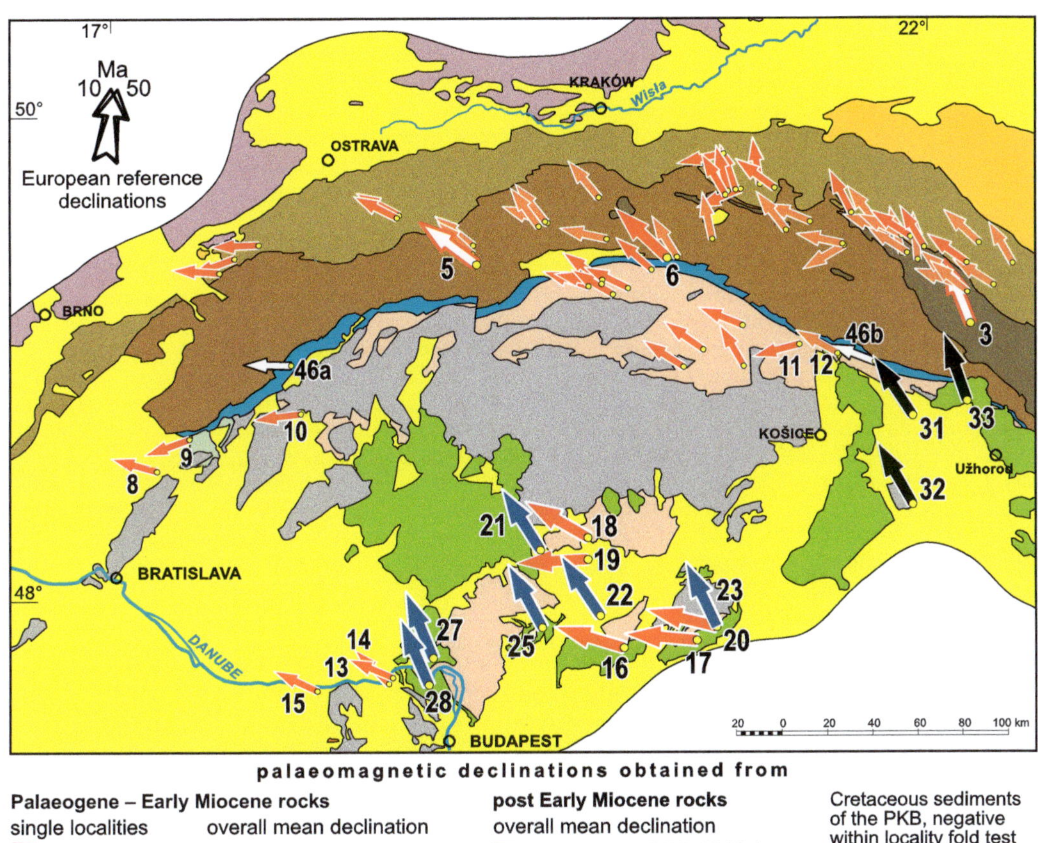


**Fig. 5.** Palaeomagnetic declinations obtained from Cenozoic rocks and two localities of the Upper Cretacous red marls from the Pieniny Klippen Belt with post-folding remanence. Numbers up to 33 refer to Table 2, 46a and 46b to Table 1.

western and central segments the characteristic magnetizations are of pre-folding age (Table 2; items 1a and 1b). In the eastern segment, five localities have primary (Table 2, 1c) and five post-folding (Table 2, item 1d) remanences, both groups exhibiting anticlockwise rotations of similar angle. The overall mean palaeomagnetic directions from the central and eastern segments show *c.* 60° anticlockwise rotation, while the palaeodeclination for the western segment is *c.* 80°.

The earlier published palaeomagnetic results from the Dukla Nappe (Table 2, item 3) are poorly documented. Judging from the original paper (Korab *et al.* 1981), the results of blanket thermal demagnetization at 350°C suggesting moderate anticlockwise rotation were interpreted in terms of tectonics. It is quite possible that this temperature was not high enough to completely remove

the possibly present overprint in the red sediments; the result should therefore be considered as indicative of rotation, although the angle remains uncertain. Table 2, item 2 summarizes the recently obtained and so-far-unpublished results by Márton and Tokarski, which suggest synfolding magnetization of the grey clastic sediments exhibiting anticlockwise rotations (Fig. 5, light arrows in the Dukla Nappe).

From the Magura Nappe of the Outer Western Carpathians, Upper Eocene and Oligocene flysch (fine-grained sandstones, siltstones and marls) was studied at 34 geographically distributed localities; 13 of these yielded statistically acceptable palaeomagnetic results (Table 2, item 4; Fig. 5, light arrows plotted on the Magura Nappe). On restoring the strata to the horizontal, the statistical parameters of the overall mean direction became worse, that is, the remanence is of post-folding/tilting age. Nevertheless, the locality mean directions are fairly consistent before tectonic correction and indicate *c.* 60° anticlockwise rotation with respect to the present north (Márton *et al.* 2009*a*).

Samples were taken from two sub-basins of the Central Carpathian Palaeogene Basin: the Podhale north of the Tatra Mountains, and the Levoča east of the Tatra Mountains. From the sampled 18 localities, 10 could be evaluated from a tectonic point of view (Fig. 5, light arrows without numbers in the Central Carpathian Palaeogene Basin). They indicate *c.* 60° anticlockwise rotation (Table 2, item 7) with positive fold/tilt test (Márton *et al.* 1999, 2009*b*). It was also documented that the palaeomagnetic directions are current independent, that is, the generally east–west-oriented sedimentary transport could not bias the palaeomagentic directions from that of the ambient Earth magnetic field (Márton *et al.* 2009*b*). Previously published palaeomagnetic directions interpreted as being of pre-tilting age for one locality from the eastern part of the Levoča basin (Fig. 5; Table 2, item 11; Túnyi & Márton 1996) and from the contact between the Levoča and East Slovak basins (Fig. 5; Table 2, item 12; Márton *et al.* 2000), and one Upper Cretaceous locality with documented secondary remanence (Fig. 5; Table 2, item 46*b*; Márton *et al.* 2013), all exhibited large anticlockwise rotations. Such a statement can also be made of the area west of the Central Carpathian Palaeogene basin (Fig. 5; Table 2, items 8–10; Túnyi & Márton 1996), interpreted as primary remanences, and the post-tilting remanence at the western end of the Pieniny Klippen Belt, measured on Upper Cretaceous red marls (Fig. 5; Table 2, item 46*a*; Márton *et al.* 2013).

In the Inner Carpathian area there is another large Palaeogene basin (Fig. 1) from which not only Palaeogene palaeomagnetic results are available, but also younger results. Three sedimentary Palaeogene localities near the river Danube (Fig. 5; Table 2, items 13–15; Túnyi & Márton 1996) and three from the southern margin of the Bükk Mountains (Márton & Márton 1996) indicate large anticlockwise rotation. Similar values were obtained from Upper Eocene andesites and thermally altered contact sediments from the Mátra Mountains (Fig. 5; Table 2, item 16), 11 Lower Miocene ignimbrite sites from the Bükk Mountains (Fig. 4; Table 2, item 20) as well as from sediments (Fig. 5; Table 2, item 18) and ignimbrites (Fig. 5; Table 2, item 19) of the Nógrád–Novohrad basin of northern Hungary (Márton & Márton 1996) and southern Slovakia (Márton *et al.* 1996). The mentioned sediments and volcanic rocks are in autochthonous position above Internal Carpathian basement units (Figs 1 & 2). At several places,

they are covered by Middle Miocene volcanic rocks and sediments which exhibit only *c.* 30° anticlockwise rotation (Fig. 5; Table 2, items 21–23, 25–28), while the Upper Miocene rocks are characterized by slight clockwise or no rotation with respect to the present north (Table 2, items 24a–f). As the volcanic horizons are well dated by the K–Ar method and most sediments by the biostratigraphic method, the age of the rotations are fairly well constrained in the area of the Inner Carpathian Palaeogene Basin. The significance of this age control for the Cenozoic displacements is discussed in the next section.

The East Slovak Basin is separated from the rest of the Pannonian Basin by a volcanic chain of the Tokaj–Slanec Mountains. In this area, including the Vihorlatské Mountains, anticlockwise-rotated palaeomagnetic directions characterize most of the magmatic rocks (Table 2, items 31–33) suggesting that the rotation is younger here than in the Central and Internal Carpathian Palaeogene basins.

## Discussion and tectonic interpretation of the palaeomagnetic results

### Cenozoic results

The Cenozoic palaeomagnetic results from the Western Carpathians represent overstep sequences for the Inner Carpathians, while in the Outer Carpathians the studied Palaeogene sediments were folded and thrusted until the Miocene. Despite the different tectonic settings, there is a remarkable consistency within and between the two areas in palaeomagnetic declinations measured on Palaeogene rocks, indicating large anticlockwise rotation (Fig. 6). As documented in the original papers (for citations see Table 2), the results were obtained using modern laboratory procedures and evaluation. Before tilt corrections, the directions of the characteristic remanences were typically far from that of Earth's the present magnetic field at the sampling areas, proving long-term stability. According to between-locality fold/tilt tests and reversal tests (where applicable), the characteristic remanences are primary or of synfolding age. In the latter case, the acquisition time is quite close to the time of deposition. The Magura Nappe (Table 2, item 4) and a population of localities in the eastern segment of the Silesian Nappe (Table 2, item 4d) are notable exceptions. However, the magnetizations of post-folding age are regionally consistent in both areas, therefore valuable in tectonic interpretation. The results summarized in Table 2 (depicted in Fig. 7a–c) suggest that during the Miocene the Inner Carpathians (an already consolidated block) accreted the sediments of the Outer Carpathians.

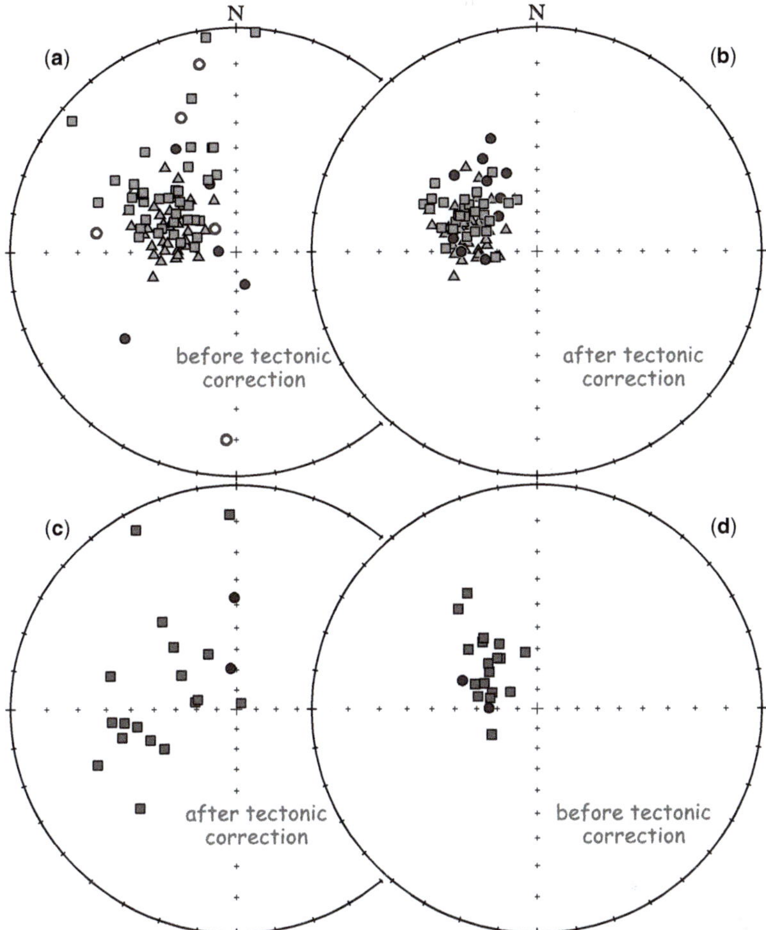



**Fig. 6.** Palaeomagnetic locality mean directions (providing the entry data for calculating overall mean palaeomagnetic directions tabulated in Table 2) from the dominantly Palaeogene sediments (squares) from the Silesian (Table 2, rows 1a–d) and Magura nappes (Table 2, row 4) of the Outer Western Carpathians, from the Central (Table 2, row 7) and Inner Carpathian (Table 2, rows 13–15, 17 and 18) Palaeogene basins, from the latest Cretaceous red marls (Table 1, rows 45 and 46) of the Pieniny Klippen Belt (dots and circles) and from the igneous rocks of the Pieniny Klippen Belt (Table 2, row 6) and of the Inner Carpathian Palaeogene basin (Table 2, rows 16, 19 and 20). Stereographic projections, full/empty symbols are vectors pointing downwards/upwards, representing vectors with positive and negative inclinations, respectively. Palaeomagnetic directions interpreted as of pre-folding/tilting age (**a**) before tilt corrections and (**b**) after tilt corrections. Note that in the Pieniny Klippen Belt, overturned strata also occur and these have negative inclinations before and positive after tectonic corrections. Palaeomagnetic directions interpreted as of post-folding/tilting age (**c**) after tilt corrections and (**d**) before tilt corrections from the PKB (Table 1, rows 46a and 46b), the Magura Nappe (Table 2, row 4) and the Silesian Nappe (Table 2, row 1d).

Relative rotations within and between nappes seem to be within the resolution of the palaeomagnetic data. Some differences in declinations within the central segment of the Silesian Nappe and in the Dukla Nappe can be attributed to post-folding uplift, a larger average rotation in the western segment of the Silesian Nappe than elsewhere to strike-slip displacements, also post-dating the folding (Márton *et al.* 2009a).

An interesting aspect of the palaeomagnetic dataset is that overall mean palaeomagnetic declinations of the vectors, both of pre- and post-folding ages (Table 2, items 1–4; Fig. 7a–c), are practically coincident. It follows that the internal deformation of the nappes must have taken place between the two acquisition times. In the Silesian and Dukla nappes, where lineations of the Anisotropy of the Magnetic Susceptibility are well defined, there is a

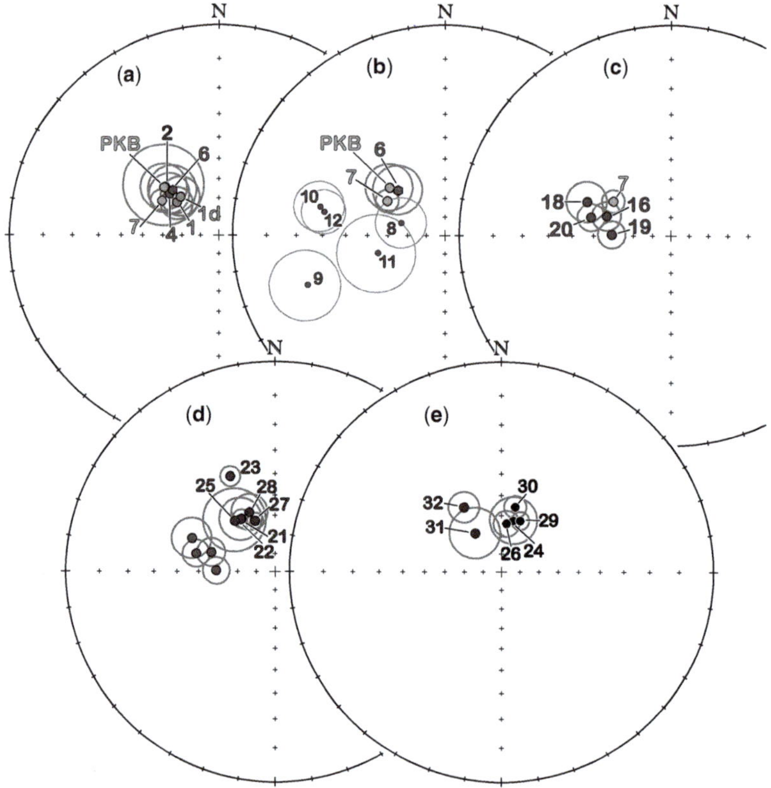

**Fig. 7.** Palaeomagnetic directions with confidence circles for the Palaeogene rocks of the Outer Carpathian nappes, the Upper Cretaceous red marls and the Miocene andesites of the Pieniny Klippen Belt and the overstep sequences of the Internal Carpathians. Stereographic projections; all directions are plotted as of normal polarity. (**a**) It is documented that the Miocene overall rotations of the Outer Carpahian nappes (Silesian, Dukla and Magura; Table 2, items 1: prefolding, 1d: post-folding, 2: synfolding, 4: post-folding, respectively), the Pieniny Klippen belt (Upper Cretaceous red marls, Table 1, item 45: prefolding, Pieniny andesites, Table 2, item 6) and the Podhale–Levoča basin (Table 2, item 7) are the same. (**b**) The Palaeogene sediments (single localities) outside of the Podhale–Levoca basin (Table 2, items 8–12) also exhibit large anticlockwise rotations. (**c**) The Podhale–Levoča basin rotated in co-ordination with the Internal Carpathian Palaeogene basin (Table 2, items 16, 18–20) during the Miocene. (**d**) The net Miocene rotation is the results of two rotational phases, since rocks from the Internal Carpathian Palogene basin younger than 17.7 Ma (Table 2, items 21–23, 25, 27 and 28) exhibit only moderate anticlockwise rotation. (**e**) When the rotations were over at that location (Table 2, items 24, 28–30), they are still ongoing (Table 2, items 31 and 32) in the East Slovak Basin.

good correlation between them and the strikes of the sampled beds (Fig. 8). In other words, the magnetic fabric was imprinted during the compressional regime which prevailed during folding and thrusting of the nappes. The order of the events which were important from palaeomagnetic and tectonic aspects is therefore: (1) acquisition of the remanent magnetizations of pre-folding age before the Miocene; (2) internal deformation of the nappes, during which remanences of synfolding age and the AMS lineations were acquired; and (3) co-ordinated rotation of the nappes together with the Inner Carpathian block.

The co-ordinated rotation of the Outer and Inner Western Carpathians must have taken place during two time periods: 18.5–17.5 and 16.0–14.5 Ma. The time constraints are due to the Internal Carpathian (north Hungarian–south Slovakian) Palaeogene Basin, where a large number of well-dated localities/sites in sedimentary and igneous rocks yielded good palaeomagnetic results (Table 2; Fig. 7d, e). Here, sediments and igneous rocks of age >18.5 Ma exhibit *c*. 60° anticlockwise rotation, those of age 17.5–16.0 Ma *c*. 30° and those of age <14.5 Ma are characterized by declinations which are close to what is expected in a stable European framework. This means that the final thrusting of the Outer Western Carpathians at *c*. 11.5 Ma post-dates the process of the docking of the Outer Carpathian nappes by anticlockwise

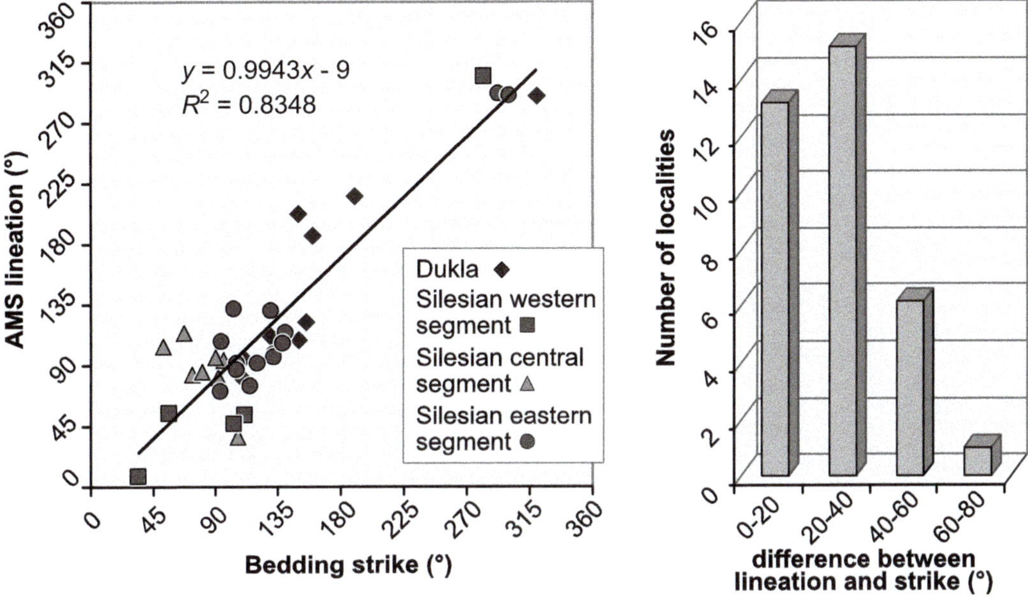



**Fig. 8.** Correlation between local tectonic strikes and AMS lineations in the Silesian and Dukla Nappes, due to compressional tectonics.

rotation to the southern margin of the European plate.

By the time the rotations were over in the Internal Carpathian Palaeogene Basin, anticlockwise rotation was still in progress in the East Slovak Basin (Fig. 7e). The tectonic implication of this result, which is corroborated by those from the Maramures area in Romania (Márton *et al.* 2007c), is still to be explored.

## Mesozoic and Palaeozoic results

Interpretation of the Mesozoic palaeomagnetic data from the Western Carpathian area poses more problems than the Cenozoic data. As can be seen from Figures 3 and 4, senses and amounts of rotations are varied, even within the same tectonic units. Regional syntectonic remagnetizations were documented in several tectonic units (such as Silica, Krížna and Choč). An additional problem is related to the complexity of the tectonic correction: Mesozoic rocks were deformed during nappe thrusting during the Late Cretaceous, and the geometry of these Eo-Alpine structures was definitely modified during the Neogene uplifts of the 'core mountains'. In order to reconstruct the position of Mesozoic nappe stacks during Late Cretaceous–Neogene time, a tectonic correction for the overlying Palaeogene rocks is applied which in the Tatra Mountains amounts to 10° (azimuth)/30° (tilt around horizontal axis) (e.g. Grabowski 1997; Grabowski *et al.*

2009; Szaniawski *et al.* 2012). In fact, all secondary magnetizations of inferred synthrusting age are of limited value for regional palaeotectonic interpretations, although they appeared to be useful in local tectonic reconstructions (e.g. position of beds during intermediate stages of thrusting; e.g. Grabowski 2000; Grabowski *et al.* 2009).

## Areas with consistent and partly consistent sense and magnitude of rotations observed on Cenozoic and older rocks

Consistent anticlockwise-rotated palaeodeclinations in both primary and secondary palaeomagnetic components are observed within the Triassic rocks of the Silica unit. The observations are in good agreement with the results from the surrounding Cenozoic basins, and indicate that the rotations observed within the Mesozoic rocks of the Silica unit are predominantly of Cenozoic age. The Silica units must have rotated together with the ALCAPA superunit and no significant rotation of Silica took place during Triassic–Miocene times.

Coincidence between the Mesozoic and Cenozoic palaeodeclinations is observed in the Malé Karpaty Mountains. The detected anticlockwise rotations of the Krížna unit and the Palaeogene cover are identical and therefore must be interpreted as Neogene. The same sense of rotation with a somewhat larger angle can be inferred for the Permian volcanic rocks of the Choč Nappe, although the almost equatorial inclination also allows an alternative interpretation of large clockwise rotation of the Choč Nappe in the Malé Karpaty.

Participation of the Pieniny Klippen Belt in the Miocene anticlockwise rotation of the Alp–Carpathian–Pannonian superunit is well constrained (Márton *et al.* 2004, 2013), since the overall mean primary magnetizatons of the Upper Cretaceous pelagic marls exhibit similar rotation to the oldest group of Cenozoic rocks from the Inner Western Carpathians (Fig. 7a). It might therefore be accepted that PKB units have not rotated significantly between the Late Cretaceous (Albian?) and Miocene, except some small-scale disturbances that might be attributed to weak oroclinal bending (Márton *et al.* 2013). The pre-Late Cretaceous dataset from the same belt clearly disagrees with the Late Cretaceous data. Callovian–Kimmeridgian results from the central part of the Pieniny Klippen Belt (Grabowski *et al.* 2008) exhibit only weak anticlockwise rotations. The data might be interpreted in favour of a (possibly) local clockwise rotation of the central part of the belt during Late Jurassic–Late Cretaceous time. However, when considering the intensive tectonic reworking of the Pieniny Klippen Belt (e.g. Birkenmajer 1986; Ratschbacher *et al.* 1993; Nemčok & Nemčok 1994; Plašienka 2012*a*), it is possible that different palaeodeclination trends observed result from local tectonic rotations and/or disharmonic tectonic deformations of hard blocks of Upper Jurassic Czorsztyn limestones and relatively more competent Upper Cretaceous marls.

Mesozoic (Krs *et al.* 1982, 1993, 1996) and Cenozoic (Márton *et al.* 2009*a*) results from the western part of the Outer Western Carpathians indicate generally consistent anticlockwise rotations of both Magura and Silesian nappes which must have occurred after the Oligocene. Non-rotated Mesozoic palaeodeclinations were reported east of 19° East from two localities of the Lower Cretaceous teschenitic rocks of the Silesian Nappe (Grabowski *et al.* 2006) and the Upper Cretaceous marls of Subsilesian Nappe (Szaniawski *et al.* 2013). At the present state of the investigations, it is too early for their robust tectonic interpretation. The apparent lack of rotations in the Mesozoic of the central part of the Silesian unit (Grabowski *et al.* 2006) needs confirmation from well-bedded sedimentary rocks. A secondary nature of magnetization cannot be excluded; although both fold and contact tests were positive, a very steep palaeo-inclination observed (between 60° and 70°, which corresponds to a latitude of *c.* 48° N) is hardly acceptable for the Early Cretaceous in the area. Although they pass the fold test, characteristic directions of Szaniawski *et al.* (2013) require additional assumptions for a proper interpretation; the inclination is far too shallow for the Late Cretaceous, and results from more geographically distributed localities are needed to verify the regional significance.

*Areas with non-consistent sense and magnitude of rotations observed on Cenozoic and older rocks*

The best-documented contrast between Palaeogene and Mesozoic palaeodeclinations occurs in the area of the High Tatra Mountains. The difference between palaeodeclinations from the Central Carpathian Palaeogene Basin (Márton *et al.* 1999, 2009*b*) and the Tatric and Fatric sequences (e.g. Kądziałko-Hofmokl & Kruczyk 1987; Grabowski 2005) amounts to 90°. The logical consequence of those observations is that the area must have rotated clockwise by an amount of *c.* 90° between the Late Cretaceous (Turonian?) and Oligocene (Grabowski & Nemčok 1999; Grabowski 2005). The nature and geographical extent of the areas involved in the rotation remains unclear. Parallelization of this rotation with that which affected the Northern Calcareous Alps between the Barremian and Danian (Mauritsch & Márton 1995; Grabowski 2005) is also hypothetical, since palaeomagnetic interpretations in the Northern Calcareous Alps have been substantially modified (Pueyo *et al.* 2007). It is noteworthy that the results from the Tatra Mountains were obtained from all major tectonic units: Tatricum (both basement rocks and sedimentary cover), Fatricum and Hronicum. All primary and most secondary Mesozoic directions reveal a NNE–E-rotated trend of palaeomagnetic declinations. The Neogene uplift was not likely to have significantly affected the primary Mesozoic palaeodeclinations (Grabowski 2005; Szaniawski *et al.* 2012) since the bedding azimuth of the studied Mesozoic beds is very close to those of the Palaeogene cover (see also Piotrowski 1978).

Most palaeomagnetic results from the Mesozoic rocks between the High Tatra and Malé Karpaty Mountains were obtained from the Krížna Nappe. It must be kept in mind that a part of these directions (Kruczyk *et al.* 1992) most probably represents syntectonic remagnetizations and a real structural context of their acquisition is not clear. However, a trend of increasing anticlockwise rotations is observed along the strike of the orogen (Grabowski *et al.* 2010), supported by the sporadic results from Tatric and Manin units (Pruner *et al.* 1998). As a whole, the data might be interpreted as an effect of oroclinal bending (Kruczyk *et al.* 1992) or radial thrusting. In this case, the difference between Palaeogene and Mesozoic palaeodeclinations would have decreased southwest of the High

Tatra Mountains. An alternative explanation is that the rotations within the Krížna Nappe might be related to the different tilting azimuths of different massifs during the Neogene (Szaniawski *et al.* 2012). Indeed, massifs uplifted around a roughly east–west-oriented axis (such as the Tatra, Nizke Tatry or Choč Mountains) reveal no rotation or only a slight clockwise rotation; those uplifted around SW–NE-aligned axes show anticlockwise rotations (Malá Fatra, Strážovske Vrchy and Malé Karpaty). A single locality in the Spišska Magura, located just at the prominent Neogene strike-slip fault (Sperner *et al.* 2002), demonstrated extreme clockwise rotation. Definite verification of either model is not possible without further integrated palaeomagnetic and field studies.

Palaeozoic and a single Mesozoic result from the central and eastern part of the Choč unit reveal large rotations. The Palaeozoic results should be verified since the results were obtained in the early days of palaeomagnetism and details of demagnetization or the interpretation of characteristic directions were not clearly presented. If we accept the reality of clockwise-rotated declinations in the Choč Nappe and correct them for the effect of the anticlockwise rotation of the entire Alp–Carpathian–Pannonian unit in the Miocene, we have to incorporate a huge clockwise rotation (*c.* 80–130°) of the Choč Nappe before Neogene time.

## *Palaeogene–Neogene(?) remagnetizations of the Mesozoic rocks in the Central Western Carpathians and the Pieniny Klippen Belt*

Documentation of secondary magnetizations of Palaeogene or Neogene age in the Mesozoic rocks is very important, since it creates a link between the Mesozoic and Cenozoic kinematics. It has already been mentioned (see 'Pieniny Klippen Belt' subsection in 'Permian and Mesozoic palaeomagnetic results' section) that the presence of consistent postfolding remagnetizations of reversed polarity and presumed Neogene age has been reported from the western, central and eastern sectors of the Pieniny Klippen Belt (Grabowski *et al.* 2008; Márton *et al.* 2013).

No similar directions have yet been reported from the rocks of the Krížna Nappe. However, reversed polarity magnetizations documented in two localities of the Tatricum unit of the Tatra Mountains (component C of Grabowski 1997) might be interpreted as overprints acquired just before the Neogene uplift. Having rotated the Tatra Mountains back to their pre-Neogene position (see section on 'Mesozoic and Palaeozoic results') the reversed overprints attain a declination of 140°, which is again close to the secondary directions

from the PKB and primary directions from the Pieniny andesites (Márton *et al.* 2004). It seems that a persistent reversed polarity remagnetization, which might have been coeval with intrusions of the Pieniny andesites (Birkenmajer & Pécskay 2000), also affected the Mesozoic rocks. Its age might be related to the so-called 'Mid-Miocene thermal event', which caused resetting of the fission track ages of some of the Oligocene rocks in the Podhale Basin (Danišik *et al.* 2012; Anczkiewicz *et al.* 2013).

### *Mesozoic palaeolatitudes*

It is evident from Table 1 that only a small amount of the available palaeomagnetic data might be suitable for palaeolatitude estimations. All syntectonic remagnetizations must be excluded since (1) their acquisition took place at different stages of the thrusting processes in the Late Cretaceous; and (2) the position of beds during acquisition of secondary magnetization is not always possible to reconstruct. It must be stressed that the data of Kądziałko-Hofmokl & Kruczyk (1987), which for a long time have been included within the reference Middle–Late Jurassic poles for the European Platform as 'Subtatric nappe sediments, Poland' (e.g. Besse & Courtillot 1991, 2002, 2003; Van der Voo 1993; Torsvik *et al.* 2012), must also be included in the group of synthrusting remagnetizations. This is supported not only by incomplete demagnetization, but also by the new results from the coeval rocks of the Pieniny Klippen Belt. Primary palaeoinclinations of the Middle–Upper Jurassic rocks of the Pieniny Klippen Belt are significantly shallower than those from the Krížna unit (Kądziałko-Hofmokl & Kruczyk 1987; *c.* 40° N latitude) and the difference is statistically significant. As it is extremely unlikely that the depositional area of the Krížna unit was situated north of the Pieniny Klippen Belt (e.g. Birkenmajer 1986; Plašienka *et al.* 1997) the secondary nature of the component from the Krížna Nappe, earlier interpreted as primary, must be accepted.

Despite the need to reject a number of data, the palaeolatitudinal drift history of the Central and Internal Western Carpathians together with the Pieniny Klippen Belt might be roughly constrained (Fig. 9). The Lower Triassic palaeo-inclinations obtained from the northern (Szaniawski *et al.* 2012) and southern periphery of the Inner Western Carpathians (Márton *et al.* 1988) are quite concordant and account for the palaeogeographic position of the area at palaeolatitude *c.* 11–13° N. For the rest of the Triassic, results are available from the Silica and Bodva nappes (Márton *et al.* 1988, 1991; Channell *et al.* 2003). They indicate a quick northwards drift from *c.* 23–25° N in the Middle

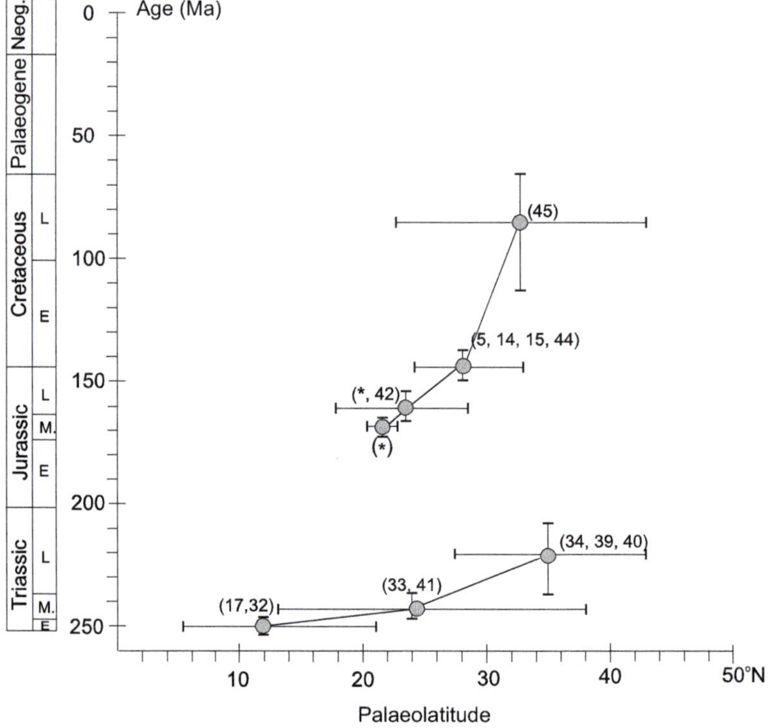



**Fig. 9.** Palaeolatitudinal drift of the Inner Carpathians and Pieniny Klippen Belt in the Mesozoic. Numbers corresponds to entries in Table 1. Asterisk (*) denote data from Jeleńska *et al.* (2011) not enclosed into Table 1.

Triassic up to 34° N in the Late Triassic. No data exist for the Lower Jurassic. Middle–Upper Jurassic results from the Polish sector of the Pieniny Klippen Belt point to *c.* 22° N (±5°) latitude during the Callovian–Kimmeridgian (Grabowski *et al.* 2008). The relatively low palaeolatitudes for the Bajocian (21.7° N ± 1.5°) and Oxfordian–Kimmeridgian (24.6° N ± 5.6°) were also reported for the western Slovak sector of the belt by Jeleńska *et al.* (2011). These observations might be interpreted in favour of a southwards drift of the Western Carpathians between the Late Triassic and Middle Jurassic. This interpretation is in good accordance with the model of the southwards drift of the Adriatic Plate during the Lower–Middle Jurassic, due to an opening of the Liguria–Piedmont ocean (Muttoni *et al.* 2005, 2013) which continued northeastwards during the Carpathian realm as the Vahic–Magura Ocean (e.g. Plašienka 2012*b*). It should be noted that southwards drift between the Middle Jurassic (Bajocian–Bathonian) and middle Oxfordian was documented in the easternmost part of the Pieniny Klippen Belt in Ukraine (Lewandowski *et al.* 2005), but its timing was apparently slightly later than in the Western Carpathians.

Abundant data from the Berriasian (Houša *et al.* 1996; Grabowski 2005; Grabowski *et al.* 2009, 2010) are very consistent and indicate a palaeolatitude of *c.* 28° N for both the Central Western Carpathians and Pieniny Klippen Belt. The northwards drift of both units in the latest Jurassic is again concordant with the similar palaeolatitudinal shift of the Lombardian Basin (Muttoni *et al.* 2005) and Ukrainian part of the Pieniny Klippen Belt (Lewandowski *et al.* 2005).

The northwards drift apparently continued throughout the Cretaceous, similar to that of the Adriatic Plate (Márton *et al.* 2010) since the mean Late Cretaceous palaeolatitude of the Pieniny Klippen Belt amounts to 33° N ± 10° (Márton *et al.* 2013).

The Mesozoic palaeolatitude data for the Outer Western Carpathians are sparse. Very high palaeolatitudes calculated from the Lower Cretaceous teschenitic rocks (*c.* 43° N; Krs *et al.* 1982, 1996; Grabowski *et al.* 2006) correspond to expected Cenozoic values (Besse & Courtillot 2002, 2003). The mean value for the four studies of the Upper Cretaceous rocks of the Magura and Silesian Nappes (Krs *et al.* 1982, 1996) is *c.* 34° N, which is not far from the coeval results from the Pieniny Klippen Belt (Márton *et al.* 2013).

## Conclusions

The present review focused on the tectonic applications of the palaeomagnetic results from the Western Carpathians, which are subdivided into an Outer and an Inner Western Carpathian nappe system, separated by the narrow, tectonically complicated zone referred to as the Pieniny Klippen Belt. The Inner Western Carpathians are further subdivided into a Central and an Internal Carpathians, which are in contact along the Meliata Suture. The nappes of the Outer Western Carpathians are rootless and were emplaced during the Early Miocene. The overstep sequences here are Miocene sediments occurring in isolated basins. The basement of the Inner Western Carpathians is built up of several nappe stacks (some thick skinned) consisting of basement and cover units, while the others are only cover units. Nappe transport terminated in the Inner Carpathians during the Late Cretaceous and the nappes are discordantly covered by latest Cretaceous–Miocene sediments and Neogene igneous rocks. The palaeomagnetic results discussed in this paper represent the above-mentioned main units, and their analysis leads to the following conclusions.

(1) The Internal Western Carpathians are characterized by a high diversity of the palaeomagnetic declinations obtained for the basement rocks. The results, interpreted partly as pre-, syn- and post-folding age, suggest important relative rotations during nappe transport and uplift of the 'core mountains' and possibly by oroclinal bending.

(2) Overall mean palaeomagnetic declinations (based on several sites/localities) in the Central Carpathians depart only slightly from the present north in the Tatra Mountains, while moderate westerly declinations occur towards the west. Exceptions are the Little Carpathians and the Silica Nappe, where large anticlockwise rotations are documented. Single sites/localities usually indicate large clockwise or anticlockwise rotations. Checking on their quality (basically applicable to Permian results) and their tectonic significance (local anomalies connected to tectonic lines?) is to be explored in the future.

(3) Overstep sequences in the Internal Western Carpathians (namely in the Central and Internal Carpathian Palaeogene basins) exhibit large and fairly consistent anticlockwise rotation up to the age of 18.5 Ma, which is similar in sense and magnitude to the net post-Late Cretaceous rotation of the Pieniny Klippen Belt. These observations prove that the named areas formed a single block during the Miocene. The block rotated first in the anticlockwise sense between 18.5 and 17.5 Ma, then during the interval of 16.0–14.5 Ma.

(4) The above-described Miocene anticlockwise rotations must have affected the basement of the Internal Western Carpathians. Consequently, the basement areas exhibiting less anticlockwise rotation than the overstep sequences must have rotated in the clockwise sense before the Late Cretaceous.

(5) Most Cretaceous and all Palaeogene data from the Outer Western Carpathians suggest an overall anticlockwise rotation. The palaeomagnetic control on the Mesozoic situation is quite poor, but the Palaeogene results are geographically distributed and represent the Magura, the Dukla and the Silesian nappes. They suggest that the Outer Carpathians rotated in co-ordination with the already consolidated Inner Carpathians and the Pieniny Klippen Belt during the Miocene. The folding of the nappes (also manifested in well-defined AMS lineations correlating in direction with the strike of the sampled beds) followed the acquisition of the primary magnetizations of the Palaeogene flysch. As the magnetization of post-folding age is aligned with that of the primary magnetization, even the first anticlockwise rotation must have been subsequent to the folding events and the Outer Carpathians only attained their present orientation during the Miocene.

We thank Dušan Plašienka for providing and discussing papers on the geological and tectonic development of the Inner Carpathians and Gábor Imre and Jan Turczynowicz for technical assistance. Careful and constructive reviews by two anonymous referees were greatly appreciated. Financial support from the Hungarian Scientific Research Fund (OTKA, Project number K105245), from Polish–Hungarian bilateral Scientific and Technological project 2011–2012 and from a joint project of the Academies of Science of Poland and Hungary (2011–2013) is acknowledged.

## References

ALEKSANDROWSKI, P. 1985. Structure of the Mt. Babia Góra region, Magura Nappe, Western Outer Carpathians: an interference of West and East Carpathian fold trends. *Annales Societatis Geologorum Poloniae*, **55**, 375–422 (in Polish with English summary).

ANCZKIEWICZ, A. A., ŚRODOŃ, J. & ZATTIN, M. 2013. Thermal history of the Podhale Basin in the Internal Western Carpathians from the perspective of apatite fission trac analyses. *Geologica Carpathica*, **64**, 141–151.

BALLA, Z. & MÁRTON-SZALAY, E. 1978. The palaeomagnetic sequence in the Börzsöny volcanic area 1. *Magyar Geofizika*, **19**, 51–59 (in Hungarian).

BAZHENOV, M., BEGAN, A., BIRKENMAJER, K. & BURTMAN, V. S. 1980. Paleomagnetic evidence of the tectonic origin of the curvature of the West Carpathian Arc. *Bulletin de L'Academie Polonaise des Sciences; Série des Sciences de la Terre*, **28**, 281–290.

BESSE, J. & COURTILLOT, V. 1991. Revised and synthetic apparenat polar wander paths of the African, Eurasian, North American and Indian Plates, and true polar wander since 200 Ma. *Journal of Geophysical Research*, **96**, 4029–4050.

BESSE, J. & COURTILLOT, V. 2002. Apparent and true polar wander and geometry of the geomagnetic field over the last 200 Myr. *Journal of Geophysical Research*, **107**, 6-1–6-31 EPM.

BESSE, J. & COURTILLOT, V. 2003. Correction to 'Apparent and true polar wander and geometry of the geomagnetic field over the last 200 Myr'. *Journal of Geophysical Research*, **108**, 3-1–3-2 EPM.

BIRKENMAJER, K. 1986. Stages of structural evolution of the Pieniny Klippen Belt, Carpathians. *Studia Geologica Polonica*, **80**, 7–32.

BIRKENMAJER, K. & NAIRN, A. E. M. 1968. Paleomagnetic studies of Polish rocks. III. Neogene igneous rocks of the Pieniny Mountains, Carpathians. *Annales de la Société Geologique de Pologne*, **38**, 475–489.

BIRKENMAJER, K. & PÉCSKAY, Z. 2000. K–Ar dating of the Miocene andesite intrusions, Pieniny Mts, West Carpathians: a supplement. *Studia Geologica Polonica*, **117**, 7–25.

BURTMAN, V. S. 1988. Kinematics of the Carpathian-Balkan region during Cretaceous and Cenozoic. *Studia Geologica Polonica*, **91**, 39–60.

CHANNELL, J. E. T., KOZUR, H. W., SIEVERS, T., MOCK, R., AUBRECHT, R. & SYKORA, M. 2003. Carnian – Norian biomagnetostratigraphy at Silická Brezová (Slovakia): correlation to Rother Tethyan sections and to Newark Basin. *Palaeogeography, Palaeoclimatology, Palaeoecology*, **191**, 65–109.

DANIŠIK, M., KOHUT, M., EVANS, N. J. & McDONALD, B. J. 2012. Eo-Alpine metamorphism and the 'mid-Miocene thermal event' in the Western Carpathians (Slovakia): new evidence from multiple thermochronology. *Geological Magazine*, **149**, 158–171.

DECKER, K., NESCIERUK, P., REITER, F., RUBINKIEWICZ, J., RYŁKO, W. & TOKARSKI, A. K. 1997. Heteroaxial shortening, strike-slip faulting and displacement transfer in the Polish Carpathians. *Przegląd Geologiczny*, **45**, 1070–1071.

ENKIN, R. J. 2003. The direction-correction tilt test: an all-purpose tilt/fold test for paleomagnetic studies. *Earth and Planetary Science Letters*, **212**, 151–166.

ENKIN, R. J. & WATSON, G. S. 1996. Statistical analysis of paleomagnetic inclination data. *Geophysical Journal International*, **126**, 495–504.

FISHER, R. A. 1953. Dispersion on a sphere. *Proceedings of the Royal Society of London, Series A*, **217**, 295–305.

FODOR, L. 2006. Tertiary tectonic evolution of the Pannonian_Carpathian-Eastern Alpine Domain: a personal view of from Pannonia in the light of the terminological question of tectonic units. *Geolines*, **20**, 34–35.

FODOR, L., FRANCU, J., KREJCI, O. & STRÁNIK, Z. 1995. From transpression to transtension: Oligocene-Miocene structural evolution of the Vienna Basin and the Eastern Alpine-Western Carpathian junction. *Tectonophysics*, **242**, 151–182.

FROITZHEIM, N., PLAŠIENKA, D. & SCHUSTER, R. 2008. Alpine tectonics of the Alps and Western Carpathians. *In*: McCANN, T. (ed) *The Geology of Central Europe. Volume 2. Mesozoic and Cenozoic*. Geological Society, London, 1141–1232.

GĄGAŁA, Ł., VÉRGES, J., SAURA, E., MALATA, T., RINGENBACH, J.-C., WERNER, P. & KRZYWIEC, P. 2012. Architecture and orogenic evolution of the northeastern Outer Carpathians from cross-section balancing and forward modeling. *Tectonophysics*, **532**, 223–241.

GALBRUN, B. 1985. Magnetostratigraphy of the Berriassian stratotype section (Berrias, France). *Earth and Planetary Science Letters*, **74**, 130–136.

GRABOWSKI, J. 1995. New paleomagnetic data from the Lower Sub-Tatric radiolarites, Upper Jurassic (Western Tatra Mts). *Geological Quaterly*, **39**, 61–74.

GRABOWSKI, J. 1997. Paleomagnetic results from the Cover (High Tatric) unit and Nummulitic Eocene in the Tatra Mts (Central West Carpathians, Poland) and their tectonic implications. *Annales Societatis Geologorum Poloniae*, **67**, 13–23.

GRABOWSKI, J. 2000. Palaeo- and rock magnetism of Mesozoic carbonate rocks in the Sub-Tatric series (Central West Carpathians) – palaeotectonic implications. *Polish Geological Institute, Special Papers*, **5**, 1–88.

GRABOWSKI, J. 2005. New Berriasian palaeopole from the Central West Carpathians (Tatra Mountains, southern Poland): does it look Apulian? *Geophysical Journal International*, **161**, 65–80.

GRABOWSKI, J. & GAWĘDA, A. 1999. Preliminary palaeomagnetic study of the High Tatra granites, Central West Carpathians, Poland. *Geological Quarterly*, **43**, 263–276.

GRABOWSKI, J. & NEMČOK, M. 1999. Summary of paleomagnetic data from the Central West Carpathians of Poland and Slovakia: evidence for the Late Cretaceous – Early Tertiary transpression. *Physics and Chemistry of the Earth*, **A24**, 681–685.

GRABOWSKI, J. & PSZCZÓŁKOWSKI, A. 2006. Magneto- and biostratigraphy of the Tithonian – Berriasian pelagic sediments in the Tatra Mountains (central Western Carpathians, Poland): sedimentary and rock magnetic changes at the Jurassic/Cretaceous boundary. *Cretaceous Research*, **27**, 398–417.

GRABOWSKI, J., KRZEMIŃSKI, L., NESCIERUK, P., SZYDŁO, A., PASZKOWSKI, M., PECSKAY, Z. & WÓJTOWICZ, A. 2003. Geochronology of the teschenitic intrusions in the Outer Western Carpathians of Poland – constraints from $^{40}$K/$^{40}$Ar ages and biostratigraphy. *Geologica Carpathica*, **54**, 385–393.

GRABOWSKI, J., KRZEMIŃSKI, L., NESCIERUK, P. & STARNAWSKA, E. 2006. Paleomagnetism of the teschenitic rocks (Lower Cretaceous) in the Outer Western Carpathians of Poland: constraints for the tectonic rotations in the Silesian unit. *Geophysical Journal International*, **166**, 1077–1094.

GRABOWSKI, J., KROBICKI, M. & SOBIEŃ, K. 2008. New palaeomagnetic results from the Polish part of the

Pieniny Klippen Belt, Carpathians – evidence for the palaeogeographic position of the Czorsztyn Ridge in Mesozoic. *Geological Quarterly*, **52**, 31–44.

GRABOWSKI, J., MICHALÍK, J., SZANIAWSKI, R. & GROTEK, I. 2009. Synthrusting remagnetization of the Krížna nappe: high resolution palaeo- and rock magnetic study in the Strážovce section, Strážovské vrchy Mts, Central West Carpathians (Slovakia). *Acta Geologica Polonica*, **59**, 137–155.

GRABOWSKI, J., MICHALIK, J., PSZCZÓŁKOWSKI, A. & LINTNEROVÁ, O. 2010. Magneto- and isotope stratigraphy around the Jurassic/Cretaceous boundary in the Vysoka unit (Male Karpaty Mountains): correlations and tectonic implications. *Geologica Carpathica*, **61**, 309–326.

GRADSTEIN, F. M., OGG, J. G., SCHMITZ, M. D. & OGG, G. M. (eds) 2012. *The Geologic Time Scale 2012*. 1st edn. Elsevier, Amsterdam.

HOUŠA, V., KRS, M., KRSOVA, M. & PRUNER, P. 1996. Magnetostratigraphy of Jurassic-Cretaceous limestones in the Western Carpathians. *In*: MORRIS, A. & TARLING, D. H. (eds) *Paleomagnetism and Tectonics of the Mediterranean Region*. Geological Society, London, Special Publications, **105**, 185–194.

JELEŃSKA, M., TÚNYI, I. & AUBRECHT, R. 2011. Low latitude Oxfordian position of the Oravic crustal segment (Pieniny Klippen Belt, Western Carpathians): palaeogeographic implications. *Palaeogeography, Palaeoclimatology, Palaeoecology*, **302**, 338–348.

KĄDZIAŁKO-HOFMOKL, M. & KRUCZYK, J. 1987. Paleomagnetism of middle-late Jurassic sediments from Poland and implications for the polarity of the geomagnetic field. *Tectonophysics*, **139**, 53–66.

KARÁTSON, D., MÁRTON, E. ET AL. 2000. Volcanic evolution and stratigraphy of the Miocene Börzsöny Mountains, Hungary: an integrated study. *Geologica Carpathica*, **51**, 325–343.

KARÁTSON, D., OLÁH, I., PÉCSKAY, Z., MÁRTON, E., HARANGI, SZ., DULAI, A. & ZELENKA, T. 2007. Miocene volcanism in the Visegrád Mountains (Hungary): an integrated approach to regional volcanic stratigraphy. *Geologica Carpathica*, **58**, 541–563.

KISS, D., MÁRTON, E. & TOKARSKI, A. K. 2015. The role of compressional tectonics, sedimentary transport and mineral composition on AMS and AARM fabrics. A case study of the flysch from the Dukla nappe, Outer Western Carpatians, Poland. *Geophysical Research Abstracts*, **17**, EGU2015-11586-1.

KORAB, T., KRS, M., KRSOVÁ, M. & PAGÁČ, P. 1981. Paleomagnetic investigations of Albian (?)-Paleocene to Lower Oligocene sediments from the Dukla unit, East Slovakian Flysch, Czechoslovakia. *Zapadne Karpaty, ser. Geologia*, **7**, 127–149.

KOTÁSEK, J. & KRS, M. 1965. Palaeomagnetic study of tectonic rotation in the Carpathian Mountains of Czechoslovakia. *Palaeogeography, Palaeoclimatology, Palaeoecology*, **1**, 39–49.

KRS, M. & ŠMÍD, B. 1979. Palaeomagnetism of Cretaceous rocks of the teschenite association, Outer West Carpathians, Czechoslovakia. *Sbornik Geologickych Ved. Uzita Geofyzika*, **16**, 7–22.

KRS, M., MUŠKA, P. & PAGÁČ, P. 1982. Review of paleomagnetic investigations in the West Carpathians of Czechoslovakia. *Geologicke prace, Spravy, Geological Institute of Dionyz Stur, Bratislava*, **78**, 39–58.

KRS, M., KRSOVÁ, M., CHVOJKA, R. & POTFAJ, M. 1991. Paleomagnetic investigations of the flysch belt in the Orava region, Magura unit, Czechoslovak Western Carpathians. *Geologicke prace, Dionýz Stúr Geological Institute, Bratislava*, **92**, 135–151.

KRS, M., CHVOJKA, R. & POTFAJ, M. 1993. Paleomagnetic investigations in the Biele Karpaty Mts unit, flysch belt of the Western Carpathians. *Geologica Carpathica*, **45**, 35–43.

KRS, M., KRSOVA, M. & PRUNER, P. 1996. Paleomagnetism and paleogeography of the Western Carpathians from the Permian to the Neogene. *In*: MORRIS, A. & TARLING, D. H. (eds) *Paleomagnetism and Tectonics of the Mediterranean Region*. Geological Society, London, Special Publications, **105**, 175–184.

KRUCZYK, J., KĄDZIAŁKO-HOFMOKL, M., LEFELD, J., PAGAČ, P. & TÚNYI, I. 1992. Paleomagnetism of Jurassic sediments as evidence for oroclinal bending of the Inner West Carpathians. *Tectonophysics*, **206**, 315–324.

KRUCZYK, J., KĄDZIAŁKO-HOFMOKL, M., TÚNYI, I., PAGAČ, P. & MELLO, J. 1998. Paleomagnetic study of Triassic sediments from the Silica nappe in the Slovak Karst, a new approach. *Geologica Carpathica*, **49**, 33–43.

KSIĄŻKIEWICZ, M. 1960. Pre-orogenic sedimentation in the Carpathian geosyncline. *Geologische Rundschau*, **50**, 8–31.

KSIĄŻKIEWICZ, M. 1977. Tectonics of the Carpathians. *In*: POZARYSKI, W. (ed.) *Geology of Poland. vol. IV, Tectonics*. Wydawnictwa Geologiczne, Warsaw, 476–604.

LEWANDOWSKI, M., KROBICKI, M., MATYJA, B. A. & WIERZBOWSKI, A. 2005. Palaeogeographic evolution of the Pieniny Klippen Basin using stratigraphic and palaeomagnetic data from the Veliky Kamenets section (Carpathians, Ukraine). *Palaeogeography, Palaeoclimatology, Palaeoecology*, **216**, 53–72.

LEXA, J., BEZÁK, V., ELEČKO, M., MELLO, J., POLÁK, M., POTFAJ, M. & VOZÁR, J. (eds) 2000. *Geological Map of Western Carpathians and Adjacent Areas 1: 500,000*. Ministry of Environment of Slovak Republic, Geological Survey of Slovak Republic, Bratislava.

LUCIŃSKA-ANCZKIEWICZ, A., VILLA, I. M., ANCZKIEWICZ, R. & ŚLĄCZKA, A. 2002. $^{40}Ar/^{39}Ar$ Dating of Alkaline Lamprophyres from the Polish Western Carpathians. *Geologica Carpathica*, **53**, 45–52.

MÁRTON, E. & MÁRTON, P. 1996. Large scale rotations in North Hungary during the Neogene as indicated by palaeomagnetic data. *In*: MORRIS, A. & TARLING, D. H. (eds) *Palaeomagnetism and Tectonics of the Mediterranean Region*. Geological Society, London, Special Publications, **105**, 153–173.

MÁRTON, E. & PÉCSKAY, Z. 1998. Correlation and dating of the Miocene ignimbritic volcanics in the Bükk foreland, Hungary: complex evaluation of paleomagnetic and K/Ar isotope data. *Acta Geologica Hungarica*, **41**, 467–476.

MÁRTON, E., MÁRTON, P. & LESS, Gy. 1988. Paleomagnetic evidence of tectonic rotations in the Southern margin of the Inner West Carpathians. *Physics of the Earth and Planetary Interiors*, **52**, 256–266.

MÁRTON, E., PAGÁČ, P. & TÚNYI, I. 1992. Paleomagnetic investigations on late Cretaceous-Cenozoic sediments from the NW part of the Pannonian Basin. *Geologica Carpathica*, **43**, 363–368.

MÁRTON, E., VASS, D. & TÚNYI, I. 1996. Rotation of the South Slovak Paleogene and Lower Miocene rocks indicated by paleomagnetic data. *Geologica Carpathica*, **47**, 31–41.

MÁRTON, E., MASTELLA, L. & TOKARSKI, A. K. 1999. Large counterclockwise rotation of the Inner West Carpathian Paleogene Flysch – evidence from paleomagnetic investigation of the Podhale Flysch (Poland). *Physics and Chemistry of the Earth*, **24**, 645–649.

MÁRTON, E., VASS, D. & TÚNYI, I. 2000. Counterclockwise rotations of the Neogene rocks in the East Slovak Basin. *Geologica Carpathica*, **51**, 159–168.

MÁRTON, E., TOKARSKI, A. K. & HALÁSZ, D. 2004. Late Miocene counterclockwise rotation of the Pieniny andesites at the contact of the Inner and Outer West Carpathians. *Geologica Carpathica*, **55**, 411–419.

MÁRTON, E., VASS, D., TÚNYI, I., MÁRTON, P. & ZELENKA, T. 2007a. Paleomagnetic age assignment of the ignimbites from the famous fossil footprints site, Ipolytarnóc (close to the Hungarian-Slovakian boundary). *Geologica Carpathica*, **58**, 531–540.

MÁRTON, E., ZELENKA, T. & MÁRTON, P. 2007b. Paleomagnetic correlation of Miocene pyroclastics of the Bükk Mts and their forelands. *Central European Geology*, **50**, 47–57. http://doi.org/10.1556/Ceu Geol.50.2007.1.4

MÁRTON, E., TISCHLER, M., CSONTOS, L., FÜGENSCHUH, B. & SCHMID, S. M. 2007c. The contact zone between ALCAPA and Tisza-Dacia megatectonic units of Northern Romania in the light of new paleomagnetic data. *Eclogae Geologicae Helvetiae*, **100**, 109–124.

MÁRTON, E., RAUCH-WŁODARSKA, M., KREJČI, O., TOKARSKI, A. K. & BUBIK, M. 2009a. An integrated palaeomagnetic and AMS study of the Tertiary flysh from the Outer Western Carpathians. *Geophysical Journal International*, **177**, 925–940. http://doi.org/10.1111/j.1365-246X.2009.04104x

MÁRTON, E., JELEŃSKA, M., TOKARSKI, A. K., SOTÁK, J., KOVÁČ, M. & SPIŠIAK, J. 2009b. Current-independent paleomagnetic declinations in flysch basins: a case study from the Inner Carpathians. *Geodinamica Acta*, **22**, 73–82. http://doi.org/10.3166/ga.22.73-82

MÁRTON, E., ZAMPIERI, D., GRANDESSO, P., ĆOSOVIĆ, V. & MORO, A. 2010. New Cretaceous paleomagnetic results from the foreland of the Southern Alps and the refined apparent polar wander path for stable Adria. *Tectonophysics*, **480**, 57–72.

MÁRTON, E., TOKARSKI, A. K., KREJČI, O., RAUCH, M., OLSZEWSKA, B., PETROVÁ, P. T. & WÓJCIK, A. 2011. 'Non-European' palaeomagnetic directions from the Carpathian Foredeep at the southern margin of the European plate. *Terra Nova*, **23**, 134–144, http://doi.org/10.1111/j.1365-3121.2011.00993.x

MÁRTON, E., GRABOWSKI, J., PLAŠIENKA, D., TÚNYI, I., KROBICKI, M., HAAS, J. & PETHE, M. 2013. New paleomagnetic results from the Upper Cretaceous red marls of the Pieniny Klippen Belt, Western Carpathians: evidence for general CCW rotation and implications for the origin of the structural arc formation. *Tectonophysics*, **592**, 1–13.

MÁRTON, P., ROZLOŽNIK, L. & SASVARI, T. 1991. Implications of paleomagnetic study of the Silica nappe, Slovakia. *Geophysical Journal International*, **107**, 67–75.

MASTELLA, L. & KONON, A. 2002. Jointing in the Silesian Nappe (Outer Carpathians, Poland) – Paleostress reconstruction. *Geologica Carpathica*, **53**, 315–325.

MAURITSCH, H. J. & MÁRTON, E. 1995. Escape models of the Alpine-Carpathian-Pannonian region in the light of paleomagnetic observations. *Terra Nova*, **7**, 44–50.

McCABE, C. & ELMORE, R. D. 1989. The occurrence and origin of late Paleozoic remagnetization in the sedimentary rocks of North America. *Reviews of Geophysics*, **27**, 471–494.

McFADDEN, P. L. & McELHINNY, M. W. 1990. Classification of the reversal test in paleomagnetism. *Geophysical Journal International*, **103**, 725–729.

MUTTONI, G., ERBA, E., KENT, D. V. & BACHTADSE, V. 2005. Mesozoic Alpine facies deposition as a result of past latitudinal plate motion. *Nature*, **434**, 59–63.

MUTTONI, G., DALLANAVE, E. & CHANNELL, J. E. T. 2013. The drift history of Adria and Africa from 280 Ma to Present, Jurassic true polar wander and zonal climate control on Tethyan sedimentary facies. *Palaeogeography, Palaeoclimatology, Palaeoecology*, **386**, 415–435.

NEMČOK, M. & NEMČOK, J. 1994. Late Cretaceous deformations of the Pieniny Klippen Belt, Western Carpathians. *Tectonophysics*, **239**, 81–109.

NEMČOK, M., POSPISIL, L., LEXA, J. & DONELICK, R. A. 1998. Tertiary subduction and slab break-off model of the Carpathian–Pannonian region. *Tectonophysics*, **295**, 307–340.

NEMČOK, M., NEMČOK, J. *ET AL.* 2000. Results of 2D balancing along 20° and 21°30′ langitude and pseudo-3D in the Smilno tectonic window: implications for shortening mechanisms of the West Carpathian accretionary wedge. *Geologica Carpathica*, **51**, 281–300.

NEMČOK, M., KRZYWIEC, P., WOJTASZEK, M., LUDHOVÁ, L., KLECKER, R. A., SECOMBE, W. J. & COWARD, M. P. 2006. Tertiary development of the Polish and eastern Slovak parts of the Carpathian accretionary wedge: insights from balanced cross-sections. *Geologica Carpathica*, **57**, 355–370.

OPDYKE, N. D. & CHANNELL, J. E. T. 1996. *Magnetic Stratigraphy*. Academic Press, San Diego.

ORLICKÝ, O. 1996. Paleomagnetism of Neovolcanics of the East-Slovak lowlands and Zemplínske Vrchy Mts: a study of the tectonics applying the paleomagnetic data (Western Carpathians). *Geologica Carpathica*, **47**, 13–20.

OSZCZYPKO, N. & OSZCZYPKO-CLOWES, M. 2006. Evolution of the Magura Basin (in Polish, English abstract). *In*: OSZCZYPKO, N., UCHMAN, A. & MALATA, E. (eds) *Palaeotectonic Evolution of the Outer Carpathians and Pieniny Klippen Belt Basins*. Institute of Geological Sciences, Jagiellonian University, Cracow, 133–164.

OSZCZYPKO, N. & ŚLĄCZKA, A. 1985. An attempt at palinspastic reconstruction of Neogene basin in the Carpathian foredeep. *Annales Societatis Geologorum Poloniae*, **55**, 55–75.

OSZCZYPKO, N., KRZYWIEC, P., POPADYUK, I. & PERYT, T. 2005. Carpathian Foredeep Basin (Poland and Ukraine): its sedimentary, structural and geodynamic evolution. *In*: GOLONKA, J. & PICHA, J. (eds) *The Carpathians and Their Foreland. Geology and Hydrocarbon Resources.* AAPG, Memoirs, Tulsa, Oklahoma, **84**, 293–350.

PICHA, F., STRANIK, Z. & KREJČI, O. 2005. Geology and hydrocarbon resources of the Outer Western Carpathians and their foreland, Czech Republic. *In*: GOLONKA, J. & PICHA, F. (eds) *The The Carpathians and Their Foreland: Geology and Hydrocarbon Resources.* AAPG, Memoirs, Tulsa, Oklahoma, **84**, 49–175.

PIOTROWSKI, J. 1978. Mesostructural analysis of the main tectonic units of the Tatra Mountains along the Kooecieliska Valley, (in Polish, English summary). *Studia Geologica Polonica*, **55**, 3–90.

PLAŠIENKA, D. 1995. Passive and active margin history of the northern Tatricum (Western Carpathians, Slovakia). *Geologische Rundschau*, **84**, 748–760.

PLAŠIENKA, D. 2012a. Early stages of structural evolution of the Carpathian Klippen Belt (Slovakian Pieniny sector). *Mineralia Slovaca*, **44**, 1–16.

PLAŠIENKA, D. 2012b. Jurassic syn-rift and Cretaceous syn-orogenic, coarse grained deposits related to opening and closure of the Vahic (South Penninic) Ocean in the Western Carpathians – an overview. *Geological Quarterly*, **56**, 601–628.

PLAŠIENKA, D., GRECULA, P., PUTIŠ, M., KOVÁČ, M. & HOVORKA, D. 1997. Evolution and structure of the Western Carpathians: an overview. *In*: GRECULA, P., HOVORKA, P. & PUTIŠ, M. (eds) *Geological Evolution of the Western Carpathians.* Mineralia Slovaca Monograph, Bratislava, 1–24.

PÓKA, T., ZELENKA, T., SEGHEDI, I., PÉCSKAY, Z. & MÁRTON, E. 2004. Miocene volcanism of the Cserhát Mts. (N. Hungary): integrated volcano-tectonic, geochronologic and pertochemical study. *Acta Geologica Hungarica*, **47**, 221–246.

PROKEŠOVA, R., PLAŠIENKA, D. & MILOVSKY, R. 2012. Structural pattern and emplacement mechanism of the Križna cover nappe (Central Western Carpathians). *Geologica Carpathica*, **63**, 13–32.

PRUNER, P., VENHODOVÁ, D. & SLEPIČKOVÁ, J. 1998. Palaeomagnetic and palaeotectonic studies of selected areas of Tatricum, Manin and Križna units. *In*: RAKUS, M. (ed.) *Geodynamic Development of the Western Carpathians.* Geological Survey of Slovak Republic, Bratislava, 281–290.

PUEYO, E. L., MAURITSCH, H. J., GAWLICK, H.-J., SCHOLGER, R. & FRISCH, W. 2007. New evidence for block and thrust sheet rotations in the central northern Calcareous Alps deduced from two pevasive remagnetization events. *Tectonics*, **26**, TC5011, http://doi.org/10.1029/2006TC001965

RATSCHBACHER, L., FRISCH, W. *ET AL.* 1993. The Pieniny Klippen Belt in the Western Carpathians of northeastern Slovakia: structural evidence for transpression. *Tectonophysics*, **226**, 471–483.

RÖGL, F. 1996. Stratigraphic correlation of the Paratethys Oligocene and Miocene. *Mitteilungen der Gesellschaft der Geologie- und Bergbaustudenten in Österreich*, **41**, 65–73.

ŚLĄCZKA, A., KRUGŁOW, S., GOLONKA, J., OSZCZYPKO, N. & POPADYUK, I. 2005. Geology and hyrocarbon resources of the Outer Carpathians, Poland, Slovakia and Ukraine. *In*: GOLONKA, J. & PICHA, J. (eds) *The Carpathians and Their Foreland, Geology and Hydrocarbon Resources.* AAPG, Memoirs, Tulsa, Oklahoma, **84**, 221–258.

SPERNER, B., RATSCHBACHER, L. & NEMČOK, M. 2002. Interplay between subduction retreat and lateral extrusion: tectonics of the Western Carpathians. *Tectonics*, **21**, 1051–1075.

ŚRODOŃ, J., KOTARBA, M., BIRON, A., SUCH, P., CLAUER, N. & WÓJTOWICZ, A. 2006. Diagenetic history of the Podhale-Orava basin and the underlying Tatra Sedimentary units (Western Carpathians): evidence from XRD and K–Ar of illite-smectite. *Clay Minerals*, **41**, 751–774.

ŚWIERCZEWSKA, A. & TOKARSKI, A. K. 1998. Deformation bands and the history of folding in the Magura Nappe, Western Outer Carpathians (Poland). *Tectonophysics*, **297**, 73–90.

SZANIAWSKI, R., LUDWINIAK, M. & RUBINKIEWICZ, J. 2012. Minor counter-clockwise rotation of the Tatra Mountains (Central West Carpathians) as derived from paleomagnetic results achieved in hematite bearing Lower Triassic sandstones. *Tectonophysics*, **560–561**, 51–61.

SZANIAWSKI, R., MAZZOLI, S., JANKOWSKI, L. & ZATTIN, M. 2013. No large-magnitude rotations of the Subsilesian Unit of the Outer Western Carpathians: evidence from primary magnetization recorded in hematite-bearing Węglówka Marls (Senonian to Eocene). *Journal of Geodynamics*, **71**, 14–24.

TOKARSKI, A. K., SWIERCZEWSKA, A. & ZUCHIEWICZ, W. 2007. Fractured clasts in neotectonic reconstructions: an example from the Nowy Sacz Basin, Western Outer Carpathians, Poland. *Studia Quaternaria*, **24**, 47–52.

TOMEK, Č. & HALL, J. 1993. Subducted continental margin imaged in the Carpathians of Czechoslovakia. *Geology*, **21**, 535–538.

TOMINAGA, M., SAGER, W. W., TIVEY, M. A. & LEE, S.-M. 2008. Deep tow magnetic anomaly study of the Pacific Jurassic Quiet Zone and implications for the geomagnetic polarity reversal timescale and geomagnetic field behaviour. *Journal of Geophysical Research*, **113**, B07110, http://doi.org/10.1029/2007JB005527

TORSVIK, T. H., VAN DER VOO, R. *ET AL.* 2012. Phanerozoic polar wander, palaeogeography and dynamics. *Earth Science Reviews*, **114**, 325–368.

TÚNYI, I. & MÁRTON, E. 1996. Indications for large Tertiary rotation in the Carpathian–Northern Pannonian region outside the North Hungarian Paleogene Basin. *Geologica Carpathica*, **47**, 43–49.

TÚNYI, I., MÁRTON, E., VASS, D. & ŽEC, B. 2004. Results of paleomagnetic study in the Vihorlatské vrchy Mts. (East Slovakia). *Scripta Facultatis Scientiarium Naturalium Universitatis Masarykianæ Brunensis*, **31–32**, 129–132.

VAN DER VOO, R. 1993. *Paleomagnetism of the Atlantic, Tethys and Iapetus Oceans.* Cambridge University Press, Cambridge.

WÓJCIK, A. & JUGOWIEC, M. 1998. The youngest members of the folded Miocene in the Andrychów region

(Southern Poland). *Przegląd Geologiczny*, **46**, 763–770.

ZATTIN, M., ANDREUCCI, B., JANKOWSKI, L., MAZZOLI, S. & SZANIAWSKI, R. 2011. Neogene exhumation in the Outer Western Carpathians. *Terra Nova*, **23**, 283–291.

ZELENKA, T., PÓKA, T., MÁRTON, E. & PÉCSKAY, Z. 2005. New data on stratigraphic position of the Tar Dacite Tuff Formation (in Hungarian). *Magyar Állami Földtani Intézet Évi Jelentése*, **2004**, 73–84.

ZUCHIEWICZ, W., TOKARSKI, A. K., JAROSIŃSKI, M. & MÁRTON, E. 2002. Late Miocene to present day structural development of the Polish segment of the Outer Carpathians. *Stephan Mueller Special Publication Series*, **3**, 185–202.

# Formation of arc-shaped orogenic belts in the Western and Central Mediterranean: a palaeomagnetic review

FRANCESCA CIFELLI[1]*, CHIARA CARICCHI[1,2] & MASSIMO MATTEI[1]

[1]*Dipartimento di Scienze, Università degli Studi di Roma TRE, Largo San Leonardo Murialdo 1, 00146, Rome, Italy*

[2]*Istituto Nazionale di Geofisica e Vulcanologia, Via di Vigna Murata 605, 00143, Rome, Italy*

**Corresponding author (e-mail: francesca.cifelli@uniroma3.it)*

**Abstract:** During the past few decades, palaeomagnetism has been used as a powerful tool for constraining kinematic models of curved orogenic systems, because of its great potential in quantifying vertical axis rotations and in discriminating between primary and secondary (orocline *s.l.*) arcs. The Mediterranean area has represented an attractive region to apply palaeomagnetic analysis, as it shows a large number of narrow arcs, defining an irregular and rather diffuse plate boundary. This paper is intended to be an updated review on the contribution of palaeomagnetism to the reconstruction of the Neogene geodynamic evolution of the arc-shaped orogenic belts in the Western and Central Mediterranean Basin. The Gibraltar Arc, the Northern Apennines and the Calabria Arc are here described, underlining the common and the different features that characterize these arcuate mountain chains. In particular, the mechanisms that lead to the present-day shape of these arcs (the subduction process) will be discussed, in the general framework of the geometry and space–time evolution of the Mediterranean subduction system.

Based on the spatio-temporal relationships between deformation and vertical axis rotation, curved orogens can be subdivided as primary or secondary (oroclines *s.l.*), if they formed respectively in a self-similar manner without undergoing important variations in their original curved shape or if their curvature in map-view is the result of a bending about a vertical axis of rotation (see Lowrie & Hirt 1986; Weil & Sussman 2004, among others). Among the oroclines we can further distinguish progressive and secondary arcs. In progressive oroclines, the curvature is acquired during orogenic deformation and in response to the same stress-field responsible for orogen, while secondary arcs (oroclines *s.s.*) develop after initial orogenesis and in response to a change in the orientation of the stress field responsible for the orogen development (see Johnston *et al.* 2013 for a recent review). In addition to the kinematics of the arc and the timing of its curvature, a crucial factor for understanding the origin of belts curvature is knowledge of the geodynamic process governing arc formation. In this context, the detailed reconstruction of the rotational history, which is mainly based on palaeomagnetism, represents a fundamental prerequisite to fully understand the origin of orogenic arcs and to unravel the geodynamic processes responsible for their curvature.

The Mediterranean area represents a classical test for the study of orogenic arcs, being formed by a large number of narrow arcs, which define an irregular and rather diffuse plate boundary (Fig. 1). In particular, the Alpine chain forms a 'snake-shaped' defined by tight arcs with an eastward (Gibraltar Arc, Western Alpine Arc), westward (Calabrian Arc, Northern Apennines, Carpathian Arc) or northeastward (Hellenic Arc) concavity (Fig. 1). Because of their structural orientation, these arcs are kinematically inconsistent with the relative motion of the Africa plate with respect to Eurasia, as derived by plate tectonics reconstructions (Dewey *et al.* 1989) or GPS data (McClusky *et al.* 2003; D'Agostino & Selvaggi 2004; Biggs *et al.* 2006). Slab roll-back (e.g. Frizon de Lamotte *et al.* 1991; Royden 1993; Lonergan & White 1997; Morales *et al.* 1999; Faccenna *et al.* 2004), breakoff of a subducting lithospheric slab (Blanco & Spakman 1993), delamination of lithospheric mantle (e.g. Comas *et al.* 1992; García-Dueñas *et al.* 1992; Seber *et al.* 1996) and removal of the mantle lithosphere (e.g. Platt & Vissers 1989) are all possible mechanisms proposed to explain the formation of the orogenic arcs developed at high angle with respect to the convergence direction. However, since Malinverno & Ryan (1986) first presented their interpretation on the formation of the Calabrian Arc as a consequence of trench retreat owing to the presence of the narrow Ionian oceanic lithosphere, most later geodynamic models (among others, Royden 1993; Faccenna *et al.* 1997, 2014; Lonergan & White 1997; Guegen *et al.* 1998; Wortel & Spakman 2000; van Hinsbergen *et al.*

*From*: Pueyo, E. L., Cifelli, F., Sussman, A. J. & Oliva-Urcia, B. (eds) 2016. *Palaeomagnetism in Fold and Thrust Belts: New Perspectives*. Geological Society, London, Special Publications, **425**, 37–63.
First published online February 17, 2016, http://doi.org/10.1144/SP425.12

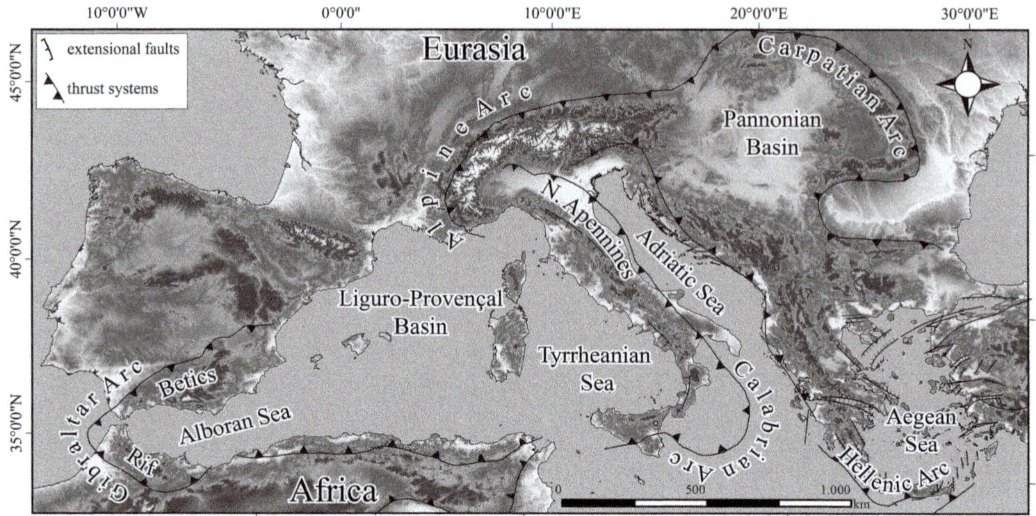



**Fig. 1.** Schematic map of the Mediterranean area with the main curved orogenic systems traced.

2014) proposed that the Mediterranean arcs formed as a consequence of the progressive retreat of the subducting slab and can be, therefore, considered secondary features achieved through large and opposite rotations along the limbs, shaping the arcs (see Rosenbaum & Lister 2004 for a detailed review of the different tectonic models proposed for the Mediterranean arcs).

The large amount of palaeomagnetic data collected in recent decades may allow further revision of this general model, informing whether the curved shape of the Mediterranean orogenic arcs was acquired by means of opposite rotations along opposing limbs of the arcs, and they are therefore oroclines whose shape has been acquired through bending about a vertical axis.

In this paper, we review palaeomagnetic data available for the Western and Central Mediterranean, from the Gibraltar Arc to the Italian Peninsula, where two arcuate orogenic belts shape the Africa–Eurasia boundary: the Calabrian Arc and the Northern Apennines. A quantitative analysis of palaeomagnetic data is used to constrain the tectonic and kinematic evolution of these arcs, in order to define their progressive or secondary nature and their relationships with first order geodynamic processes in the Mediterranean area.

## The Calabrian Arc

The Calabrian Arc defines a mountain belt encircling the Tyrrhenian Sea, from the Southern Apennines to Sicilian Maghrebides with an arcuate

trend defined by the regional variation in the strike of fold axes, striking from NW–SE in Southern Apennines to east–west across Sicily (Fig. 1). The area is defined by two sectors: the Calabria–Peloritane Domain (CPD) in the central part, and the Southern Apennines and Sicilian Maghrebides thrust-and-fold belts at the edges of the curved-shape belt (Fig. 2). These two sectors are characterized by different tectonic evolution, structural architecture and deep lithospheric structure (Wortel & Spakman 2000; Bonardi *et al.* 2001, among many others). The Southern Apennines and Sicilian Maghrebides are mainly composed of Mesozoic to Cenozoic shallow-platform and deep basin sedimentary units of the African and Adriatic continental margins, which form the Apennine fold-and-thrust system together with sediments from Neogene–Quaternary foredeep and thrust-top basins. The main compressional phases in Southern Apennines and Sicily started in the early Miocene and were almost finished during the early Pleistocene (Argnani *et al.* 1987; Butler & Grasso 1993; Patacca & Scandone 2001, 2004). The emplacement of the thrust nappe in Southern Apennines was followed by extensional tectonics, which progressively disrupted the fold-and-thrust belt. Extensional processes started during late Pliocene–early Pleistocene (Ascione & Romano 1999) along the Tyrrhenian coast and migrated during the Quaternary to reach the axial part of the chain where, today, deformation is mostly accomplished by NW–SE-oriented active normal faults (e.g. Valensise & Pantosti 2001).

The core of the Calabrian Arc is formed by the CPD domain that mostly consists of a Hercynian

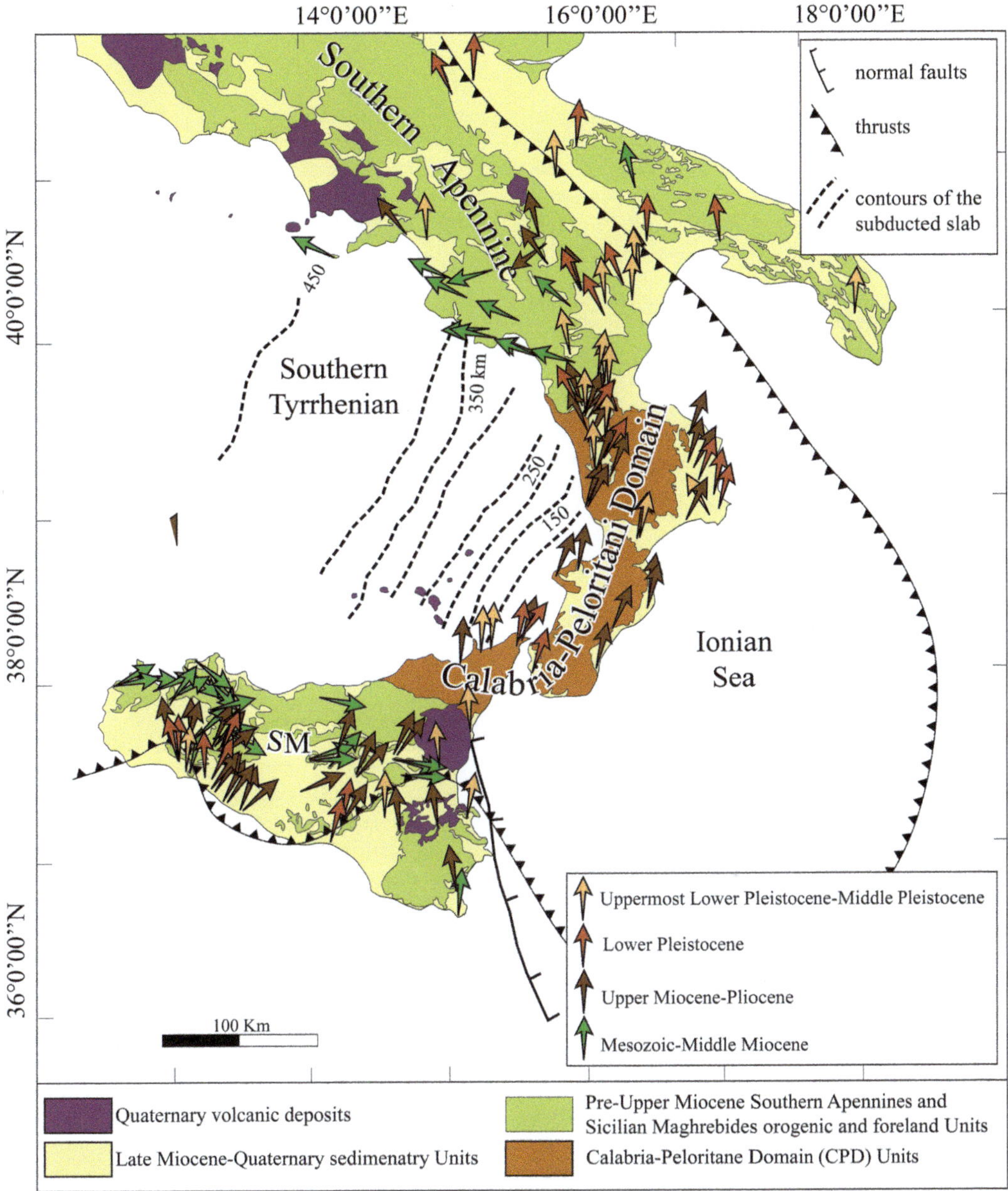


**Fig. 2.** Simplified geological map of the Calabrian Arc and estimated palaeomagnetic rotations based on data from the Southern Apennines, Calabrian Arc and Sicily (including results from this work). Each arrow represents results from one site or group of sites. Mesozoic to Palaeogene and Neogene rotations have been calculated comparing the obtained palaeodeclinations with the coeval expected African reference directions (Besse & Courtillot 2002) and with the geocentric axial dipole (GAD) field direction, respectively. Data come from Aifa *et al.* (1988), Barberi *et al.* (1974), Besse *et al.* (1984), Butler *et al.* (1999), Catalano *et al.* (1976), Channell *et al.* (1980), Channell *et al.* (1990), Channell *et al.* (1992), Cifelli *et al.* (2004, 2007*b*), Duermejier *et al.* (1998), Gattacceca & Speranza 2002, Gregor *et al.* (1975), Jackson (1990), Maffione *et al.* (2013), Manzoni (1975); Marton & Nardi (1994), Mattei *et al.* (2002), Mattei *et al.* (2004), Nairn *et al.* (1985), Sagnotti (1992), Scheepers (1992), Scheepers (1994), Scheepers & Langereis (1993), Scheepers & Langereis (1994), Scheepers *et al.* (1993), Scheepers *et al.* (1994), Schult (1976), Speranza *et al.* (1999, 2000, 2003, 2011), Tauxe *et al.* (1983). SM, Sicilian Maghrebides; CPD, Calabria Peloritani Domain. A hypothetical axis of the chain (0–100%) was traced across the arc and tectonic rotations were represented as the projection of the sampled sites on this axis in Figure 5.

basement, Alpine polymetamorphic rock successions related to the deformation of the Southern Tethyan margin and Mesozoic sedimentary units that record the early phases of Apennine orogenic construction and subsequent extension-related exhumation (e.g. Rossetti *et al.* 2004). These units are unconformably overlain by Cenozoic sedimentary sequences, deposited in different tectonic provinces of the Calabrian Arc. In particular, Neogene-to-Quaternary forearc basin sedimentary sequences developed along the Ionian side of the CPD (Cavazza *et al.* 1997; Bonardi *et al.* 2001) at the rear of an accretionary wedge located offshore in the Ionian Sea (Cernobori *et al.* 1996; Minelli & Faccenna 2010). Conversely, Miocene-to-Quaternary deposits cropping out along the Tyrrhenian side are interpreted as extensional back-arc basins, associated with opening and expansion of the Southern Tyrrhenian Sea (Sartori 1990; Mattei *et al.* 2002; Cifelli *et al.* 2007*a*). Since the middle to late Miocene, the CPD, originally located in the innermost Apennine chain sector, was affected by post-orogenic extension and drifted away from Sardinia to be finally assembled to the Southern Apennine–Maghrebide orogenic system (Alvarez *et al.* 1974; Malinverno & Ryan 1986; Faccenna *et al.* 1997; Bonardi *et al.* 2001; Mattei *et al.* 2002).

Starting from the 1970s, in the Calabrian Arc more than 500 sites were sampled either for palaeomagnetic or for magnetostratigraphic investigations (Fig. 2). The age of the investigated rocks range from middle Jurassic to Pleistocene (among others, Channell *et al.* 1980, 1990; Besse *et al.* 1984; Aifa *et al.* 1988; Sagnotti 1992; Scheepers & Langereis 1993, 1994; Scheepers *et al.* 1993, 1994; Duermeijer *et al.* 1998; Butler *et al.* 1999; Speranza *et al.* 1999, 2000, 2003, 2011; Gattacceca & Speranza 2002; Mattei *et al.* 2002, 2004; Cifelli *et al.* 2004; Cifelli *et al.* 2007*b*). In the Southern Apennines and in Sicily, palaeomagnetic data from Mesozoic and Palaeogene sedimentary sequences (today incorporated in the fold-and-thrust belt) indicate that such units record large magnitude clockwise (CW) and counterclockwise (CCW) rotations with respect to the coeval reference pole, respectively (Fig. 2). Palaeomagnetic data from upper Miocene and Pleistocene deposits come instead from foredeep and piggyback basins lying in the external part of the Maghrebian and Southern Apennines thrust belts. In Sicily, most of the data show a significant magnitude of CW rotation until lower Pleistocene (Butler *et al.* 1992; Scheepers & Langereis 1993; Duermejier *et al.* 1998; Speranza *et al.* 1999, 2003). In the Southern Apennines, most available palaeomagnetic data from middle–upper Miocene foredeep siliciclastic sediments outcropping along the Tyrrhenian side of the chain, show CCW rotations 40° up to 125° (Scheepers & Langereis

1994; Maffione *et al.* 2013). Similarly, data from Pliocene strata collected in claystones from the Potenza, Calvello, and Sant'Arcangelo foredeep and piggyback basins show CCW rotations (Sagnotti 1992; Scheepers *et al.* 1993). Data from lower Pleistocene sediments of the Sant'Arcangelo basin and Bradanic foredeep indicate a CCW rotation of 25° (Sagnotti 1992; Scheepers *et al.* 1993). Finally, both in Sicily and Southern Apennines, uppermost lower Pleistocene and middle Pleistocene sediments are not rotated (Scheepers *et al.* 1994; Cifelli *et al.* 2004; Mattei *et al.* 2004) (Fig. 2).

In the CPD, palaeomagnetic data have been mainly collected in upper Miocene to Pleistocene sedimentary sequences cropping out both in the post-orogenic basins along the Tyrrhenian coast and in the forearc basins located along the Ionian side of the region (Fig. 3). No data are available from Mesozoic and Palaeogene units, hindering the possibility to reconstruct the older rotational history of the CPD crustal block. Upper Miocene to Pleistocene sediments show a general CW rotational pattern along the entire CPD. Serravallian to upper Tortonian sediments sampled from the Tyrrhenian and Ionian side of Calabria show similar 20° CW rotations. This same value has been obtained from Pliocene to lower Pleistocene strata in the Calabria and Peloritani areas (Tauxe *et al.* 1983; Aifa *et al.* 1988; Scheepers *et al.* 1994; Speranza *et al.* 2000; Mattei *et al.* 2002; Cifelli *et al.* 2004). Similar to the Southern Apennines and Sicily, in the Calabria-Peloritane Domain the uppermost lower Pleistocene–middle Pleistocene sediments generally do not show appreciable rotations (Cifelli *et al.* 2007*b*). This is well documented in the Crati Basin, where palaeomagnetic data from the uppermost lower Pleistocene–middle Pleistocene sedimentary strata are not affected by tectonic rotations. Exceptions to this are the northwestern sector of the Crati basin (Cifelli *et al.* 2007*b*), and uppermost lower Pleistocene deposits of the Crotone forearc Basin (Speranza *et al.* 2011), where CCW rotations are recorded (Fig. 2).

# Northern Apennines

The Northern Apennines consist of several palaeogeographic domains, organized in a fold-and-thrust belt forming a broad curved structure displaced towards the Adriatic foreland (Fig. 3). The curvature of the arc follows the Adriatic margin of the belt, with the main structures ranging from NW–SE/WNW–ESE in the north to almost north–south in the south (e.g. Barchi *et al.* 1998). The uppermost (internal) nappes of the Northern Apennines are represented by oceanic units (Ligurian) characterized by the occurrence of Jurassic ophiolites and

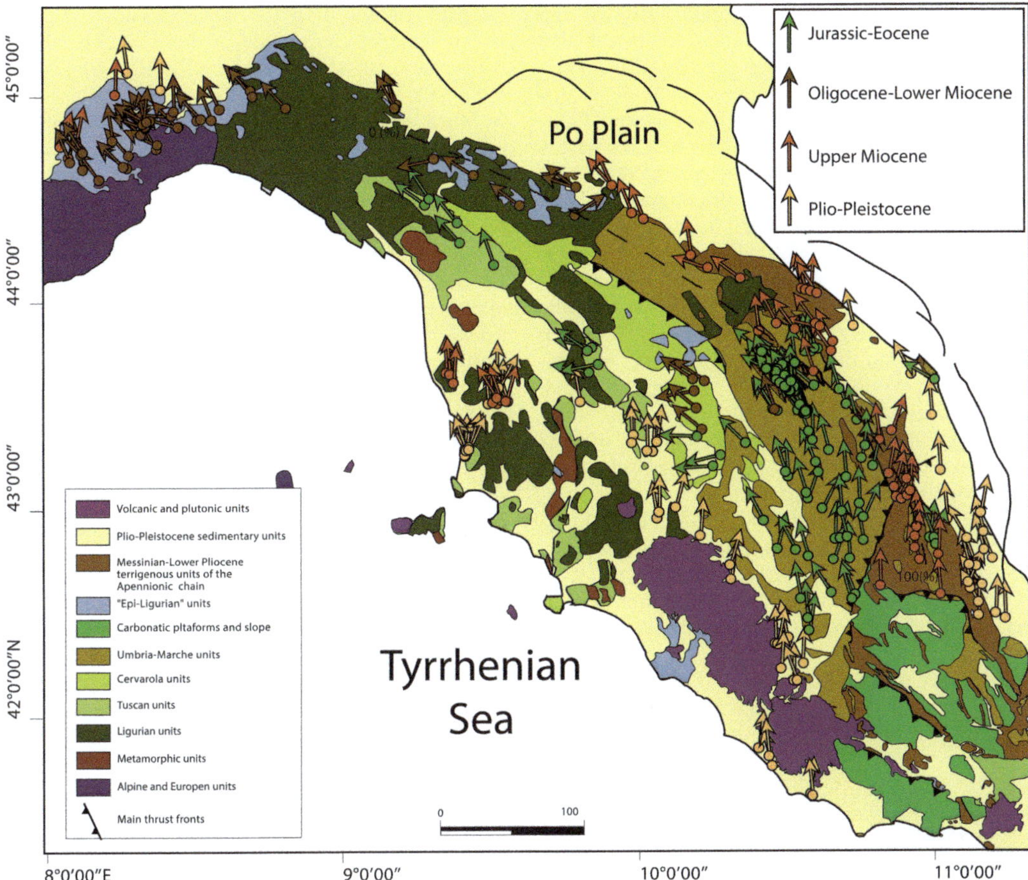



**Fig. 3.** Simplified geological map of Northern Apennines. Each arrow represents results from one site, group of sites or magnetostratigraphic sections. Tectonic rotations have been calculated comparing the obtained palaeodeclinations with the coeval expected African reference directions (Besse & Courtillot 2002). Data come from Aiello & Hagstrum (2001), Alvarez & Lowrie (1978), Alvarez & Lowrie (1984), Caricchi *et al.* (2014), Carrapa *et al.* (2003), Channell & Tarling (1975), Channell (1992), Cirilli *et al.* (1984), Dela Pierre *et al.* (1992), Hirt & Lowrie (1988), Klootwijk & VandenBerg (1975), Lanci & Wezel (1995), Latal *et al.* (2000), Lowrie & Alvarez (1975, 1977, 1979), Lowrie *et al.* (1980, 1982), Maffione *et al.* (2008), Marton & D'Andrea (1992), Mattei *et al.* (1996), Muttoni *et al.* (1998), Napoleone *et al.* (1983), Roggenthen & Napoleone (1977), Sagnotti *et al.* (1994, 2000), Sarti *et al.* (1995), Satolli *et al.* (2007, 2008), Speranza & Parisi (2007), Speranza *et al.* (1997), Tarduno *et al.* (1992), Thio (1988), VandenBerg *et al.* (1978). (Modified from Cifelli & Mattei 2010.) A hypothetical axis of the chain (0–100%) was traced across the arc and tectonic rotations were represented as the projection of the sampled sites on this axis in Figure 5.

their Middle Lias–early Cretaceous sedimentary cover, which are overlain by Cretaceous-Oligocene flysch sequences. The Ligurian domain units overthrust eastward the Tuscan domain units, formed by Upper Triassic to Eocene marine carbonates and by Oligocene–lower Miocene foredeep sequences, that are deformed in an array of thrust sheets and ultimately thrust over the units of the Umbria-Marche-Romagna domain. The latter consists of sedimentary sequences deposited on a continental margin with basal Late Triassic evaporites, platform carbonates (Lias), and pelagic sequences (Jurassic–Eocene) enriched upward in terrigenous deposits (Oligocene) and flysch sequences (Miocene).

The tectonic evolution of the Northern Apennines was characterized by the progressive migration of the orogenic front towards the Adriatic foreland. This evolution was marked by the onset, starting from late Cretaceous in the Ligurian Domain, of siliciclastic deposits getting progressively younger towards the Adriatic-Ionian foreland (e.g. Carmignani *et al.* 1994). Starting from the Oligocene onwards, foredeep basins migrated eastward and formed on the top of continental sequences that belonged to the passive margin of Adria. Their incorporation into the Apennines orogenic wedge marked the subduction of Adriatic continental lithosphere underneath Europe. Afterwards, during the Neogene, foredeep basins further migrated towards the Apulia foreland in front of the migrating thrust nappes. In the Northern Apennines this process is well documented by stratigraphic and seismic studies, which precisely constrain the Neogene evolution of the foredeep basins in the front of the Apenninic chain (e.g. Patacca *et al.* 1992). Foredeep basins formed on the top of progressively easternmost (external) units, up to the Adriatic foreland, where the foredeep Quaternary deposition is no longer active. The formation and evolution of foredeep basins were driven by the loading of the adjacent thrust belt and related to subduction processes, such as the flexural retreat of the subducting lithosphere (Royden *et al.* 1987). This process was particularly severe during Plio-Pleistocene times as evidenced by the presence, in the external part of the Apenninic chain, of a foredeep-basin system, that contains up to 8 km of Pliocene–Quaternary sedimentary rocks (Royden *et al.* 1987). An active tectonic regime is represented by NW–SE-trending normal faults, with well documented historical and recent seismicity, mostly located in the internal sector of the Umbria-Marche-Romagna region, at the edges of intramontane sedimentary basins (among others, Chiaraluce *et al.* 2004; Sani *et al.* 2009).

From a palaeomagnetic point of view, the Northern Apennines are one of the most famous sampling localities in the world, being the type-section for late Cretaceous-Palaeogene Scaglia Formation is one of the most studied sedimentary sequences for palaeomagnetism and magnetostratigraphy (e.g. Channell & Tarling 1975; Klootwijk & VandenBerg 1975; Lowrie & Alvarez 1975; Vandenberg *et al.* 1978). In the following years, other Authors documented in the Cretaceous-Palaeogene rocks from other regional localities widespread vertical-axis rotations associated with thrusting and folding in Northern Apennines. These new data suggested that results from the Umbria arc could not be extrapolated to the entire Italian peninsula (e.g. Channell *et al.* 1978, 1992; Van der Voo 1993). In the 1990s palaeomagnetic data came from Messinian to Pleistocene sediments of the external front of the chain (e.g. Dela Pierre *et al.* 1992; Lanci & Wezel 1995; Speranza *et al.* 1997). These results better constrained the timing of tectonic rotation, indicating that Plio-Pleistocene rotations affected the external part of the arc contributing to the overall present-day geometry. These rotations occurred with different senses and variable amount along the arc. In particular, the post-Messinian 20° CCW rotation measured in the Marche-Romagna area are of the same amplitude of those calculated for Mesozoic sequences in northern Umbria, suggesting a Plio-Pleistocene age for the rotations measured in older sequences (Speranza *et al.* 1997; Sagnotti *et al.* 2000). Caricchi *et al.* (2014) reported results from a palaeomagnetic study on pre-orogenic and syn-orogenic rocks from the internal sector of the Northern Apennines Arc. They analysed Eocene pelagic foreland ramp deposits (Scaglia Toscana Formation) and Oligocene–lower Miocene siliciclastic turbidites (Cervarola and Falterona Formations), named hereafter TCF units. Palaeomagnetic results show that the internal sector of the Northern Apennines underwent large CCW rotations (up to 110°) with respect to the Adria–Africa foreland. In particular, results show that the Tuscan units underwent a larger amount of CCW rotations in the southern part of the arc than in the northern one, a trend that is opposite to that observed in the more external units.

Moving northwestward along the chain, Upper Oligocene–middle Miocene Epiligurian units record a 52° CCW rotation with respect to Africa (Muttoni *et al.* 1998; Fig. 3). Muttoni *et al.* (2000) suggested that rotation was Oligocene–Miocene in age (e.g. Montigny *et al.* 1981; Speranza *et al.* 2002; Gattacceca *et al.* 2007) while the remaining part, observed in upper Miocene to Pliocene sediments, was due to Pliocene shortening episodes occurring along the Apennine front, which may have (re)activated thrust planes in the Apennines structures below the Ligurian wedge. Palaeomagnetic data by Muttoni *et al.* (2000) supported the results from Speranza *et al.* (1997), who found a post Messinian CCW rotation of *c.* 20° in the northern part of the Northern Apennines. Along the extensional Tyrrhenian margin, palaeomagnetic data have been obtained from post-orogenic Late Miocene–Pleistocene units, which were not involved in the Apennine orogenic deformation and show no evidence of tectonic rotation (Alvarez & Lowrie 1978; Sagnotti *et al.* 1994; Mattei *et al.* 1996; Fig. 3).

## The Gibraltar Arc

The Betic (southern Spain) and the Rif (North Africa) chains form the Gibraltar Arc, an arcuate orogenic belt that surrounds the Alboran Basin, in the westernmost Mediterranean Sea (Fig. 1). These orogenic systems are characterized by north-directed thrusting in the Betics and south-directed thrusting in the Rif, which took place simultaneously since late Oligocene with extensional deformation in internal parts of the orogen (e.g. Platt & Vissers 1989; Crespo-Blanc *et al.* 1994). Both the Betic and Rif systems contain three distinct structural domains, commonly referred to as the Internal Zones, the Flysch Domain and the External Zones (Fig. 4). The Internal Zones, or Alboran Domain, are located onshore and offshore in the inner part of the arc. They consist mostly of metamorphic units, which constitute the remnants of a Palaeogene orogen (Azañón & Crespo-Blanc 2000) that thrust onto the margins of Southern Spain and Northern Morocco during the early Miocene (e.g. Comas et al. 1992; Lonergan & White 1997). The Flysch Domain is constituted by siliciclastic sediments of Cretaceous to early Miocene age, deposited in a deep marine or oceanic setting near the northern African margin (Didon et al. 1973). The External Zones are characterized by thin-skinned fold-and-thrust belts composed of Mesozoic to Cenozoic continental margin strata deformed continuously from early Miocene to early Pliocene times (Lonergan & White 1997; Platt et al. 2003). In the Betic Cordillera, the Subbetic and Prebetic Zones represent the outer margin of the chain and are formed by Mesozoic and Cenozoic sediments deposited in basinal (Subbetic) and shelf (Prebetic) environments of the rifted palaeomargin of South Iberia. These rocks were detached from their hercynian Hercynian basement from early Miocene onwards, and formed a NNE–SSW- to NE–SW-trending fold-and-thrust belt, with associated foredeep and foreland basins along its outer margin (Guadalquivir Complex and Guadalquivir Basins, respectively). In Morocco, the External Rif mostly consists of Mesozoic–Eocene, pre-orogenic sediments from the rifted continental margin of Africa and it is detached along the upper Triassic evaporitic redbeds and thrust towards the Atlas–Meseta foreland. The different tectonic units that form the Rif chain are arranged in a complex fold-and-thrust architecture that is characterized by out-of-sequence and synchronous thrust propagation, which is responsible for most of the anomalous tectonic contacts recognized throughout the Rif chain (Morley 1992; Platt et al. 2003; Frizon de Lamotte et al. 2004). The outer front of the Rif chain appears to be sealed by upper Miocene sediments along the Gharb plain, and in the Atlantic abyssal plain (Meghraoui et al. 1996; Frizon de Lamotte et al. 2004), which indicates that the main phases of nappe emplacement in the Gibraltar Arc ended during late Miocene.

First palaeomagnetic sampling in the Gibraltar Arc was carried out in Mesozoic rocks (Table 1). In the Betics, palaeomagnetic samples were taken from Jurassic to Late Cretaceous sedimentary units, mainly from external Betics (Fig. 4). In particular most of the data come from western Sub-Betics (Ogg et al. 1984; Steiner et al. 1987; Osete et al. 1988, 2004; Platzman & Lowrie 1992; Allerton et al. 1993; Allerton 1994; Platzman 1994; Villalain et al. 1994; Platt et al. 1995, 2003; Kirker & McClelland 1996; Mayfield 1999) with only a few sites from the eastern Sub-Betics (Mazaud et al. 1986; Ogg et al. 1988; Allerton et al. 1993, 1994; Allerton 1994; Osete et al. 2004) and Pre-Betics (Allerton et al. 1993; Osete et al. 2004). In general, Mesozoic rocks from the Betic chain show mostly CW rotations (Fig. 4b). In the Rif orogenic wedge and Internal Domain, there are fewer palaeomagnetic results from Mesozoic rocks compared with the Betics (Platzman 1992; Platt et al. 2003), yet available data indicate that large vertical-axis rotations occurred and most of them are CCW (Fig. 4b).

In recent years an increasing amount of palaeomagnetic data were collected from younger rocks, both from the Betics and Rif Chain. Oligo-Miocene units were collected in the Internal Domain (Calvo et al. 1994; Saddiqi et al. 1995; Feinberg et al. 1996; Villasante-Marcos et al. 2003; Platzman & Platt 2004; Cifelli et al. 2008; Berndt et al. 2015) (Table 1). Similar to Mesozoic rocks, palaeomagnetic results from these units show the same opposite pattern of vertical axis rotations along the two arms of the Arc. Rotations are supposed to be early Miocene in age (Feinberg et al. 1996).

In the Betics, CW rotations were also measured in the Granada and Alcalà la Real upper-Miocene intramontane basins of Central Betics (Martín-Suarez et al. 1998; Mattei et al. 2006). On the other hand, in the Eastern Betics, Calvo et al. (1997) measured more complex vertical axis rotations (mainly CCW) in upper Miocene to Pliocene sedimentary and volcanic rocks. Mattei et al. (2006) measured CCW vertical axis rotations in the middle Miocene to lower Pliocene sites in the Lorca and Vera basins and, locally, in the Tortonian units of the Huercal–Overa Basin (Fig. 4). These CCW rotations have been observed in the different sedimentary basins located along a deformation belt dominated by Late Miocene–Quaternary left-lateral strike-slip faults, which characterize the whole South Iberian continental margin in the region of Almería–Murcia. In this region CCW rotations do not extend to sedimentary basins, which are located far away from the main strike-slip faults (Calvo et al. 1997; Krijgsman & Garcés 2004), and indicate the occurrence of small, fault-bounded blocks that rotate about the vertical axis as a consequence of movement along these faults. It is worth noting that some of these faults are supposed to be currently active (Booth-Rea et al. 2004). Magnetostratigraphic investigation and palaeomagnetic sampling have been carried out on Late Miocene sedimentary sections in the foreland basin both from the Betics and Rif chain (Table 1). Palaeomagnetic data suggest that no rotations occurred in the Late Tortonian to Messinian sediments in the Guadalquivir foreland basin (Krijgsman et al. 2004; Mattei et al. 2006). In the Rif Chain, palaeomagnetic studies have been

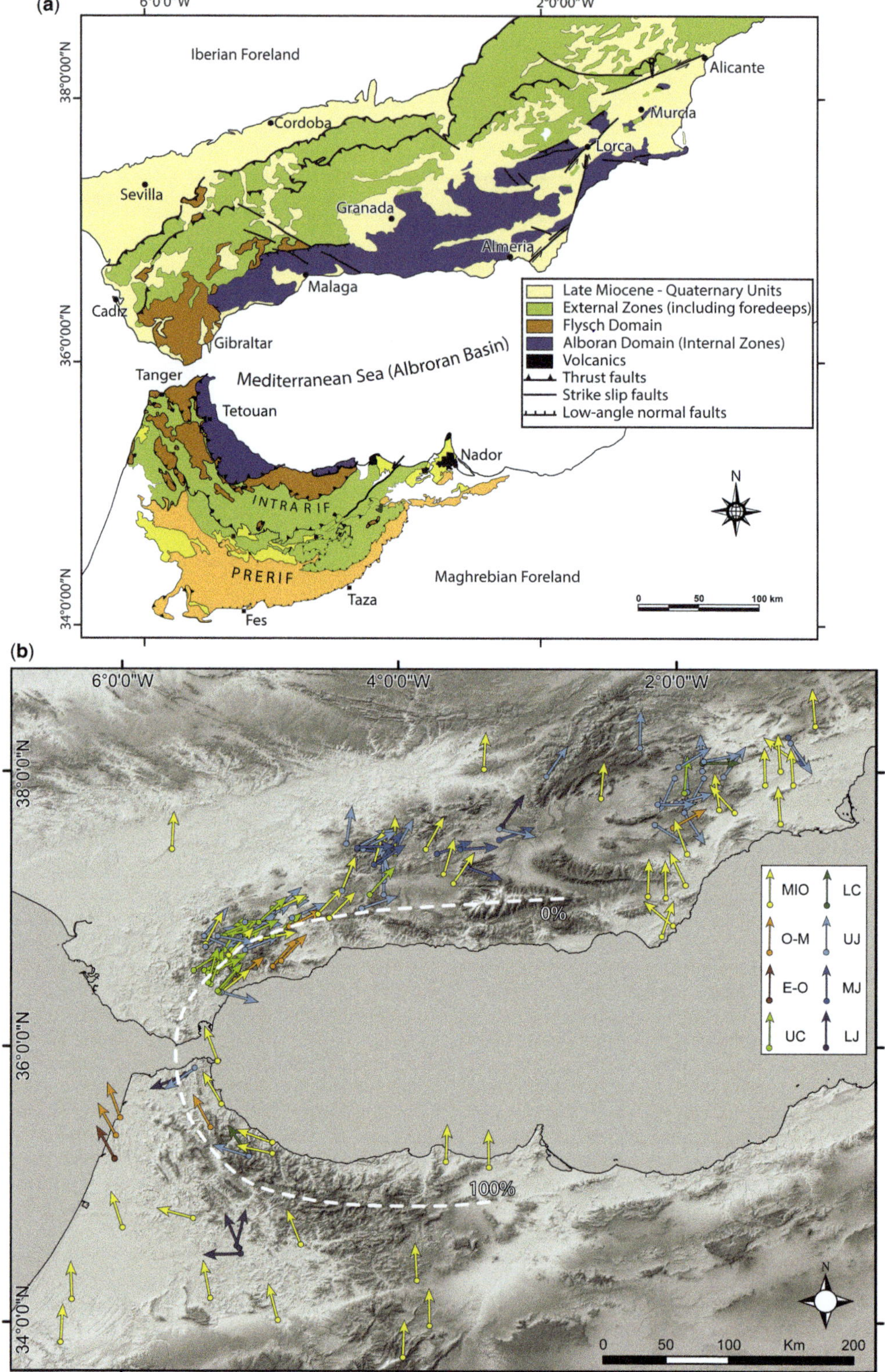

(a)
6°0'0"W
2°0'00"W
Iberian Foreland
38°00'N
Cordoba
Alicante
Murcia
Lorca
Sevilla
Granada
Almeria
36°00'N
Malaga
Cadiz
Gibraltar
Late Miocene - Quaternary Units
External Zones (including foredeeps)
Flysch Domain
Alboran Domain (Internal Zones)
Volcanics
Thrust faults
Strike slip faults
Low-angle normal faults
Tanger
Mediterranean Sea (Alboran Basin)
Tetouan
Nador
INTRA RIF
N
34°00'N
PRERIF
Maghrebian Foreland
0    50    100 km
Taza
Fes

(b)
6°0'0"W
4°0'0"W
2°0'0"W
38°0'0"N
0%
MIO    LC
O-M    UJ
E-O    MJ
36°0'0"N
UC    LJ
100%
34°0'0"N
N
0    50    100    Km    200

carried out on Neogene basins of the Rif fore-land domain for magnetostratigraphic investigations (Krijgsman *et al.* 1999; Hilgen *et al.* 2000; Krijgsman & Garcés 2004; Van Assen *et al.* 2006). Palaeomagnetic data from Tortonian to Messinian deposits indicate that no major rotations have occurred since deposition of these sediments. Based on these results, Krijgsman & Garcés (2004) concluded that the Neogene basins of the Rif chain have not experienced any rotations about vertical axes since late Tortonian. Cifelli *et al.* (2008) investigated the upper Miocene sediments from the Rif orogenic wedge. Palaeomagnetic results indicate that the upper Tortonian–Messinian sedimentary units sampled in the thrust-top Taounate–Tefrannt–Dhar Souk basins underwent about 20° CCW rotation, suggesting that CCW rotations, previously measured in the Mesozoic units of the Rif chain, extended into the upper Miocene units of the chain. Cifelli *et al.* (2008) also sampled foreland domain deposits from the Taza Basin. Their data are consistent with the palaeomagnetic results of Krijgsman & Garcés (2004) obtained in the same basin, indicating that, in the foreland basin, upper Miocene deposits did not undergo tectonic rotation. Similar directions were observed in the post-nappe basins, which are located at the northern edge of the Rif (Van Assen *et al.* 2006; Cifelli *et al.* 2008; Fig. 4).

## Discussion

The large quantity of palaeomagnetic data collected in the past years along the Mediterranean arcs enables a quantitative description of the distribution of tectonic rotations in space and time, and a detailed reconstruction of their rotational history. In the following sections we will describe the geographic and space distribution of palaeomagnetic rotations in the studied arcs, focusing on the common and different features that characterize these arcuate mountain chains, in relation to their geological complexity and geodynamic history.

For the discussion on the Calabrian Arc and Northern Apennines we consider data summarized in some recent papers (Cifelli *et al.* 2008; Caricchi *et al.* 2014). In these works, palaeomagnetic rotations have been calculated, according to Demarest (1983), by comparing Northern and Southern Apennines palaeomagnetic results with coeval expected African palaeopoles computed by Besse & Courtillot (2002). The use of the African apparent polar wander path (APWP) as a proxy for Adriatic and

Southern Italy geodynamics seems reasonable, as palaeomagnetism in the Tethyan domain has been robustly shown to have a post-Permian Africa–Adria palaeomagnetic coupling (e.g. Channell & Tarling 1975; Muttoni *et al.* 2001).

In the Gibraltar Arc, the choice of reference palaeomagnetic poles for the evaluation of tectonic rotations is more problematic. In fact, during the Mesozoic and early-middle Cenozoic times, the Iberian plate was located at the western boundary of the Tethys Ocean and formed an independent microplate between Africa and Europe (Vergés & Fernàndez 2012). In particular, the kinematics of the Iberia plate, with respect to Europe, has been modified by the late Cretaceous opening of the Bay of Biscay and the CCW rotation of Iberia with respect to stable Europe, whose precise timing and kinematics are still not well defined (see Neres *et al.* 2012 for a recent discussion of this issue). As a consequence, at least for Mesozoic times, palaeomagnetic data from the Betics cannot be directly compared with available European APWP, and Iberia reference poles have to be considered in order to calculate the precise amount of palaeomagnetic rotations. In this work, the Jurassic–Cenozoic palaeomagnetic data from the Betic and Rif chains have been compared with the Iberia palaeopoles and with the Eurasia and South African reference APWP curves to determine the amount of vertical axis rotations for each locality. For the Rif chain we used the Besse & Courtillot (2002) South Africa APWP for the time interval comprised between 185 and 5 Ma. For the Betic chain we used: the Besse & Courtillot (2002) Eurasian APWP for ages between 25 and 5 Ma; the recently calculated Iberia reference poles by Neres *et al.* (2012) for ages comprised 70 and 88 Ma (Late Cretaceous) and 136–160 Ma (Late Jurassic–early Cretaceous); and the two Iberia reference poles of 158.5 and 179.5 Ma, recently calculated by Osete *et al.* (2007), for middle and late Jurassic sites. Palaeomagnetic rotations computed for the Betic and Rif Chain are reported in Table 1. In our review of data from Gibraltar Arc, we did not consider those data that presented high $\alpha_{95}$ values or that probably underwent recent remagnetization or that could not be located in the map because information from the original papers was missing.

### Geographic distribution of palaeomagnetic rotations

*Orogenic wedge (orogenic units) v. foreland domain.* Palaeomagnetic rotations have been measured

---

**Fig. 4.** (**a**) Simplified geological map of the Gibraltar Arc area. (**b**) Palaeomagnetic declinations (and relative confidence limits) from sedimentary rocks in the Gibraltar Arc. See Table 1 for references. Rotations have been calculated comparing the obtained palaeodeclinations with the coeval expected Europe and African reference directions (see text and Table 1 for details).

**Table 1.** *Palaeomagnetic rotations in the Gibraltar Arc*

| Tectonic domain | Name | $N$ | Age | Age (Ma) | Latitude (°N) | Longitude (°W) | Palaeomagnetic declination | Palaeomagnetic inclination | $\alpha_{95}$ | Pole age | $R$ | Error (±) | Reference |
|---|---|---|---|---|---|---|---|---|---|---|---|---|---|
| *BETICS* | | | | | | | | | | | | | |
| *Western Subbetics* | | | | | | | | | | | | | |
| | CM | 1 | LJ | 176 | 37.60 | 3.25 | 21.2 | 56.6 | 10.2 | 179.5* | 34.5 | 15 | 1 |
| | ALJ | 2 | MJ | 174 | 37.32 | 4.17 | 43.6 | 53.1 | 11.3 | 179.5* | 57 | 15 | 1 |
| | AC | 1 | MJ | 173.6 | 37.34 | 4.13 | 25.8 | 43 | 7.6 | 179.5* | 39 | 9 | 2 |
| | AB | 1 | MJ | 171.6 | 37.34 | 4.13 | 22.1 | 17.4 | 10.4 | 179.5* | 35.5 | 9 | 2 |
| | CYB | 1 | MJ | 168.5 | 37.44 | 4.29 | 53.9 | 54.8 | 3.6 | 158.5* | 88 | 6 | 1 |
| | BH | 1[†] | MJ | 168.5 | 37.30 | 3.50 | 75.8 | 46.6 | 5.2 | 158.5* | 110 | 7 | 3 |
| | BC | 1[†] | MJ | 168.5 | 37.43 | 4.29 | 59.4 | 42.9 | 4 | 158.5* | 94 | 5 | 3 |
| | PNS | 2 | MJ | 168.5 | 37.43 | 3.55 | 51.4 | 48.8 | 11.7 | 158.5* | 86 | 14 | 1 |
| | BJ | 1[†] | MJ | 168.5 | 37.44 | 4.28 | 53.8 | 47 | 2.8 | 158.5* | 88 | 4 | 3 |
| | BB | 1 | MJ | 168 | 37.40 | 3.72 | 35.9 | 32.4 | 5.4 | 158.5* | 70.3 | 6 | 2 |
| | GNC1 | 1 | MJ | 168 | 37.50 | 3.27 | 36.3 | 38.9 | 12.5 | 158.5* | 70.5 | 13 | 2 |
| | CYK | 1 | UJ | 153.5 | 37.44 | 4.29 | 26.1 | 37.9 | 2.7 | 158.5* | 60.6 | 4 | 1 |
| | A | 1[†] | UJ | 153.5 | 37.20 | 4.10 | 322.2 | 27.9 | 2.8 | 158.5* | −3 | 4 | 4 |
| | B | 1[†] | UJ | 153.5 | 37.43 | 4.25 | 23 | 35.5 | 3.9 | 158.5* | 57 | 5 | 4 |
| | V | 4 | UJ | 153.2 | 36.72 | 5.39 | 20 | 29 | 4.6 | 158.5* | 55 | 5 | 5 |
| | CJ | 2 | UJ | 153.2 | 36.40 | 5.30 | 70.6 | 54 | 8.8 | 158.5* | 105.2 | 12 | 6 |
| | RJ01 | 1 | UJ | 153.2 | 36.70 | 5.30 | 39.9 | 48.9 | 6.5 | 158.5* | 74.6 | 8 | 6 |
| | RJ02 | 1 | UJ | 153.2 | 36.80 | 5.10 | 20.7 | 47.4 | 7 | 158.5* | 55.4 | 9 | 6 |
| | RJ03 | 1 | UJ | 153.2 | 37.00 | 4.90 | 43.8 | 49.7 | 5.7 | 158.5* | 78.4 | 8 | 6 |
| | RJ04 | 1 | UJ | 153.2 | 36.80 | 5.40 | 77 | 34.9 | 4.2 | 158.5* | 111.7 | 5 | 6 |
| | RJ05 | 1 | UJ | 153.2 | 36.80 | 5.30 | 65.7 | 44 | 6.5 | 158.5* | 100.4 | 8 | 6 |
| | RJ06 | 1 | UJ | 153.2 | 36.90 | 5.10 | 58.3 | 45.5 | 3.2 | 158.5* | 93 | 5 | 6 |
| | RJ07 | 1 | UJ | 153.2 | 36.90 | 5.00 | 25.3 | 57.7 | 10.6 | 158.5* | 60 | 16 | 6 |
| | RJ08 | 1 | UJ | 153.2 | 36.80 | 5.40 | 6.2 | 41.6 | 7.7 | 158.5* | 40.9 | 9 | 6 |
| | RJI0 | 1 | UJ | 153.2 | 36.70 | 5.30 | 29.4 | 44.2 | 8.2 | 158.5* | 64 | 9 | 6 |
| | RJ12 | 1 | UJ | 153.2 | 36.70 | 5.10 | 33.7 | 37.7 | 9.1 | 158.5* | 68.3 | 9 | 6 |
| | CH | 10 | UJ | 153.2 | 36.93 | 4.76 | 33 | 41 | 9 | 158.5* | 67 | 10 | 7 |
| | ERJ | 2 | UJ | 153.2 | 37.47 | 4.37 | 333.6 | 38.2 | 6.7 | 158.5* | 8 | 7 | 1 |
| | AJ | 4 | UJ | 153.2 | 37.00 | 4.26 | 36.1 | 43.3 | 15.9 | 158.5* | 70.5 | 17 | 6 |
| | SG | 1 | UJ | 153.2 | 37.10 | 4.10 | 328.3 | 36 | 7.8 | 158.5* | 3 | 8 | 8 |
| | BRJu | 1 | UJ | 153.2 | 37.41 | 4.02 | 353.8 | 38.8 | 11.6 | 158.5* | 28.3 | 12 | 1 |
| | MM | 5 | UJ | 153.2 | 37.61 | 2.15 | 88 | 41 | 12 | 158.5* | 122 | 13 | 7 |
| | GAB | 10 | UJ | 153.2 | 37.76 | 2.13 | 55 | 37 | 8 | 158.5* | 89 | 8 | 7 |
| | JCA | 2 | UJ | 150.7 | 36.39 | 5.27 | 4 | 44.9 | 4.9 | 136–160[†] | 38 | 6 | 9 |

| JAT | 2 | UJ | 150.7 | 36.66 | 5.22 | 12.9 | 36.4 | 8.6 | 136–160[†] | 47 | 9 | 9 |
|---|---|---|---|---|---|---|---|---|---|---|---|---|
| JBU | 2 | UJ | 150.7 | 36.78 | 4.99 | 44 | 33.2 | 5.6 | 136–160[†] | 78 | 6 | 9 |
| JTE | 2 | UJ | 150.7 | 36.98 | 4.90 | 17.3 | 45 | 4.4 | 136–160[†] | 51 | 6 | 9 |
| JGO | 2 | UJ | 150.7 | 37.13 | 4.40 | 349.1 | 40 | 4 | 136–160[†] | 23 | 5 | 9 |
| A | 1[†] | UJ | 148.2 | 37.20 | 4.10 | 324.4 | 32.5 | 2.5 | 136–160[†] | −4 | 4 | 4 |
| B | 1[†] | UJ | 148.2 | 37.43 | 4.25 | 30.4 | 41.6 | 1.6 | 136–160[†] | 62 | 4 | 4 |
| GO | 1 | UJ | 145.5 | 37.58 | 3.27 | 68.4 | 11.8 | 7.3 | 136–160[†] | 102 | 7 | 2 |
| B304 | 1 | UC | 83 | 36.55 | 5.40 | 62 | 35 | 4 | 70–88[†] | 61 | 5 | 10 |
| B305 | 1 | UC | 83 | 36.55 | 5.40 | 74 | 46 | 5 | 70–88[†] | 73 | 6 | 10 |
| B306 | 1 | UC | 83 | 36.55 | 5.48 | 59 | 22 | 16 | 70–88[†] | 58 | 14 | 10 |
| B307 | 1 | UC | 83 | 36.46 | 5.37 | 16 | 53 | 10 | 70–88[†] | 15 | 14 | 10 |
| B308 | 1 | UC | 83 | 36.45 | 5.35 | 21 | 45 | 10 | 70–88[†] | 20 | 11 | 10 |
| B309 | 1 | UC | 83 | 36.53 | 5.37 | 67 | 50 | 10 | 70–88[†] | 66 | 13 | 10 |
| RC10 | 1 | UC | 83 | 36.54 | 5.31 | 59 | 40 | 8 | 70–88[†] | 58 | 9 | 10 |
| cc02 | 1 | UC | 83 | 36.39 | 5.29 | 39 | 51 | 4 | 70–88[†] | 38 | 6 | 10 |
| CC | 2 | UC | 83 | 36.40 | 5.30 | 42.3 | 49.2 | 14.8 | 70–88[†] | 41.5 | 18.2 | 6 |
| RC01 | 1 | UC | 83 | 36.80 | 5.10 | 64.9 | 41.6 | 4.8 | 70–88[†] | 64 | 6 | 6 |
| RC02 | 1 | UC | 83 | 36.70 | 5.250 | 31.2 | 38.1 | 5.6 | 70–88[†] | 30 | 6 | 6 |
| RC03 | 1 | UC | 83 | 37.00 | 4.90 | 73.1 | 20.7 | 5.9 | 70–88[†] | 72 | 6 | 6 |
| RC04 | 1 | UC | 83 | 36.80 | 5.25 | 65.3 | 30.3 | 5.6 | 70–88[†] | 64 | 6 | 6 |
| RC05 | 1 | UC | 83 | 36.80 | 5.30 | 69 | 29.7 | 5.2 | 70–88[†] | 68 | 6 | 6 |
| RC06 | 1 | UC | 83 | 36.80 | 5.30 | 80.6 | 34.1 | 3.7 | 70–88[†] | 80 | 5 | 6 |
| RC07 | 1 | UC | 83 | 36.80 | 4.90 | 53.3 | 36.7 | 6.5 | 70–88[†] | 52 | 7 | 6 |
| RC08 | 1 | UC | 83 | 36.90 | 5.00 | 48.9 | 48.2 | 5.4 | 70–88[†] | 48 | 7 | 6 |
| RC09 | 1 | UC | 83 | 36.70 | 5.20 | 55.2 | 41.5 | 5.1 | 70–88[†] | 54 | 6 | 6 |
| SG | 1 | UC | 83 | 37.10 | 4.20 | 42.6 | 54.2 | 6 | 70–88[†] | 41.5 | 8.7 | 8 |
| ALJ | 2 | MIO | 15 | 37.32 | 4.17 | 18 | 51.5 | 11.9 | 15[‡] | 15.6 | 15 | 1 |
| BRJ | 2 | MIO | 15 | 37.40 | 4.02 | 3 | 60.3 | 7.7 | 15[‡] | 0 | 13 | 1 |
| JAT | 1 | MIO | 15 | 36.66 | 5.22 | 44.1 | 49.1 | 4 | 15[‡] | 41.8 | 6 | 8 |
| JBU | 1 | MIO | 15 | 36.78 | 4.99 | 81.9 | 62.8 | 9.1 | 15[‡] | 79.6 | 16 | 8 |
| JCA | 1 | MIO | 15 | 36.39 | 5.27 | 53.4 | 69.7 | 2.7 | 15[‡] | 51.1 | 4 | 8 |
| JGO | 1 | MIO | 15 | 37.13 | 4.40 | 23.2 | 44.9 | 3.9 | 15[‡] | 20.8 | 5 | 8 |
| JTE | 1 | MIO | 15 | 36.98 | 4.90 | 66.3 | 63.7 | 2.7 | 15[‡] | 64 | 6 | 8 |
| JTO | 4 | MIO | 15 | 36.96 | 4.57 | 45.1 | 48.3 | 3.4 | 15[‡] | 43 | 5 | 8 |
| JUB | 1 | MIO | 15 | 36.74 | 5.39 | 31.3 | 53.3 | 2.2 | 15[‡] | 29 | 4 | 8 |
| JAB | 1 | MIO | 15 | 36.93 | 4.49 | 46.1 | 43.8 | 5.3 | 15[‡] | 44 | 6 | 8 |

*Eastern Subbetics*

| SL | 1 | MJ | 168.5 | 38.25 | 1.20 | 117.2 | 53 | 9.6 | 158.5* | 151 | 13 | 1 |
|---|---|---|---|---|---|---|---|---|---|---|---|---|
| MAR | 1 | UJ | 153.2 | 37.70 | 1.94 | 103 | 31 | 9 | 158.5* | 147 | 9 | 11 |

(Continued)

**Table 1.** *Palaeomagnetic rotations in the Gibraltar Arc  (Continued)*

| Tectonic domain | Name | N | Age | Age (Ma) | Latitude (°N) | Longitude (°W) | Palaeomagnetic declination | Palaeomagnetic inclination | $\alpha_{95}$ | Pole age | R | Error (±) | Reference |
|---|---|---|---|---|---|---|---|---|---|---|---|---|---|
| | ARC | 5 | UJ | 153.2 | 38.03 | 1.98 | 23.1 | 40.4 | 24 | 158.5* | 57 | 25 | 11 |
| | CAR | 3 | UJ | 153.2 | 37.95 | 2.01 | 168.6 | 42.2 | 44 | 158.5* | −157 | 54 | 11 |
| | ENG | 2 | UJ | 153.2 | 37.77 | 2.00 | 40.9 | 28.1 | 37.9 | 158.5* | 74.7 | 35 | 11 |
| | QPR | 5 | UJ | 153.2 | 38.07 | 1.80 | 45.2 | 46 | 23 | 158.5* | 79 | 27 | 12 |
| | BRT | 3 | UJ | 153.2 | 38.00 | 1.81 | 323 | 35 | 15.3 | 158.5* | −3 | 15 | 12 |
| | ALM | 3 | UJ | 153.2 | 37.85 | 1.92 | 321.9 | 41.7 | 16.6 | 158.5* | −4 | 18 | 11 |
| | GON | 1 | UJ | 153.2 | 37.95 | 1.8 | 174 | 23 | 16.5 | 158.5* | −152 | 14 | 11 |
| | PON | 2 | UJ | 153.2 | 37.96 | 1.66 | 1.6 | 35.6 | 53.9 | 158.5* | 35 | 65 | 11 |
| | KSL | 1_m | UJ | 148 | 38.21 | 1.18 | 111.7 | 46.6 | 3 | 136–160[†] | 145 | 5 | 13 |
| | VCY.VCZ | 2_m | LC | 140 | 38.06 | 1.81 | 56.4 | 47.7 | 4.4 | 136–160[†] | 89.6 | 6 | 14 |
| | AS | 6 | UC | 73 | 37.84 | 1.94 | 6 | 39 | 14 | 70–88[†] | 4.3 | 14.5 | 15 |
| *Prebetics* | | | | | | | | | | | | | |
| | CAZ[†] | 2 | UJ | 153.2 | 37.96 | 2.93 | 0.3 | 29.2 | 7.3 | 158.5* | 34 | 7 | 1 |
| | B49 | 1 | UJ | 153.2 | 38.17 | 2.26 | 327 | 48 | 6.4 | 158.5* | 1 | 8 | 11 |
| *Neogene Basins* | | | | | | | | | | | | | |
| Foreland | CAR | 1_m | MIO | 6.3–6.0 | 37.43 | 5.64 | 3 | 52 | 4 | 5[‡] | 3 | 6 | 16 |
| | GUA | 4 | MIO | 8 | 38.01 | 3.38 | 4.8 | 54.8 | 11.4 | 10[‡] | 3 | 16 | 17 |
| Central Betics | ALR | 5 | MIO | 13 | 37.43 | 3.80 | 31.1 | 47.7 | 12.3 | 15[‡] | 29 | 15 | 17 |
| | PUR | 1_m | MIO | 6 | 37.25 | 3.67 | 196.5 | −39.7 | 5.9 | 5[‡] | 16 | 6 | 18 |
| | GRA | 9 | MIO | 6 | 37.18 | 3.6 | 34.9 | 20.6 | 11 | 5[‡] | 34 | 9 | 17 |
| Eastern Betics | MUR | 7 | MIO | 6 | 38.00 | 1.25 | 356.5 | 51.1 | 8.9 | 5[‡] | −4 | 11 | 19 |
| | H-O | 5 | MIO | 7 | 37.40 | 1.92 | 160.1 | −31.1 | 34.6 | 5[‡] | −20 | 32 | 17 |
| | CR | 6 | MIO | 5 | 37.62 | 1.25 | 356.9 | 46.8 | 11.3 | 5[‡] | −4 | 13 | 19 |
| | CO | 1_m | MIO | 7.8–7.5 | 38.33 | 1.00 | 357 | 48 | 4 | 10[‡] | −5 | 5 | 20 |
| | GAL | 1_m | MIO | 3.5–1.7 | 37.80 | 2.54 | 5 | 45 | 11 | 5[‡] | 5 | 12 | 21 |
| | CH-SM | 1_m | MIO | 7.6–6.0 | 38.08 | 1.16 | 309 | 27 | 4 | 5[‡] | −52 | 4 | 22 |
| | SL | 1_m | MIO | 6.8–4.6 | 37.90 | 1.36 | 359 | 44 | 3 | 5[‡] | −1.5 | 4 | 22 |
| | LS | 1_m | MIO | 7.8–7.5 | 37.72 | 1.69 | 354 | 52 | 4 | 10[‡] | −8 | 6 | 23 |
| | ZORR | 1_m | MIO | 5.6–5.3 | 37.09 | 2.2 | 181 | −40 | 5 | 5[‡] | 1 | 6 | 24 |
| | VV | 1_m | MIO | 7.6–7.3 | 37.90 | 1.16 | 357.8 | 48.4 | 3.3 | 10[‡] | −4 | 5 | 25 |
| | L-T | 10 | MIO | 10 | 37.7 | 1.58 | 320 | 37.9 | 14 | 10[‡] | −42 | 14 | 17 |
| | Vera | 3 | MIO | 5 | 37.17 | 1.93 | 335.5 | 31.3 | 4.4 | 5[‡] | −25 | 5 | 17 |
| | ABAD | 1_m | MIO | 6.8–6 | 37.08 | 2.07 | 359 | 57 | 5 | 5[‡] | −1.5 | 8 | 26 |

| | | | | | | | | | | | | |
|---|---|---|---|---|---|---|---|---|---|---|---|---|
| *Internal Domain* | | | | | | | | | | | | |
| | RON | 11 | O-M | 20–18 | 36.48 | 5.17 | 54.8 | 42 | 10.1 | 20[‡] | 50.6 | 11 | 27 |
| | OJEN | 7 | O-M | 20–18 | 36.62 | 4.84 | 48.5 | 46 | 14.2 | 20[‡] | 44 | 17 | 27 |
| | CA3 | 1 | O-M | 20–18 | 36.87 | 4.78 | 67.7 | 69.9 | 16.3 | 20[‡] | 63.5 | 43 | 27 |
| | | 19 | O-M | 20–18 | 36.58 | 4.90 | 46.1 | 47.5 | 6.6 | 20[‡] | 41.9 | 8 | 28 |
| | EST | 7 | O-M | 23 | 37.59 | 2.02 | 243 | −27 | 9 | 25[‡] | 60 | 9 | 29 |
| | A | 7 | MIO | 12.5 | 36.80 | 2.10 | 20 | 50 | 11 | 10[‡] | 18 | 14 | 30 |
| | B | 4 | MIO | 11 | 36.88 | 2.02 | 312 | 49 | 20 | 10[‡] | −50 | 25 | 30 |
| *MOROCCO* *Foreland* | | | | | | | | | | | | |
| | OA/Gh | 1_m | MIO | 6.3–6 | 34.16 | 6.37 | 358 | 37 | 4 | 5[§] | −2 | 4 | 31 |
| | AeB/Gh | 1_m | MIO | 6.5–5.5 | 33.85 | 6.45 | 3 | 43 | 4 | 5[§] | 3 | 5 | 16 |
| | Z/Gu | 1_m | MIO | 7.6–7.0 | 33.97 | 3.77 | 360 | 47 | 4 | 10[§] | −1 | 5 | 32 |
| | KZ/Gu | 1_m | MIO | 6.0–4.8 | 33.74 | 3.96 | 2 | 49 | 5 | 5[§] | 2 | 6 | 32 |
| | MR49/52 | 4 | MIO | **6** | 34.30 | 3.86 | 357.1 | 48.9 | 11 | 5[§] | −3 | 13 | 33 |
| | RI03 | 1 | MIO | **7** | 34.68 | 5.99 | 341.7 | 53.1 | 9.9 | 5[§] | −18 | 13 | 33 |
| | MR30 | 1 | MIO | **7** | 34.01 | 4.86 | 344.4 | 33.4 | 6.2 | 5[§] | −15 | 6 | 33 |
| *Alboran Domain* | | | | | | | | | | | | |
| | 14 | 1 | LJ | 187 | 35.79 | 5.54 | 222.6 | 43.9 | 13.5 | 185[§] | −112 | 16 | 34 |
| | 13 | 1 | UJ | 153.2 | 35.84 | 5.47 | 198.5 | 40.8 | 8.6 | 155[§] | −125 | 9 | 34 |
| | 18 | 2 | UJ | 153.2 | 35.20 | 5.07 | 251.4 | 35.5 | 8.8 | 155[§] | −73 | 9 | 34 |
| | 17 | 1 | LC | 83 | 35.20 | 5.07 | 299.9 | 35.3 | 12.8 | 85[§] | −34 | 15 | 34 |
| | BB | 7 | MIO | 20 | 35.30 | 4.90 | 291 | 58 | 13.1 | 20[§] | −70 | 20 | 35 |
| | SA | 7 | MIO | 20 | 35.89 | 5.30 | 340.2 | 49.2 | 4.4 | 20[§] | −21 | 7 | 35 |
| | ST | 6 | MIO | 18 | 35.22 | 4.90 | 283.8 | 58.6 | 10.3 | 20[§] | −78 | 16 | 36 |
| | RI04 | 1 | MIO | **7** | 35.58 | 5.27 | 328.6 | 45.5 | 11.8 | 5[§] | −31 | 13 | 33 |
| *External Rif* | | | | | | | | | | | | |
| | R8 | 1 | LJ | 187 | 34.49 | 5.13 | 243 | 50 | 11.3 | 185[§] | −92 | 15 | 37 |
| | R9.20 | 2 | LJ | 187 | 34.55 | 5.16 | 349 | 33 | 4.8 | 185[§] | 14 | 8 | 37 |
| | R7 | 1 | LJ | 187 | 34.53 | 5.14 | 312 | 39 | 7.7 | 185[§] | −23 | 10 | 37 |
| | RI01 | 1 | E-O | **39** | 35.18 | 6.05 | 326 | 58.2 | 10.5 | 40[§] | −30 | 17 | 33 |
| | MR03 | 1 | O-M | **22** | 35.48 | 6.01 | 339.9 | 21 | 17.8 | 20[§] | −21 | 15 | 33 |
| | MR06 | 1 | O-M | **22** | 35.35 | 6.04 | 330.8 | 30.8 | 13 | 20[§] | −30 | 12 | 33 |
| | MR08 | 1 | MIO | **14** | 34.75 | 5.48 | 285 | 50.8 | 7 | 15[§] | −75 | 9 | 33 |
| | RI05 | 1 | O-M | **23** | 35.41 | 5.35 | 330.1 | 58.2 | 8.6 | 25[§] | −29 | 14 | 33 |
| | MR28 | 1 | MIO | **7.5** | 34.17 | 5.35 | 349.4 | 64.7 | 10.7 | 10[§] | −11 | 20 | 33 |

(Continued)

**Table 1.** *Palaeomagnetic rotations in the Gibraltar Arc (Continued)*

| Tectonic domain | Name | $N$ | Age | Age (Ma) | Latitude (°N) | Longitude (°W) | Palaeomagnetic declination | Palaeomagnetic inclination | $\alpha_{95}$ | Pole age | $R$ | Error (±) | Reference |
|---|---|---|---|---|---|---|---|---|---|---|---|---|---|
| *Neogene Basins* | | | | | | | | | | | | | |
| Alboran Dom | If/M | 1_m | MIO | 6.5–6.0 | 35.12 | 3.34 | 359 | 49 | 8 | 5[§] | −1 | 10 | 38 |
| | BOU | 6 | MIO | 6 | 35.16 | 3.65 | 181.7 | −56.6 | 8.6 | 5[§] | 1.8 | 12 | 33 |
| External Rif | T-T-DS | 12 | MIO | 7 | 34.56 | 4.69 | 338.2 | 44.2 | 10 | 5[§] | −22 | 11 | 33 |

$N$, number of sites; $\alpha_{95}$, a statistical parameter; $R$, tectonic rotation; LJ, Lower Jurassic; MJ, Middle Jurassic; UJ, Upper Jurassic; LC, Lower Cretaceous; UC, Upper Cretaceous; E-O, Eocene–Oligocene; O-M, Oligocene–Miocene; MIO, Miocene.

*,[†]Iberia palaeopoles from Osete *et al.* (2007) and Neres *et al.* (2012), respectively.

‡Europe palaeopoles from Besse & Courtillot (2002).

§Africa palaeopoles from Besse & Courtillot (2002).

References are: (1) Osete *et al.* (2004), (2) Osete *et al.* (1988), (3) Steiner *et al.* (1987), (4) Ogg *et al.* (1984), (5) Kirker & McClelland (1996), (6) Platzman & Lowrie (1992), (7) Mayfield (1999), (8) Platzman (1994), (9) Villalain *et al.* (1994), (10) Platt *et al.* (2003), (11) Allerton *et al.* (1993), (12) Allerton (1994), (13) Mazaud *et al.* (1986), (14) Ogg *et al.* (1988), (15) Allerton *et al.* (1994), (16) Krijgsman *et al.* (2004), (17) Mattei *et al.* (2006), (18) Martin-Suarez *et al.* (1998), (19) Calvo *et al.* (1997), (20) Dinarès-Turell *et al.* (1999), (21) Garcés *et al.* (1997), (22) Garcés *et al.* (2001), (23) Krijgsman *et al.* (2000), (24) Krijgsman *et al.* (2001), (25) Krijgsman *et al.* (2006), (26) Sierro *et al.* (1996), (27) Villasante-Marcos *et al.* (2003), (28) Feinberg *et al.* (1996), (29) Platzman & Platt (2004), (30) Calvo *et al.* (1994), (31) Hilgen *et al.* (2000), (32) Krijgsman *et al.* (1999), (33) Cifelli *et al.* (2008), (34) Platzman (1992), (35) Berndt *et al.* (2015), (36) Saddiqi *et al.* (1995), (37) Platt *et al.* (2003), (38) Van Assen *et al.* (2006).

in the orogenic units of the Betics, Rif, Sicilian Maghrebids and Southern Apennines chains, different from rotations found in the Guadalquivir, Taza, Hyblean and Apulia foreland basins (Figs 2–4). The absence of palaeomagnetic rotations in the foreland basins is not dependent on the age of the sampled units, which age spans from Mesozoic to Pleistocene. This distribution demonstrates that palaeomagnetic rotations are confined to the orogenic wedge and do not extend to the Iberia, Africa and Apulia forelands, which remained almost unrotated during the formation of the orogens.

*Palaeomagnetic rotations along the arcs.* In the Mediterranean Arcs CCW palaeomagnetic rotations have been measured in the northern arm of the Northern Apennines Arc, in the Southern Apennines and in the Rif chain, whereas CW rotations have been measured in the southern part of the Northern Apennines Arc, in the Betic Chain, in the CPD, and in the Sicilian Maghrebids (Fig. 1). This pattern of palaeomagnetic rotations is, at first approximation, what is expected in oroclines, which are defined on the basis of a strong correlation between the trend of geological structures and palaeomagnetic rotations. This relationship has already been demonstrated for the Gibraltar Arc (Platt *et al.* 2003), the Northern Apennines Arc (Speranza *et al.* 1997) and the Calabrian Arc (Cifelli *et al.* 2007b, 2008). However, in all these arcs an oroclinal bending model is not able to explain the whole pattern of palaeomagnetic rotations, which are also influenced by other tectonic processes. In particular, divergences from a simple oroclinal bending model for in the Mediterranean arcs include: (a) the CPD in the Calabrian Arc; (iib) the Tuscan units in the Northern Apennines Arc; and (iiic) the Eastern Betics in the Gibraltar Arc.

In order to test the orocline origin for the Mediterranean Arcs, we plotted palaeomagnetic rotations as a function of their location along the length of the Calabrian, Northern Apennines and Gibraltar Arc, analysing palaeomagnetic rotations in relation to their distance, in percentage, along the arcs (Fig. 5). A hypothetical axis of the chains was traced across the arcs and tectonic rotations were represented as the projection of the sampled sites on this axis (see Figs 2–4 for location).

Data from the Calabrian Arc, the Southern Apennines and Sicily indicate that a large amount of tectonic rotation has occurred. Within these areas palaeomagnetic rotations show a very large scatter, which is due to the presence of local tectonic structures, to the large age range of sampled lithology and to the different structural position (internal and external) of the sampled units (shown with different symbols). The distribution of palaeomagnetic rotations along the arc shows that an abrupt change

in palaeomagnetic directions occurs within a narrow region located between the Southern Apennines and Calabria. This region is characterized by a complex pattern of palaeomagnetic rotations (Cifelli *et al.* 2007b), which change from CCW rotations in the north to CW rotations in the south (Fig. 2). This type of pattern is not expected in a typical orocline, where a systematic change in the magnitude of along-strike palaeomagnetic rotations should occur. Furthermore, no significant differences in palaeomagnetic rotations are observed along the entire segment of the arc represented by the CPD, which indicates that this sector behaved as an almost homogeneous rigid block (Fig. 5).

In the Northern Apennines Arc, the Umbria–Marche–Romagna units show a gradual change (with a large scatter) in the magnitude of along-strike palaeomagnetic rotations (Fig. 5b), as expected in an orocline. This trend is well evident in the pre-orogenic Mesozoic–Eocene units, outcropping in the axial part of the chain (green full squares in Fig. 5b), and in upper Miocene foredeep units, that outcrop in the external part of the arc (orange full squares in Fig. 5b). Both these units show CW rotations in the SE part of the arc and a progressive increase of CCW rotations in the NW sector of the arc. A different pattern is observed in the Tuscan Domain units, which crop out in the internal part of the chain (empty squares in Fig. 5b). In these units, the amount of CCW vertical axis rotations decreases from the southern sector of the arc, where a very large amount of CCW rotations has been measured, towards the northern part of the arc, where CCW rotations are significantly less (Caricchi *et al.* 2014). This distribution cannot be explained using a simple oroclinal bending model, and needs a more complex interplay of different tectonic processes (Caricchi *et al.* 2014). This point will be discussed later in detail.

In the Gibraltar Arc palaeomagnetic rotations are CW in the Western Betics and CCW in the Rif chain. Notwithstanding this general pattern, palaeomagnetic data do not show a clear trend in the magnitude of along-strike palaeomagnetic rotations. In the Western Betics, moving from west to east, no significant variation in the amount of palaeomagnetic rotation is observed in pre-orogenic (green and orange full squares in Fig. 5), intermountain (orange empty triangles in Fig. 5) or internal (red full crosses in Fig. 5) units. The eastern Betics (not reported in Fig. 5) show a different distribution of palaeomagnetic rotations with respect to the western and central Betics (Fig. 4). This distribution seems to be influenced by the tectonic structures that characterize this area. CCW rotations have been observed in different sedimentary basins located along a deformation belt dominated by Late Miocene–Quaternary left lateral strike-slip faults,

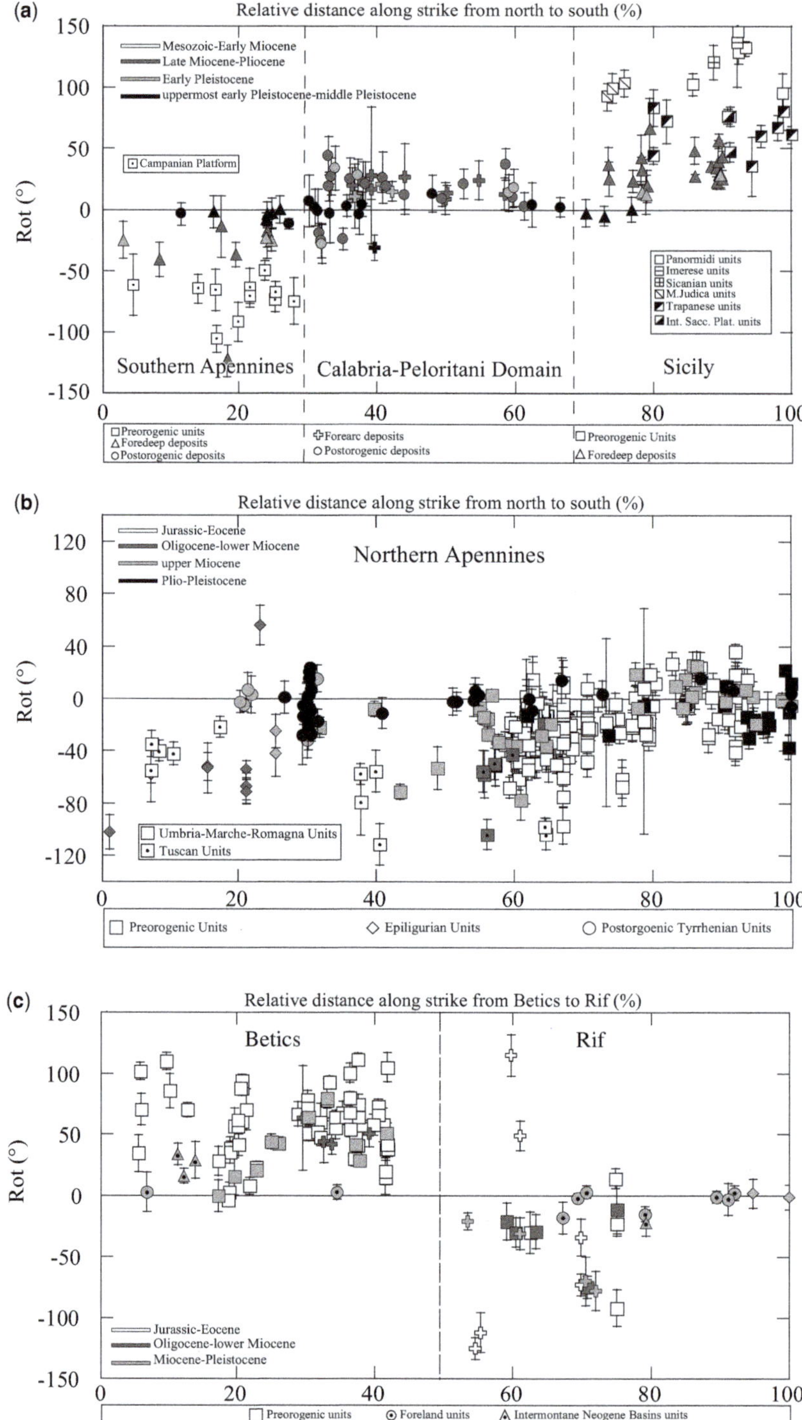


**Fig. 5.** Distribution of palaeomagnetic rotations as a function of distance along the chain (from north to south) along the Gibraltar Arc. A hypothetical axis of the chain (0–100%) was traced across the arcs (see Figs 2–4) and tectonic rotations were represented as the projection of the sampled sites on this axis.

which characterize the South Iberian continental margin in the region of Almeria–Murcia. In this region, CCW rotations do not extend to sedimentary basins, which are located far away from the main strike-slip faults (Calvo *et al.* 1997; Krijgsman *et al.* 2004). In the Rif chain the few reliable data from pre-orogenic units show a wide range of rotation values, without a significant trend in palaeomagnetic rotations along this sector of the arc.

*Palaeomagnetic rotations from the inner to the outer sectors of the arcs.* A common characteristic of the investigated Mediterranean arcs is that the amount of palaeomagnetic rotation tends to decrease moving from the more internal units towards the more external ones. This is particularly well documented in the Southern Apennines and Sicily, at the opposite edges of the Calabrian Arc (Fig. 2). The distribution of palaeomagnetic declinations indicates a gradual decrease in the amount of rotation away from internal sectors (Tyrrhenian side) towards external sectors (Channell *et al.* 1990; Oldow *et al.* 1990; Gattacceca & Speranza 2002; Speranza *et al.* 2003; Cifelli *et al.* 2007b). In Sicily the amount of palaeomagnetic rotations decreases, from sites of the same age, moving from the internal units (up to 145° CW), to the intermediate units (50° CW), whereas the external, undeformed units are unrotated (Figs 2 & 5). Similarly, in the Southern Apennines a gradual reduction in the amount of palaeomagnetic rotations from the internal units is observed, which record up to 125° CCW rotation, to external sectors of the chain, which record smaller amounts of CCW rotation. A peculiar pattern is observed in the CPD where, except for results from Speranza *et al.* (2011), no differences exist between palaeomagnetic rotations measured in the Tyrrhenian post-orogenic deposits and the Ionian forearc basins deposits (Figs 2 & 5).

In the Northern Apennines a different distribution of palaeomagnetic rotations is observed in the southern and northern sectors of the arc. In the southern sector the amount of palaeomagnetic rotations decreases from the Internal Tuscan Units (where up to 100° CCW rotations have been measured) towards the external Umbria–Marche–Romagna Units (where rotations are 30° CCW in average). In the northern sector of the arc the distribution of palaeomagnetic rotations from the internal to the external sectors of the chain is not well constrained, even if the few available palaeomagnetic directions seem to suggest that palaeomagnetic rotations decrease moving from the internal Epiligurian Units (30–90° CCW), to the Tuscan Units (about 40° CCW) to the external Umbria–Marche–Romagna units (20–30° CCW).

In the Gibraltar Arc the variation of palaeomagnetic rotations from the internal to the external sectors of the arc is less defined because of the important episodes of magnetic overprint that occurred in the area during the Neogene (e.g. Villalain *et al.* 1994). However, in the Rif chain a progressive decrease of palaeomagnetic rotations is observed from the Internal Zone, where large amounts of tectonic rotations, mostly CCW (except Tetouan area), were measured (Platzman 1992; Platzman *et al.* 1993), in respect to the external units that record minor amounts of CCW rotations and to external intramontane basins that underwent 20° CCW rotations (Fig. 4). In the western and central Betics a large amount of data have been collected in the external domains, whereas few data have been obtained form the internal units, which have been affected by magnetic overprints that seem to hide a part of rotational history of these units (e.g. Villalain *et al.* 1994).

## Timing of palaeomagnetic rotations

The distribution of palaeomagnetic rotations v. age in the Mediterranean arcs is shown in Figure 6. In these diagrams all the data for each limb of the arcs are reported, independently from their position along the arc, whereas distinctive symbols are used for the different tectonic units that form the chains. The large dispersion of palaeomagnetic rotations for data of similar ages mostly reflects the different tectonic position of the sampled units (e.g. internal v. external position) and the relative position of palaeomagnetic rotations along the trace of the arc (Fig. 5). Notwithstanding this oversimplification, all diagrams show a common pattern that characterizes the two different trends: (a) a first segment, which spans Mesozoic to Neogene time, has values of palaeomagnetic rotations that are high, but remain almost constant during the entire time interval for each structural unit; and (b) a second segment, which corresponds to the Neogene time interval, where a progressive decrease in palaeomagnetic rotations with time occurs. These different trends define a change in the geodynamic setting, and mark the progressive incorporation of different palaeogeographic domains into the orogenic wedge (Fig. 6).

In Sicily and in the Southern Apennines, the first segment spans Jurassic to Oligocene–middle Miocene time, with values of palaeomagnetic rotations that, for the same structural unit, remain almost constant during the entire time interval. In the second segment, which corresponds to the middle–late Miocene to Pleistocene time interval, a progressive decrease in palaeomagnetic rotations with time occurs (Cifelli *et al.* 2007b; Fig. 6a). The first segment of the diagrams shows that the internal Apennine and Sicilian units did not rotate with respect to Africa during this time span, when they were still

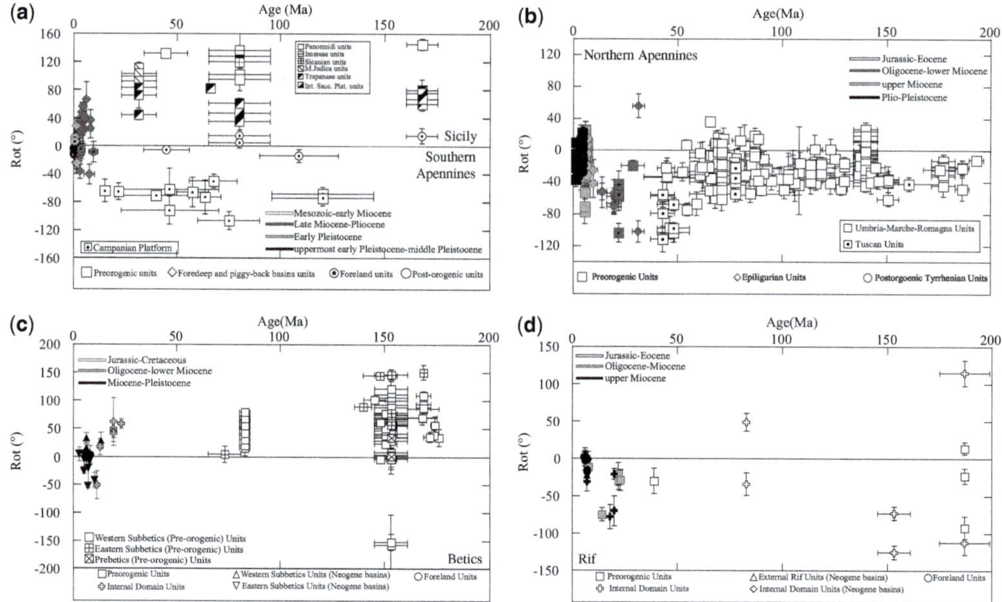

**Fig. 6.** Units mean rotations (in degrees) v. age for palaeomagnetic rotations from Calabrian Arc (**a**), Northern Apennines (**b**) and Gibraltar Arc (**c** and **d**).

part of the passive margin of the African–Apulia plate or represented the foreland of the former Apennine and Maghrebian chain, before their palaeogeographic realms were incorporated in the fold-and-thrust belt. The transition between the two segments of the diagrams occur during middle–late Miocene and allows defining the beginning of rotational processes in Sicily and in the Southern Apennines, when internal Apennine and Maghrebian units started to be incorporated in the orogenic wedge (Gattacceca & Speranza 2002; Speranza *et al.* 2003; Cifelli *et al.* 2007*b*). Sicily and the Southern Apennines show a progressive decrease in the magnitude of rotations from late Miocene to early Pleistocene, when thrust sequences forming the Southern Apennines and Maghrebian chains were progressively emplaced and translated towards the Apulia and the Hyblean foreland, respectively. It is also worth noting that lower Pleistocene sediments in Sicily and in the Southern Apennines show a large amount of opposite sense rotations, whereas palaeomagnetic declinations from uppermost lower to middle Pleistocene sediments indicate no significant vertical axis rotations occurred. Together these data suggest that, in a large part of the Calabrian Arc, vertical axis rotations occurred during the early Pleistocene and were almost finished by the middle Pleistocene (Fig. 2).

In the Northern Apennines the rotation v. age diagram is very well constrained because a large amount of palaeomagnetic data are available from

pre-, syn- and post-orogenic units. In the external part of the chain, palaeomagnetic data have been obtained from Lower Jurassic to Oligocene– Lower Miocene Umbria–Marche–Romagna pre-orogenic units and from Late Miocene–early Pliocene syn-orogenic deposits from external fore-deep basins. Even if some variability of palaeomagnetic rotations along the arc is taken into account, the external units of the arc yield 30° CCW rotation from Lower Jurassic to Late Miocene, implying that these units have not rotated during this entire time span, and that this rotation occurred after Late Miocene. In the more internal domain large CCW rotations have been measured in Palaeogene pre-orogenic and Lower Miocene syn-orogenic Tuscan units, whereas no palaeomagnetic rotations have been measured in Late Miocene–Pliocene post-orogenic units (Figs 3 & 6b). This distribution of palaeomagnetic rotations v. age implies that rotations become younger from the internal towards the more external units. In fact, the Tuscan internal units underwent a large phase of CCW rotation between the early Miocene (age of the youngest rotated sediments) and the Late Miocene (age of the non-rotated post-orogenic units), whereas the external Umbria–Marche–Romagna units rotated after Late Miocene onward.

In the Gibraltar Arc the age distribution of palaeomagnetic data is very inhomogeneous, as data have been collected almost exclusively from Late Jurassic, Late Cretaceous and Neogene units

(Fig. 6c, d). However, two distinct trends in the rotation v. age diagram are recognized: an early segment, which spans Lower Jurassic to Oligocene–Lower Miocene time, where values of palaeomagnetic rotations remain almost constant during the entire time interval; and a second segment, which corresponds to the Neogene time interval, where a progressive decrease in palaeomagnetic rotations with time occurs. In the Western Betics, in particular Early Jurassic, Late Cretaceous and Oligocene–Lower Miocene data show the same amount of CW rotations (50–60° on average), whereas in the Late Miocene Granada and Alcalá-la-Real intramontane sedimentary basins the amount of CW rotation is about 30–35° (Fig. 6c).

## *The geodynamic setting for the formation of the Mediterranean Arcs*

The Mediterranean arcs revised in this study developed at the front of the Alboran, Liguro–Provençal and Tyrrhenian back-arc basins, and on top of narrow subducting slabs. These slabs are well defined using deep seismic data and seismic tomography (Wortel & Spakman 2000) (Fig. 7). Both the arcs and back-arc basins formed, contemporaneously, during the Cenozoic and Quaternary, causing the progressive dismembering of the Mediterranean Alpine chain that previously formed an almost rectilinear belt along the Europa–Africa plate boundary (see Faccenna *et al.* 2014 for a recent review). Palaeomagnetic data show that the Mediterranean arcs achieved their curvature mostly during the main phases of back-arc opening. In some cases, these phases were accompanied by drifting and rotation of almost rigid crustal blocks, as observed in the CPD and in the internal part of the Northern Apennines (Fig. 7). In particular, the internal units of the Northern Apennines arc record a large amount of CCW rotations. This can only be explained if these units first rotated CCW along with Corsica–Sardinia during its Lower Miocene rotational

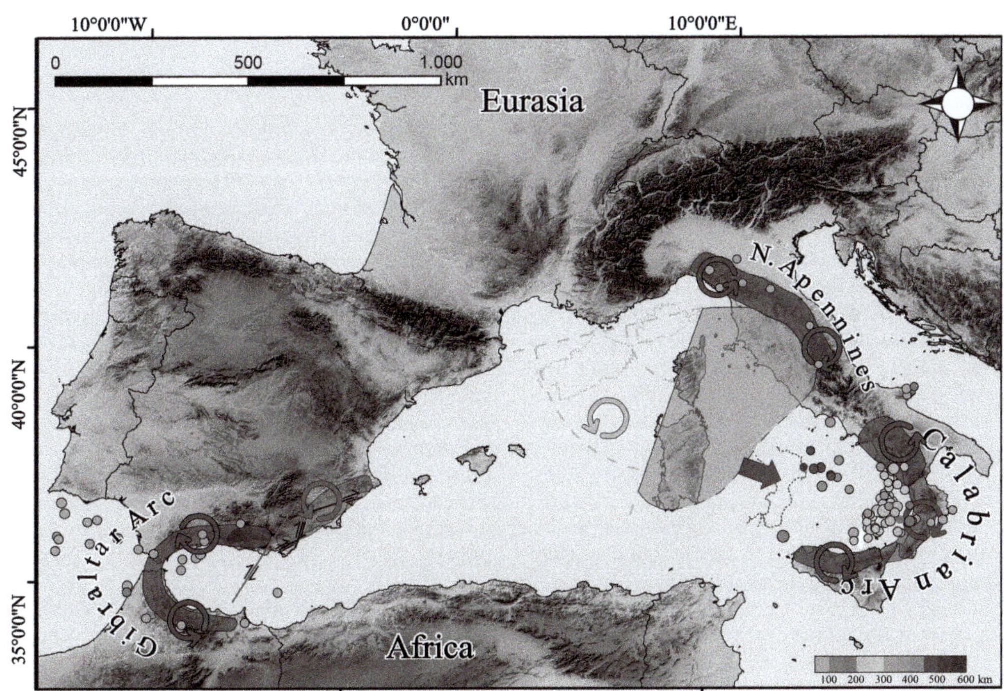


**Fig. 7.** Schematic sketch representing the main tectonic events that occurred in the Western and Central Mediterranean Basin. Brown coloured areas represent areas where palaeomagnetic rotations were strictly coeval to thrust emplacement and ceased together with the ending of compressional tectonics. Violet coloured areas represent the SE drift of the CPD and its impingement into the core of the Calabrian Arc. Yellow coloured areas represent the CCW rotation and drifting of the Corsica–Sardinia block and the contemporaneous CCW rotation of the internal units along with Corsica–Sardinia during its rotational drifting (i.e. the opening of the Liguro–Provencal Basin). Green coloured areas represent the Eastern Betics and the CCW rotations observed along a deformation belt dominated by Late Miocene–Quaternary left lateral strike-slip faulting. Coloured dots are hypocentral depth (International Seismological Centre catalogue, magnitude range >4).

drifting (i.e. the opening of the Liguro–Provencal Basin), implying that in this phase the Adriatic–Ionian subduction slab roll-back caused the CCW rotation of the whole upper plate of the Central Mediterranean subduction system (Caricchi *et al.* 2014). This scenario changed with the end of the Corsica–Sardinia drifting and CCW rotation, when back-arc extensional processes were transferred to the Northern Tyrrhenian Basin. Since that time the vertical axis rotations were confined to the external part of the Apennines orogenic wedge, which progressively acquired its present-day curvature as a consequence of slab roll-back of the subducting Adriatic plate, whereas the post-orogenic basins located along the Northern Tyrrhenian Basin did not rotate during the same time interval (Fig. 7).

In the Calabrian Arc the bending process has been mainly controlled by the presence of a small oceanic lithosphere plate (the Ionian Sea), intervening between the Apulia and Africa continental margins (Malinverno & Ryan 1986). This configuration of the subducting plate caused a differential southeastward roll-back of the trench during the Neogene and Quaternary, and the progressive drifting of a continental region (CPD), now impinged in the core of the Calabrian Arc. Palaeomagnetic data support this model, indicating a different behaviour between the CPD, located on the top of the Ionian slab, and the Southern Apennines and Sicilian Maghrebides units, which represent the part of the Calabrian Arc where continental lithosphere was partially subducted (Fig. 7).

In the Gibraltar Arc, the bending of the arc started during the Neogene and continued after the late Miocene with significant (about 20–30°) vertical axis rotations in the western Betics (CW) and in the Rif (CCW) (Fig. 6). Part of the curvature was acquired during the opening of the Alboran back-arc basin as a consequence of the westward retreat and fragmentation of the subducting slab, whose remnants are well imaged by seismic tomography below the Gibraltar arc. In this context it is worth noting that the Rif chain in northern Morocco shows a southwestern translation with respect to the Africa plate at rates exceeding $5 \text{ mm a}^{-1}$ (Fadil *et al.* 2006). This motion is compatible with southwestward roll-back of the subducted Africa plate beneath the Alboran Sea and Rif Mountains, suggesting the presence of a small remnant of the subducting slab (Fadil *et al.* 2006; Koulali *et al.* 2011).

## Conclusion

Data described in this paper indicate that vertical axis rotations with respect to bounding Africa, Europe and Iberia plates played a key role in the Neogene and Quaternary geodynamic evolution in the Western and Central Mediterranean Basin. In particular, the Mediterranean arcs share the following common features: (a) vertical axis rotations are confined to the orogenic wedge, and do not extend to the undeformed forelands, which are non rotated; (b) the magnitude of palaeomagnetic rotations decreases from the inner to the external sectors of the arcs; (c) within the same tectonic unit of the orogenic wedge, the amount of palaeomagnetic rotations remains almost constant in the pre-orogenic units, independently of their ages, whereas syn-orogenic units show a progressive decreasing of palaeomagnetic rotations with time; and (d) palaeomagnetic rotations are coeval with thrust emplacement and ceased together with the ending of compressional tectonics.

The curvature of the arcs was achieved during orogenic deformation through (a) opposite vertical axis rotations on the two limbs of the arcs, and (b) rotation of rigid crustal blocks (e.g. Corsica–Sardinia and CPD), which rotated during the opening of the back-arc basins (e.g. Liguro–Provençal and Southern Tyrrhenian basins). Rotation of crustal blocks and oroclines represents the upper plate response to the evolution of the Mediterranean subduction system. These systems are characterized by slab roll-back, segmentation and separation of an originally continuous subduction zone that led to the development of different subducting slabs, each characterized by a different tectonic evolution of the subduction process. The Mediterranean arcs acquired their curvature progressively, following the segmentation, narrowing and retreating of the Africa subducting slab. This mechanism is most evident in the Northern Apennines and in the Calabrian Arc, where the main phases of vertical axis rotation correspond with fast episodes of slab retreat. These episodes are characterized by the opening of back-arc basins and rapid outward migration of foredeep basins, whereas the end of vertical axis rotation corresponds with the cessation of slab retreat and outward propagation of the external thrust front (Mattei *et al.* 2007; Caricchi *et al.* 2014).

Described data clearly indicate that the curvature of the Mediterranean arcs was acquired progressively and that most of their curvature was attained during the same geodynamic process, which is the progressive retreat of narrow subducting slabs. On this basis, they can be considered progressive oroclines according to recent classification of oroclines (Weil & Sussman 2004; Johnston *et al.* 2013).

Finally, we would like to stress the historical (e.g. Carey 1955) and present day importance of the Mediterranean arcs for understanding the origin and evolution of oroclines. In fact, the large amount of palaeomagnetic data available in the

Mediterranean arcs represent an important dataset to successfully reconstruct the velocity, geographic distribution and timing of vertical axis rotations that occurred during the formation of the arcs. At the same time, the enormous amount of geological, stratigraphic and geophysical data that exist from the area allows for precise reconstruction of the tectonic processes that occurred along the European–African margin during the Mediterranean arcs evolution. We see it as necessary that these datasets be combined in order to fully develop general models about the formation of orogenic arcs during the evolution of the Mediterranean region.

# References


AIELLO, I.W. & HAGSTRUM, J.T. 2001. Paleomagnetism and paleogeography of Jurassic radiolarian cherts from the northern Apennines of Italy Paleomagnetism and paleogeography of Jurassic radiolarian cherts from the northern Apennines of Italy. *Geological Society of America Bulletin*, **113**, 469–481.

AIFA, T., FEINBERG, H. & POZZI, J.P. 1988. Plio-Pleistocene evolution of the Tyrrhenian arc: paleomagnetic determination of uplift and rotational deformation. *Earth and Planetary Science Letters*, **87**, 438–452.

ALLERTON, S. 1994. Vertical axis rotation associated with folding and thrusting: an example from the Eastern Subbetics of Southern Spain. *Geology*, **22**, 1039–1042.

ALLERTON, S., LONERGAN, L., PLATT, J., PLATT, J., PLATZMAN, E.S. & MCCLELLAND, E. 1993. Paleomagnetic Rotations in the Eastern Betic Cordillera, Southern Spain. *Earth and Planetary Science Letters*, **119**, 225–241

ALLERTON, S., REICHERTER, K. & PLATT, J.P. 1994. A structural and paleomagnetic study of a section through the eastern Subbetic, southern Spain. *Journal of the Geological Society, London*, **151**, 659–668.

ALVAREZ, W. & LOWRIE, W. 1978. Upper Cretaceous paleomagnetic stratigraphy at Moria (Umbrian Apennines, Italy); verification of the Gubbio section? Geophys. *Journal of the Royal Astronomy Society*, **55**, 1–17.

ALVAREZ, W. & LOWRIE, W. 1984. Magnetic stratigraphy apllied to synsedimentary slumps, turbidites and basin analysis: the Scalgia limestone at Furlo (Italy). *Geological Society American Bulletin*, **95**, 324–336.

ALVAREZ, W., COCOZZA, T. & WEZEL, F.C. 1974. Fragmentation of the Alpine orogenic belt by microplate dispersal. *Nature*, **248**, 309–314.

ARGNANI, A., CORNINI, S., TORELLI, L. & ZITELLINI, N. 1987. Diachronous foredeep-system in the Neogene-Quaternary of the Strait of SIcily. *Memorie Società Geologica Italiana*, **38**, 407–417.

ASCIONE, A. & ROMANO, P. 1999. Vertical movements on the eastern margin of the Tyrrhenian extensional basin. New data from Mt. Bulgheria (Southern Apennines, Italy). *Tectonophysics*, **315**, 337–356.

AZAÑÓN, J.M. & CRESPO-BLANC, A. 2000. Exhumation during a continental collision inferred from the tectonometamorphic evolution of the Alpujarride Complex in the central Betics (Alboran Domain, SE Spain). *Tectonics*, **19**, 549–565.

BARBERI, F., INNOCENTI, F., FERRARA, G., KELLER, J. & VILLARI, L. 1974. Evolution of Eolian arc volcanism (Southern Tyrrhenian Sea). *Earth and Planetary Science Letters*, **21**, 269–276.

BARCHI, M.R., DE FEYTER, A., MAGNANI, B., MINELLI, G., PIALLI, G. & SOTERA, B. 1998. The structural style of the Umbria-Marche fold and thrust belt. *Memorie della Società Geologica Italiana*, **52**, 557–578.

BERNDT, T., RUIZ-MARTINEZ, V.C. & CHALOUAN, A. 2015. New constraints on the evolution of the Gibraltar Arc from paleomagnetic data of the Ceuta and Beni Bousera peridotites (Rif, northern Africa). *Journal of Geodynamics*, **84**, 19–39.

BESSE, J. & COURTILLOT, V. 2002. Apparent and true polar wander and the geometry of the geomagnetic field over the last 200 Myr. *Journal of Geophysical Research*, **107**, http://doi.org/10.1029/2000JB000050

BESSE, J., POZZI, J.P., MASCLE, G. & FEINBERG, H. 1984. Paleomagnetic study of Sicily: consequences for the deformation of Italian and African margins over the last 100 million years. *Earth and Planetary Science Letters*, **67**, 377–390.

BIGGS, J., BERGMAN, E., EMMERSON, B., FUNNING, G.J., JACKSON, J., PARSONS, B. & WRIGHT, T.J. 2006. Fault identification for buried strike-slip earthquakes using InSAR: the 1994 and 2004 Al Hoceima, Morocco earthquakes. *Geophysical Journal International*, **166**, 1347–1362.

BLANCO, M.J. & SPAKMAN, W. 1993. The P-wave velocity structure of the mantle below the Iberian Peninsula: evidence for subducted lithosphere below southern Spain. *Tectonophysics*, **221**, 13–34.

BONARDI, G., CAVAZZA, W., PERRONE, V. & ROSSI, S. 2001. Calabria Peloritani Terrane and Northern Ionian Sea. *In*: VAI, G.B. & MARTINI, I.P. (eds) *Anatomy of an Orogen: The Apennines and Adjacent Mediterranean Basins*. Kluwer Academic, Dordrecht, 287–306.

BOOTH-REA, G., AZAÑON, J.M. & GARCÍA-DUEÑAS, V. 2004. Extensional tectonics in the northeastern Betics (SE Spain): case study of extension in a multilayered upper crust with contrasting rheologies. *Journal of Structural Geology*, **26**, 2009–2058.

BUTLER, R.W.H. & GRASSO, M. 1993. Tectonic controls on base level variations and depositional sequences within thrust-top and foredeep basins: examples from the Neogene thrust belt of central Sicily. *Basin Research*, **5**, 137–151.

BUTLER, R.W.H., GRASSO, M. & LA MANNA, F. 1992. Origin and deformation of the Neogene-Recent Maghrebian foredeep at the Gela Nappe, SE Sicily. *Journal of Geological Society, London*, **149**, 547–556.

BUTLER, R.W.H., MCCLELLAND, E. & JONES, R.E. 1999. Calibrating the duration and timing of the Messinian salinity crisis in the Mediterranean: linked tectonoclimatic signals in thrust-top basins of Sicily. *Journal of the Geological Society, London*, **156**, 827–835.

CALVO, M., OSETE, M.L. & VEGAS, R. 1994. Palaeomagnetic Rotations in Opposite Senses in Southeastern Spain. *Geophysical Research Letters*, **21**, 761–764.

CALVO, M., VEGAS, R. & OSETE, M.L. 1997. Paleomagnetic results from Upper Miocene and Pliocene rocks from the Internal Zone of the eastern Betic Cordilleras (southern Spain). *Tectonophysics*, **277**, 271–283.

CAREY, S.W. 1955. The orocline concept in geotectonics. *Proceedings of the Royal Society of Tasmania*, **89**, 255–288.

CARICCHI, C., CIFELLI, F., SAGNOTTI, L., SANI, F., SPERANZA, F. & MATTEI, M. 2014. Paleomagnetic evidence for a post-Eocene 90° CCW rotation of internal Apennine units: a linkage with Corsica–Sardinia rotation? *Tectonics*, **33**, 374–392.

CARMIGNANI, L., DECANDIA, A., FANTOZZI, L., LAZZAROTTO, A., LIOTTA, D. & MECCHERI, M. 1994. Tertiary extensional tectonics in Tuscany (Northern Apennines, Italy). *Tectonophysics*, **238**, 295–315.

CARRAPA, B., BERTOTTI, G. & KRIJGSMAN, W. 2003. Subsidence, stress regime and rotation(s) of a tectonically active sedimentary basin within the western Alpine Orogen: the Tertiary Piedmont Basin (Alpine domain, NW Italy). *In*: MCCANN, T. & SAINTOT, A. (eds) *Tracing Tectonic Deformation Using the Sedimentary Record*. Geological Society, London, Special Publications, **208**, 205–227.

CATALANO, R., CHANNELL, J.E.T., D'ARGENIO, B. & NAPOLEONE, G. 1976. Mesozoic paleo- geography of Southern Apennines and Sicily – Problems of paleotectonics and paleo- magnetism. *Memorie Società Geologica Italiana*, **15**, 95–118.

CAVAZZA, W., BLENKINSOP, J., DECELLES, P., PATTERSON, R.T. & REINHARDT, E. 1997. Stratigrafia e sedimentologia della sequenza sedimentaria oligocenico-quaternaria del bacino calabro-ionico. *Bollettino della Società Geologica Italiana*, **116**, 51–77.

CERNOBORI, L., HIRN, A., MCBRIDE, J.H., NICOLICH, R., PETRONIO, L., ROMANELLI, M. & STREAMERS/PROFILES WORKING GROUPS 1996. Crustal image of the Ionian and its Calabrian margins. *Tectonophysics*, **264**, 175–189.

CHANNELL, J.E.T. 1992. Paleomagnetic data from Umbria (Italy): implications for the rotation of Adria and Mesozoic apparent polar wander paths. *Tectonophysics*, **216**, 365–378.

CHANNELL, J.E.T. & TARLING, D.H. 1975. Paleomagnetism and the rotation of Italy. *Earth and Planetary Science Letters*, **25**, 177–188.

CHANNELL, J.E.T., LOWRIE, W., MEDIZZA, F. & ALVAREZ, W. 1978. Paleomagnetism and tectonics in Umbria, Italy. *Earth and Planetary Science Letters*, **39**, 199–210.

CHANNELL, J.E.T., CATALANO, R. & D'ARGENIO, B. 1980. Paleomagnetism and deformation of the Mesozoic continental margin of Sicily. *Tectonophysics*, **61**, 391–407.

CHANNELL, J.E.T., OLDOW, J.S., CATALANO, R. & D'ARGENIO, B. 1990. Paleomagnetically determined rotations in the western Sicilian fold and thrust belt. *Tectonics*, **9**, 641–660.

CHANNELL, J., DI STEFANO, E. & SPROVIERI, R. 1992. Calcareous Plankton Biostratigraphy, magnetostratigraphy and paleoclimatic history of the Plio-Pleistocene Monte San Nicola Section (Southern Sicily). *Bollettino della Società Geologica Italiana*, **31**, 351–382.

CHIARALUCE, L., AMATO, A. ET AL. 2004. Complex normal faulting in the Apennines thrust-and-fold. *Bulletin of the Seismology Society of America*, **94**, 99–116.

CIFELLI, F. & MATTEI, M. 2010. Curved orogenic systems in the Italian Peninsula: a paleomagnetic review. *In*: BELTRANDO, M., PECCERILLO, A., MATTEI, M., CONTICELLI, S. & DOGLIONI, C. (eds) *The Geology of Italy: Tectonics and Life Along Plate Margins. Journal of the Virtual Explorer*. Electronic Edition 36, paper 17, http://doi.org/10.3809/jvirtex.2010.00239

CIFELLI, F., ROSSETTI, F., MATTEI, M., HIRT, A.M., FUNICIELLO, R. & TORTORICI, L. 2004. An AMS, structural and paleomagnetic study of quaternary deformation in eastern Sicily. *Journal of Structural Geology*, **26**, 29–46.

CIFELLI, F., ROSSETTI, F. & MATTEI, M. 2007a. The architecture of brittle postorogenic extension: results from an integrated structural and paleomagnetic study in north Calabria (southern Italy). *Geological Society of America Bulletin*, **119**, 221–239.

CIFELLI, F., MATTEI, M. & ROSSETTI, F. 2007b. The tectonic evolution of arcuate mountain belts on top of a retreating subduction slab: the example of the Calabrian Arc. *Journal of Geophysical Research*, **112**, B09101, http://doi.org/10.1029/2006JB004848

CIFELLI, F., MATTEI, M. & DELLA SETA, M. 2008. Calabrian Arc oroclinal bending: the role of subduction. *Tectonics*, **27**, TC5001.

CIRILLI, S., MÀRTON, P. & VIGLI, L. 1984. Implications of a combined biostratigraphic and palaeomagnetic study of the Umbrian Maiolica Formation. *Earth and Planetary Science Letters*, **71**, 203–214.

COMAS, M.C., GARCÍA-DUEÑAS, V. & JURADO, M.J. 1992. Neogene tectonic evolution of the Alboran Sea from Mcs data. *Geo-Marine Letters*, **12**, 157–164.

CRESPO-BLANC, A., OROZCO, M. & GARCÍA-DUEÑAS, V. 1994. Extension v. compression during the Miocene tectonic evolution of the Betic chain. Late folding of normal fault systems. *Tectonics*, **13**, 78–88.

D'AGOSTINO, N. & SELVAGGI, G. 2004. Crustal motion along the Eurasia-Nubia plate boundary in the Calabrian Arc and Sicily and active extension in the Messina Straits from GPS measurements. *Journal of Geophysical Research*, **109**, B11402, http://doi.org/11410.11029/12004JB002998

DELA PIERRE, F., GHISETTI, F., LANZA, R. & VEZZANI, L. 1992. Paleomagnetic and structural evidence of Neogene tectonic rotation of the Gran Sasso racne (central Apennines, Italy). *Tectonophysics*, **215**, 335–348.

DEMAREST, H.H. 1983. Error analysis for the determination of tectonic rotation from paleomagnetic data. *Journal of Geophysical Research*, **88**, 4321–4328.

DEWEY, J.F., HELMAN, M.L., KNOTT, S.D., TURCO, E. & HUTTON, D.H.W. 1989. Kinematics of the western Mediterranean. *In*: COWARD, M.P., DIETRICH, D. & PARK, R.G. (eds) *Alpine Tectonics*. Geological Society, London, Special Publications, **45**, 265–283.

DIDON, J., DURAND-DELGA, M. & KORNPROBST, J. 1973. Homologies géologiques entre les deux rives du détroit de Gibraltar. *Bulletin de la Société Géologique de France*, **7**, 77–105.

DINARÈS-TURELL, J., ORTI, F., PLAYA, E. & ROSELL, L. 1999. Palaeomagnetic chronology of the evaporitic sedimentation in the Neogene Fortuna Basin (SE Spain): early restriction preceding the 'Messinian Salinity Crisis'. *Palaeogeography Palaeoclimatology Palaeoecology*, **154**, 161–178.

DUERMEJIER, C.E., VAN VUGT, N., LANGEREIS, C.G., MEU-LENKAMP, J.E. & ZACHARIASSE, W.J. 1998. A major late Tortonian rotation phase in the Crotone basin using AMS as tectonic tilt correction and timing of the opening of the Tyrrhenian basin. *Tectonophysics*, **287**, 233–249.

FACCENNA, C., MATTEI, M., FUNICIELLO, R. & JOLIVET, L. 1997. Styles of back-arc extension in the Central Mediterranean. *Terra Nova*, **9**, 126–130.

FACCENNA, C., PIROMALLO, C., CRESPO-BLANC, A., JOLIVET, L. & ROSSETTI, F. 2004. Lateral slab deformation and the origin of the western Mediterranean arcs. *Tectonics*, **23**, TC1012, http://doi.org/1010.1029/2002TC001488

FACCENNA, C., BECKER, T.W. *ET AL.* 2014. Mantle dynamics in the Mediterranean. *Review of Geophysics*, **52**, 283–332.

FADIL, A., VERNANT, P. *ET AL.* 2006. Active tectonics of the western Mediterranean: geodetic evidence for rollback of a delaminated subcontinental lithospheric slab beneath the Rif Mountains, Morocco. *Geology*, **34**, 529–532.

FEINBERG, H., SADDIQI, O. & MICHARD, A. 1996. New constraints on the bending of the Gibraltar arc from palaeomagnetism of the Ronda peridotites (Betic Cordilleras, Spain). *In*: MORRIS, A. & TARLING, D.H. (eds) *Palaeomagnetism and Tectonics of the Mediterranean Region*. Geological Society, London, Special Publications, **105**, 43–52.

FRIZON DE LAMOTTE, D., ANDRIEUX, J. & GUÉZOU, J.C. 1991. Cinématique des chevauchements néogènes dans l'Arc bético-rifain: discussion sur les modèles géodynamiques. *Bulletin de la Société Géologique de France*, **162**, 611–626.

FRIZON DE LAMOTTE, D., CRESPO-BLANC, A. *ET AL.* 2004. TRANSMED Transect I: Iberian Meseta–Guadalquivir Basin–Betic Cordillera–Alboran Sea–Rif–Moroccan Meseta–High Atlas–Sahara Platform. *In*: *Proceedings 32nd International Geological Congress*, Florence, **SO6-01**.

GARCÉS, M., AGUSTI, J. & PARÉS, J.M. 1997. Late Pliocene continental magnetochronology in the Guadix-Baza Basin (Betic Ranges, Spain). *Earth and Planetary Science Letters*, **146**, 677–687.

GARCÉS, M., KRIJGSMAN, W. & AGUSTI, J. 2001. Chronostratigraphic framework and evolution of the Fortuna basin (Eastern Betics) since the late Miocene. *Basin Research*, **13**, 199–216.

GARCÍA-DUEÑAS, V., BALANYÁ, J.C. & MARTÍNEZ-MARTÍNEZ, J.M. 1992. Miocene extensional detachments in the outcropping basement of the Northern Alboran basin (Betics) and their tectonic implications. *Geo-Marine Letters*, **12**, 88–95.

GATTACCECA, J. & SPERANZA, F. 2002. Paleomagnetism of Jurassic to Miocene sediments from the Apenninic carbonate platform (southern Apennines, Italy): evidence for a 60° counterclockwise Miocene rotation. *Earth and Planetary Science Letters*, **201**, 19–34.

GATTACCECA, J., DEINO, A., RIZZO, R., JONES, D.S., HENRY, B., BEAUDOIN, B. & VADEBOIN, F. 2007. Miocene rotation of Sardinia: new paleomagnetic and geochronologic constraints and geodynamic implications. *Earth and Planetary Science Letters*, **258**, 359–377.

GREGOR, A.G.B., NAIRN, E.M. & NEGENDANK, J.F.W. 1975. Paleomagnetic investigations of the Tertiary and Quaternary rocks: IX. The Pliocene of southeast Sicily and some Cretaceous rocks from Capo Passero. *Geologische Rundschau*, **64**, 948–957.

GUEGEN, E., DOGLIONI, C. & FERNANDEZ, M. 1998. On the post 25 Ma geodynamic evolution of the western Mediterranean. *Tectonophysics*, **298**, 259–269.

HILGEN, F.J., BISSOLI, L., IACCARINO, S., KRIJGSMAN, W., MEIJER, R., NEGRI, A. & VILLA, G. 2000. Integrated stratigraphy and astrochronology of the Messinian GSSP at Oued Akrech (Atlantic Morocco). *Earth and Planetary Science Letters*, **182**, 237–251.

HIRT, A.M. & LOWRIE, W. 1988. Paleomagnetism of the Umbria–Marche orogenic belt. *Tectonophysics*, **146**, 91–103.

JACKSON, K.C. 1990. A palaeomagnetic study of Apennine thrusts, Italy: Monte Maiella and Monte Raparo. *Tectonophysics*, **178**, 231–240.

JOHNSTON, S.T., WEIL, A.B. & GUTIÉRREZ-ALONSO, G. 2013. Oroclines: thick and thin. *Geological Society of America Bulletin*, **125**, 643–663.

KIRKER, A. & MCCLELLAND, E. 1996. Application of net tectonic rotations and inclination analysis to a high-resolution paleomagnetic study in the Betic Cordillera. *In*: MORRIS, A. & TARLING, D.H. (eds) *Paleomagnetism and Tectonics of the Mediterranean Region*. Geological Society, London, Special Publications, **105**, 19–32.

KLOOTWIJK, C.T. & VANDENBERG, J. 1975. The rotation of Italy: preliminary paleomagnetic data from the Umbrian sequence: Northern Apennines. *Earth and Planetary Science Letters*, **25**, 263–273.

KOULALI, A., OUAZAR, D. *ET AL.* 2011. New GPS constraints on active deformation along the Africa-Iberia plate boundary. *Earth and Planetary Science Letters*, **308**, 211–217.

KRIJGSMAN, W. & GARCÉS, M. 2004. Pleomagnetic constrains on the geodynamic evolution of the Gibraltar Arc. *Terra Nova*, **16**, 281–287.

KRIJGSMAN, W., LANGEREIS, C.G. *ET AL.* 1999. Late Neogene evolution of the Taza-Guercif Basin (Rifian Corridor, Morocco) and implications for the Messinian salinity crisis. *Marine Geology*, **153**, 147–160.

KRIJGSMAN, W., GARCÉS, M., AGUSTI, J., RAFFI, I., TABERNER, C. & ZACHARIASSE, W.J. 2000. The 'Tortonian salinity crisis' of the eastern Betics (Spain). *Earth and Planetary Science Letters*, **181**, 497–511.

KRIJGSMAN, W., FORTUIN, A.R., HILGEN, F.J. & SIERRO, F.J. 2001. Astrochronology for the Messinian Sorbas basin (SE Spain) and orbital (precessional) forcing for evaporite cyclicity. *Sedimentary Geology*, **140**, 43–60.

KRIJGSMAN, W., GABOARDI, S., HILGEN, F.., IACCARINO, S., DE KAENEL, E. & VAN DER LAAN, E. 2004. Revised astrochronology for the Ain el Beida section (Atlantic Morocco): no glacio-eustatic control for the onset of the Messinian Salinity Crisis. *Stratigraphy*, **1**, 87–101.

KRIJGSMAN, W., LEEWIS, M.E., GARCÉS, M., KOUWENHOVEN, T.J., KUIPER, K.F. & SIERRO, F.J. 2006. Tectonic control for evaporate formation in the Eastern Betics. *Sedimentary Geology*, **188–189**, 155–170.

LANCI, L. & WEZEL, F.C. 1995. Rotazioni tettoniche di età Messiniana nell'Appennino Marchigiano. *In*:

*Proceedings Geodinamica e Tettonica Attiva del Sistema Tirreno-Appennino*. Studi Geologici Camerti, Università di Camerino, Camerino, 9–10 February 1995, 330–332.

LATAL, C., SCHOLGERAAND, R. & PREISINGERB, A. 2000. Paleomagnetic investigations imply rotations within the Cretaceous–Tertiary transition section at Cerbara (Italy). *Physics and Chemistry of the Earth*, **25**, 499–503.

LONERGAN, L. & WHITE, N. 1997. Origin of the Betic–Rif mountain belt. *Tectonics*, **16**, 504–522.

LOWRIE, W. & ALVAREZ, W. 1975. Paleomagnetic evidence for rotation of the Italian Peninsula. *Journal of Geophysical Research*, **80**, 1579–1592.

LOWRIE, W. & ALVAREZ, W. 1977. Late Cretaceous geomagnetic polarity sequence: detailed rock and paleomagnetism studies of Scaglia Rossa limentone at Gubbio, Italy. *Geophysical Journal of the Royal Astronomy Society*, **51**, 2561–2581.

LOWRIE, W. & ALVAREZ, W. 1979. Paleomagnetism and Rock magnetism of the Pliocene Rhyolite at San Vincenzo, Tuscany, Italy. *Journal of Geophysics*, **45**, 417–432.

LOWRIE, W. & HIRT, A.M. 1986. Paleomagnetism in arcuate mountain belts. *In*: WEZEL, F.C. (ed.) *The Origin of the Arcs*. Elsevier, New York, 141–158.

LOWRIE, W., CHANNELL, J.E.T. & ALVAREZ, W. 1980. A review of magnetic stratigraphy investigations in Cretaceous pelagic carbonate rocks. *Journal of Geophysical Research*, **85**, 3597–3605.

LOWRIE, W., ALVAREZ, W., NAPOLEONE, G., PERCH-NIELSEN, K., SILVA, I.P. & TOUMARKINE, M. 1982. Paleogene magnetic stratigraphy in Umbrian pelagic carbonate rocks: the Contessa sections, Gubbio. *Geological Society of America Bulletin*, **93**, 414–432.

MAFFIONE, M., SPERANZA, F., FACCENNA, C., CASCELLA, A., VIGNAROLI, G. & SAGNOTTI, L. 2008. A synchronous Alpine and Corsica-Sardinia rotation. *Journal of Geophysical Research*, **113**, 1–2.

MAFFIONE, M., SPERANZA, F., CASCELLA, A., LONGHITANO, S.G. & CHIARELLA, D. 2013. A ~125° post-early Serravallian counterclockwise rotation of the Gorgoglione Formation (Southern Apennines, Italy): new constraints for the formation of the Calabrian Arc. *Tectonophysics*, **590**, 24–37.

MALINVERNO, A. & RYAN, W.B.F. 1986. Extension in Tyrrhenian sea and shortening in the Apennines as result of arc migration driven by sinking of the lithosphere. *Tectonics*, **5**, 227–254.

MANZONI, M. 1975. Paleomagnetic evidene for rotation of northern Calabria. *Geophysical Reserach Letters*, **2**, 427–429.

MARTÍN-SUAREZ, E., OMS, O., FREUDENTHAL, M., AUGUSTI, J. & PARÉS, J.M. 1998. Continental Mio-Pliocene transition in the Granada Basin. *Lethaia*, **31**, 161–166.

MARTON, E. & NARDI, G. 1994. Cretaceous Paleomagnetic results from Murge (Apulia, southern, Italy): tectonic implications. *Geophysical Journal International*, **19**, 842–856.

MARTON, P. & D'ANDREA, M. 1992. Paleomagnetically inferred tectonic rotation of the Abruzzi and northwestern Umbria. *Tectonophysics*, **202**, 43–53.

MATTEI, M., KISSEL, C. & FUNICIELLO, R. 1996. No tectonic rotation of the Tuscan Tyrrhenian margin (Italy) since late Miocene. *Journal of Geophysical Research*, **101**, 2835–2845.

MATTEI, M., CIPOLLARI, P., COSENTINO, D., ARGENTIERI, A., ROSSETTI, F. & SPERANZA, F. 2002. The Miocene tectonic evolution of the Southern Tyrrhenian Sea: stratigraphy, structural and paleomagnetic data from the on-shore Amantea basin (Calabrian Arc, Italy). *Basin Research*, **14**, 147–168.

MATTEI, M., PETROCELLI, V., LACAVA, D. & SCHIATTARELLA, M. 2004. Geodynamic implications of Pleistocene ultrarapid vertical-axis rotations in the Southern Apennines, Italy. *Geology*, **32**, 789–792.

MATTEI, M., CIFELLI, F., ROJAS, I.M., CRESPO-BLANC, A., COMAS, M.C., FACCENNA, C. & PORRECA, M. 2006. Neogene tectonic evolution of the Gibraltar Arc: new paleomagnetic constrains from the Betic chain. *Earth and Planetary Science Letters*, **250**, 522–540.

MATTEI, M., CIFELLI, F. & D'AGOSTINO, N. 2007. The evolution of the Calabrian Arc: evidence from paleomagnetic and GPS observations. *Earth and Planetary Science Letters*, **263**, 259–274.

MAYFIELD, A. 1999. *Paleomagnetic and kinematic constraint on deformation during oblique convergence, Betic Cordillera, southern Spain*. PhD thesis, University of London.

MAZAUD, A., GALBRUN, B., AZÉMA, J., ENAY, R., E., F. & RASPLUS, L. 1986. Données magnétostratigraphiques sur le Jurassique supérieur et le Berrisien du NE des Cordillères bètiques. *Comptes Rendus de l'Académie des Sciences*, **302**, Série II, 1165–1170.

MCCLUSKY, S., REILINGER, R., MAHMOUD, S., BEN SARI, D. & TEALEB, A. 2003. GPS contraints on Africa (Nubia) and Arabia plate motions. *Geophysical Journal International*, **155**, 126–138.

MEGHRAOUI, M., MOREL, J.L., ANDRIEUX, J. & DAHMANI, M. 1996. Tectonique plioquaternaire de la chaine tello-rifaine et de la mer d'Alboran: Une zone complexe de convergence continent-continent. *Bulletin de la Societe Geologique de France*, **167**, 141–157.

MINELLI, L. & FACCENNA, C. 2010. Evolution of the Calabrian accretionary wedge (central Mediterranean). *Tectonics*, **29**, 1–21.

MONTIGNY, R., EDEL, J.B. & THUIZAT, R. 1981. Oligo-Miocene rotation of Sardinia: K–Ar ages and paleomagnetic data of Tertiary volcanic. *Earth and Planetary Science Letters*, **54**, 261–271.

MORALES, J., SERRANO, I. *ET AL.* 1999. Active continental subduction beneath the Betic Cordillera and the Alboran Sea. *Geology*, **27**, 753–758.

MORLEY, C.K. 1992. Tectonic and sedimentary evidence for synchronous and out-of-sequence thrusting Larache–Acilah area, Western Morocco, Rif. *Journal of the Geological Society, London*, **149**, 39–49.

MUTTONI, G., ARGNANI, A., KENT, D.V., ABRAHAMSEN, N. & CIBIN, U. 1998. Paleomagnetic evidence for Neogene tectonic rotations in the northern Apennines, Italy. *Earth and Planetary Science Letters*, **154**, 25–40.

MUTTONI, G., LANCI, L., ARGNANI, A., HIRT, A.M., CIBIN, U., ABRAHAMSEN, N. & LOWRIE, W. 2000. Paleomagnetic evidence for a Neogene two-fase counterclockwise rotation in the Northern Apennines (Italy). *Tectonophysics*, **326**, 241–253.

MUTTONI, G., GARZANTI, E., ALFONSI, L., CIRILLI, S., GERMANI, D. & LOWRIE, W. 2001. Motion of Africa and Adria since Permian: paleomegnetic and paleoclimatic constrains from Northern Libya. *Earth and Planetary Science Letters*, **192**, 159–174.

NAIRN, A.E.M., NARDI, G., GREGOR, C.B. & INCORONATO, A. 1985. Coherence of the Trapanese units during tectonic emplacement in western Sicily. *Bollettino della Società Geologica Italiana*, **104**, 267–272.

NAPOLEONE, G., SILVA, I.P., HELLER, F., CHELI, P., COREZZI, S. & FISCHER, A.G. 1983. Eocene magnetic stratigraphy at Gubbio, Italy, and its implications for Paleogene. *Geological Society of America Bulletin*, **94**, 181–191.

NERES, M., FONT, E., MIRANDA, J.M., CAMPS, P., TERRINHA, P. & MIRÃO, J. 2012. Reconciling Cretaceous paleomagnetic and marine magnetic data for Iberia: New Iberian paleomagnetic poles. *Journal of Geophysical Research*, **117**, http://doi.org/10.1029/2011JB009067

OGG, J.G., STEINER, M.B., OLORIZ, F. & TAVERA, J.M. 1984. Jurassic magnetostratigraphy, 1. Kimmeridgian–Tithonian of Sierra Gorda and Carcabuey, southern Spain. *Earth and Planetary Science Letters*, **71**, 147–162.

OGG, J.G., STEINER, M.B., COMPANY, M. & TAVERA, J.M. 1988. Magnetostratigraphy across the Berriasian–Valanginian stage boundary (Early Cretaceous), at Cehegin (Murcia Province, southenr Spain). *Earth and Planetary Science Letters*, **87**, 205–215.

OLDOW, J.S., CHANNELL, J., CATALANO, R. & D'ARGENIO, B. 1990. Contemporaneous thrusting and large-scale rotations in the western Sicilian fold and thrust belt. *Tectonics*, **9**, 661–681.

OSETE, M.L., FREEMAN, R. & VEGAS, R. 1988. Preliminary paleomagnetic results form the Subbteic Zone (Betic Cordillera, southern Spain): kinematic and structural implications. *Physics of the Earth and Planetary Interiors*, **52**, 283–300.

OSETE, M.L., VILLALAIN, J.J., OSETE, C., SANDOVAL, J. & GARCÍA-DUEÑAS, V. 2004. New palaeomagnetic data from the Betic Cordillera: constraints on the timing and the geographical distribution of tectonic rotations in Southern Spain. *Pure and Applied Geophysics*, **161**, 701–722.

OSETE, M.L., GÓMEZ, J.J., PAVÓN-CARRASCO, F.J., VILLALAÍN, J.J., PALENCIA-ORTAS, A., RUIZ-MARTÍNEZ, V.C. & HELLER, F. 2007. The evolution of Iberia during the Jurassic from palaeomagnetic data. *Tectonophysics*, **502**, 105–120.

PATACCA, E. & SCANDONE, P. 2001. Late thrust propagation and sedimentary response in the thrust-belt-foredeep system of the Southern Apennines (Pliocene-Pleistocene). *In*: VAI, G.B. & MARTINI, L.P. (eds) *Anatomy of an Orogen: The Apennines and Adjacent Mediterranean Basins*. Springer, Netherlands, 401–440.

PATACCA, E. & SCANDONE, P. 2004. The Plio-Pleistocene thrust belt-foredeep system in the Southern Apennines and Sicily (Italy). *Bollettino della Società Geologica Italiana*, **Vol. Spec.**, 93–129.

PATACCA, E., SARTORI, R. & SCANDONE, P. 1992. Tyrrhenyan Basin and Apenninic arcs: kinematics relations since Late Tortonian times. *Memorie della Società Geologica Italiana*, **45**, 425–451.

PLATT, J. & VISSERS, R.L.M. 1989. Extensional collapse of thickened continental lithosphere: a working hypothesis for the Alboran Sea and Gibraltar Arc. *Geology*, **17**, 540–543.

PLATT, J.P., ALLERTON, S., KIRKER, A. & PLATZMAN, E.S. 1995. Origin of the western Subbetic arc, southern Spain: palaeomagnetic and structural evidence. *Journal of Structural Geology*, **17**, 765–775.

PLATT, J.P., ALLERTON, S., KIRKER, A., MANDEVILLE, C., MAYFIELD, A., PLATZMAN, E.S. & RIMI, A. 2003. The ultimate arc: differential displacement, oroclinal bending, and vertical axis rotation in the External Betic-Rif arc. *Tectonics*, **22**, 1017, http://doi.org/1010.1029/2001TC001321

PLATZMAN, E.S. 1992. Paleomagnetic rotations and the kinematics of the Gibraltar arc. *Geology*, **20**, 311–314.

PLATZMAN, E.S. 1994. East-West thrusting and anomalous magentic declinations in the Sierra Corda, Betic Cordillera, southern Spain. *Journal of Structural Geology*, **16**, 11–20.

PLATZMAN, E. & LOWRIE, W. 1992. Paleomagnetic evidence for rotation of the Iberian peninsula and the external Betic Cordillera, southern Spain. *Earth and Planetary Science Letters*, **108**, 45–60.

PLATZMAN, E.S. & PLATT, J.P. 2004. Kinematics of a twisted core complex: oblique axis rotation in an extended terrane (Betic Cordillera, southern Spain). *Tectonics*, **23**, TC6010, http://doi.org/10.1029/2003TC001549

PLATZMAN, E.S., PLATT, J.P. & OLIVIER, P. 1993. Palaeomagnetic rotations and fault kinematics in the Rif arc of Morocco. *Journal of Geological Society, London*, **150**, 707–718.

ROGGENTHEN, W.M. & NAPOLEONE, G. 1977. Upper Cretaceous–Paleocene magnetic stratigraphy at Gubbio, Italy IV. Upper Maastrichtian–Paleocene magnetic stratigraphy. *Geological Society of America Bulletin*, **88**, 378–382.

ROSENBAUM, G. & LISTER, G.S. 2004. Formation of arcuate orogenic belts in the western Mediterranean region. *In*: WEIL, A. & SUSSMAN, A.J. (eds) *Orogenic Curvature*. The Geological Society of America, Boulder, CO, **383**, 41–56.

ROSSETTI, F., GOFFÉ, B., MONIÉ, P., FACCENNA, C. & VIGNAROLI, G. 2004. Alpine orogenic *P–T–t*-deformation history of the Catena Costiera area and surrounding regions (Calabrian Arc, southern Italy): the nappe edifice of north Calabria revised with insights on the Tyrrhenian-Apennine system formation. *Tectonics*, **23**, TC6011, http://doi.org/6010.1029/2003TC001560

ROYDEN, L.H. 1993. Evolution of retreating subduction boundaries formed during continental collision. *Tectonics*, **12**, 629–638.

ROYDEN, L.E., PATACCA, E. & SCANDONE, P. 1987. Segmentation and configuration of subducted lithosphere in Italy: an important control on thrust-belt and foredeep-basin evolution. *Geology*, **15**, 714–717.

SADDIQI, O., FEINBERG, H., ELAZZAB, D. & MICHARD, A. 1995. Paléomagnétisme des péridotites des Beni Bousera, Rif interne, Maroc: conséquences pour l'évolution miocène de l'arc de Gibraltar. *Comptes Rendus Académie des Sciences Paris*, **321**, 361–368.

SAGNOTTI, L. 1992. Paleomagnetic evidence for a Pleistocene counterclockwise rotation of the Sant'Arcangelo basin, Southern Italy. *Geophysical Research Letters*, **19**, 135–138.

SAGNOTTI, L., MATTEI, M., FACCENNA, C. & FUNICIELLO, R. 1994. Paleomagnetic evidence for no tectonic rotation of the Central Tyrrhenian Margin Since Upper Pliocene. *Geophysical Research Letters*, **21**, 481–484.

SAGNOTTI, L., WINKLER, A., ALFONSI, L., FLORINDO, F. & MARRA, F. 2000. Paleomagnetic constraints on the Plio-Pleistocene geodynamic evolution of the external central-northern Apennines (Italy). *Earth and Planetary Science Letters*, **180**, 243–257.

SANI, F., BONINI, M. ET AL. 2009. Late PlioceneQuaternary evolution of outermost hinterland basins of the Northern Apennines (Italy), and their relevance to active tectonics. *Tectonophysics*, **476**, 336–356.

SARTI, G., FLORINDO, F. & SAGNOTTI, L. 1995. Risultati preliminari di un'indagine interdisciplinare (analisi di facies, biostratigrafia, magnetostratigrafia) svolta su due sezioni mioceniche superiori plioceniche inferiori della val di Fine (Toscana, Pisa). *Tettonica Attiva e Geodinamica del Sistema Tirreno-Appennino*, **378-380**.

SARTORI, R. 1990. The main results of ODP Leg 107 in the frame of Neogene to Recent geology of peri-Tyrrhenian areas. *Proceedings of the Ocean Drilling Program, Scientific Results*, **107**, 715–730.

SATOLLI, S., BESSE, J., SPERANZA, F. & CALAMITA, F. 2007. New 125–150 Ma high-resolution Apparent Polar Wander Path for Adria from magnetostratigraphic sections in Umbria–Marche (Northern Apennines, Italy): timing and duration of the global Jurassic–Cretaceous hairpin turn. *Earth and Planetary Science Letters*, **257**, 329–342.

SATOLLI, S., BESSE, J. & CALAMITA, F. 2008. Paleomagnetism of Aptian–Albian sections from the Northern Apennines (Italy): implications for the 150–100 Ma apparent polar wander of Adria and Africa. *Earth and Planetary Science Letters*, **276**, 115–128.

SCHEEPERS, P.J.J. 1992. No tectonic rotation for the Apulia–Gargano foreland in the Pleistocene. *Geophysics Research Letters*, **19**, 2275–2278.

SCHEEPERS, P.J.J. 1994. Tectonic Rotation in the Tyrrhenian arc system during Quaternary and Late Tertiary. *Geologica Ultraiectina*, **112**, 352, http://doi.org/10.1029/2004JB002998

SCHEEPERS, P.J.J. & LANGEREIS, C.G. 1993. Analysis of NRM directions from the Rossello composite: implications for tectonic rotations of the Caltanissetta basin, Sicily. *Earth and Planetary Science Letters*, **119**, 243–258.

SCHEEPERS, P.J.J. & LANGEREIS, C.G. 1994. Magnetic fabric of the Pleistocene clays from the Tyrrhenian arc: a magnetic lineation induced in the final stage of the middle Pleistocene compressive event. *Tectonics*, **13**, 1190–1200.

SCHEEPERS, P.J.J., LANGEREIS, C.G. & HILGEN, F. 1993. Counter-clockwise rotations in the Southern Apennines during the Pleistocene: paleomagnetic evidence from the Matera area. *Tectonophysics*, **225**, 379–410.

SCHEEPERS, P.J.J., LANGEREIS, C.G., ZIJDERVELD, J.D.A. & HILGEN, F. 1994. Paleomagnetic evidence for counter-clockwise rotations in the Southern Apennines fold-and-thrust belt during the late Pliocene and middle Pleistocene. *Tectonophysics*, **239**, 43–59.

SCHULT, A. 1976. Palaeomagnetism of Cretaceous and Jurassic volcanic rocks in Western Sicily. *Earth and Planetary Science Letters*, **31**, 454–457.

SEBER, D., BARAZANGI, M., IBENBRAHIM, A. & DEMNATI, A. 1996. Geophysical evidence for lithospheric delamination beneath the Alboran Sea and Rif-Betic mountains. *Nature*, **379**, 785–790.

SIERRO, F.J., GONZÁLEZ-DELGADO, J.A., DABRIO, C.J., FLORES, J.A. & CIVIS, J. 1996. Late Neogene depositional sequence in the foreland basin of Guadalquivir (SW Spain). *In*: FRIEND, P.F. & DABRIO, C.J. (eds) *Tertiary Basins of Spain. The Stratigraphic Record of Crustal Kinematics*. Cambridge University Press, Cambridge, 399–345.

SPERANZA, F. & PARISI, G. 2007. High-resolution magnetic stratigraphy at Bosso Stirpeto (Marche, Italy): anomalous geomagnetic field behaviour during early Pliensbachian (early Jurassic) times? *Earth and Planetary Science Letters*, **256**, 344–359.

SPERANZA, F., SAGNOTTI, L. & MATTEI, M. 1997. Tectonics of the Umbria-Marche-Romagna Arc (central northern Apennines, Italy): new paleomagnetic constraints. *Journal of Geophysical Research*, **102**, 3153–3166.

SPERANZA, F., MANISCALCO, R., MATTEI, M., DI STEFANO, A., BUTLER, R.W.H. & FUNICIELLO, R. 1999. Timing and magnitude of rotations in the frontal thrust systems of southwestern Sicily. *Tectonics*, **18**, 1178–1197.

SPERANZA, F., MATTEI, M., SAGNOTTI, L. & GRASSO, F. 2000. Rotational differences between the northern and southern Tyrrhenian domains: paleomagnetic constraints from the Amantea basin (Calabria, Italy). *Journal of the Geological Society, London*, **157**, 327–334.

SPERANZA, F., VILLA, I.M., SAGNOTTI, L., FLORINDO, F., COSENTINO, D., CIPOLLARI, P. & MATTEI, M. 2002. Age of the Corsica-Sardinia rotation and Liguro-Provençal Basin spreading: new paleomagnetic and Ar/Ar evidences. *Tectonophysics*, **347**, 231–251.

SPERANZA, F., MANISCALCO, R. & GRASSO, M. 2003. Pattern of orogenic rotations in central-eastern Sicily: implications for the timing of spreading in Tyrrhenian Sea. *Journal of the Geological Society, London*, **106**, 183–195.

SPERANZA, F., MACRÌ, P., RIO, D., FORNACIARI, E. & CONSOLARO, C. 2011. Paleomagnetic evidence for a post 1.2 Ma disruption of the Calabria terrane: consequences of slab breakoff on orogenic wedge tectonics. *Bulletin of the Geological Society of America*, **123**, 925–933.

STEINER, M.B., OGG, J.G. & SANDOVAL, J. 1987. Jurassic magnetostratigraphy, 3. Bathonian-Bajovician of Carcabuey, Sierra Harana and Campillo de Arenas (Subbetic Cordillera, southern Spain). *Earth and Planetary Science Letters*, **82**, 357–372.

TARDUNO, J.A., LOWRIE, W., SLITER, W.V., BRALOWER, T.J. & HELLER, F. 1992. Reversed Polarity Characteristic Magnetizations in the Albian Contessa Section, Umbrian Apennines, Italy: Implications for the Existence of a Mid-Cretaceous Mixed Polarity Interval originally derived from the interpretation of sections from the Umbrian. *Journal of Geophysical Research*, **97**, 241–271.

TAUXE, L., OPDYKE, N.D., PASINI, G. & ELMI, C. 1983. Age of the Plio-Pleistocene boundary in the Vrica section, southern Italy. *Nature*, **304**, 125–129.

THIO, H.K. 1988. Magnetotectonics in the Piemont Tertiary Basin. *Physics of the Earth and Planetary Interiors*, **52**, 308–319.

VALENSISE, G. & PANTOSTI, D. 2001. Seismogenic faulting, moment release patterns and seismic hazard along the central and southern Apennines and the Calabrian Arc. *In*: VAI, G.B. & MARTINI, I.P. (eds) *Anatomy of an Orogen: the Apennines and Adjacent Mediterranean Basins*. Kluwer Academic, Dordrecht, 495–512.

VAN ASSEN, E., KUIPER, K.F., BARHOUN, N., KRIJGSMAN, W. & SIERRO, F.J. 2006. Messinian astrochronology of the Melilla basin: stepwise restriction of the Mediterranean–Atlantic connection through Morocco. *Palaeogeography, Palaeoclimatology, Palaeoecology*, **238**, 15–31.

VANDENBERG, J., KLOOTWIJK, C.T. & WONDERS, A.A.H. 1978. Late Mesozoic and Cenozoic movements of the Italian Peninsula: further paleomagnetic data from the Umbrian sequence. *Geological Society of American Bulletin*, **89**, 133–150.

VAN DER VOO, R. 1993. *Paleomagnetism of the Atlantic, Tethys and Iapetus Oceans*. Cambridge University Press, Cambridge.

VAN HINSBERGEN, D.J.J., VISSERS, R.L.M. & SPAKMAN, W. 2014. Origin and consequences of western Mediterranean subduction, rollback, and slab segmentation. *Tectonics*, **33**, 393–419.

VERGÉS, J. & FERNÀNDEZ, M. 2012. Tethys–Atlantic interaction along the Iberia–Africa plate boundary: the Betic–Rif orogenic system. *Tectonophysics*, **579**, 144–172.

VILLALAIN, J.J., OSETE, M.L., VEGAS, R., GARCÍA-DUEÑAS, V. & HELLER, F. 1994. Widespread Neogene Remagnetization in Jurassic Limestones of the South-Iberian Paleomargin (Western Betics, Gibraltar Arc). *Physics of the Earth and Planetary Interiors*, **85**, 15–33.

VILLASANTE-MARCOS, V., OSETE, M.L., GERVILLA, F. & GARCIA-DUENAS, V. 2003. Paleomagnetic study of the Ronda peridotites (Betic Cordillera, southern Spain). *Tectonophysics*, **377**, 119–141.

WEIL, A. & SUSSMAN, A.J. 2004. Classifying curved orogens based on timing relationships between structural development and vertical-axis rotations. *In*: WEIL, A. & SUSSMAN, A.J. (eds) *Orogenic Curvature*. The Geological Society of America, Boulder, **383**, 1–15.

WORTEL, M.J.R. & SPAKMAN, W. 2000. Subduction and slab detachment in the Mediterranean-Carpathian region. *Science*, **290**, 1910–1917.

# Palaeomagnetism of Mesozoic magmatic bodies of the Fuegian Cordillera: implications for the formation of the Patagonian Orocline

AUGUSTO E. RAPALINI[1]*, JAVIER PERONI[2], TOMÁS LUPPO[1], ALEJANDRO TASSONE[1], MARÍA ELENA CERREDO[1], FEDERICO ESTEBAN[1], HORACIO LIPPAI[1] & JUAN FRANCISCOVILAS[1]

[1]*Instituto de Geociencias Básicas, Aplicadas y Ambientales de Buenos Aires (IGEBA), Facultad de Ciencias Exactas y Naturales, Universidad de Buenos Aires, Conicet*

[2]*Área Geofísica, Dirección de Recursos Geológico-Mineros, Servicio Geológico y Minero Argentino (SEGEMAR), Buenos Aires, Argentina*

**Corresponding author (e-mail: rapalini@gl.fcen.uba.ar)*


**Abstract:** It is not known whether the Patagonian Orocline, the major bend of the southern Andes at the southern tip of South America, is a primary or secondary feature. Palaeomagnetic data along the Patagonian Orocline are still too scarce to provide a reliable and unambiguous answer to this question. New palaeomagnetic results on Late Jurassic–Late Cretaceous magmatic units along the central segment of the Fuegian Cordillera are reported. Data from four Late Cretaceous small intrusions and three sites on Late Jurassic–Early Cretaceous metabasalts and metagabbros showed anticlockwise declination deviations between 21° and 46° with respect to South America. From these and previous data, a picture of a nearly homogeneous post-Late Cretaceous regional rotation of the central Fuegian Cordillera is suggested. This supports a model of nearly 30° of anticlockwise secondary bending of the Patagonian Orocline since the Late Cretaceous (72 Ma). Lack of rotation of post-50 Ma sedimentary rocks exposed to the north of our study region, and larger rotations (of *c.* 90°) reported to the south of it suggest that a geographical and/or temporal progression of rotation values from south to north in the Fuegian part of the Patagonian Orocline should be investigated.

The nearly 90° change in the trend of the southern Andes at *c.* 53° S is known as the 'Patagonian Orocline' (Carey 1955). This major bend turns a roughly north–south-aligned mountain chain in southern Patagonia into a nearly east–west orogen in the island of Tierra del Fuego (Fig. 1c). Whether this bend is original and dates back to Cretaceous times or is a whole or partial secondary feature, that is, a 'true orocline', is a matter of debate. Dalziel & Elliot (1973) proposed an oroclinal origin for this bend in a model in which an originally rectilinear orogen, comprising the southern Patagonian–Fuegian Andes and the Antarctic Peninsula, was subsequently curved during the opening of the Drake Passage in Tertiary times. More recent tectonic models have involved both whole or partial oroclinal bending (e.g. Cunningham *et al.* 1991; Kraemer *et al.* 2002) or an originally curved orogen (Diraison *et al.* 2000; Ghiglione & Cristallini 2007). Palaeomagnetism has proved to be a very powerful tool to determine whether the curvature of an orogen is original or secondary (Morris & Anderson 1998). As such, it has been used since the work of Dalziel *et al.* (1973) by several authors to test if the Patagonian Orocline is a true orocline (Burns *et al.* 1980; Cunningham *et al.* 1991). Rapalini (2007) presented a review of the available palaeomagnetic data

in southern Patagonia at that time, suggesting that data were still too scarce and in most cases of dubious quality to reliably determine the origin of the bend. More recently, Maffione *et al.* (2010) presented the first palaeomagnetic data from Tertiary sedimentary rocks in the foreland of the Fuegian fold and thrust belt. They interpreted their data as evidence of no oroclinal bending since *c.* 50 Ma.

In order to provide more and better palaeomagnetic data to answer the question about the origin of the Patagonian Orocline, a systematic palaeomagnetic study was carried out on magmatic units exposed in the central part of the Fuegian Andes, that is, between the Fagnano Lake in the north and the Navarino Island in the south. Our results indicate the presence of a widespread and relatively constant value of anticlockwise declination deviations across the region, suggesting that it underwent a partial oroclinal bending of *c.* 30° since Late Cretaceous time (*c.* 75 Ma).

## Geological framework and palaeomagnetic sampling

The island of Tierra del Fuego (Fig. 1) shows a complex geological and tectonic evolution. On top of a

*From*: PUEYO, E. L., CIFELLI, F., SUSSMAN, A. J. & OLIVA-URCIA, B. (eds) 2016. *Palaeomagnetism in Fold and Thrust Belts: New Perspectives*. Geological Society, London, Special Publications, **425**, 65–80.
First published online July 22, 2015, http://doi.org/10.1144/SP425.3

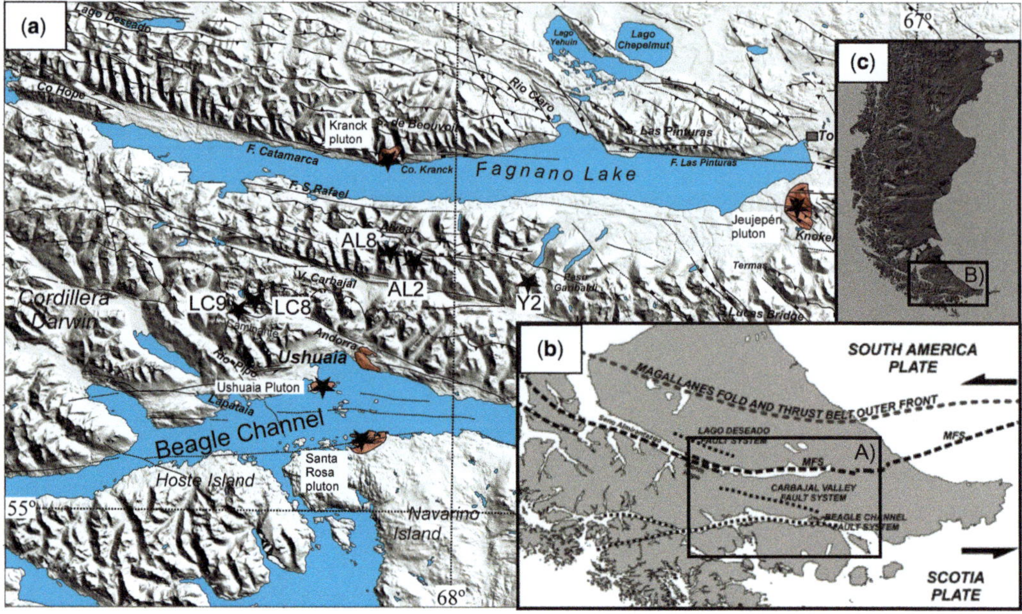



**Fig. 1.** (**a**) Digital elevation model (DEM) of the study region with main structural features and location of the sampling localities (black stars) from which consistent palaeomagnetic data were obtained, modified from Peroni (2012). (**b**) Principal structural systems of Tierra del Fuego island and location of the study region (modified from Lodolo *et al.* 2006). (**c**) DEM of Southern Patagonia and Tierra del Fuego. Note the change in trend of the Andean orogen.

highly deformed metamorphic basement of Late Palaeozoic age (e.g. Olivero & Martinioni 2001; Hervé *et al.* 2010) lies unconformably a succession of meta-volcanic, volcano-sedimentary and sedimentary rocks of Late Jurassic–Tertiary age. A major tectonic event that affected southern Patagonia was a widespread extensional regime that led to the formation of the Rocas Verdes marginal basin (Dalziel *et al.* 1974; Suárez & Pettigrew 1976; Hanson & Wilson 1991; Calderón *et al.* 2007) during the Late Jurassic. Remains of the ocean-type floor of the basin, with its pillow-basalt, basic dykes and gabbros, are now exposed along the southern Patagonia and Fuegian Andes (e.g. Cordillera Sarmiento, Tortuga Island, etc). In the study area, this extensional regime is represented by the acidic to mesosilicic volcano-pyroclastic rocks and basalts intercalated with clastic sedimentary rocks, grouped into the Lemaire Formation (known as Tobífera Formation in Chile). This has been characterized as a submarine complex developed on the marginal basin (Bruhn 1979; Hanson & Wilson 1991; Olivero & Martinioni 2001). The age of this unit is more likely Middle Jurassic on the basis of scarce radiometric datings and regional correlations (see Menichetti *et al.* 2008). The Lemaire Formation is highly deformed in the study area. Stratification is only preserved there in coarse-grained facies, while a pervasive tectonic cleavage

developed in almost all rocks of this unit but especially in fine-grained facies such as shales and siltstones (Olivero & Martinioni 2001; Esteban *et al.* 2011). Numerous basaltic sills, concordant with stratification, are intercalated in the succession of the Lemaire Formation. Most are not thicker than a few metres; however, some are several tens of metres thick. The Lemaire Formation is covered by *c.* 6 km of Lower Cretaceous deep-marine black mudstones, volcaniclastic turbidites and tuffs, corresponding to the Yahgán Formation (Bruhn 1979; Caminos *et al.* 1981; Olivero *et al.* 1999; Olivero & Martinioni 2001; Olivero & Malumián 2008). A more marginal setting in the basin is represented by the Beauvoir Formation, which consists of black shales and marls and is approximately correlative of the Yahgán Formation (Olivero & Martinioni 2001). Sedimentation into progressively shallower environments took place throughout the whole Cretaceous, synchronous with the uplift of the Fuegian Cordillera. The first-proven Andean-derived sediments were deposited during the Danian. Progression of the Andean fold and thrust belt towards the foreland (N–NE) took place during most of the Tertiary. At some time during this time span, compressive deformation was partially to largely replaced by a strike-slip deformational regime associated with the relative movements of the South America and Scotia tectonic plates (Olivero & Martinioni 2001;

Menichetti *et al.* 2008). The main deformational phase of the rocks exposed in the Fuegian Cordillera occurred during the Late Cretaceous and would have initiated by *c.* 100 Ma (Klepeis *et al.* 2010). It involved ductile deformation including isoclinal folds reaching the higher metamorphic peak – amphibolite grade – in the innermost part of the Fuegian Andes (Darwin Cordillera, see Cunningham 1995; Klepeis *et al.* 2010; Maloney *et al.* 2011). This phase has been associated with tectonic inversion and closure of the Rocas Verdes marginal basin. Its end was marked by the emplacement of intrusive bodies ('The Beagle granitic suite', Mukasa & Dalziel 1996) with ages ranging from *c.* 90 to 70 Ma. This was also the time of major uplift and cooling and was immediately followed by the development of the Magallanes fold and thrust belt, involving thick-skin and thin-skin tectonics in the inner and outer parts of the orogen, respectively (Menichetti *et al.* 2008).

Palaeomagnetic studies were carried out on 37 sites located on four, mainly undeformed, small intrusions of Late Cretaceous age, that is, the Jeujepén, Kranck, Ushuaia and Santa Rosa bodies plus several Late Jurassic basaltic sills intercalated in the Lemaire Formation. However, positive results were not obtained from all sites, as described in 'Palaeomagnetic results' section.

## Laboratory procedures

Sampling was carried out over several field trips and samples collected with a gasoline-powered portable drill in most cases. Some sills in the Lemaire Formation were sampled with block samples due to the difficulties of carrying a drill to very remote areas in the Fuegian Cordillera. Cores were oriented with both magnetic and sun compasses whenever possible. No systematic discrepancies were found between both types of measurements, when considering the local magnetic field declination of *c.* 12° E. Cores were sliced into 2.2-cm-long specimens. Palaeomagnetic measurements and demagnetizations were carried out at the Laboratorio de Paleomagnetismo Daniel A. Valencio from IGEBA (Instituto de Geociencias Básicas, Aplicadas y Ambientales de Buenos Aires, University of Buenos Aires). Measurements were performed with a 2G (DC squids) cryogenic magnetometer. Demagnetization was performed with a static three-axis degausser attached to the cryogenic magnetometer and an ASC two-chamber palaeomagnetic furnace, with an internal field of <10 nT. Two to four pilot specimens per site were submitted to stepwise alternating magnetic fields and thermal demagnetizations (e.g. Butler 1992). Assessment of magnetic behaviour of pilot specimens allowed the best stepwise demagnetization procedure to be

determined for the remaining specimens at each site. Typical demagnetization sequences were: 3, 6, 9, 12, 15, 20, 25, 30, 40, 50, 60, 70, 85 and 100 mT or 100, 150, 200, 250, 300, 350, 400, 450, 500, 525, 550, 575 and 600°C. Steps of >600°C were only necessary occasionally. In most sites that carry a characteristic remanence, AF demagnetization proved to be a more efficient technique (with the only exception of the Santa Rosa pluton); it was therefore more widely used in the whole collection of samples than the thermal method. Super-IAPD (Torsvik *et al.* 2000) software was used to visually analyse and compute the magnetic components and calculate their means. Principal component analysis (Kirschvink 1980) and Fisherian statistics (Fisher 1953) were used for those tasks, respectively.

## Palaeomagnetic results

### The Jeujepén pluton

This is a small monzodioritic pluton located to the southeast of the eastern limit of the Fagnano Lake (Fig. 1) and emplaced in the laminated to massive black shales of the Beauvoir Formation (Olivero *et al.* 1999). The Jeujepén pluton is an epizonal intrusion with <10 km$^2$ of exposure, encompassing a lithological variation from gabbro to monzonite within the shoshonite series (Cerredo *et al.* 2000*a*, *b*, 2005). The region is characterized by north-verging ESE–WNW thrusts (Lodolo *et al.* 2000, 2001, 2003) and the presence of the major tectonic feature of the area, the Magallanes–Fagnano strike-slip fault zone, located a few kilometres to the north of the Jeujepén body. Cerredo *et al.* (2011) recently published a U–Pb SHRIMP age on magmatic zircons of the pluton of 72.0 ± 0.8 Ma, much younger than a previous whole-rock K–Ar age of 93 ± 4 Ma. (Acevedo *et al.* 2000). A total of 31 cores from 5 different sites presented a characteristic remanence that was determined by AF stepwise demagnetization (Fig. 2a). A viscous random component was erased in most specimens after the first demagnetization steps. Unblocking temperatures of <600°C and effective AF cleaning suggest magnetite as the likely magnetic carrier. This was confirmed by microscopic observations, both optical and by scanning electronic microscope (SEM, Fig. 2b).

Averaging at site or at sample level yielded virtually identical mean remanence directions for the Jeujepén pluton. At site level the mean direction is declination $D$ of 313.7°, inclination $I$ of −33.7°, $\alpha_{95} = 7.5°$ and $N = 5$, where $N =$ number of sites. At sample level it is $D = 314.5°$, $I = −33.6°$, $\alpha_{95} = 4.5°$ and $n = 31$, where $n =$ number of samples (Fig. 2c, d).

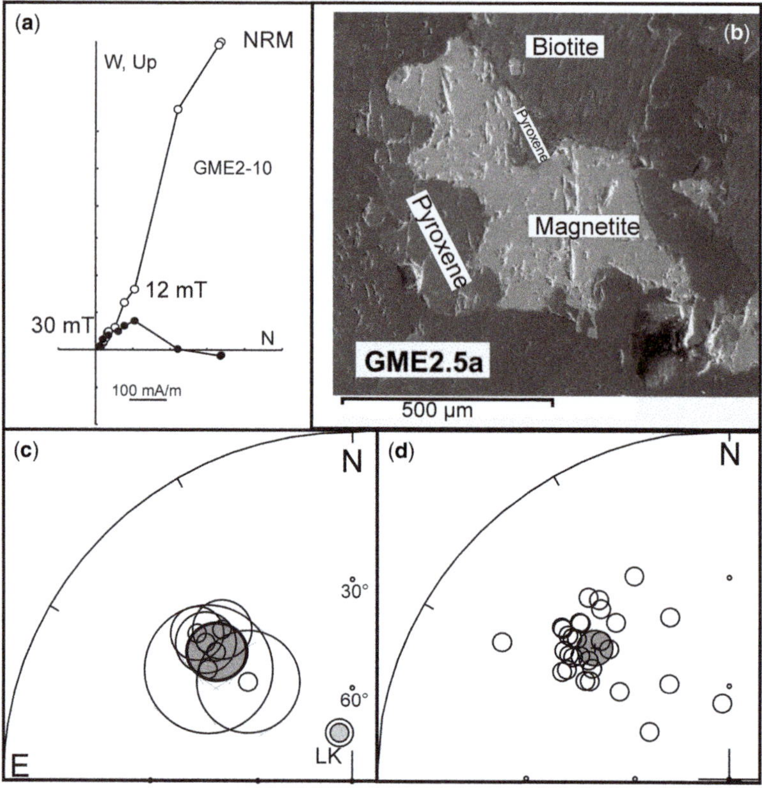



**Fig. 2.** (**a**) Representative demagnetization diagram of a sample of the Jeujepén pluton. Open (full) symbols correspond to vector representations on the vertical (horizontal) plane. (**b**) Scanning electron microscope image of a sample of the Jeujepén pluton with large magnetite crystals. Pyroxene and biotite crystals are also observed. (**c**) *In situ* site-mean characteristic remanence directions for the Jeujepén pluton. Each site direction is presented with the corresponding $\alpha_{95}$ (large circles). The overall mean $\alpha_{95}$ is formatted in grey. (**d**) Sample characteristic remanence directions for the Jeujepén pluton and its overall mean $\alpha_{95}$ (grey circle). In (c) and (d) open symbols indicate upwards (negative) inclinations.

According to the age of the Jeujepén pluton, its (sample-based) mean direction was compared with the expected direction for the sampling locality according to the South American reference palaeomagnetic pole for 70 Ma ($D = 343.7°$; $I = -69.5°$; $\alpha_{95} = 3.2°$; Besse & Courtillot 2002). Using a more recent reference apparent polar wander path for South America (Torsvik *et al.* 2008) provides a very similar reference direction ($D = 346.8°$, $I = -72.4°$) in this and the following localities. This comparison shows significant deviations both in declination ($29.2 \pm 6.1°$) and inclination ($35.9 \pm 3.9°$).

## The Kranck pluton

The Kranck pluton is a small epizonal intrusion exposed on the northern margin of the central part of the Fagnano Lake (Fig. 1). It has an outcrop area of *c.* 11 km$^2$ (Peroni 2012) and a large lithologic variation from gabbro to monzonite with late syenite dykes and veins. As well as the Jeujepén pluton, it intrudes the black shales of the Beauvoir Formation. A K–Ar cooling age on amphibole of $95.1 \pm 2.9$ Ma was recently obtained for this pluton (Cerredo *et al.* 2011).

Thermal stepwise demagnetization of pilot specimens from this pluton proved useless as remanence became unstable after a few demagnetization steps. AF cleaning, on the other hand, allowed the characteristic remanence to be determined after erasing a large but very soft random component interpreted as a viscous overprint. Demagnetization proceeded up to 50–60 mT when less than 5–10% of the original remanence remained (Fig. 3a, b). Magnetic behaviour combined with microscopic observations (both optical and by SEM; Fig. 3c) suggest magnetite as the main magnetic carrier in

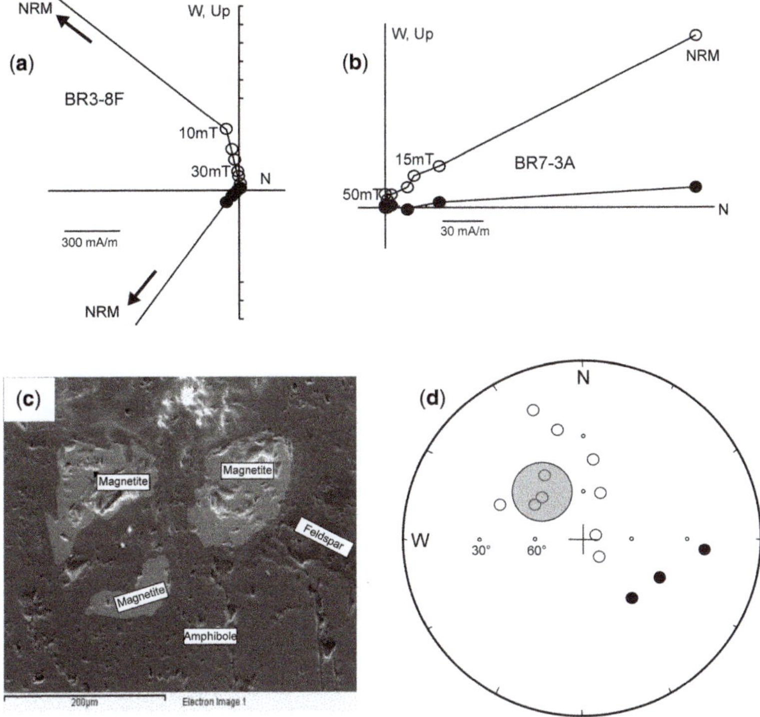



**Fig. 3.** (**a, b**) Representative demagnetization diagrams for the Kranck pluton. Symbols as in Figure 2a. (**c**) Scanning electron microscope image of a sample of the Kranck pluton with large magnetite crystals. Amphibole and feldspar crystals are also observed. (**d**) Sample characteristic remanence directions for the Kranck pluton and its overall mean $\alpha_{95}$ (grey circle). Open (full) symbols indicate upwards (downwards) inclinations.

this body. A preliminary study yielded characteristic remanence directions from 14 samples (Fig. 3d), 11 of normal polarity and 3 of reversed polarity, suggesting that remanence acquisition spanned at least one reversal of the Earth Magnetic Field and that a sample-based mean direction is more appropriate. The presence of both polarities suggests remanence acquisition after the Cretaceous Normal Superchron that ended prior to 83 Ma. This suggests magnetization much younger than the intrusion itself or a younger age for the Kranck pluton. Considering that a previous K–Ar age of 93 Ma for the nearby Jeujepén pluton has recently been superseded by a SHRIMP U–Pb age on zircons of 72 Ma (see previous section), and that K–Ar ages of 104–77 Ma for the Ushuaia pluton have also been superseded by a U–Pb SHRIMP age of 75 Ma (see next section), the possibility of a significantly younger age for the Kranck pluton consistent with the recording of both polarities of the Earth Magnetic Field is likely. In any case, the remanence age is certainly younger than 83 Ma. The mean direction after inverting the reversed polarity directions is $D = 322.9°$; $I = -51.8°$ ($n = 14$; $\alpha_{95} = 16.5°$). When compared with the expected direction for the sampling locality according to the 70 Ma reference pole for South America (Besse & Courtillot 2002), significant anticlockwise rotation (20.8 ± 16.5°) and flattening (17.7 ± 13.3°) are observed.

## The Ushuaia pluton

A small epizonal pluton hosted in the turbiditic sequence of the Yahgán Formation is exposed close to the city of Ushuaia (Fig. 1), and is referred to as the Ushuaia pluton (Peroni *et al.* 2009). This is mainly exposed to the east of the city and, although most of the body is covered by forest, scarce outcrops and magnetometric surveys (Peroni 2012) allow a minimum exposure surface of 11 km$^2$ to be inferred. Magnetometric modelling yielded a laccolithic body with an estimated volume of *c.* 140 km$^3$ (Peroni *et al.* 2009; Peroni 2012). Monzonite to monzodiorite facies dominate the exposures, although monzogabbroic to hornblenditic terms are also significant. Near the margins they grade into porphyritic dacite to andesite, which in some cases also appear as dykes. A small outcrop of this pluton is observed in the Ushuaia Peninsula (González Guillot *et al.* 2011) south of the city. A quarry at the old airport of Ushuaia exposes the porphyritic dacite and diorite facies and a well-defined contact-aureole made up of hornfels. Magnetometric modelling and field relations suggest that the Ushuaia pluton is not affected by thrusting, although it was affected by minor brittle deformation associated with the activity of the Beagle Channel Fault System (Menichetti *et al.* 2007). Ramos *et al.* (1986) published a whole-rock K–Ar age of 77 ± 3 Ma for a porphyritic dacite. More recently, Acevedo *et al.* (2002) produced another whole-rock K–Ar age of 100 ± 6 for this unit, and Peroni *et al.* (2009) reported K–Ar ages in hornblende of 101.9 ± 2.8 and 104.2 ± 3.1 Ma. However, U–Pb SHRIMP dating of zircons from the syenitic facies yielded a younger age, in this case 74.7 + 2.2/−2.0 Ma (Barbeau *et al.* 2009), which is considered the most reliable crystallization age for the pluton. Seven sites (49 cores and 14 block samples) were located in the outcrops of this

composite body at Peninsula Ushuaia (Fig. 1). Four sites were located in a diorite (Y12, GMA2, GMA3 and GMA4), two on hornfels of the contact-aureole (GMA1 and GMA5) and one on a dacite (Y11). AF demagnetization proved to be more efficient to determine the characteristic remanence (Fig. 4). Most samples were carriers of a secondary viscous overprint with random directions that could be erased by fields under 20 mT. At higher magnetic fields, most samples show a linear decay towards the origin of coordinates. However, the hornfels tend to show higher coercitivities with 30–50% of the remanence remaining after 100 mT, suggesting the presence of antiferromagnetic carriers. All sites showed good to moderate directional consistency of the characteristic remanence. Sites from the diorite, however, present poor between-site consistency (Fig. 4e). This is different for the dacitic and the hornfels sites (Fig. 4f), which show a WNW-upwards-directed remanence with high inclination. It is interpreted that the dioritic sites may be affected by undetected tectonic disturbance and are not used for any tectonic

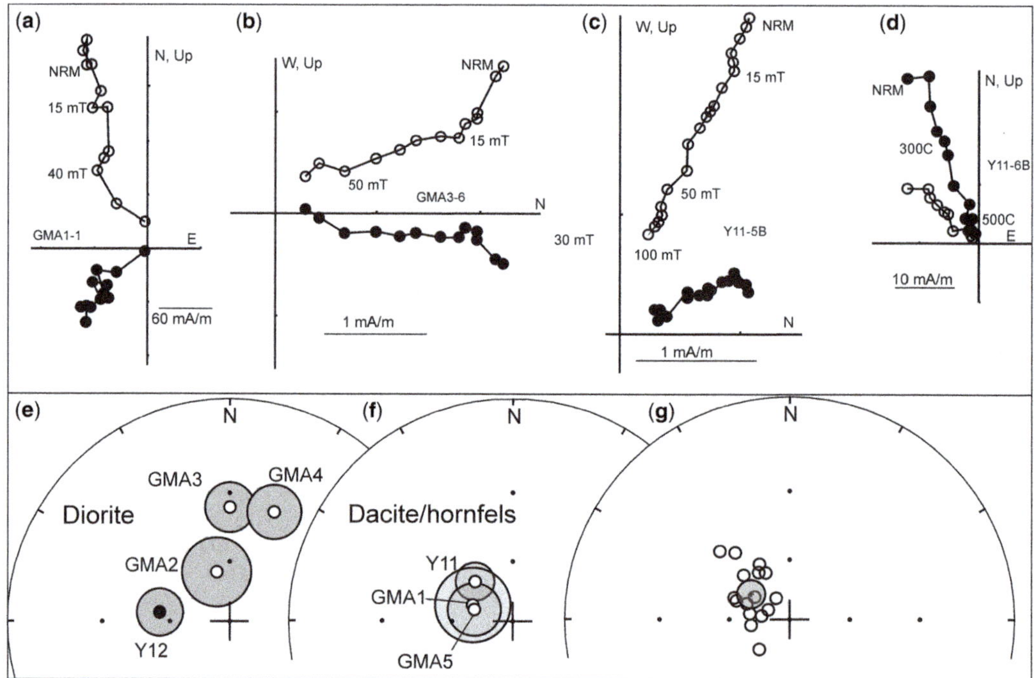



**Fig. 4.** (**a**) Representative demagnetization diagram of a sample from a hornfel in the contact-aureole of the Ushuaia pluton. (**b**) As for (a) for a sample of the main body of the intrusion (diorite) exposed at the Peninsula Ushuaia. (**c**) As for (a) for a sample of a dacite submitted to AF demagnetization. (**d**) As for (c) submitted to thermal demagnetization. Symbols as in Figure 2a. (**e**) *In situ* site mean directions for the main body (diorite) of the Ushuaia intrusion exposed at Peninsula Ushuaia (note the unconsistent directions). (**f**) As for (e) for sites located in the hornfels and the dacite. (**g**) As for (f) for each sample characteristic remanence direction. Symbols as in Figure 2c and d.

interpretation. The mean-site direction ($D = 300.3°$; $I = -67.5°$; $\alpha_{95} = 11.3°$; $N = 3$) is not significantly different from a sample-based mean ($D = 303.5°$; $I = -66.9°$; $\alpha_{95} = 6.9°$; $n = 16$). Comparison of the sample-based mean direction for the Ushuaia pluton with that expected from the South American mean palaeomagnetic pole of 70 Ma (Besse & Courtillot 2002) yielded an anticlockwise rotation of $40.2 \pm 7.9°$ with a non-significant flattening of $2.6 \pm 5.7°$.

## The Santa Rosa pluton

The Santa Rosa pluton is a small body exposed on the north-western margin of the Navarino Island, Chile (Figs 1 & 5). It is composed of a range of lithologies from gabbro to quartz-diorite, plus andesitic to granodioritic dykes. It intrudes deformed and metamorphosed shales and sandstones assigned to the Early Cretaceous Yahgán Formation. It was previously mapped and studied by Suárez

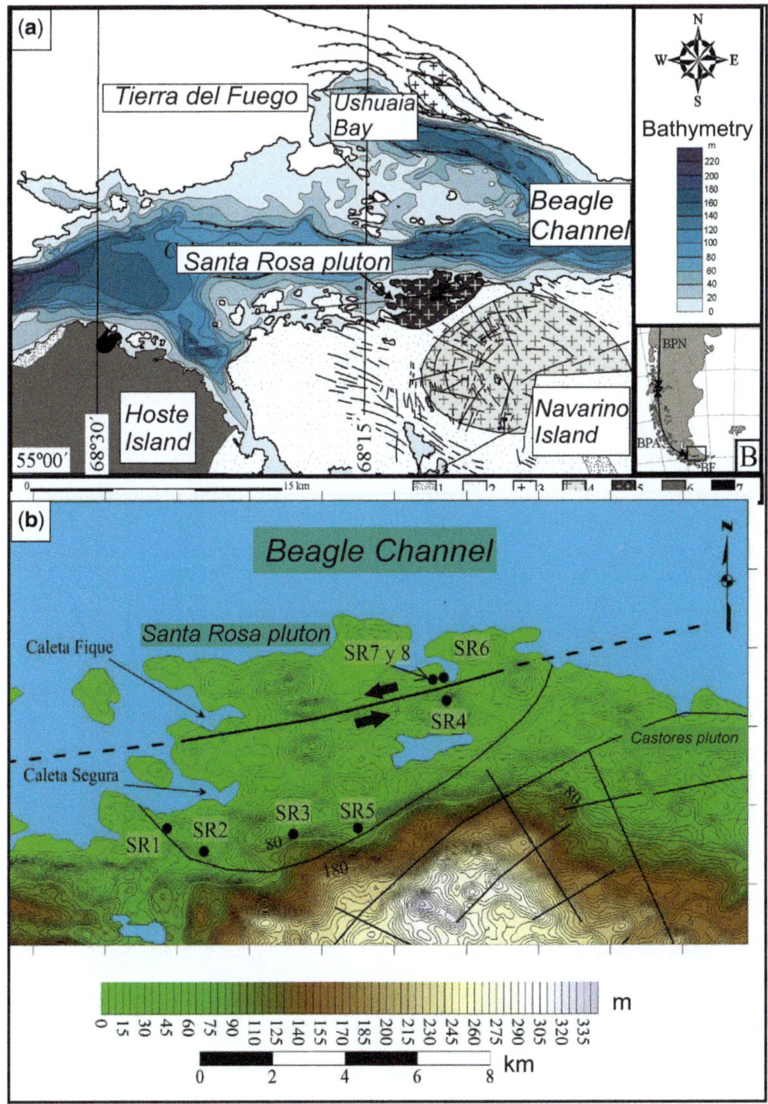


**Fig. 5.** (**a**) General location of the Santa Rosa pluton in the northern margin of the Navarino Island. Bathymetry of the Beagle Channel is shown. (**b**) Distribution of sampling sites on the Santa Rosa pluton. Note the important strike-slip fault affecting the body. Topographic contours of the northern Navarino Island are shown. Modified from Peroni (2012).

*et al.* (1985*a*, *b*, 1987) who also produced a set of cooling ages by K–Ar dating, both in biotite (88.3 ± 2.3 and 88.7 ± 2.4 Ma) and hornblende (81.8 ± 6.7, 83.7 ± 6.1, 83.8 ± 4.9 and 88.6 ± 6.5 Ma). All these datings are consistent within their uncertainties.

The Santa Rosa pluton shows a well-defined sub-vertical foliation (Suárez *et al.* 1987), which is also shown in the magnetic fabric (Peroni 2012). The outcrops of the pluton are dissected by a rectilinear ENE–WSW-oriented sinistral strike-slip fault (Peroni 2012; Menichetti, pers. comm. 2009, Fig. 5b).

A total of 44 cores were studied from 8 sites distributed on the south-western and north-eastern margins of the pluton (Fig. 5). Sites SR1 and SR6 did not yield consistent remanence directions. Thermal demagnetization proved to be a more efficient

methodology to isolate and define the characteristic remanence of the pluton (Fig. 6a–c). Discrete unblocking temperatures between 500°C and 600°C indicate that Ti-poor titanomagnetite is the likely carrier of the remanence. This was confirmed by SEM observations (Fig. 6d). Characteristic remanent directions for the Santa Rosa pluton are distributed into two groups; those from sites SR2, 3 and 5 located in the SW corner of the body have WNW-upwards-directed directions, while those on the NE margin present WSW-upwards-directed directions (Fig. 7b, c). Independent means for both areas were then calculated (Table 1, Santa Rosa pluton A and B, respectively). Both were compared with the expected direction for the Santa Rosa pluton according to the 80 Ma mean palaeomagnetic pole for South America (Besse & Courtillot

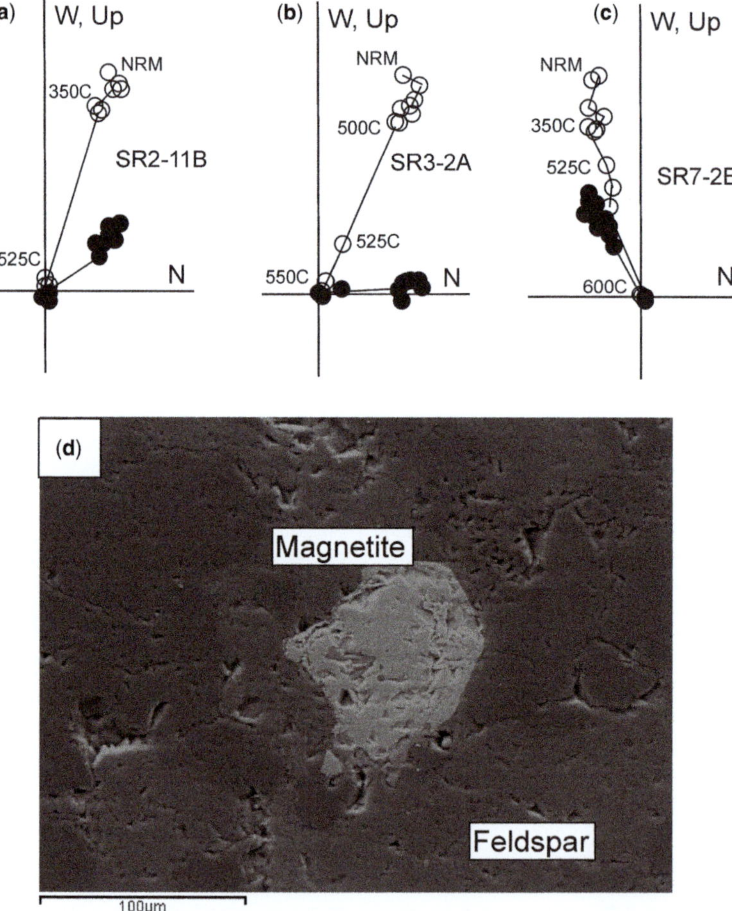



**Fig. 6.** (**a–c**) Representative demagnetization diagrams of samples of the Santa Rosa pluton submitted to thermal cleaning. Symbols as in Figure 2a. (**d**) Scanning electron microscope image of a sample of the Santa Rosa pluton with a large magnetite crystal. Feldspar crystals are also observed.

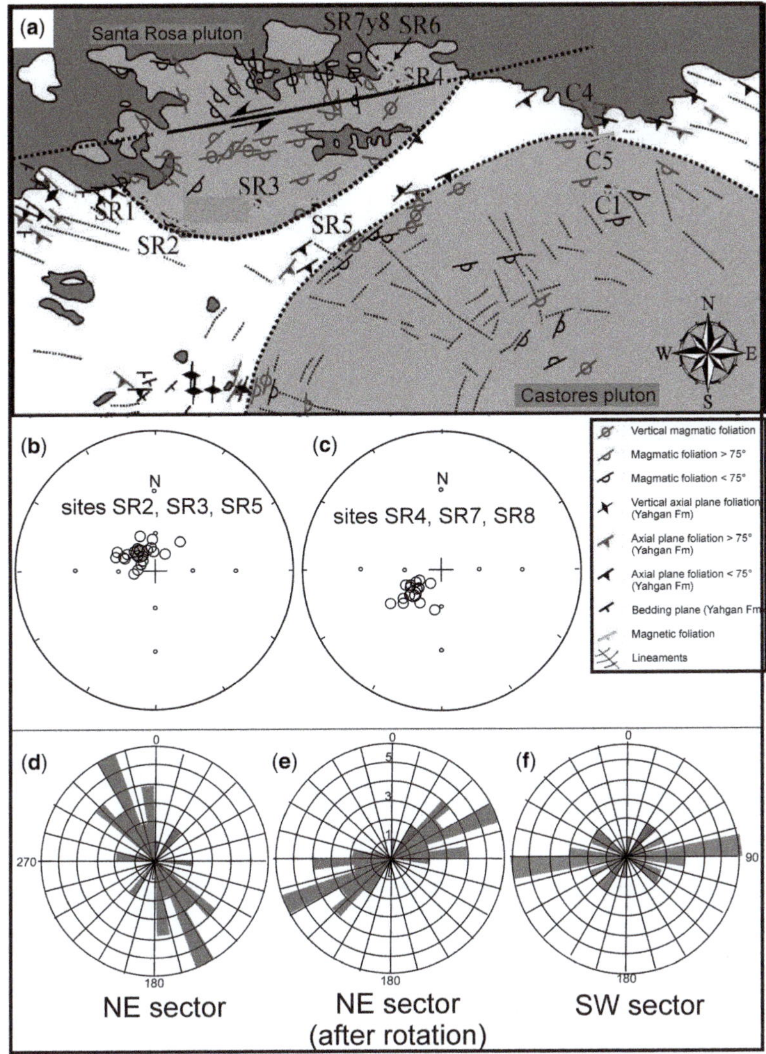



**Fig. 7.** (**a**) Sketch of the magmatic and magnetic foliations of the Santa Rosa and the nearby Castores plutons. Location of sampling sites is also indicated (modified from Peroni 2012). (**b**) Characteristic remanence directions for samples of sites SR2, SR3 and SR5, located to the south of the strike-slip fault. (**c**) As for (b) for sites SR4, SR7 and SR8, located to the north of the fault. Symbols as in Figure 2d. (**d**) Frequency diagram of the trend of magmatic foliations for the NE sector of the Santa Rosa pluton. (**e**) As for (d) after relative rotation of the northern domain respect to the southern (88°). (**f**) As for (d) for the SW sector of the pluton. Note the much better agreement with foliations after relative rotation of the northern domain.

2002; $D = 345.5°$; $I = -69.6°$) yielding anticlockwise rotation values of $28.5 \pm 9.7°$ and $116.5 \pm 9.5°$ and very minor to insignificant inclination anomalies of $-3.3 \pm 6.3°$ and $8.7 \pm 5.9°$ for the A and B zones, respectively. These results indicate a large (nearly 90°) anticlockwise relative rotation of the north-eastern sites with respect to the south-western sites. Figure 7 shows a map of the Santa Rosa pluton with an indication of the trend of the

well-defined foliations (Suárez *et al.* 1987). It is evident from this that two domains can be clearly separated in the pluton into a southern domain (where sites 2, 3 and 5 are located) which encompasses most of the pluton with ENE-trending foliations, and a northern domain (where sites 4, 7 and 8 are located) with foliations that systematically trend towards the NNW. Both domains are separated by a prominent ENE-trending strike-slip fault,

**Table 1.** *Mean characteristic remanence directions for each locality described in this paper*

| Geologic unit/locality | Age (Ma) | N (no. sites) | n (no. samples) | ChRM | | | Rotation angle (°) | Flattening (°) |
|---|---|---|---|---|---|---|---|---|
| | | | | D (°) | I (°) | $\alpha_{95}$ (°) | | |
| Jeujepén Pluton | $72 \pm 1$ | 5 | 31 | 314.5 | −33.6 | 4.5 | $29.2 \pm 6.1$ | $35.9 \pm 3.9$ |
| Kranck Pluton | <83 | 7 | 14 | 322.9 | −51.8 | 16.5 | $20.8 \pm 16.5$ | $17.7 \pm 13.3$ |
| Ushuaia Dacite + hornfels | $74 \pm 2$ | 3 | 16 | 303.5 | −66.9 | 6.9 | $40.2 \pm 7.9$ | $2.6 \pm 5.7$ |
| Santa Rosa Pluton A | $85 \pm 2$ | 3 | 21 | 317.0 | −72.9 | 5.7 | $28.5 \pm 9.7$ | $−3.3 \pm 6.3$ |
| Santa Rosa Pluton B | $85 \pm 2$ | 3 | 15 | 229.0 | −60.9 | 5.2 | $116.5 \pm 9.5$ | $8.7 \pm 5.9$ |
| *Lemaire sills* | | | | | | | | |
| Y2 | L.Jur | 1 | 7 | 301.0 | −59.8 | 9.4 | $44.6 \pm 12.0$ | $9.7 \pm 8.0$ |
| LC8 | L.Jur | 1 | 6 | 303.9 | −69.8 | 16.4 | $41.7 \pm 17.7$ | $−0.3 \pm 13.4$ |
| LC9 | L.Jur | 1 | 11 | 165.4 | 11.6 | 4.9 | $0.1 \pm 9.2$ | $57.9 \pm 4.8$ |
| AL2 | L.Jur | 1 | 3 | 114.5 | 15.1 | 11.8 | $51.1 \pm 13.8$ | $54.4 \pm 9.8$ |
| AL8 | L.Jur | 1 | 10 | 320.2 | −58.4 | 5.4 | $25.4 \pm 9.5$ | $11.1 \pm 5.1$ |

The rotation and flattening values, with their uncertainties computed according to Beck (1989), are presented. Positive values of rotation refer to anticlockwise. Positive (negative) flattening values indicate observed inclinations lower (higher) than expected. Reference directions are those computed for each locality according to the 70 Ma mean South American reference pole for the Jeujepén, Kranck and Ushuaia plutons and the 80 Ma for the Santa Rosa body and the Lemaire sills.

probably associated with the transform system along the Beagle Channel (Cunningham 1993). Figure 7 also shows a frequency diagram of foliation trends for both domains. The different trends of the foliations of the NE and SW sectors are clear. However, after rotation of the former by 88° clockwise (to compensate for the relative rotation between both domains, according to the palaeomagnetic data), a much better agreement is observed supporting the local rotation of the northern margin of the Santa Rosa pluton as a consequence of the strike-slip deformation in the area.

## Sills of the Lemaire Formation

The Lemaire Formation is exposed along a large area between the Fagnano Lake to the north and the Beagle Channel to the south, and from Paso Garibaldi in the east to the Cordillera Darwin in the west (Fig. 1). The exposures in this area consist of a volcano-sedimentary complex that has been interpreted as the record of the Late Jurassic extensional regime that led to the formation of the Rocas Verdes marginal basin (Dalziel *et al.* 1974). The complex includes sedimentary rocks (conglomerates, chert and black radiolarian and carbonaceous mudstones), acidic volcanics and volcaniclastic rocks and basaltic facies associated with the generation of the oceanic crust of the basin (Hanson & Wilson 1991; Olivero & Martinioni 2001; Cerredo *et al.* 2007; González Guillot *et al.* 2010). Many of the basalts appear as concordant sills with variable thickness from <1 m up to 100 m. Two deformational events affected the sequence during the inversion of the Rocas Verdes basin during the Late Cretaceous (Dalziel & Palmer 1979;

Menichetti *et al.* 2008). The first event produced $F_1$ folds and slaty cleavage $S_1$ with general vergence to the northeast. During this deformation, the rocks were affected by a prehnite-pumpellyite to greenschist facies metamorphism (Kohn *et al.* 1993, 1995). The second deformational event is associated with the formation and emplacement of NNE-verging fold and thrust systems (Menichetti *et al.* 2008) which originated large $F_2$ folds with a well-developed crenulation cleavage $S_2$. U–Pb zircon age obtained by laser ablation multicollector inductively coupled plasma mass spectrometry (LA-MC-ICPMS) in a rhyolite lava of the Lemaire Formation in central Tierra del Fuego yielded $163.9 \pm 3.6$ Ma (Palotti *et al.* 2012). Although no independent dating of the basic volcanic bodies intercalated in the Lemaire Formation are available, field relations and petrographic analysis suggest that they are contemporaneous with sedimentation and underwent the same metamorphic and tectonic processes that affected the sediments of the Lemaire Formation (see Esteban *et al.* 2011).

Sampling sites across the Fuegian Andes were located along road cuts, quarries or natural exposures. They were sampled either with a portable drill (sites Y1-Y5, Y8, Y9) in which cases 6–10 independent cores were obtained or as one large block sample at each site (LC5, LC8, LC9, PT4, PT6, AL2 and AL8). Sites comprised basic metavolcanic sills and lavas. A few samples from the encasing metasediments were also collected at some sites. In all cases, measurements of the cleavage and bedding planes were obtained at each site along the margins of the magmatic bodies. The thickness of the sampled bodies varied from 1–2 m (Y1, LC8) to several tens of metres (Y2, Y10, Y11).

LC9 was located 3 m from the base of a very large basic sill which may be >100 m thick. Whether this is a single sill or a composite body is not clear, due to the inaccessible nature of these outcrops.

Sampled rocks are generally porphyritic with clinopyroxene and plagioclase phenocrysts immersed in a subophitic groundmass. Primary magmatic associations display variable overprinting by static seafloor metamorphic assemblages (clinoamphibole-zoisite), later affected by oriented deformation-related paragenesis (chlorite-prehnite ± pyrite). More details on the petrographic character and microstructures of these rocks can be found in Esteban *et al.* (2011).

Many of the sampled sites on these rocks were either unstable or showed within-site inconsistent remanence directions. Of the 15 sites, consistent directions were only found in sites Y2, LC8, LC9, AL2 and AL8. Figure 8 illustrates typical magnetic behaviours of these samples. In general, they showed low–moderate coercivities with characteristic remanence being destroyed at fields between 20 and 60 mT. Unblocking temperatures were mostly *c.* 300°C, suggesting pyrrhotite as the magnetic carrier, although temperatures *c.* 500°C consistent with Ti-poor titanomagnetites were also observed. The widespread presence of pyrrhotite in these rocks was recently reported by Esteban *et al.* (2011). Figure 8 shows the mean directions from the above-mentioned sites, both *in situ* and after correction for the bedding attitude of the encasing lithology. It is evident that *in situ* directions are more consistent and suggest that magnetization is mainly post-tectonic. *In situ* directions also show higher inclinations, more appropriate to those expected at these latitudes. However, mean directions from LC8 show a better consistency with the remaining sites and those from the previously presented units, as well as a coherent inclination value, after bedding correction. In this case, the bedding correction direction has been used. Two sites (LC9, AL2) show positive inclinations,

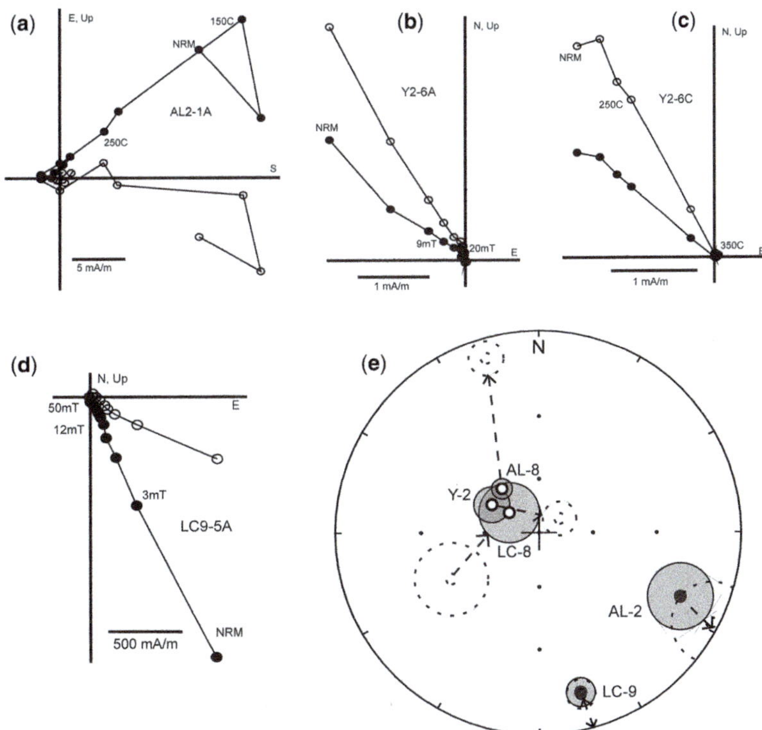

**Fig. 8.** (**a–d**) Representative demagnetization diagrams of samples from different sites of sills of the Lemaire Formation. (**b**) and (**c**) show the similar results provided by AF and thermal demagnetizations. Symbols as in Figure 2a. (**e**) Mean characteristic directions for different sites at the Lemaire Formation sills. Directions are presented with their $\alpha_{95}$ confidence circles. Arrows indicate displacements of the mean directions due to untilting of the successions in which the sills are intercalated. Dashed circles show directions not considered (generally after bedding correction). Note low inclinations of sites AL-2 and LC-9 and consistent directions of sites Y-2 (*in situ*), AL-8 (*in situ*) and LC-8 (after correction). Symbols as in Figure 2c.

suggesting a reversed polarity magnetic field during remanence acquisition. As already mentioned no independent age determination is available for these rocks, but stratigraphic, petrographic and microstructural considerations suggest they are coetaneous with sedimentation of the Middle–Upper Jurassic Lemaire Formation. Clear signs of low-grade metamorphism affecting these volcanic rocks and post-tectonic nature of the remanence in all but one site, plus pyrrhotite as the main carrier, indicate a much younger age for the remanence. The presence of both polarities suggests a post-83 Ma magnetization (end of the Cretaceous Normal Superchron, Gradstein *et al.* 2012).

Mean directions for each site were compared with the reference palaeomagnetic pole for South America for 80 Ma (Besse & Courtillot 2002). Although the age of remanence acquisition is not firmly established, a Late Cretaceous age is likely as the remanence is carried in most cases by pyrrhotite which has been determined as a by-product of the Late Cretaceous metamorphic overprint of the sequence (Menichetti *et al.* 2008; Esteban *et al.* 2011).

Rotation and flattening values are presented in Table 1. All cases, except LC9, show significant anticlockwise rotations. Sites AL2 and LC9 show very large inclination anomalies indicating undetected tectonic tilting since remanence acquisition. Both sites consistently show the largest departures from rotation values of the whole group.

## Discussion and interpretation

Figure 9a shows the distribution of declination anomalies of the studied rocks along the central part of the Fuegian Cordillera. Sites LC9 and AL2, which showed very large inclination anomalies, have been excluded. The overall picture is one of a systematic anticlockwise rotation of *c.* 30° of this section of the Fuegian Cordillera, from the Fagnano Lake in the north to the Navarino Island in the south. A much larger rotation as shown by the northern sector of the Santa Rosa pluton is assigned to local tectonic rotations, in this case associated with a major strike-slip fault dissecting the magmatic body. The fact that palaeohorizontal control is either ambiguous or absent from the different sites may explain some dispersion in the rotation values, due to minor undetected tectonic tilting. However, minor local tectonic rotations cannot be ruled out. In any case, the consistency of rotations found at so many different localities indicates a systematic rotation pattern or a rigid-body rotation of the central Fuegian Cordillera.

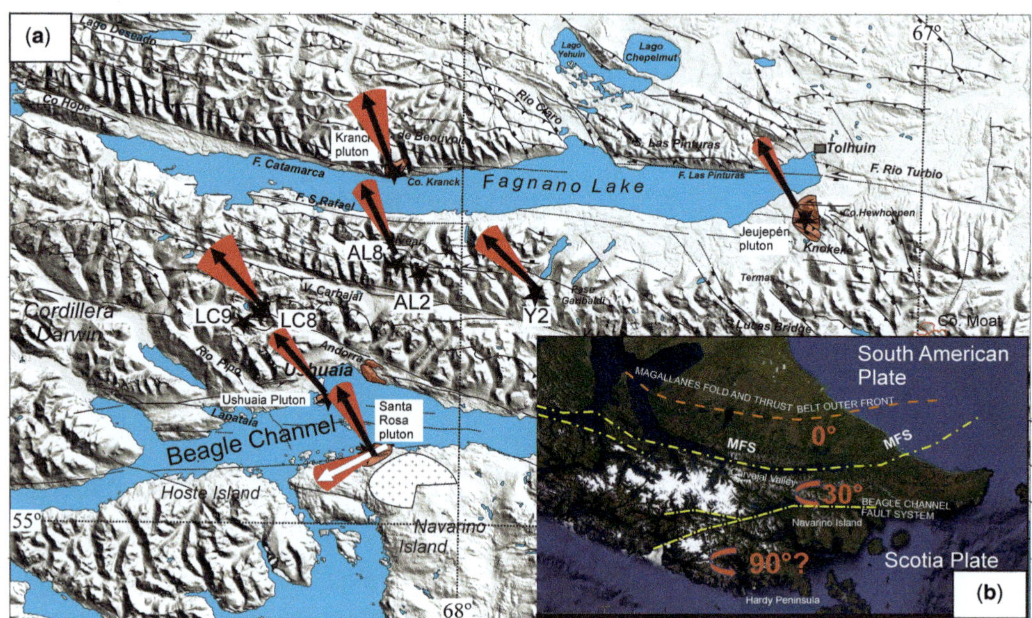


**Fig. 9.** (**a**) DEM of Figure 1 with the mean declination anomalies (rotation values of Table 1) for each locality in the Fuegian Andes reported in this study. Uncertainties are represented as red triangles around the arrow. Values for sites LC-9 and AL-2 are not presented due to the very large inclination anomalies of these sites. Rotation values for the NE sector of the Santa Rosa pluton is represented by a white arrow. (**b**) Hypothetical geographical progression of oroclinal bending of the Fuegian Andes.

Very important Cenozoic regional sinistral strike-slip systems affecting the study area (e.g. Cunningham 1993; Lodolo *et al.* 2003) may produce systematic anticlockwise rotations of crustal blocks in a domino-type style, which in turn could yield similar values of declination deviations at different localities as found in our study. However, there is a lack of significant faults transversal to the main structural trends of the Cordillera (see Fig. 1; Menichetti *et al.* 2008; Peroni 2012) which must be present in such a model as the blocks limiting faults. Furthermore, since the strike-slip deformation post-dates the main compressional phases, it is expected that in such a scenario main thrust systems should be disrupted in several localities by transversal strike-slip faults accommodating relative displacements between neighbouring rigid blocks. None of these features have yet been identified in the central areas of the Fuegian Cordillera. Finally, the overall WNW trend of the thrusting system structures (see Fig. 1) would imply that if these structures underwent *c.* 30° anticlockwise local rotations, they were originally formed as NNW structures which is difficult to reconcile with the east–west trend of the Fuegian Andes. These considerations make a systematic rotation pattern of independent crustal blocks unlikely. A *c.* 30° anticlockwise rigid body rotation of the central Fuegian Cordillera is therefore accepted as the simpler and more likely explanation for the palaeomagnetic data reported.

Several of the units studied are well dated between 85 and 72 Ma; a post-72 Ma rotation of the Cordillera is therefore implied. Maffione *et al.* (2010) recently presented palaeomagnetic data from Palaeogene sedimentary rocks of the Magallanes fold and thrust belt (MFTB), located to the north of our study localities. Mean remanence directions at five sites showed no significant declination deviation, despite very large inclination flattening that was interpreted by the authors as due to compaction of a primary remanence. Taken at face value, these data suggest that no rotation of the MFTB occurred after 50 Ma. Very minor rotations (less than 20°) since 60 Ma along the same belt have been reported by Poblete *et al.* (in press). Comparison of our data with those from Maffione *et al.* (2010) and Poblete *et al.* (in press) suggest that either the central Fuegian Cordillera rotated 30° anticlockwise between 72 and 50 Ma or its rotation did not affect the thin-skinned MFTB.

Cunningham *et al.* (1991) reported a nearly 90° anticlockwise rotation of outcrops of the Hardy Formation in Hardy Peninsula, to the south of our study localities. The Hardy Formation is correlative to the Lower Cretaceous Yahgán Formation (Olivero & Malumián 2008). This study showed that these rocks were remagnetized after tilting. The authors

interpreted a remagnetization at *c.* 90 Ma, during the Cretaceous Normal Superchron. The large anticlockwise rotation was interpreted as evidence of the oroclinal bending of the Fuegian Andes since the Middle Cretaceous. Poblete *et al.* (2013) have reported large anticlockwise rotations (*c.* 90°) in metasediments of the Hardy and Yahgán formations as well as from pillow basalts in the latter, consistent with earlier results from Cunningham *et al.* (1991). Sites from this study are more widely distributed, from the Navarino Island in the north to the Hardy Peninsula in the south. Meanwhile, Paleocene–Eocene sedimentary rocks show much smaller rotations of *c.* 25°. Unfortunately, these results have only been communicated as an extended abstract and no detailed description has yet been published.

Several different models have been proposed to explain the almost 90° curvature of the Patagonian Andes in Tierra del Fuego and neighbouring islands. They range from fully oroclinal bending (Dalziel *et al.* 1973) to partial bending (Kraemer 2003) and from a non-rotational orocline (Cunningham 1993) to a primary curvature with systematic local rotations (Diraison *et al.* 2000) or without them (Ghiglione & Cristallini 2007). Rapalini (2007) stated that the palaeomagnetic database obtained up to that date was too small and unreliable for a valid test of an oroclinal bending of the region. Despite their limitations, the studies of Maffione *et al.* (2010) and those presented here started to fill significant gaps in the database, both temporal and geographic. Diraison *et al.* (2000) presented an analogue model in which an originally curved continental margin submitted to subduction developed a fold and thrust belt in which systematic rotations of *c.* 30° developed without any oroclinal bending. However, the systematic rotations found in the central Fuegian Cordillera cannot be easily explained by such a model; most rotations have been found in late- to post-tectonic intrusives that are younger in age than the Middle–Late Cretaceous main deformational phase of the Fuegian Andes (Menichetti *et al.* 2008).

If the recent results by Poblete *et al.* (2013) are considered, then a geographic pattern of systematic rotations appears (Fig. 9b). While the central Fuegian Cordillera shows a rigid body rotation of *c.* 30°, the outer MFTB shows very minor to insignificant rotations and the inner part of the orogen *c.* 90° anticlockwise rotations. This conclusion is virtually identical to the original interpretation of Burns *et al.* (1980) over 30 years ago with the first palaeomagnetic data obtained in the region. These data, however, do not pass present-day reliability criteria. The pattern described is consistent with an oroclinal bending of the whole Fuegian Andes since initial closing of the Rocas Verdes Basin during the Middle Cretaceous, as proposed by Kraemer (2003). The fact that the maximum age of rotation decreases from south to north, that is, 90 Ma in the extreme south, 72 Ma in the central areas and 50 Ma in the northern parts, also permits a temporal progression of the bending that should have occurred between 90 and 50 Ma. Precise discrimination between temporal and geographical patterns of oroclinal bending awaits further palaeomagnetic results.

## Conclusions

Palaeomagnetic studies on the late- to post-tectonic intrusions of the Jeujepén (*c.* 72 Ma), Kranck (70–80 Ma?), Ushuaia (*c.* 74 Ma) and Santa Rosa (*c.* 85 Ma) plutons show that all these bodies were rotated *c.* 30° anticlockwise after their intrusion. The Jeujepén pluton and, to a minor extent the Kranck pluton, also underwent significant northwards tilting. The systematic pattern found in these plutons is confirmed by isolated sites on thick metabasaltic sills intercalated in the Late Jurassic Lemaire Formation that carry a post-tectonic remanence. The systematic rotation pattern found in different localities of the central Fuegian Cordillera suggest partial bending of the Patagonian Orocline since *c.* 72 Ma. Previous palaeomagnetic results on Palaeogene sedimentary rocks of the Magallanes fold and thrust belt that do not show significant rotations may be interpreted as bending ended by 50 Ma. Around 90° of anticlockwise rotation in the Hardy Peninsula to the south of the study area, and unpublished results by Chilean colleagues in several sites distributed in the southern Chilean islands, suggest the presence of a decreasing angle of rotation from south to north along the Patagonian Orocline. Whether this pattern is temporal or geographic, or a combination of both, awaits further palaeomagnetic results.

This study was supported by different grants to AER, JFV and AT by the University of Buenos Aires, CONICET and ANPCyT (Argentina). Numerous and useful discussions with Marco Menichetti were very helpful at different stages of these studies. We acknowledge all people who contributed and/or facilitated our field work in Tierra del Fuego. In particular, special thanks go to José Luis Hormaechea, Gerardo Connon, Luís Barbero and Carlos Ferrer of the Estación Astronómica Río Grande for their support during data acquisition in the field, and to Parques Nacionales of Argentina for providing access and facilities for the fieldwork at the National Park of Tierra del Fuego. Special thanks also go to Francisco Hervé for collaborative research and for helping us to obtain authorization by DIFROL (Chile) to work in the Navarino Island. Sampling in areas of difficult access at Laguna del Caminante and Sierra de Alvear was performed with the help of M. González-Guillot.

## References

ACEVEDO, R. D., ROIG, C. E., LINARES, E., OSTERA, H. A., VALÍN-ALBERDI, M. L. & QUEIROGA-MAFRA, J. M. 2000. La intrusión plutónica del Cerro Jeu-Jepén. Isla Grande de Tierra del Fuego, República Argentina. *Cuadernos do Laboratorio Xeolóxico de Laxe*, **25**, 357–359.

ACEVEDO, R. D., LINARES, E., OSTERA, H. A. & VALÍN-ALBERDI, M. L. 2002. La Hornblendita Ushuaia (Tierra del Fuego): Geoquímica y Geocronología. *Revista de la Asociación Geológica Argentina*, **57**, 133–142.

BARBEAU, D. L., GOMBOSI, D. J., ZAHID, K. M., BIZIMIS, M., SWANSON-HYSELL, N., VALENCIA, V. & GEHRELS, G. E. 2009. U-Pb zircon constraints on the age and provenance of the Rocas Verdes basin-fill, Tierra del Fuego, Argentina. *Geochemistry, Geophysics, Geosystems*, **10**, http://doi.org/10.1029/2009 GC002749

BECK, M. E., Jr., 1989. Paleomagnetism of continental North America; implications for displacement of crustal blocks within the Western Cordillera, Baja California to British Columbia. *In*: PAKISER, L. C. & MOONEY, W. D. (eds) *Geophysical Framework of the Continental United States*. Geological Society of America, Boulder, Colorado, Memoirs, **172**, 471–492.

BESSE, J. & COURTILLOT, V. 2002. Apparent and true polar wander and the geometry of the geomagnetic field over the last 200 Myr. *Journal of Geophysical Research*, **107**, http://doi.org/10.1029/2000JB000050

BRUHN, R. L. 1979. Rock structures formed during back-arc basin deformation in the Andes of Tierra del Fuego. *Bulletin of the Geological Society of America*, **90**, 998–1012.

BURNS, K. L., RICKARD, M. J., BELBIN, L. & CHAMALAUN, F. 1980. Further paleomagnetic confirmation of the Magellanes orocline. *Tectonophysics*, **63**, 75–90.

BUTLER, R. F. 1992. *Paleomagnetism: Magnetic Domains to Geologic Terranes*. Blackwell Scientific Publications, Boston.

CALDERÓN, M., HERVÉ, F., MASSONE, H. J., TASSINARI, C. G., PANKHURST, R. J., GODOY, E. & THEYE, T. 2007. Petrogenesis of the Puerto Eden igneous and Metamorphic complex, Magallanes, Chile: Late Jurassic syn-deformational anatexis of metapelites and granitoid magma genesis. *Lithos*, **93**, 17–38.

CAMINOS, R., HALLER, M., LAPIDO, O., LIZUAIN, A., PAGE, R. & RAMOS, V. 1981. Reconocimiento geológico de los Andes Fueguinos. Territorio Nacional de Tierra del Fuego. *Proceedings 8th Congreso Geológico Argentino*, San Luis, 20–26 September 1981, **3**, 759–786.

CAREY, S. W. 1955. The orocline concept in geotectonics. *Proceedings of the Royal Society of Tasmania*, **89**, 255–288.

CERREDO, M. E., TASSONE, A., COREN, F., LODOLO, E. & LIPPAI, H. 2000*a*. *Postorogenic Alkaline Magmatism in the Fuegian Andes: the Jeujepen Intrusive (Tierra del Fuego Islan)*. Proceedings 9th Congreso Geológico Chileno, Puerto Varas, Chile, 31 August–4 July 2000, **29**, 192–196.

CERREDO, M. E., TASSONE, A. A., COREN, F., LODOLO, E. & LIPPAI, H. 2000*b*. *Hewhoepen Intrusive: an Alkaline Pluton Along the Strike of the Magallanes Fagnano*

*Fault System, Tierra del Fuego Island, Argentina*. Proceedings 31st International Geological Congress, 6–17 August 2000, Río de Janeiro, Brazil.

CERREDO, M. E., REMESAL, M. B., TASSONE, A. & LIPPAI, H. 2005. *The Shoshonitic Suite of Hewhoepen Pluton, Tierra del Fuego, Argentina*. Proceedings 16th Congreso Geológico Argentino, La Plata, Argentina, 20–24 September 2005, **1**, 539–544.

CERREDO, M. E., REMESAL, M. B., TASSONE, A. & MENICHETTI, M. 2007. The ocean crust of Rocas Verdes Basin in Tierra del Fuego, Argentina. *In: Simposio Argentino del Jurásico*, No. 3, Actas 1: 33. Mendoza, 2–5 May 2007.

CERREDO, M. E., TASSONE, A., RAPALINI, A., HERVÉ, F. M. & FANNING, M. 2011. *Campanian Magmatism in the Fuegian Andes: new SHRIMP age of Jeujepen Pluton, Argentina*. Proceedings 18th Congreso Geológico Argentino, Neuquen, Argentina, 2–6 May 2011.

CUNNINGHAM, W. D. 1993. Strike-slip faults in the southernmost Andes and the development of the Patagonian orocline. *Tectonics*, **12**, 169–186.

CUNNINGHAM, W. D. 1995. Orogenesis at the southern tip of the Americas: the structural evolution of the Cordillera Darwin metamorphic complex, southernmost Chile. *Tectonophysics*, **244**, 197–229.

CUNNINGHAM, W. D., KLEPEIS, K. A., GOSE, W. A. & DALZIEL, I. W. 1991. The Patagonian Orocline: new paleomagnetic data from the Andean magmatic arc in Tierra del Fuego, Chile. *Journal of Geophysical Research*, **96**, 16061–16069.

DALZIEL, I. W. D. & ELLIOT, D. H. 1973. The Scotia Arc and Antarctic margin. *In*: NAIRN, A. E. M. & STEHLI, F. G. (eds) *The Ocean Basins and Margins, The South Atlantic*. Plenum Press, New York, **1**, 171–245.

DALZIEL, I. W. D. & PALMER, K. F. 1979. Progressive deformation and orogenic uplift at the southern extremity of the Andes. *Bulletin of the Geological Society of America*, **90**, 259–280.

DALZIEL, I. W. D., KLIGFIELD, R., LOWRIE, W. & OPDYKE, N. D. 1973. Paleomagnetic data from the southernmost Andes and the Antarctandes. *In*: TARLING, D. H. & RUNCORN, S. K. (eds) *Implications of Continental Drift to Earth Sciences*. Academic Press, San Diego, California, **1**, 87–101.

DALZIEL, I. W. D., DE WIT, M. J. & PALMER, K. F. 1974. Fossil marginal basin in the southern Andes. *Nature*, **250**, 291–294.

DIRAISON, M., COBBOLD, P. R., GAPAIS, D., ROSSELLO, E. A. & LE CORRE, C. 2000. Cenozoic crustal thickening, wrenching and rifting in the foothills of the southernmost Andes. *Tectonophysics*, **316**, 91–119.

ESTEBAN, F., TASSONE, A. *ET AL.* 2011. Magnetic fabric and microstructures across the Andes of Tierra del Fuego. *Argentina Andean Geology*, **38**, 64–81.

FISHER, R. A. 1953. Dispersion on a sphere. *Proceedings of the Royal Society of London*, **217**, 295–305.

GHIGLIONE, M. & CRISTALLINI, E. 2007. Have the southernmost Andes been curved since Late Cretaceous time? An analog test for the Patagonian Orocline. *Geology*, **35**, 13–16.

GONZÁLEZ GUILLOT, M., ACEVEDO, D. & ESCAYOLA, M. 2010. El Gabro Rancho Lata: Magmatismo mesozoico off–axis de la cuenca marginal Rocas Verdes en los Andes Fueguinos de Argentina. *Revista Mexicana de Ciencias Geológicas*, **27**, 431–448.

GONZÁLEZ GUILLOT, M., ESCAYOLA, M. & ACEVEDO, D. 2011. Calc-alkaline rear-arc magmatism in the Fuegian Andes: implications for the mid-cretaceous tectonomagmatic evolution of southernmost South America. *Journal of South American Earth Sciences*, **31**, 1–16.

GRADSTEIN, F., OGG, J., SCHMITZ, M. & OGG, G. 2012. *The Geologic Time Scale 2012*. Elsevier, Boston, USA.

HANSON, B. E. & WILSON, T. J. 1991. Submarine rhyolitic volcanism in a Jurasic proto-marginal basin, southern Andes, Chile and Argentina. *In*: HARMON, R. S. & RAPELA, C. W. (eds) *Andean Magmatism and its Tectonic Setting*. Geological Society of America, Boulder, Special Paper, **265**, 13–27.

HERVÉ, F., FANNING, M., PANKHURST, R. J., MPODOZIS, C., KLEPEIS, K., CALDERÓN, M. & THOMSON, S. 2010. Detrital zircon SHRIMP U/Pb age study of the Cordillera Darwin Metamorphic Complex: sedimentary sources and implications for the evolution of the Pacific margin of Gondwana. *Journal of the Geological Society, London*, **167**, 555–568, http://doi.org/10.1144/0016-76492009-124

KIRSCHVINK, J. L. 1980. The least-squares line and plane and the analysis of palaeomagnetic data. *Geophysical Journal of the Royal Astronomical Society*, **62**, 699–718.

KLEPEIS, K. A., BETKA, P. *ET AL.* 2010. Continental underthrusting and obduction during the Cretaceous closure of the Rocas Verdes rift basin, Cordillera Darwin, Patagonian Andes. *Tectonics*, **29**, http://doi.org/10.1029/2009TC002610

KOHN, M. J., SPEAR, F. S. & DALZIEL, I. W. D. 1993. Metamorphic P-T path from Cordillera Darwin, a core complex in Tierra del Fuego, Chile. *Journal of Petrology*, **34**, 519–542.

KOHN, M. J., SPEAR, F. S., HARRISON, T. M. & DALZIEL, I. W. D. 1995. $^{40}$Ar/$^{39}$Ar geochronology and P-T-t paths from the Cordillera Darwin metamorphic complex, Tierra del Fuego, Chile. *Journal of Metamorphic Geology*, **13**, 251–270.

KRAEMER, P. E. 2003. Orogenic shortening and the origin of the Patagonian orocline (56°S. Lat). *Journal of South American Earth Sciences*, **15**, 731–748.

KRAEMER, P. E., PLOSZKIEWICZ, J. V. & RAMOS, V. A. 2002. Estructura de la Cordillera Patagónica Austral entre los 46° y 52°S. *In*: HALLER, M. A. (ed.) *Geología y Recursos Naturales de la Provincia de Santa Cruz*. Proceedings of the 15th Congreso Geológico Argentino, Santa Cruz, 23–26 April 2002, 353–364.

LODOLO, E., TASSONE, A., MENICHETTI, M., COREN, F. & STERZAI, P. 2000. *Deciphering the Morphostructure of the Tierra del Fuego Region from Remote-Sensing and Geophysical Data*. Proceedings 25th European Geophysical Society, Nice, France, 25–29 April 2000.

LODOLO, E., TASSONE, A., MENICHETTI, M., LIPPAI, H., HORMAECHEA, J. L., FERRER, C. & CONNON, G. 2001. *The Magallanes-Fagnano Fault System in the Lago Fagnano, Tierra del Fuego Island: Morphology and Structure*. Proceedings 26th European Geophysical Society General Assembly, Nice, France, 25–30 March 2001.

LODOLO, E., MENICHETTI, M., BARTOLE, R. & BEN-AVRAHAM, Z. 2003. Magallanes-Fagnano continental transform fault (Tierra del Fuego, southermost South America). *Tectonics*, **22**, http://doi.org/10.1029/2003TC001500

LODOLO, E., DONDA, F. & TASSONE, A. 2006. Western Scotia Sea margins: improved constraints on the opening of the Drake Passage. *Journal of Geophysical Research*, **111**, http://doi.org/10.1029/2006JB004361

MAFFIONE, M., SPERANZA, F., FACCENNA, C. & ROSSELLO, E. 2010. Paleomagnetic evidence for a pre-early Eocene ($\pm$50 Ma) bending of the Patagonian orocline (Tierra del Fuego, Argentina): Paleogeographic and tectonic implications. *Earth and Planetary Science Letters*, **289**, 273–286.

MALONEY, K. T., CLARKE, G. L., KLEPEIS, K. A., FANNING, C. M. & WANG, W. 2011. Crustal growth during back-arc closure: Cretaceous exhumation history of Cordillera Darwin, southern Patagonia. *Journal of Metamorphic Geology*, **29**, 649–672.

MENICHETTI, M., TASSONE, A., PERONI, J. I., GONZÁLEZ GUILLOT, M. & CERREDO, M. E. 2007. Assetto strutturale, petrografia e geofisica della Bahía Ushuaia – Argentina. *Rendiconti Online Societa Geolica Italiana, Nuova Serie*, **4**, 259–262.

MENICHETTI, M., LODOLO, E. & TASSONE, A. 2008. Structural geology of the Fuegian Andes and Magallanes fold-and-thrust belt-Tierra del Fuego Island. *Geologica Acta*, **6**, 19–42.

MORRIS, A. & ANDERSON, M. W. 1998. Paleomagnetism and tectonic rotations. *Tectonophysics*, **299**, 1–253.

MUKASA, S. B. & DALZIEL, I. W. D. 1996. Southernmost Andes and South Georgia Island North Scotia Ridge: Zircon U-Pb and muscovite $^{40}Ar/^{39}Ar$ age constraints on the tectonic evolution of the southwestern Gondwanaland. *Journal of South American Earth Sciences*, **9**, 349–365.

OLIVERO, E. B. & MALUMIÁN, N. 2008. Mesozoic-Cenozoic stratigraphy of the Fuegian Andes, Argentina. *Geologica Acta*, **6**, 5–18.

OLIVERO, E. B. & MARTINIONI, D. R. 2001. A review of the geology of Argentinian Fuegian Andes. *Journal of South American Earth Sciences*, **14**, 175–188.

OLIVERO, E. B., MARTINIONI, D. R., MALUMIÁN, N. & PALAMARCZUK, S. 1999. *Bosquejo Geológico de la Isla Grande de Tierra del Fuego, Argentina*. Proceedings 14th Congreso Geológico Argentino, Salta, 19–24 September 1999, **1**, 291–294.

PALOTTI, P., MENICHETTI, M., CERREDO, M. E. & TASSONE, A. 2012. A new U/Pb zircon determination for the Lemaire Formation of Fuegian Andes, Tierra del Fuego, Argentina. *Rendiconti Online Societa Geolica Italiana*, **22**, 170–173.

PERONI, J. 2012. *Modelado Geofísico-Geológico de Plutones en las Islas Grande de Tierra del Fuego (Argentina) y Navarino (Chile)*. PhD thesis, Facultad de Ciencias Exactas y Naturales, Universidad de Buenos Aires, Argentina.

PERONI, J. I., TASSONE, A., MENICHETTI, M. & CERREDO, M. E. 2009. Geophysical modeling and structure of Ushuaia Pluton, Fuegian Andes, Argentina. *Tectonophysics*, **476**, 436–449.

POBLETE, F., ROPERCH, P., HERVÉ, F., RAMÍREZ, C. & ARRIAGADA, C. 2013. On the possibility and timing for tectonic rotations in Patagonia: new paleomagnetic and AMS data from southernmost South América. *Bollettino di Geofisica Teorica ed Aplicada*, **54**, 340–341.

POBLETE, F., ROPERCH, P., HERVÉ, F., DIRAISON, M., ESPINOZA, M. & ARRIAGADA, C. In press. The curved Magallanes fold and thrust belt: tectonic insights from a paleomagnetic and anisotropy of magnetic susceptibility study. *Tectonics*, **33**, 2526–2551.

RAMOS, V. A., HALLER, M. J. & BUTRON, F. 1986. Geología y evolución tectónica de las islas Barnevelt, Atlántico Sur. *Revista de la Asociación Geológica Argentina*, **41**, 137–154.

RAPALINI, A. E. 2007. A paleomagnetic analysis of the Patagonian Orocline. *Geologica Acta*, **5**, 287–294.

SUÁREZ, M. & PETTIGREW, T. H. 1976. An Upper Mesozoic Island-arc – back-arc system in the southern Andes and South Georgia. *Geological Magazine*, **113**, 305–400.

SUÁREZ, M., HERVÉ, M. & PUIG, A. 1985*a*. Hoja Isla Hoste e Islas Adyacentes, XII Región. *In: Carta Geológica de Chile N°65*. Servicio Nacional de Geología y Minería, Santiago, Chile.

SUÁREZ, M., HERVÉ, M. & PUIG, A. 1985*b*. *Plutonismo Diapírico del Cretácico en Isla Navarino*. Proceedings 4th Congreso Geológico Chileno, Antofagasta, Chile, 19–24 August 1985, **4**, 549–563.

SUÁREZ, M., HERVÉ, M. & PUIG, A. 1987. Cretaceous diapiric plutonismo in the southern Cordillera, Chile. *Geological Magazine*, **124**, 569–575.

TORSVIK, T. H., BRIDEN, J. C. & SMETHURST, M. A. 2000. *Super-IAPD: InteractiveAnalysis of Palaeomagnetic Data*. World Wide Web Address: http://www.geodynamics.no/Web/Content/Software/

TORSVIK, T. H., MULLER, R. D., VAN DER VOO, R., STEINBERGER, B. & GAINA, C. 2008. Global plate motion frames: toward a unified model. *Reviews of Geophysics*, **46**, RG 3004, 1–44.

# Palaeomagnetic data from the Precordillera fold and thrust belt constraining Neogene foreland evolution of the Pampean flat-slab segment (Central Andes, Argentina)

MARIA S. JAPAS[1]*, GUILLERMO H. RÉ[1], SEBASTIÁN ORIOLO[2] & JUAN F. VILAS[1]

[1]*Departamento de Ciencias Geológicas, Instituto de Geociencias Básicas, Aplicadas y Ambientales (IGeBA, CONICET-UBA), Universidad de Buenos Aires, Pabellón II, Ciudad Universitaria (1428) Ciudad Autónoma de Buenos Aires, Argentina*

[2]*Geoscience Centre of the Georg–August–Universität Göttingen, Germany*

**Corresponding author (e-mail: msjapas@gl.fcen.uba.ar)*

**Abstract:** Available and new palaeomagnetic data reveal transpressional deformation in the Argentine Precordillera fold and thrust belt contemporaneous with Juan Fernández ridge subduction. Localized changes in the orientation of palaeomagnetic directions indicate a vertical axis rotation pattern linked to local, oblique brittle-ductile shear zones that overprint the regional structure. The nearly homogeneous clockwise block rotation pattern from Western-Central Precordillera shows localized rotation nulls along NNW-trending left-lateral oblique belts, revealing that overprinting anticlockwise tectonic rotations could have balanced previous clockwise rotations. Conversely, clockwise rotations in Eastern Precordillera are only localized in the vicinity of these NNW-trending structures, and are controlled by rigid block rotations linked to basement-involved deformation.These results combined with available regional palaeomagnetic data in the forearc would indicate a regional bending of the upper plate margin between 27° S and 33° S that seems to be related to the subduction of the Juan Fernández aseismic ridge.

Representing the longest modern active-convergent margin in the world, the Andean chain has been extensively studied. Along its length of more than 7000 km, some peculiar tectonic features (e.g. changes in arc-magmatism and deformation front patterns, etc.) induced its along-strike segmentation that are primarily controlled by latitudinal differences of the subducted oceanic lithosphere, such as subduction angle, thickness and age. One of these segments is the 27–33° S Pampean flat-slab sector, located in the southern Central Andes (Ramos *et al.* 2002; Fig. 1a), and is mainly characterized by distinctive Neogene–present-day deformation and magmatism. This flat-slab segment was developed as the consequence of the Juan Fernández ridge subduction (Pilger 1981; Gutscher *et al.* 2000; Fig. 1a).

The effect of the Nazca slab flattening and the consequent changes in the deformational pattern of the South American upper plate, including migration of deformation towards the foreland (Gutscher *et al.* 2000, Ramos *et al.* 2002) and continental margin curvature (Bevis & Isacks 1984; Yáñez *et al.* 2001; Japas & Ré 2012*b*), can be used to unravel the evolution of the Argentine Precordillera fold and thrust belt (hereafter referred as Precordillera; Figs 1 & 2a). The Precordillera represents the foreland of the Andean chain during the Neogene, and has therefore recorded the sedimentation and deformational histories for the last 18 Ma (Jordan

*et al.* 1993). In addition, the Precordillera contains some oblique transpressional and transtensional megazones that could have played a significant role in controlling inland migration of deformation and tectonic rotations during the Neogene (Japas 1998; Ré *et al.* 2001; Japas & Ré 2012*a*, *b*), such that localized deformation might be important.

The Precordillera is a north–south-trending, complex fold and thrust belt (Fig. 1b; Baldis & Chebli 1969; Ortiz & Zambrano 1981; Baldis *et al.* 1982) that developed since the Neogene as the deformation front migrated towards the Andean foreland (the Bermejo foreland basin, Figs 1c & 2a). Although the Precordillera was traditionally linked to orthogonal shortening (Allmendinger *et al.* 1990; von Gosen 1992; Zapata & Allmendinger 1996*a*; Cristallini & Ramos 2000; Ramos *et al.* 2002), more recent contributions have shown that transpressional deformation may also have taken place (Japas 1998; Ré *et al.* 2001; Siame *et al.* 2005; Álvarez-Marrón *et al.* 2006). Regardless of whether transpression is partitioned or non-partitioned, all these latter authors propose a regional, dextral transpressional deformation for the Precordillera. In this setting, vertical axis rotations deduced from palaeomagnetic data can help in explaining some of these uncertainties, providing unique information to discriminate between deformation accommodated during regional deformation and thickening

*From*: PUEYO, E. L., CIFELLI, F., SUSSMAN, A. J. & OLIVA-URCIA, B. (eds) 2016. *Palaeomagnetism in Fold and Thrust Belts: New Perspectives.* Geological Society, London, Special Publications, **425**, 81–105.
First published online August 18, 2015, http://doi.org/10.1144/SP425.9

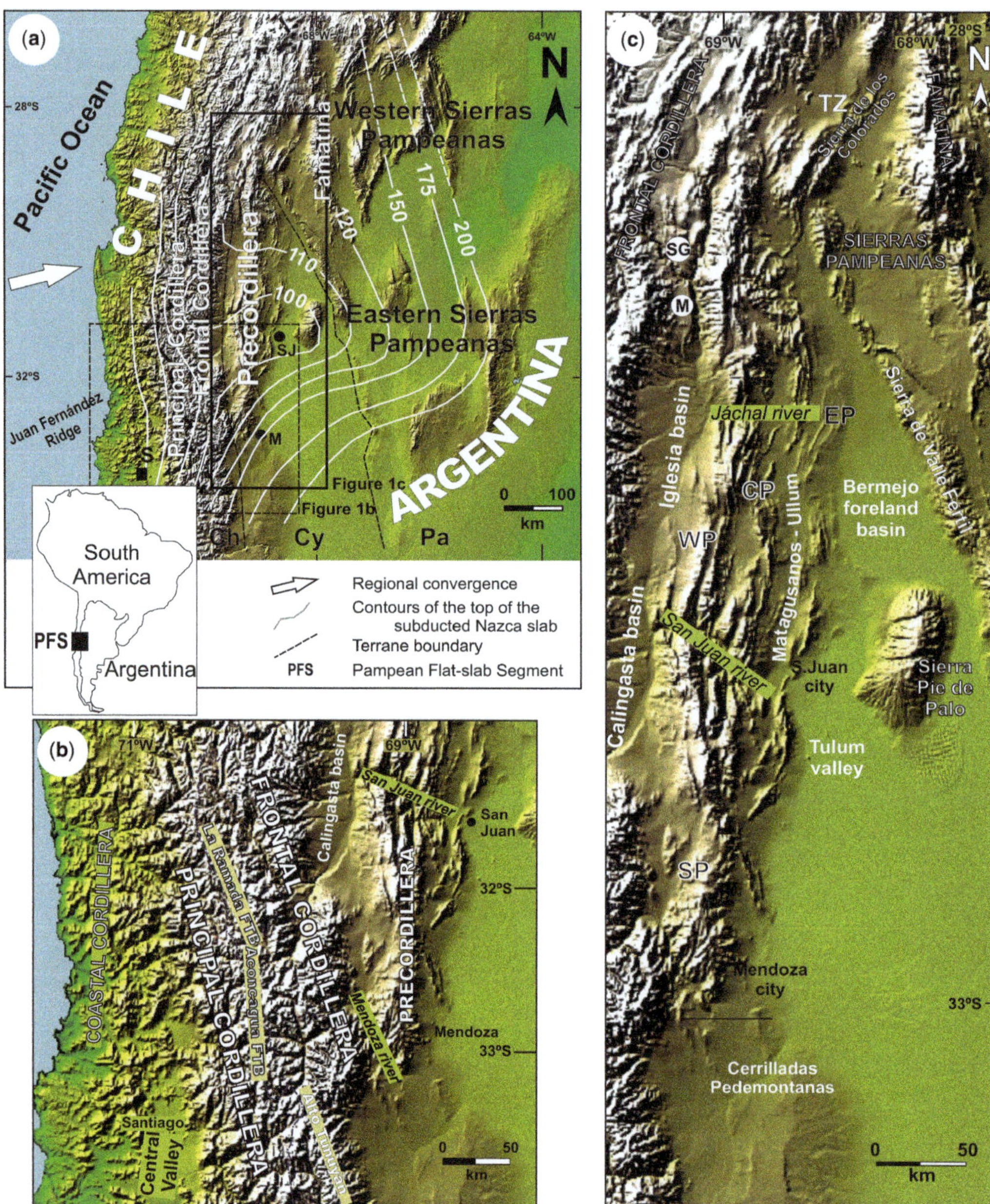



**Fig. 1.** (a) The Pampean flat-slab segment in the southern Central Andes context (Shuttle Radar Topography Mission image). White contours refer to the top of the subducted Nazca slab (dashed lines after Cahill & Isacks 1992; solid lines after Anderson *et al.* 2007). Dashed black lines separate terranes at 28–33° S: Chilenia (Ch), Cuyania (Cy), Pampia (Pa). M: Mendoza city, SJ: San Juan city, S: Santiago (Chile). (b) Principal Cordillera fold and thrust belts in the Pampean flat-slab segment: the thick-skinned La Ramada and the thin-skinned Aconcagua fold and thrust belts (SRTM image). FTB: fold and thrust belt. (c) The Argentine Precordillera fold and thrust belt (SRTM image). WP: Western Precordillera; CP: Central Precordillera; EP: Eastern Precordillera; SP: Southern Precordillera; TZ: transitional zone between Precordillera and Sierras Pampeanas; M and SG: del Médano and San Guillermo minor basins. See the geological map of this SRTM image in Figure 2. Courtesy NASA/JPL-Caltech: http://www2.jpl.nasa. gov/srtm/southAmerica.htm#PIA03388.

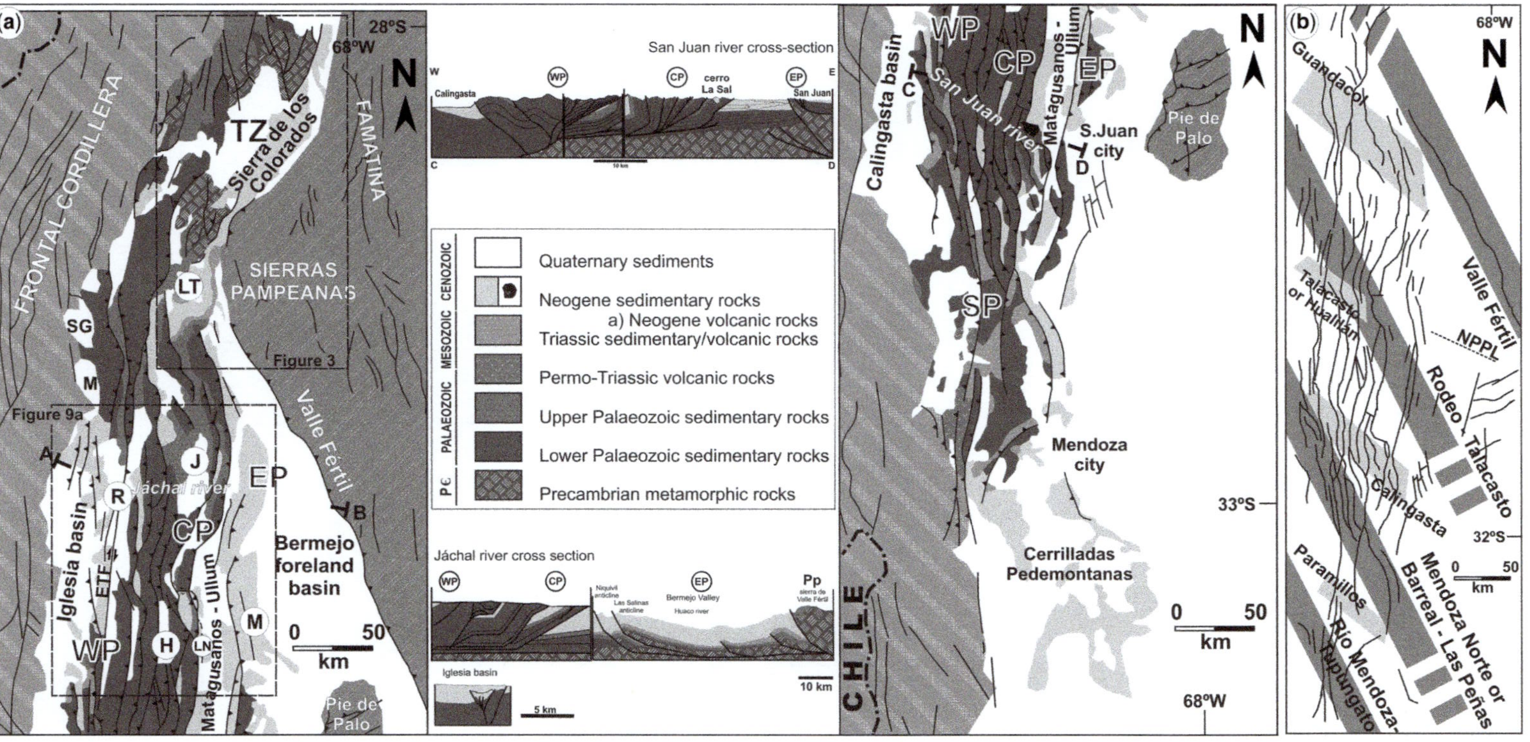

**Fig. 2.** (**a**) The Argentine Precordillera: simplified geological map (after Caminos *et al.* 1993; Ragona *et al.* 1995; Zapata & Allmendinger 1996*a*; SEGEMAR 2012 http://sig.segemar.gov.ar/) and regional cross-sections (San Juan river section after Ramos *et al.* 1986; Cristallini & Ramos 1995; Jáchal river section after Allmendinger *et al.* 1990; Álvarez-Marrón *et al.* 2006; Pérez & Bagur Delpiano 2012). Abbreviations as for Figure 1 plus: LT: La Troya depocentre; SG: Pampa de San Guillermo; M: Pampa del Médano; R: Rodeo town; J: Jáchal-El Fiscal basin; H: Pampa de Hualilán; LN: Loma Negra–Matagusanos brachianticline; ETF: El Tigre fault; and Pp: Sierras Pampeanas. (**b**) Oblique (transpressional and transtensional) belts (modified from Japas *et al.* 2012) of Figure 1c.

(limited, homogeneous vertical axis rotations) and that acquired through heterogeneous (localized) deformation on major lithospheric faults (Dupont-Nivet *et al.* 2003).

The main purpose of this study is twofold. First, this investigation constrains Precordillera Cenozoic structural evolution based on new and available palaeomagnetic data. Second, this work proposes a regional geodynamic model of the Pampean flatslab segment based on comparisons with existing palaeomagnetic data from the Andean forearc (Arriagada *et al.* 2008, 2009) to test the validity of the Central Andes Rotation Pattern (CARP, Somoza *et al.* 1996) along this Central Andean segment.

## Geological setting

Composed of different tectonostratigraphic units, the 27–33° S Andean segment shows major crustal-scale anisotropies (Alvarado *et al.* 2007; Gans *et al.* 2011; Ammirati *et al.* 2013) that resulted from the ancient accretion and amalgamation of different terranes (Pampia, Cuyania, Chilenia; Ramos 1999*a*; Ramos *et al.* 2002; Fig. 1a)

In Argentina, the Pampean flat-slab segment comprises four main morphotectonic units: the Principal Cordillera, the Frontal Cordillera, the Precordillera and the Sierras Pampeanas (Fig. 1a). The Principal Cordillera embraces the northern thick-skinned La Ramada (Cristallini 1996) and the southern thin-skinned Aconcagua (Cegarra & Ramos 1996) fold and thrust belts (Fig. 1b). The Frontal Cordillera and the Sierras Pampeanas represent dominantly Palaeozoic basement inliers with thick-skinned deformation (Heredia *et al.* 2002; Ramos *et al.* 2002), whereas the Precordillera is a thin- and thick-skinned fold and thrust belt (Allmendinger *et al.* 1990; Cristallini & Ramos 2000).

### *The Precordillera*

The Precordillera is considered to be an allochthonous terrane (Cuyania, Ramos *et al.* 1996; Occidentalia, Dalla Salda 2005; Ramos & Dalla Salda 2011) that was accreted to the continental SW margin of Gondwana during the Early Palaeozoic (Astini *et al.* 1996; Fig. 1a).

Although unexposed, the Precordillera basement is considered to be Grenvillian in age, as revealed by metamorphic xenoliths in Miocene volcanic rocks (Leveratto 1968; Abbruzzi *et al.* 1993; Kay *et al.* 1996). The Palaeozoic rocks of the Precordillera record the evolution from a Cambrian–Ordovician passive-margin to a Silurian–Devonian active-margin, culminating in a post-collisional history of Carboniferous–Permian active subduction. From the Cambrian to the Early Permian, sediments were deposited and successively deformed during different orogenic events (the Middle–Late Ordovician Famatinian–Ocloyic orogeny, the Late Devonian Famatinian–Chanic orogeny and the Early Permian San Rafael orogeny). Geodynamic changes linked to plate rearrangements resulted in Triassic rifting. After the Late Cretaceous inception of the Andean orogeny, this region evolved as part of the Bermejo foreland basin until it began to be disrupted at *c.* 19 Ma (Jordan *et al.* 1993; Alonso *et al.* 2011) during the Neogene eastwards migration of the orogenic front; the latter was a consequence of the flattening of the Nazca plate beneath South America (Pampean flat-slab stage; Table 1).

Major structural differences allow for the discrimination between the Western (WP), Central (CP) and Eastern Precordillera (EP) (Ortiz & Zambrano 1981; Allmendinger *et al.* 1990; Cristallini & Ramos 2000) and the Southern Precordillera (SP) (Cortés *et al.* 2005, 2006; Figs 1c & 2a).

The Western Precordillera region is defined by thin-skinned east-verging thrusts detached at the deep-marine Ordovician beds. A major structure of this sector is the El Tigre fault (ETF in Fig. 2a; Bastías *et al.* 1984; Bastías 1985; Siame *et al.* 1997; Terrizzano *et al.* 2009; Fazzito 2011), a N10° E-striking strike-slip fault that accommodates right-lateral motions (at least since the Pliocene; Siame *et al.* 1997, 2005; Cortés *et al.* 1999). To the west, the Iglesia-Calingasta piggyback basin separates the Western Precordillera from the Frontal Cordillera (Allmendinger *et al.* 1990; Beer *et al.* 1990; Figs 1c & 2a). To the east, the Central Precordillera also shows thrusts verging to the east, with the detachment level localized on Cambrian and Early Ordovician carbonates. In this sector, outcrops of Late Palaeozoic, Late Mesozoic and Cenozoic units are common. East of the Matagusanos–Ullum valley (Figs 1c & 2a), the Eastern Precordillera shows thick-skinned west-verging thrusts and folds. Folded Neogene rocks crop out to the north, whereas south of San Juan city Cambrian platform units are exposed through thrusts (Pérez *et al.* 2012). A thick-skinned triangular zone appears between both thin- and thick-skinned belts (Zapata & Allmendinger 1996*a*; Fig. 2a). The southern Precordillera (Fig. 2a) is dominated by NNW-striking structures that resulted from inversion and reactivation of Palaeozoic and Triassic major structures (Cortés *et al.* 2005, 2006; Terrizzano *et al.* 2010).

The existence of heterogeneous deformation linked to localized, cross-strike and oblique deformational zones was highlighted by Japas (1998), Ré *et al.* (2001), Japas *et al.* (2002), Cortés & Cegarra (2004), Cortés *et al.* (2005, 2006), Japas & Ré (2012*a, b*) and Oriolo *et al.* (2014, 2015). These deformational belts were considered as

**Table 1.** *Cenozoic stratigraphy of the Precordillera and the Sierra de los Colorados transitional zone*

| Era | Period | Epoch | Stage | Iglesia basin | Western Precordillera | Central Precordillera | Eastern Precordillera | Southern Precordillera | Transitional Zone | Tectonic setting Rock sequence | Cycle | Tectonic events |
|---|---|---|---|---|---|---|---|---|---|---|---|---|
| CENOZOIC | Quaternary | Holocene | | Alluvial, colluvial, eolian and evaporitic sediments | | | | | | Retroarc foreland basin Synorogenic deposits and arc magmatism | Andean | Andean Orogeny — Flat-slab stage |
| | | Pleistocene | | Alluvial, colluvial, eolian and evaporitic sediments | | | | | | | | |
| | Neogene | Pliocene | Late | Rodeo Fm | | Mogna Formation | Mogna Formation | Mogotes Formation | Santa Florentina Formation | | | |
| | | | Early | | | | Río Jáchal Formation Quebrada del Cura Formation | Cerro Redondo andesites La Pilona Formation Mariño Formation | Toro Negro Formation; Vinchina Formation (upper section) | | | |
| | | Miocene | Late | | Iglesia Group and eq. | El Corral and Loma de las Tapias formations | Huachipampa Formation Quebrada del Jarillal Formation | | Vinchina Formation (lower section) | | | |
| | | | Middle | | andesites and dacites | Cuculí and Albarracín formations; andesites and dacites | Río Salado Formation | | | | | |
| | | | Early | | | | | | | | | |
| | Palaeogene | Oligocene | | Las Trancas Formation | | Río Huaco and del Áspero groups | | Divisadero Largo Formation | Quebrada de la Montosa Formation Vallecito Formation Puesto La Flecha Formation | | | Normal-slab stage |
| | | Eocene | | Valle del Cura Formation | | | | | | | | |
| Mz | Cretaceous | Late | | | | Ciénaga de Huaco Formation | (Ciénaga de Huaco Formation) | | | | | |

Data compiled from Ragona *et al.* (1995); Ramos (1999*b*); Zencich *et al.* (2008); Giambiagi *et al.* (2010); Alonso *et al.* (2011); Ciccioli *et al.* (2011, 2014); Rodríguez Brizuela & Tauber (2006). Light grey shadow refers to the Sierra de los Colorados stratigraphy.

brittle-ductile (following Ramsay & Huber 1987) megashear zones (Fig. 2b), and their motions are consistent with the regional N76° E convergence direction defined by DeMets *et al.* (1990). Based on strain fabrics, localized WNW deformational zones were defined (Ré *et al.* 2001; Japas *et al.* 2002; Fig. 2b). They represent sinistral transtensional zones, comprising the Guandacol (see also Chernicoff & Nash 2002), the Talacasto (Ré *et al.* 2001), the Calingasta (Japas *et al.* 2012) or San Juan (Ré *et al.* 2001), and the Paramillos belts (Japas *et al.* 2014a). Recently the Talacasto deformational zone was considered as a cross-strike structure by Oriolo *et al.* (2011, 2012, 2014, 2015), and Oriolo (2012) named it the Hualilán belt. Further to the east, in the Tertiary Volcanic Belt in the Sierras Pampeanas broken foreland (Urbina & Sruoga 2009; Japas *et al.* 2010), Neogene magmatic rocks are confined to these WNW corridors that seem to have played an active structural role in magmatism emplacement during its migration into the foreland (Oriolo *et al.* 2014).

On the other hand, localized NNW-trending oblique belts consist of transpressional structures with components of left-lateral motions as deduced from local deflections of structures at the regional scale (Japas 1998; Ré *et al.* 2001), crustal seismicity (Ré *et al.* 2001), quaternary folding (Cortés & Cegarra 2004; Japas *et al.* 2011) and kinematic indicators (Japas *et al.* 2011). This left-lateral NNW-trending transpressional structures comprise the following four belts (Fig. 2b): the Valle Fértil (Rossello *et al.* 1996; Introcaso & Ruiz 2001); the Rodeo–Talacasto (Japas *et al.* 2011); the Mendoza Norte (Ré *et al.* 2001) or Barreal–Las Peñas (Cortés & Cegarra 2004, Cortés *et al.* 2005); and the Río Mendoza–Tupungato (Cortés *et al.* 2005).

Since the palaeomagnetic database is more robust in the vicinity of the Rodeo–Talacasto belt (Fig. 2b), our analysis will focus on these NNW-trending transpressional structures. The Rodeo–Talacasto belt was originally recognized because of the en échèlon arrangement of a set of thrust-controlled exposures of synorogenic Neogene rocks that are also aligned with the NNW-trending Loma Negra–Matagusanos brachianticline (Japas *et al.* 2011; Fig. 2). Along this NNW-trending belt, strike of Andean regional thrusts and folds changes from NNE–SSW to north–south/NNW–SSE, and NNE-oriented outcrops of Palaeozoic sedimentary rocks are deflected anticlockwise (Japas *et al.* 2011; Fig. 2).

## Sierra de los Colorados area

The thick Neogene sedimentary sequence exposed at the Sierra de los Colorados (Fig. 3) comprises the Vinchina and Toro Negro Formations (Turner 1964; Ramos 1970). These Andean foreland units from the northern Pampean flat-slab segment crop out following a continuous NNE-trending strip (Fig. 3a, b), dipping to the west with decreasing angle towards the top of the sequence.

Although variable in thickness, the Vinchina and Toro Negro formations reach *c.* 5500 m and 2000 m thick respectively at the Quebrada de La Troya section (Fig. 3c). Recent data from Ciccioli *et al.* (2014) indicate a maximum U–Pb age of 15.6 ± 0.4 Ma (detritic zircons in a tuffaceous interval of the Vinchina Formation Lower Member) and a depositional U–Pb age of 9.24 ± 0.034 Ma (volcanic zircons in a tuff of the Vinchina Formation Upper Member). The Toro Negro Formation was defined as Upper Miocene–Lower Pliocene by Rodríguez Brizuela & Tauber (2006).

Uplift of the Sierra del Espinal–Umango occurred at the time of deposition of the Upper Member of Vinchina Formation and the basal Toro Negro Formation (Ciccioli *et al.* 2010).

## Methods

### *Palaeomagnetic sampling and laboratory procedures*

New palaeomagnetic information from Sierra de los Colorados area (northern Bermejo foreland basin; Fig. 3 & Table 2) as well as available palaeomagnetic data (Table 3) were considered in order to analyse Neogene vertical axis rotations in the Precordillera fold and thrust belt. The Sierra de los Colorados area is a transitional zone that shares a common early basin history until *c.* 10 Ma with La Troya depocentre from the northern Central Precordillera (Ciccioli *et al.* 2011, 2014; Figs 2 & 3). All the units considered in this contribution comprise proximal to distal Neogene continental sedimentary sequences mainly represented by fluvial, lacustrine and alluvial fan deposits. Locally, thicknesses of Neogene strata can reach more than 6000 m (e.g. Sierra de los Colorados and Mogna localities; Fig. 2b).

Data from previous studies (Table 3) which were collected for magnetostratigraphic analysis were reprocessed following the methodology proposed by Beck (1989). In these cases, and following Reynolds (1987), the average of the stratigraphic levels sampled within a stratigraphic interval defines a site; each site contains, for example, 50 stratigraphic levels with at least 5 palaeomagnetic samples per level (Table 3).

In the case of the new data from Sierra de los Colorados area (Table 2), samples were collected for palaeomagnetic study along the Quebrada de La Troya (Fig. 3). In this case, three sections of the Vinchina Formation (Vinchina Formation

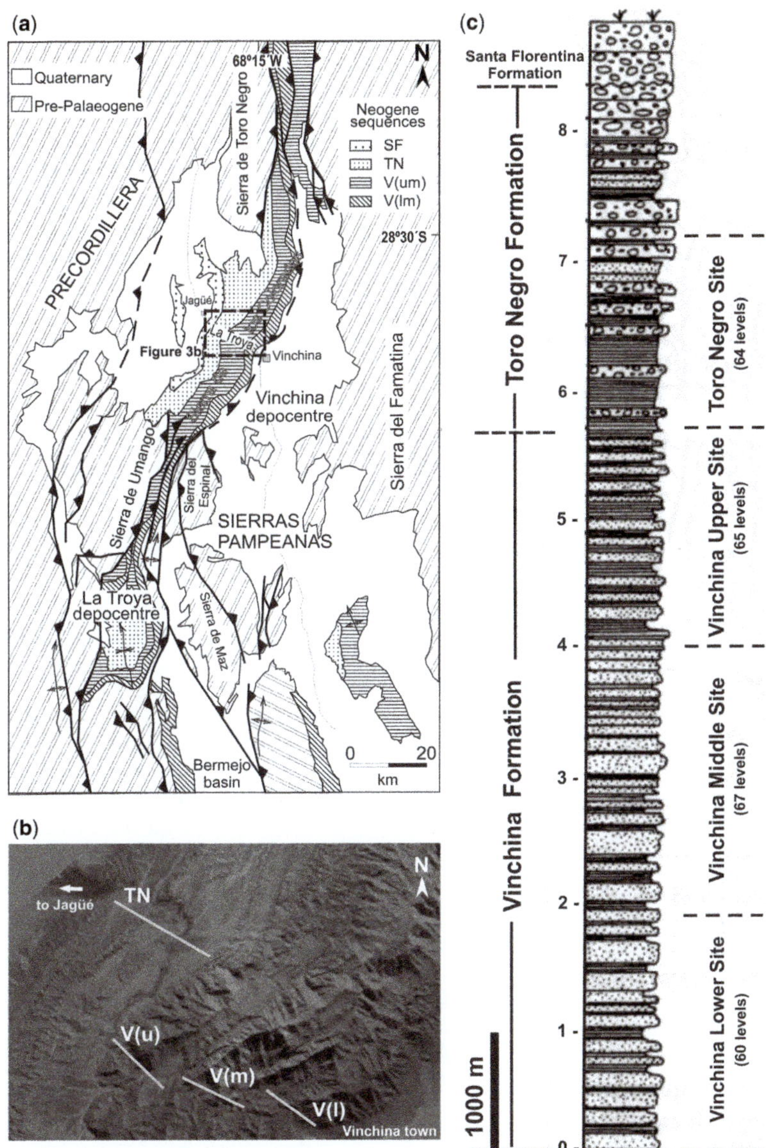



**Fig. 3.** (**a**) The Sierra de los Colorados area (after Ramos 1999b) showing the Quebrada de La Troya sampled profile. V(lm): Lower Member of the Vinchina Formation; V(um): Upper Member of the Vinchina Formation; TN: Toro Negro Formation; SF: Santa Florentina Formation. (**b**) The Quebrada de La Troya area showing the four sampled sections (palaeomagnetic sites). (**c**) Schematic section for the Vinchina and Toro Negro formations in the Quebrada de La Troya area. Number of sampled levels for each site is also indicated.

Upper, Middle and Lower sites) and one in the lower member of the Toro Negro Formation were sampled (Table 1). Oriented hand samples from 256 new levels were taken using sun and magnetic compasses. Vinchina Formation sampling involved 576 specimens from 192 levels with a level interval of 30 m. Sampling comprised 180 samples (60 levels), 201 samples (67 levels) and 195 samples

(65 levels) from the basal, intermediate and upper sections, respectively. On the other hand, a total of 192 specimens were obtained for the Toro Negro Formation (three specimens for all 64 levels) with a level interval of 30 m.

Samples were cut into standard specimens of 2.2 cm length. At this stage the original sampling was reduced by *c.* 15% as some of the block

**Table 2.** *Palaeomagnetic poles from the Vinchina Formation and the Lower Member of Toro Negro Formation*

| Site | $n/N$ | Before tectonic correction | | | | After tectonic correction | | | |
|---|---|---|---|---|---|---|---|---|---|
| | | $I$ | $D$ | $\alpha_{95}$ | $k$ | $I$ | $D$ | $\alpha_{95}$ | $k$ |
| Lower Vinchina Formation | 158/129 | 45.3 | 265.3 | 16 | 13.2 | 41.1 | 208.7 | 8 | 19.9 |
| Middle Vinchina Formation | 163/137 | 49.2 | 265.1 | 21 | 17.8 | 50.6 | 206.9 | 9 | 19.1 |
| Upper Vinchina Formation | 173/153 | 47.1 | 188.6 | 13 | 10.6 | 49.8 | 193.8 | 5 | 28.9 |
| Lower Toro Negro Formation | 148/131 | 55.5 | 201.9 | 12 | 9.2 | 45.0 | 187.3 | 5 | 27.1 |

$n$, original specimen number; $N$, selected specimen number; $k$, Fisher statistical parameter; $\alpha_{95}$, semi-angle of 95% confidence cone

samples were broken in the process of drilling. Remanent magnetization was measured using a three-axis 2G DC squid cryogenic magnetometer. Alternating field demagnetization (AF) was carried out to a maximum of 110 mT (peak) using a static 2G600 demagnetizer attached to the magnetometer. Step-wise thermal demagnetization up to 650°C was performed using either a two-chambers ASC or Schonstedt TSD-1 oven at the Laboratorio de Paleomagnetismo Daniel Valencio (Departamento de Ciencias Geológicas of the Universidad de Buenos Aires).

A total of 642 specimens belonging to 217 levels (Fig. 3) from the Vinchina and Toro Negro formations were subjected to stepwise demagnetization to examine the coercivity and unblocking temperature spectra of the natural remanent magnetization (NRM). Bulk magnetic susceptibility was measured after each thermal step by using a MS2W Bartington susceptometer in order to monitor possible magnetic mineral changes.

The magnetic behaviour of each specimen was analysed by visual inspection of Zijderveld plots, stereographic projections and intensity demagnetization curves. Magnetic components were determined by using principal component analysis (PCA, Kirschvink 1980). Components were determined with at least four consecutive demagnetization steps and only those defined by a maximum angular deviation (MAD) value lower than 15° were accepted. Programs SuperIAPD (Torsvik 1992) and MAG88 (Oviedo 1989), both applying Fisher (1953) statistics, were used.

## Palaeomagnetic results

Data from the Sierra de los Colorados area were recalculated from Ré (2008) in order to obtain four independent palaeomagnetic poles (three for the different sections of the Vinchina Formation and one for the Toro Negro lower member), and contrasted them to the present-day pole. Magnetic susceptibility of the Vinchina Formation ranges from 8 to $170 \times 10^{-5}$ (SI). Although higher in some levels (up to 110 mA/m), the NRM varies between 0.3 and 60.0 mA/m. Each member is dominated by NRM values <11 mA/m, with geometrical mean of 3–4 mA/m.

Most of the samples showed only one characteristic component of remanence that, in some cases, is accompanied by a softer component easily erased after the first steps of demagnetization. Alternating field and thermal demagnetization procedures allowed samples to be classified as follows (Fig. 4):

(1) Type A: Magnetic remanence was erased after applying AF peaks higher than 40 mT, usually around 90 mT, with a geometrical mean of 55 mT. Thermal demagnetization of these samples showed unblocking temperatures of 580°C, with no changes in magnetic susceptibility during the procedure, pointing to (pure) magnetite as the magnetic mineral. A total of 295 specimens (40% of the processed samples) belong to this group (Fig. 4a).

(2) Type B: Magnetic remanence showed coercive forces lower than Type A, as AF procedures eliminated the NRM after peaks of 25–50 mT (geometrical mean of 30 mT). Thermal demagnetization showed unblocking temperatures varying over the range 300–400°C (up to 550°C), with no significant changes in the magnetic susceptibility during the procedure (Fig. 4b). Considering both, the oxidant depositional environment and the obtained unblocking temperatures, (Ti) magnetite is considered to be the magnetic carrier in these samples. This includes 255 specimens (*c.* 35%).

(3) Type C: Samples of this group became unstable with applied AF peaks lower than 30 mT and temperatures at *c.* 200°C. This behaviour characterizes 116 specimens (*c.* 25%) which were discarded from further analysis (Fig. 4c).

Demagnetization and IRM curves reveal magnetite in most of the samples, whereas hematite was present at the lower section of the Vinchina (Fig. 5).

Many of the sites contain specimens showing all three kinds of behaviour, indicating that the

**Table 3.** *Tectonic rotations measured for Neogene rocks in the Precordillera fold and thrust belt*

| Sites | Age (Ma) | Latitude (°S) | Longitude (°E) | Calculated pole (latitude) | Calculated pole (longitude) | Calculated inclination | Calculated declination | $\alpha_{95}$ | $R$ | $\Delta R$ (error) | $P$ | $\Delta P$ (error) | References |
|---|---|---|---|---|---|---|---|---|---|---|---|---|---|
| Santo Domingo (SD) | 37 | −28.5 | 291.2 | −75.0 | 184.5 | NA | NA | 6 | 16 | NA | NA | NA | Vizán *et al.* (2013) |
| Las Juntas (J) | 10–18 | −29.4 | 291.4 | −76.7 | 22.6 | 47.9 | 15.2 | 5 | 15 | 6 | 0 | 5 | Reynolds (1987)* |
| Río Mañero(Mñ) | 6–15 | −29.7 | 292.2 | −84.7 | 180.3 | 54.1 | 2.4 | 4 | 2 | 6 | 5 | 5 | Reynolds (1987)* |
| Mogna (M) | 2–18 | −30.5 | 291.5 | −85.2 | 322.7 | 54.0 | 357.0 | 4 | 3 | 6 | 4 | 5 | Milana (1991)* |
| Mendoza (Mz) | 14–16 | −32.5 | 291.1 | −66.1 | 229.8 | 34.7 | 337.9 | 6 | 22 | 7 | 13 | 6 | Costa (1994)* |
| Tucumanesas (Tu) | Neogene | −29.2 | 291.9 | −81.0 | 260.9 | 56.2 | 5.8 | 8 | 6 | 9 | 8 | 7 | Puente (1992)* |
| Angualasto (A) | 5–9 | −30.0 | 290.9 | −82.7 | 2.3 | 46.1 | 7.8 | 7 | 8 | 8 | 3 | 7 | Ré & Barredo (1993)* |
| Rodeo (R) | c. 8 | −30.1 | 291.0 | −83.9 | 208.7 | 48.1 | 353.1 | 11 | 7 | 11 | 1 | 9 | Ré & Barredo (1993)* |
| Río Azul (RA) | 8–17 | −30.4 | 291.4 | −84.2 | 278.8 | 55.5 | 1.1 | 6 | 2 | 7 | 6 | 6 | Jordan *et al.* (1990)* |
| Las Flores (F) | c. 16 | −30.2 | 290.9 | −49.6 | 24.3 | 42.5 | 45.4 | 12 | 45 | 12 | 6 | 10 | Stein (1994)* |
| Tudcum (T) | c. 7 | −30.1 | 290.8 | −74.9 | 47.7 | 55.4 | 16.7 | 17 | 17 | 17 | 6 | 14 | Balbi (1995)* |
| Site 20 (Vinchina) (V) | c. 17 | −28.7 | 292.0 | −69.2 | 23.9 | 56.3 | 202.9 | 5 | 23 | 7 | 8 | 6 | Urreiztieta (1996), Aubry *et al.* (1996)[†] |
| Site 21 (Vinchina) (V) | c. 17 | −28.7 | 292.0 | −74.6 | 31.5 | 49.2 | 197.6 | 8 | 18 | 9 | 1 | 8 | Urreiztieta (1996), Aubry *et al.* (1996)[†] |
| Lower Vinchina Formation (V) | c. 16 | −28.5 | 291.8 | −63.8 | 197.6 | 41.1 | 208.7 | 8 | 29 | 8 | 5 | 7 | Ré (2008), recalculated in this work |
| Middle Vinchina Formation (V) | c. 12 | −28.5 | 291.8 | −66.6 | 215.4 | 50.6 | 206.9 | 9 | 27 | 9 | 3 | 8 | Ré (2008), recalculated in this work |
| Upper Vinchina Formation (V) | c. 10 | −28.5 | 291.8 | −77.7 | 35.1 | 49.8 | 193.8 | 5 | 14 | 7 | 2 | 6 | Ré (2008), recalculated in this work |
| Toro Negro Formation (V) | c. 6 | −28.5 | 291.8 | −83.2 | 6.8 | 45.0 | 187.3 | 5 | 7 | 6 | 2 | 5 | Ré (2008), recalculated in this work |
| Site 22 (22) | 8–10 | −30.3 | 291.6 | −88.4 | 152.7 | 48.0 | 181.2 | 6 | 1 | 7 | 1 | 6 | Urreiztieta (1996)– Aubry *et al.* (1996)[†] |
| Site 23 (23) | 10–14 | −30.8 | 291.6 | −77.1 | 143.9 | 35.5 | 187.3 | 7 | 7 | 7 | 11 | 7 | Urreiztieta (1996), Aubry *et al.* (1996)[†] |
| Site 24 (24) | 10–18 | −30.8 | 291.6 | −77.7 | 158.8 | 39.0 | 189.7 | 8 | 10 | 8 | 9 | 8 | Urreiztieta (1996), Aubry *et al.* (1996)[†] |
| Site 25 (25) | 10–18 | −30.8 | 291.6 | −65.4 | 172.4 | 31.3 | 203.3 | 7 | 22 | 7 | 14 | 7 | Urreiztieta (1996), Aubry *et al.* (1996)[†] |
| Huaco N (HN) | 3–10 | −30.1 | 291.6 | −56.2 | 205.7 | 45.3 | 218.9 | 8 | 38 | 8 | 3 | 7 | Beer (1990)* |
| Huaco C (HC) | 3–10 | −30.2 | 291.6 | −79.1 | 191.9 | 46.5 | 192.1 | 4 | 12 | 6 | 2 | 5 | Johnson *et al.* (1984)* |
| Huaco S (HS) | 3–10 | −30.3 | 291.6 | −77.9 | 188.9 | 45.4 | 193.3 | 13 | 13 | 13 | 3 | 11 | Beer (1990)* |
| Lomas del Inca (LI) | 5–10 | −31.5 | 290.8 | −74.8 | 126 | 39.6 | 193.7 | 8 | 13 | 5 | 11 | 5 | Basile (2004) |
| Loc 20-24 (TN2) | 5–11 | −30.4 | 290.8 | −70.9 | 193.8 | −44.7 | 21.3 | 8 | 21 | 9 | −4 | 7 | Fazzito (2011) |
| Loc 18-17-19-21 (TN1) | 5–11 | −30.4 | 290.8 | −68.6 | 309.7 | −67.4 | 349.4 | 7 | −10 | 16 | 17 | 6 | Fazzito (2011) |
| Loc A (TC) | 5–11 | −30.8 | 290.8 | −24.7 | 244.8 | −66.6 | 83.8 | 8 | 83 | 17 | 16 | 7 | Fazzito (2011) |
| Loc D (TC) | 5–11 | −30.8 | 290.8 | 40.6 | 313 | −29.3 | 17.3 | 13 | 17 | 12 | −20 | 11 | Fazzito (2011) |
| Loc G8 (TC) | 5–11 | −30.8 | 290.8 | −17.3 | 225.9 | −45.1 | 286.8 | 8 | −73 | 10 | −4 | 7 | Fazzito (2011) |

Letters in parentheses indicate locality. For additional information (number of samples, etc.) the reader is referred to the original contribution. Reference: actual-pole (90–360°). Dark-grey cells: new results from the Sierra de los Colorados area, comprising the Lower, Upper and Middle Vinchina Formation and the Toro Negro Formation. Pale-grey cells: data obtained from magnetostratigraphic analysis. *Reprocessed by Ré (2008); [†]Recalculated in Ré (2008). NA, not available in Vizán *et al.* (2013).

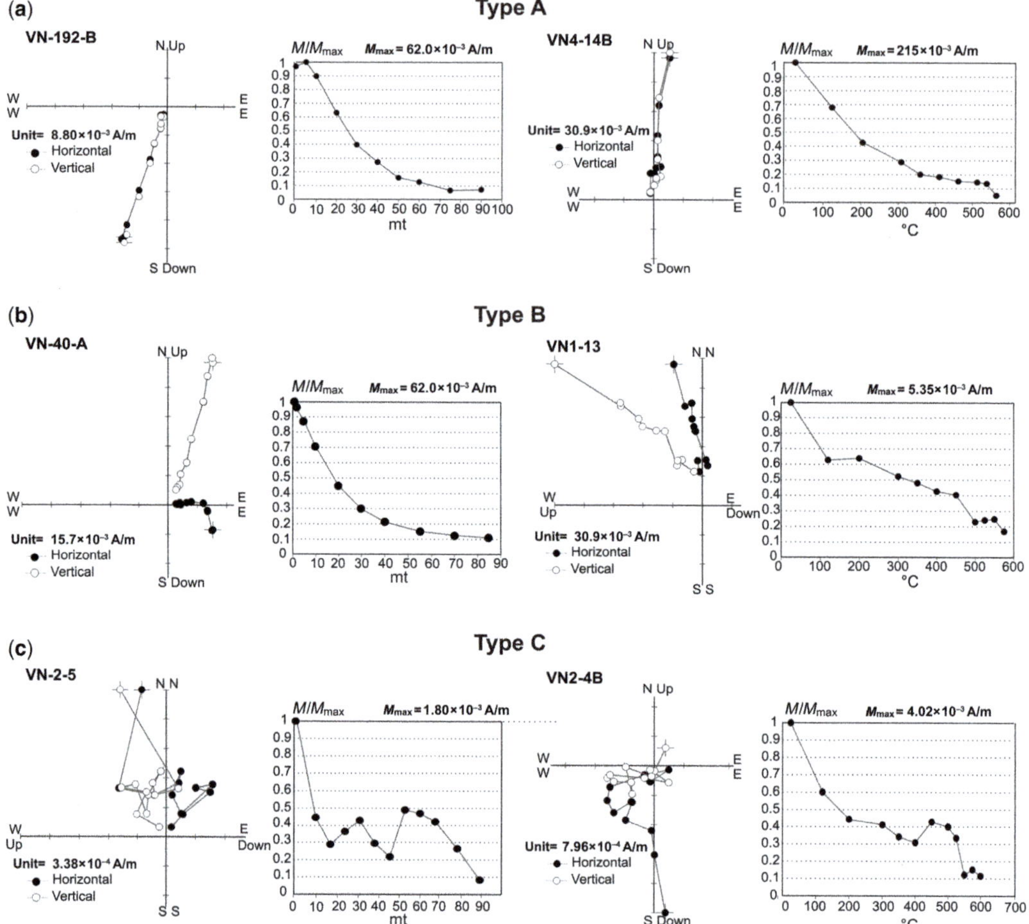

**Fig. 4.** Typical magnetic behaviour of Vinchina and Toro Negro Formation samples, illustrated in Zijderveld orthogonal plots and demagnetization curves. The left and right sides show the response to AF and thermal demagnetization, respectively. Open (solid) symbols: vertical (horizontal) projection. (**a**) Type A behaviour (normally magnetized and reversely magnetized) sample; (**b**) Type B behaviour; and (**c**) Type C behaviour. mt: millitesla.

different behaviours respond to local variation in the nature of the magnetic minerals (i.e. Ti content, degree of oxidation and effective grain size).

As Type A behaviour shows the highest coercive force for the magnetic minerals, these specimens are considered the most reliable record of the past magnetic field. Consequently, only Type A specimens were used to calculate site mean directions, termed 'Class A'. For sites where Type A specimens were absent, the site mean direction was calculated using Type B specimens and consequently termed 'Class B'.

The four sampled sections show site mean directions of positive and negative inclination (i.e. reversed and normal polarity; Fig. 6). Magnetization was considered as pre-tectonic by Ré (2008)

because the best grouping of mean directions arises after structural correction. Neither folding nor conglomerate tests could be performed because the nearly monoclinal westwards-dipping sequence of mud- and sandstones was unsuitable for running the stability test. However, the occurrence of the two polarities of the magnetic field and the mean inclination observed in the four sites indicate an absence of regional remagnetization.

Virtual geomagnetic poles (VGP) were calculated from the site mean directions. The palaeomagnetic pole for each member was calculated as the mean of Class A sites VGPs, with a cutoff angle (45°) established by the method of Vandamme (1994). VGPs showing an angular deviation to the mean pole greater than the cutoff were considered

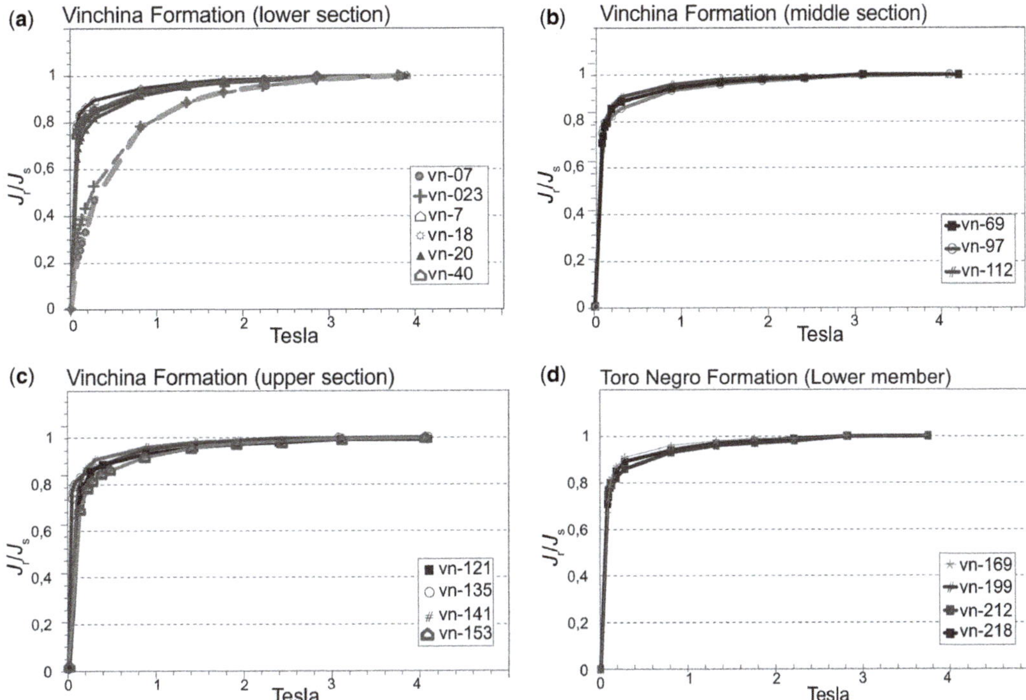


**Fig. 5.** Typical isothermal magnetization behaviour of Vinchina and Toro Negro Formation samples: (**a**) Vinchina Formation lower section; (**b**) Vinchina Formation middle section; (**c**) Vinchina Formation upper section; and (**d**) Toro Negro Formation lower member. vn-number refers to original sample identification.

oblique and were not considered further. The remaining VGPs were averaged to compute the mean palaeomagnetic poles (Table 2).

## Precordillera palaeomagnetic database

Vertical axis rotation data for the Precordillera sampled localities are shown in Table 3 and Figure 7, where the reprocessed data from previous magnetostratigraphic studies and the palaeomagnetic rotation data were plotted.

Deposits younger than 2.5 Ma are mainly conglomerates related to proximal synorogenic facies that are inadequate for palaeomagnetic measurements; rotation parameters linked to deformation of <2.5 Ma are therefore scarce.

The northern sector of the Precordillera was more extensively sampled, so the vertical axis rotation analysis will be centred on this area. The palaeomagnetic data in Figure 7 reveal a strong heterogeneous rotation pattern: Neogene units ranging from >18 Ma to 2.6 Ma display variable tectonic rotations (Fig. 7b). As the highly variable spectrum shown by the palaeomagnetic data from the central sector of the El Tigre locality seems to be indicative of proximity to the El Tigre fault

(TC sites − 73–83°; Table 3 & Fig. 7a); these data will be not consider in the analysis.

From this database, the following considerations can be made.

- When compared with other Andean foreland segments, palaeomagnetic results at the 28–33° S reveal a rather unique pattern, confirming the hypothesis of Somoza *et al.* (1996). Instead of a homogeneous Neogene clockwise rotation pattern, the foreland at the Pampean flat-slab segment shows some localized areas registering no significant block rotation (Fig. 7a).
- When assembled by rock age, Neogene Precordillera palaeomagnetic data can be grouped into an old Early Miocene A population (*c.* 18–11 Ma), an intermediate Late Miocene–Early Pliocene B population (*c.* 11–4 Ma) and a young Late Pliocene–Early Pleistocene C population (*c.* 4–2 Ma) (Fig. 7b). Tectonic block rotation values for the A group reveal clockwise rotations ranging from 45° to *c.* 0°. The 11–4 Ma B group shows variable, smaller vertical axis rotation (30 to − 10°). On the other hand, younger strata show no significant palaeomagnetic rotation.

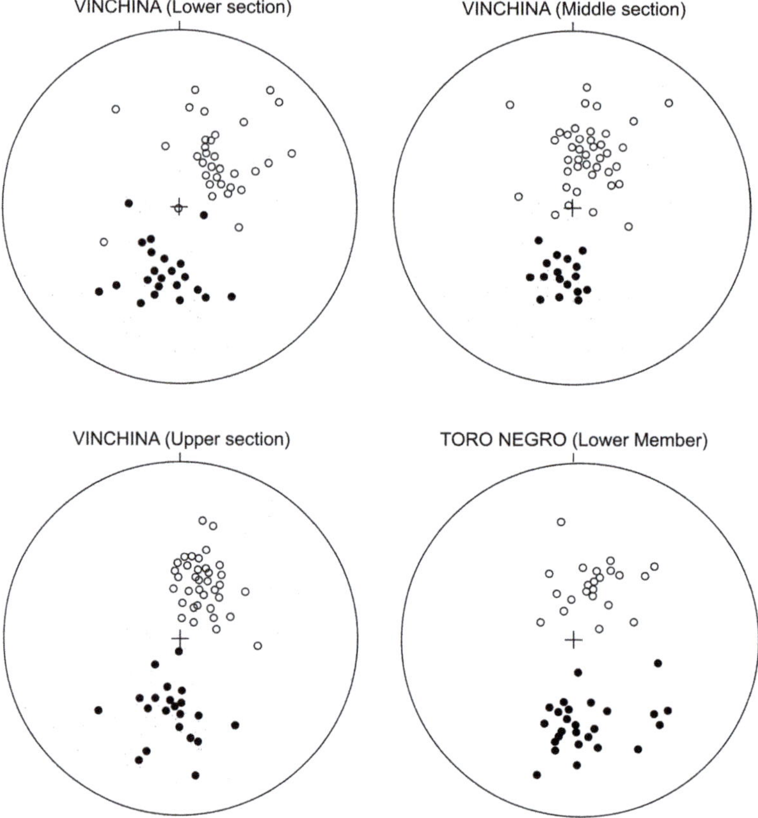

**Fig. 6.** Site mean directions for members of the Vinchina and Toro Negro formations (stereonet plots). Open/solid symbols: negative/positive inclination. Data after tectonic correction.

- When compared to each other, vertical rotation patterns from the Western-Central and Eastern Precordillera show some significant differences. In the Western-Central Precordillera, clockwise rotations are representative of those localities showing NNE-trending structures; nearly null rotations were obtained for those localized regions showing *c.* north–south-trending structures (Fig. 8a, b). Conversely, in the Eastern Precordillera, clockwise rotations were recognized in regions affected by *c.* north–south-trending structures and negligible rotations were detected at those areas showing NNE-striking structures (Fig. 8a, b).

## Interpretation and discussion

### Vertical axis rotations in the Pampean flat-slab foreland

Analysing the vertical axis rotations in the Precordillera oblique belt context, some conclusions can be made (Fig. 7). Instead of a non-systematic rotation distribution as described by Ré & Rapalini (1995), an organized, yet complex, pattern appears to be evident. In the Western and Central Precordillera, those localities showing no Neogene rotation are confined to the NNW-trending transpressional belts, whereas in the Eastern Precordillera those localities undergoing null rotation are localized out of the NNW-trending belts. Two distinctive regional palaeomagnetic domains can therefore be defined: the Western-Central Precordillera and the Eastern Precordillera.

*Western-Central Precordillera.* Palaeomagnetic data from the Rodeo area allows for the recognition of three local subdomains (Fig. 8a): a northern subdomain showing non-significant vertical axis rotation (Angualasto and Rodeo localities); a central subdomain (Tudcum) indicative of low but significant clockwise vertical axis rotation; and a southern subdomain (Las Flores and TN2: Loc 20–24; Table 3; Figs 7 & 8a) with the highest clockwise

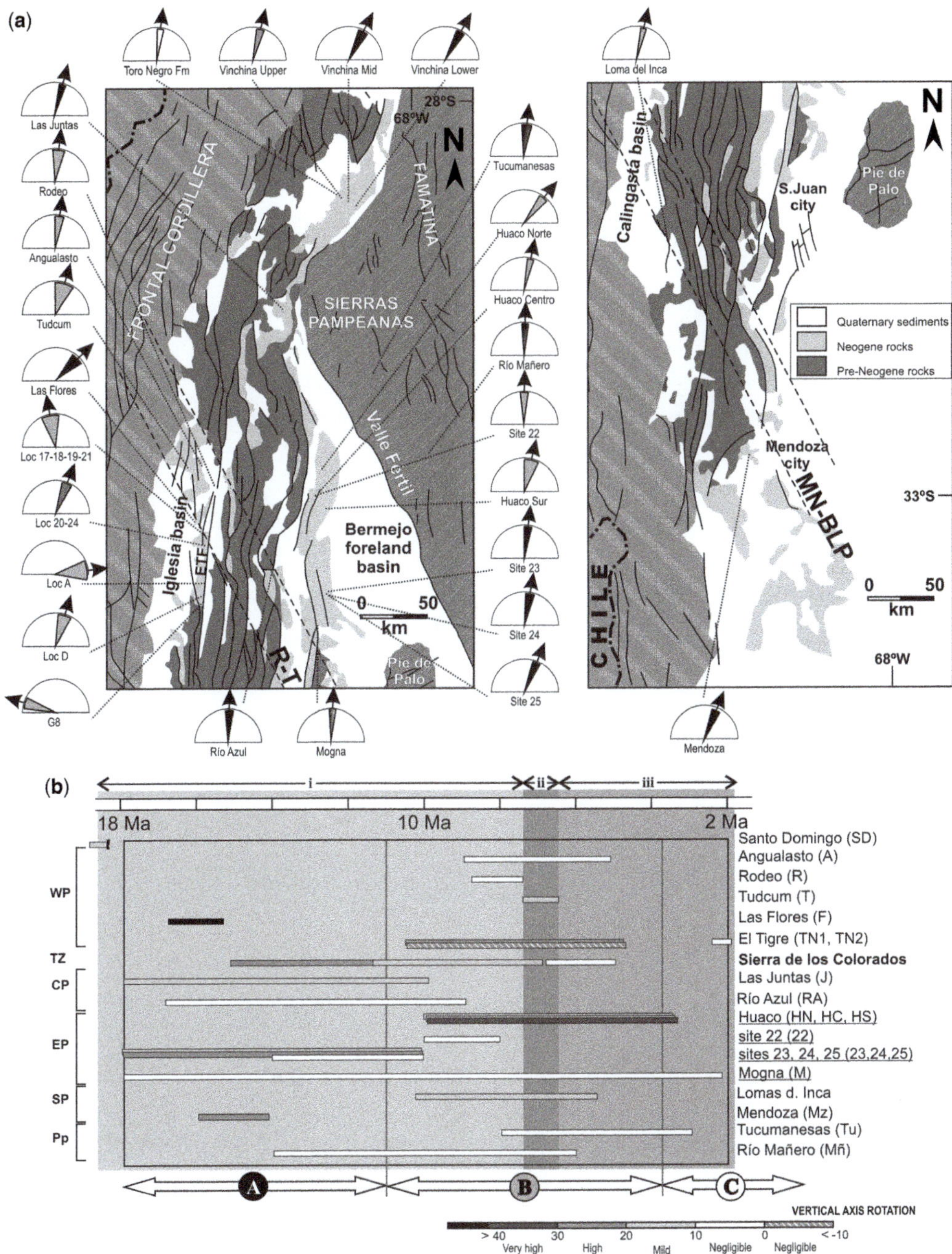


**Fig. 7.** Palaeomagnetic data from Table 3 discriminated by age of the considered strata and rotation magnitude. (**a**) Vertical axis rotation data in the geological context. Data from sites 20 and 21 are omitted from this figure because they are equivalent to the new results from the lower Vinchina Formation sector in this study. Colours in cones representing the declination error of the mean directions indicate A (black), B (grey) and C (white) populations. Dashed lines denote the Rodeo–Talacasto (R-T) and the Mendoza Norte/Barreal–Las Peñas (MN-BLP) belts. (**b**) Vertical axis rotation data in the foreland sedimentary record context. Grey fields i, ii, iii refer to the foreland basin stages defined by Jordan *et al.* (1999). A, B, C indicate the different palaeomagnetic populations identified (Early Miocene population, Late Miocene–Early Pliocene and Late Pliocene–Early Pleistocene, respectively). Eastern Precordillera localities are underlined. Abbreviations as for Figures 1 and 2.

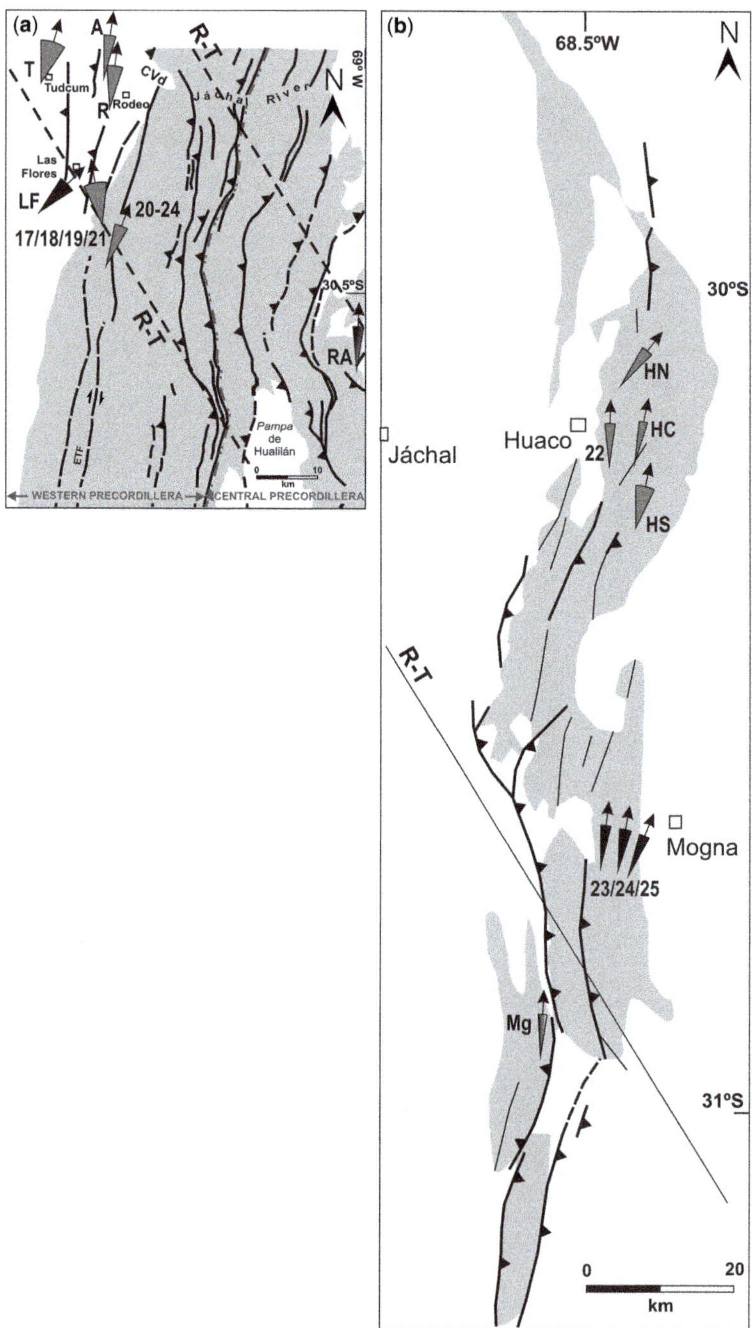



**Fig. 8.** Local maps from the NW and SW Rodeo–Talacasto belt areas showing vertical axis rotation data. Cones represent the declination error of the mean directions; colours in these cones indicate A (black), B (grey) and C (white) populations. Dashed lines denote the Rodeo–Talacasto (R-T) belt. (**a**) Western-Central Precordillera: Rotations revealed by palaeomagnetism are lower inside the Rodeo–Talacasto belt. Structure modified from Cardó *et al.* (1998). ETF: El Tigre fault, interpreted here as the fault segment of the El Tigre Fault System (Bastías *et al.* 1984) located between the Rodeo–Talacasto and the Mendoza Norte/Barreal–Las Peñas belts. CVD: Cuesta del Viento dam. (**b**) Vertical axes rotation data corresponding to Eastern Precordillera showing that the lowest rotation magnitude is to the north of the Rodeo–Talacasto belt and that the highest values are located near this and the Valle Fértil belt (see Fig. 7).

vertical axis rotation values (45° and 22°, respectively). The NNW segment of the Rodeo–Talacasto belt shows transpressional deformation in Neogene sediments, revealed by the NNW-trending positive flower structure described from seismic data (Álvarez-Marrón *et al.* 2006; Fig. 2a) as well as some other localized structural features that are restricted to the Rodeo–Talacasto belt (Cerro Negro de Iglesia and Cuesta del Viento anticlines, en-échèlon-arranged faults with Quaternary activity, particular fracture fabric; Jordan *et al.* 1993; Ragona *et al.* 1995; Costa *et al.* 2000; Perucca & Martos 2009; Alonso *et al.* 2011; Pérez & Costa 2011). All these structures allow us to interpret a localized NNW-trending sinistral transpression (Fig. 9a, b). The tectonic significance of the overprint of the Rodeo–Talacasto belt in this sector could also be recognized by the segmentation imposed on two major structural elements in Western Precordillera; the *c.* 800-km-long El Tigre Fault System (Bastías *et al.* 1984) and the Iglesia basin (Fig. 2a). In the case of the Central Precordillera,

the negligible vertical axis rotation obtained for the Río Azul area seems to be compatible with its position inside the Rodeo–Talacasto belt. To the southeast, the presence of out-of-sequence thrusts in the Talacasto area (Zapata & Allmendinger 1996*a*; Fig. 9a, c) and the strain gradients detected across the Rodeo–Talacasto belt, by the different shortening values, would also indicate overprinted sinistral transpression.

Considering the proposed transpressional conditions, differences in vertical axis rotation values could be linked to the overprint of anticlockwise rotations controlled by the NNW-trending sinistral Rodeo–Talacasto belt. This interpretation is consistent with the very local anticlockwise value obtained by Fazzito (2011) at the southern Rodeo–Talacasto bounding structure (TN1: Loc 17-18-19-21 in Table 3 & Figs 7 & 8a). In this context, the general absence of anticlockwise rotation values should be regarded as the consequence of late overprint of the Rodeo–Talacasto belt on a previous clockwise rotation pattern, and

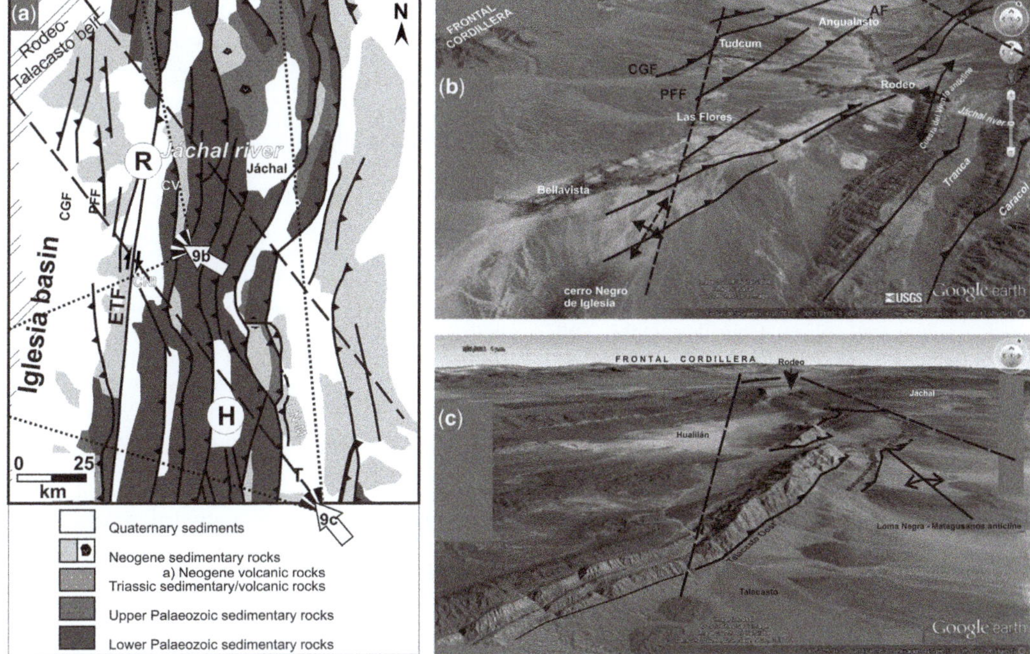


**Fig. 9.** (**a**) The Rodeo–Talacasto belt (see location in Fig. 2a). CGF: Colangüil–Guañizuil fault; PFF: Pismanta–Las Flores fault; AF: Angualasto fault (adapted from Perucca & Martos 2009). Dashed lines define the Rodeo–Talacasto belt. Localized folds: CNI: Cerro Negro de Iglesia; CV: Cuesta del Viento; LN-M: Loma Negra–Matagusanos. (**b**) The northern Iglesia basin (oblique view to the NNW as shown by the white arrow in (a). The Rodeo–Talacasto belt is recognized here based on the local development of Cenozoic anticlines (Cerro Negro de Iglesia and Cuesta del Viento) and en échèlon arrangements of faults with Quaternary activity, among others. (**c**) Panoramic view showing the main aligned out-of-sequence thrusts (view from the SSE as shown by the white arrow in (a)). Dashed lines converging into the horizon represent the Rodeo–Talacasto belt boundaries.

also because rocks younger than 2.5 Ma suitable for palaeomagnetic rotation studies are lacking.

To the south of the Rodeo–Talacasto belt, palaeomagnetic data confirm that rotations from the Late Miocene–Early Pliocene B population are variable but the minimum values arise at those localities within the NNW-trending belts (e.g. Loma del Inca at the Mendoza Norte/Barreal–Las Peñas belt). Although showing higher clockwise rotation than the B population, the Early Miocene A population follows the same relation.

*Eastern Precordillera.* The vertical axis rotation pattern is significantly different from that observed in the Western-Central Precordillera as it is characterized by negligible rotation linked to NNE-trending structures and clockwise rotations linked to *c.* north–south-trending structures (Figs 7b & 8b). The nearly north–south-trending structures are closely related to the Rodeo–Talacasto belt. As in the Central Precordillera, this belt is also characterized by the presence of localized out-of-sequence thrusts that deformed Neogene strata (Pontón Grande thrust; Zapata & Allmendinger 1996*a*).

NNE-trending structures from the Eastern Precordillera are interpreted as guiding the thrust component of the partitioned regional transpression recognized by Siame *et al.* (2005) since the Pliocene, which is in agreement with the low magnitude of rotations detected by palaeomagnetic analysis. Conversely, the *c.* north–south-trending segments of the main Eastern Precordillera structures show a clockwise rotation of *c.* 20°. As these segments seem to be deformed by the sinistral NNW-trending transpressional belts and deformation is basement-involved, it can be concluded that localized rigid block rotation took place. Considering the NNE-striking basement fabric reported by Zapata (1998), these NNW-trending sinistral belts would have triggered right-lateral displacements between NNE-bounded basement blocks. This domino-like mechanism induced local clockwise vertical rotations (see Ron *et al.* 1984, 1986). This is also coincident with the geometry of the décollement described by Zapata (1998) that show strike changes from NNE (near Jáchal) to NNW (close to the Valle Fértil transpressional belt). Furthermore, shortening values provided by Zapata & Allmendinger (1996*a*) indicating higher shortening in NNE–SSW- than in north–south-trending structures would also support this interpretation. The Eastern Precordillera rotation pattern can also be recognized to the west, in the Sierras Pampeanas (Tucumanesas and Río Mañero sites, both showing local NNE–SSW-trending structures; Fig. 7a).

The south-eastern sector of the Rodeo–Talacasto belt could be extended to the south of Sierra Pie de Palo (Fig. 2b) where the presence of a basement high south of Sierra Pie de Palo reported by Ortiz & Zambrano (1981) and Martínez *et al.* (2008) would reveal basement-block uplift segmented by the Rodeo–Talacasto belt.

The present interpretation agrees with the kinematic scenario depicted for Precordillera by Siame *et al.* (2005) and Oriolo *et al.* (2014), indicating deformation partitioned in both motion and time. The two kinematic fields recognized in the Hualilán area by Oriolo *et al.* (2014; Fig. 2a) support an axis reorientation from NE- to WNW-shortening during Precordillera Neogene evolution, and are consistent with as-yet unpublished kinematic results obtained in other areas from Precordillera by the authors. This clockwise reorientation of kinematic axes was previously identified by Siame *et al.* (2005) and determined to have started in the Pliocene (5–2.5 Ma).

## Time constraints for the activation of the Rodeo–Talacasto belt

In the Loma Negra–Matagusanos anticline, structures from the two kinematic stages described in the previous section affected the upper section of the Río Jáchal Formation (*c.* 2.75 Ma, following Furque *et al.* 2003), revealing that this overprinted Rodeo–Talacasto megazone would have been activated from the latest Pliocene onwards, nearly the time of Eastern Precordillera uplift.

Considering a 2.75 Ma age for the activation of the Rodeo–Talacasto belt, the non-significant rotation deduced for the youngest 4–2.0 Ma rocks should be regarded as the result of the overprinting of anticlockwise- to previous slight clockwise rotation. This activation age seems to be concordant with thermochronological data recently obtained for the northern Sierra del Valle Fértil uplift (Ortiz *et al.* 2014).

Overall, the Western-Central Precordillera fold and thrust belt indicates two deformational stages. The first stage (*c.* 18–2.75 Ma) was accommodated through distributed deformation and thickening (clockwise rotation) controlled by the NNE orogenic front, and changed when the sinistral NNW-trending transpressional set of structures were activated during the uplift of the Sierras Pampeanas (broken-foreland stage, Jordan *et al.* 1999). From that time onwards, localized anticlockwise rotations would have locally overprinted the previous clockwise pattern, yielding to localized non-significant rotation or smaller clockwise rotation magnitudes (Fig. 10). Scarce anticlockwise values have been recognized in Neogene rocks: (a) the southern NNW-trending structure that defines the Rodeo–Talacasto belt near Las Flores (TN1; Fazzito 2011); and (b) the central section of the El

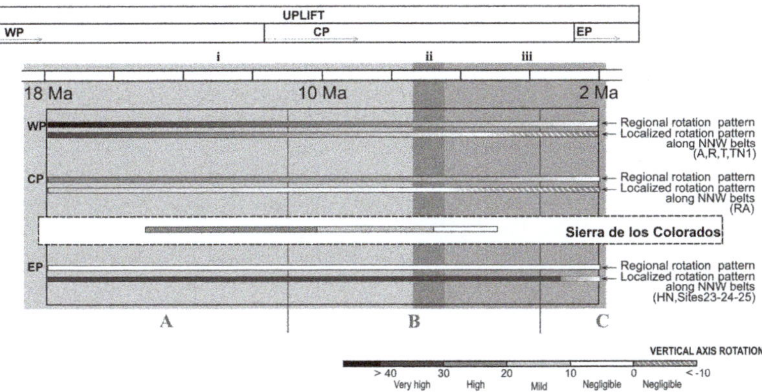

**Fig. 10.** Schematic diagram to explain the observed Precordillera rotation pattern as the result of regional dextral transpressional deformation and late oblique NNW-trending sinistral transpressional belts, in the context of diachronic uplift linked to propagation of deformation towards the foreland. The horizontal top bar for each Precordillera sector (Western, Central, Eastern) indicates rotation linked to the early regional dextral transpressional regime (distributed deformation and thickening), whereas the horizontal bottom bar represents the expected pattern when modified by the overprint of the localized oblique NNW-trending sinistral transpressional belts. Instead of full greys, gradient-fill greys were used in these bars in order to represent the rotation variation trend. Notice the significant correlation between the reference top and bottom bars for each sector of Precordillera with local data from different localities shown in Figure 7b according to their position relative to the late NNW-trending belts. Notice that the Sierra de los Colorados area is coincident with the regional Central Precordillera pattern (compare the Sierra de los Colorados rotation bar with the Central Precordillera top bar that represent the regional rotation pattern without overprint of late oblique NNW-trending belts). Abbreviations as for Figure 1. Age of WP, CP and EP uplifts are indicated following Jordan *et al.* (1993, 2001), Zapata & Allmendinger (1996*a*, *b*), Siame *et al.* (2005) and Alonso *et al.* (2011). For references see Figure 7b. A: Angualasto, R: Rodeo, T: Tudcum, TN1: El Tigre Loc 18-17-19-21; RA: Río Azul; HN: Huaco Norte.

Tigre fault (Fazzito 2011; Fazzito *et al.* 2011; Table 3). These results are partially similar to previous results published by Maffione *et al.* (2009) for the 22–24° S segment of the Eastern Cordillera. Unlike the Eastern Cordillera, oblique structures of the Precordillera comprise regional brittle-ductile megashear zones and not single faults. On the other hand, deformation at the Eastern Precordillera began during the broken-foreland stage defined by Jordan *et al.* (1999), shortly after 2.6 Ma (Zapata & Allmendinger 1996*a*), and was strongly controlled by these transpressional NNW-trending belts. The partitioning of deformation depicted by Siame *et al.* (2005) for the Precordillera would therefore have started in the Late Pliocene and would have been controlled by the NNW-trending transpressional belts. While the Western-Central Precordillera accommodated deformation between the NNW-trending belts through earlier NNE-trending dextral strike-slip structures (e.g. the NNE-striking El Tigre fault that is delimited by the Rodeo–Talacasto and the Mendoza Norte/Barreal–Las Peñas belts), the Eastern Precordillera accommodated deformation through NNE-thrusting. This implies that while the Rodeo–Talacasto belt overprints structures from Western-Central Precordillera, it guides the structural configuration of the Eastern Precordillera. This model seems to be reasonably consistent with the stress regime picture presented by Siame *et al.* (2005) and the recent GPS-derived velocity field proposed by Brooks *et al.* (2003). In this palaeomagnetic domain scenario, vertical axis rotations of the Sierra de los Colorados area (Vinchina and Toro Negro localities) show strong similarities to that of the Western-Central Precordillera domain (Fig. 7b). Our interpretations are also supported by preliminary results from the northern sector of the Sierra de los Colorados where north–south-trending strata reveal null rotation (Ré *et al.*, unpub. data).

## Further tectonic implications

In the Sierra de los Colorados area, deformation began at *c.* 10–12 Ma since both the lower and middle sections of the Vinchina Formation show the same amount of rotation (*c.* 28° clockwise). This allows us to interpret that the early evolution of this region is equivalent to that of the Central Precordillera (Fig. 10), at least until the NNW-trending Valle Fértil belt overprinted and stamped the Sierras Pampeanas structural grain. This overprint would have occurred at *c.* 4–6 Ma (e.g. Ciccioli *et al.* 2010) and is recognizable south of the Quebrada de La Troya area, in the Sierra de Espinal–Umango region (Fig. 3a).

In the Precordillera context, overprinting relationships allow us to consider these NNW-trending belts as late structures; consequently, they could have segmented the El Tigre Fault System as previously mentioned, but also some other main features such as the prime Iglesia piggyback basin that could have been fragmented into a series of minor sub-basins (del Médano–San Guillermo, Iglesia, Calingasta; Figs 1c & 2). In this context, the Pampa de Hualilán and the Jáchal-El Fiscal basins (two along-strike, uplifted small Neogene basins; Fig. 2a) could be derived from a relatively small piggyback basin that was disconnected during the Pliocene by the Rodeo–Talacasto oblique megashear zone.

There seems to be a direct link between Andean NNW-striking megashear zones and the development of out-of-sequence thrusts at the Pampean flat-slab scale. Examples are not only restricted to the Precordillera (Allmendinger *et al.* 1990; Jordan *et al.* 1993; Cristallini & Ramos 1995; Zapata & Allmendinger 1996*a*; Giampaoli & Cegarra 2003), since out-of-sequence thrusts reported in the Principal Cordillera (Cristallini 1996; Cristallini & Ramos 2000; Pérez 2001) and in the Alto Tunuyán foreland basin between Principal Cordillera and Frontal Cordillera (Giambiagi *et al.* 2001) show the same localized setting relative to major NNW-trending shear zones. The out-of-sequence localized belts in the Principal and Frontal Cordillera are aligned with the NNW-trending Vacas–Tupungato transpressional belt recognized by Japas *et al.* (2012). This empirical link could be explained by the presence of a structural obstacle hampering thrust translation and backwards propagation of deformation (see Hayward & Graham 1989; Cristallini & Ramos 2000), producing a more complex pattern than that of a single thin-skinned fold and thrust belt. The uplift of basement blocks of the Sierras Pampeanas and the control of extensional faults linked to the Cuyana basin could have triggered the development of out-of-sequence thrusts constrained to the Rodeo–Talacasto and Mendoza Norte/Barreal–Las Peñas transpressional belts, respectively. This is in agreement with previous proposals (Cortés *et al.* 2005) and recent analogue models (Yagupsky *et al.* 2008; Oriolo *et al.* 2015) in that the structure of pre-Neogene rocks would have controlled the development of the Neogene oblique belts. As mentioned, the Rodeo–Talacasto oblique belt is coincident with a main ancient regional structure exposed in the Frontal Cordillera that seems to have been active during the Late Palaeozoic–Triassic such that it controlled Permian Choiyoi volcanism in the Frontal Cordillera (see Kleiman & Japas 2009). In the case of the Valle Fértil and Mendoza Norte/Barreal–Las Peñas belts, they were controlled by the Pampia–Cuyania terrane boundary and the Cuyana basin border, respectively (Cortés *et al.* 2005).

Additionally, another significant coincidence linked to the presence of a major NNW-trending zone at the Pampean flat-slab segment comprises the reorientation of kinematic axes during the Neogene. A clockwise reorientation can be interpreted from the Principal Cordillera data in Cristallini & Ramos (2000), which show WNW-directed compression along a late NNW-trending belt, in contrast with the ENE-trending axes out of this belt.

## Vertical axis rotation pattern, brittle-ductile megashear zones and margin curvature

At the Pampean flat-slab segment, the rotation pattern in the foreland contrasts with that in the forearc (Japas & Ré 2012*a*, *b*). Between 27° S and south of 33° S, some recognized tectonic rotations in Chile allow Arriagada *et al.* (2009) to partition the forearc region into four segments (Fig. 11a): the north-of-28° S and south-of-33° S segments showing *c.* 30–39° clockwise rotations; the 28–30° S region recording *c.* 18° clockwise rotation; and the 30–33° S sector revealing very low clockwise rotation of *c.* 4°. It should be regarded that, irrespective of the considered segment, there is a constant *c.* 15° relationship between the palaeomagnetic vector and the trend of the mountain front (Japas & Ré 2012*a*, *b*; Fig. 11a).

Based on the Western and Central Precordillera domain pattern, and considering the regional Palaeogene scenario of nearly homogeneous clockwise *c.* 35° rotations in line with the Central Andes Rotation Pattern in the forearc (Arriagada *et al.* 2008), the overprinting of a late stage of WNW-trending sinistral deformation could have induced anticlockwise tectonic rotations reducing Palaeogene clockwise rotation values (Japas & Ré 2012*a*, *b*; Arriagada *et al.* 2013). Based on this and on the presence of younger rotations to the south than to the north (post-Late Miocene and post-Oligocene, respectively; Arriagada *et al.* 2009), as well on differences in foreland–forearc tectonic rotations patterns, Japas & Ré (2012*a*, *b*) interpreted that the arrival of the Juan Fernández ridge at 28° S after the Oligocene could have induced the margin to curve. This southwards migration of deformation induced by the Juan Fernández ridge subduction and the higher curvature localized south of 30° S are also compatible with the foreland Holocene to recent deformation demonstrated by Casa *et al.* (2011), concentrated from the Rodeo–Talacasto belt to the south.

This interpretation is consistent with that previously proposed by Japas & Ré (2005) about the conditions controlling the margin deformation and ridge migration along the active boundary.

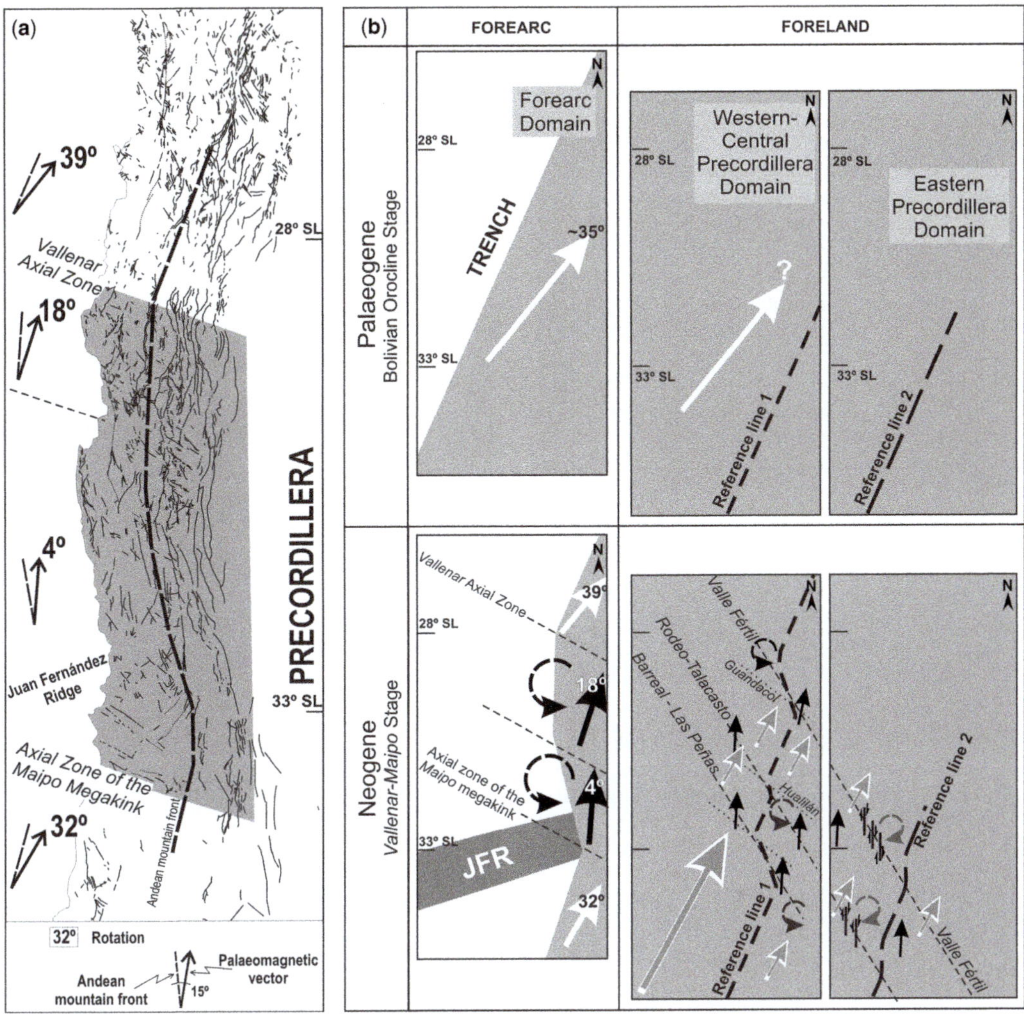


**Fig. 11.** (**a**) Fracture fabric in the forearc (after Sernageomin 2003). Notice that WNW-trending structures are mainly constrained to the 28–33° S (e.g. Rivera & Yáñez 2009; Sruoga *et al.* 2012; Japas *et al.* 2012, 2014*b*). The numbers indicate the amount of vertical axis rotation obtained by Arriagada *et al.* (2009). The Vallenar axial zone and the axial zone of the Maipo Megakink were defined by these authors. Irrespective of the considered segment, and as shown by Japas & Ré (2012*a*, *b*), there is a constant *c.* 15° angular relationship between the palaeomagnetic vector and the western highest mountain front. (**b**) Sketch showing the palaeomagnetic constraints for bending of the continental upper plate margin (palaeomagnetic data from Chile: Arriagada *et al.* 2009). Black arrows indicate present-day rotation values; white arrows (forearc) represent the regional clockwise rotation pattern previous to the arrival of the Juan Fernández ridge (JFR); grey arrows (foreland) symbolize regional clockwise rotation pattern linked to the early subduction of the Juan Fernández ridge and consequent transfer of deformation into the foreland. Modified from Japas & Ré (2012*a*, *b*). See Figures 7 and 8 for more information.

Convergence direction (trend N76° E; DeMets *et al.* 1990) oblique to ridge axis (trending N70° E), and both oblique to the margin trend (N10° E), would have conditioned the 28–33° S margin and Pampean flat-slab evolution. Whereas ridge axis trend would have controlled the strike of the deformed margin, convergence/ridge obliquity would have imposed the width of the rotating continental margin zone and would have defined the migration sense of the aseismic ridge along the active plate boundary.

The presented data and the interpretation proposed in this work (similar to that by Japas & Ré 2012*a*, *b*), are significantly different from previous proposals; they indicate a late overprint stage in the forearc and in the foreland that the authors linked to the Juan Fernández ridge subduction stage. This would reveal a different geodynamic scenario than that suggested by Isacks (1988) since, in the present interpretation, the foreland curved due to margin bending that in turn resulted from the penetration of the Juan Fernández ridge (Fig. 11b). Moreover, as in the Western and Central Precordillera domain, the absence of Miocene rotations in the forearc could be the result of the overprinting of late sinistral transpressional structures (WNW-trending in the forearc and NNW-trending in the foreland) cancelling slight clockwise rotations (Fig. 11b). In this context, the hypothesis of a scenario characterized by two independent steps (Eocene in the forearc, Miocene in the backarc; Somoza *et al.* 1996) should be reconsidered, since comparisons between regions undergoing diachronic heterogeneous deformation overprinted by a late stage of lateral-displacement structures would generate a complex mosaic of vertical axis rotation magnitudes.

## Conclusions

Based on available palaeomagnetic data, three regional domains were recognized in the Pampean flat-slab segment:

(1) The Forearc domain shows a segmented homogeneous vertical axis rotation pattern, interpreted as the result from the overprint of anticlockwise rotations induced by WNW structures on the regional clockwise Central Andes rotation pattern (CARP, Somoza *et al.* 1996).

(2) The Western-Central Precordillera domain, represented by a heterogeneous vertical axis rotation pattern linked to a first stage of deformation (regional clockwise vertical axis rotations concordant with CARP) lately overprinted by localized anticlockwise rotation induced by the NNW-trending sinistral transpressional belts.

(3) The Eastern Precordillera domain displays a heterogeneous vertical axis rotation pattern linked to the presence of the NNW-trending transpressional belts, suggesting both partitioning of regional deformation and active involvement of basement structures since *c.* 2.75 Ma.

The coincidence between the subduction of the Juan Fernández ridge, the development of the Pampean flat-slab segment, the margin curvature at 28–33° S and the presence of localized vertical axis rotation patterns and kinematic reorientation linked to overprinted oblique megashears allow us to propose a scenario of ridge indentation style controlling the foreland deformation.

The vertical axis rotation pattern reveals overprinting relations defined by Pliocene NNW–SSE-trending left-lateral oblique belts deforming the regional Miocene north–south/NNE–SSW-trending dextral transpressional fabric. The southern end of the Palaeogene CARP should therefore be extended further south of 28° S.

This research was funded by CONICET (PIP 6411 and PIP 11420100100334 to MSJ) and by Universidad de Buenos Aires (UBACyT N°20020120200157 to GHR). Authors extend their gratitude to *Dirección de Vialidad Nacional* and *Dirección Provincial de Vialidad* (San Juan province) for their support during the field work and to Haroldo Vizán for valuable suggestions on an earlier draft of the manuscript. Ruth Soto and Nuria Carrera are thanked for their critical and helpful reviews, as well as editors Emilio Pueyo and Aviva Sussman for their assistance in clarifying the presentation.

## References

ABBRUZZI, J. M., KAY, S. M. & BICKFORD, M. E. 1993. *Evidence of a Grenville age island arc basament in Precordillera, San Juan Province, Argentina.* Geological Society of America, Boulder, Abstracts with Programs, **25**, 232–233.

ALLMENDINGER, R. W., FIGUEROA, D., SNYDER, D., BEER, J., MPODOZIS, C. & ISACKS, B. L. 1990. Foreland shortening and crustal balancing in the Andes at 30°S latitude. *Tectonics*, **9**, 789–809.

ALONSO, M. S., LIMARINO, C. O., LITVAK, V. D., POMA,S. M., SURIANO, J. & REMESAL, M. B. 2011. Paleogeographic, magmatic and paleoenvironmental scenarios at 30° S during the Andean orogeny: cross sections from the volcanic-arc to the orogenic front (San Juan, Argentina). *In*: SALFITY, J. A. & MARQUILLAS, R. A. (eds) *Cenozoic Geology of the Central Andes of Argentina.* SCS Publisher, Salta, Argentina, 23–45.

ALVARADO, P., BECK, S. & ZANDT, G. 2007. Crustal structure of the south-central Andes Cordillera and backarc region from regional waveform modeling. *Geophysical Journal International*, **170**, 858–875, http://doi.org/10.1111/j.1365-246X.2007.03452.x

ÁLVAREZ-MARRÓN, J., RODRÍGUEZ-FERNÁNDEZ, R., HEREDIA, N., BUSQUETS, P., COLOMBO, F. & BROWN, D. 2006. Neogene structures overprinting Palaeozoic thrust systems in the Andean Precordillera at 30° S latitude. *Journal of the Geological Society, London*, **163**, 949–964.

AMMIRATI, J. B., ALVARADO, P., PERARNAU, M., SÁEZ, M. & MONSALVO, G. 2013. Crustal structure of the Central Precordillera of San Juan, Argentina (31° S) using teleseismic receiver functions. *Journal of South American Earth Sciences*, **46**, 100–109.

ANDERSON, M., ALVARADO, P., ZANDT, G. & BECK, S. 2007. Geometry and brittle deformation of the subducting Nazca plate, central Chile and Argentina. *Geophysical Journal International*, **171**, 419–434.

ARRIAGADA, C., ROPERCH, P., MPODOZIS, C. & COBBOLD, P. R. 2008. Paleogene building of the Bolivian Orocline: tectonic restoration of the Central Andes in 2.D map view. *Tectonics*, **27**, TC6014, http://doi.org/10.1029/2008TC002269

ARRIAGADA, C., MPODOZIS, C., YÁNEZ, G., CHARRIER, R., FARÍAS, M. & ROPERCH, P. 2009. *Rotaciones tectónicas en Chile central: El oroclino de Vallenar y el 'megakink' del Maipo.* 12° Congreso Geológico Chileno, Santiago de Chile, 1–4.

ARRIAGADA, C., FERRANDO, R., CÓRDOVA, L., MORATA, D. & ROPERCH, P. 2013. The Maipo Orocline: a first scale structural feature in the miocene to recent geodynamic evolution in the central Chilean Andes. *Andean Geology*, **40**, 419–437.

ASTINI, R. A., RAMOS, V. A., BENEDETTO, J. L., VACCARI, N. E. & CAÑAS, F. L. 1996. *La Precordillera: un terreno exótico a Gondwana.* 13° Congreso Geológico Argentino y III.° Congreso de Exploraciones e Hidrocarburos, Buenos Aires, **5**, 293–324.

AUBRY, L., ROPERCH, P., URREIZTIETA, M., ROSSELLO, E. A. & CHAUVIN, A. 1996. Paleomagnetic study along the southeastern edge of the Altiplano–Puna Plateau: neogene tectonic rotations. *Journal of Geophysical Research*, **101**, 17 883–17 899.

BALBI, A. 1995. *Estudio magnetoestratigráfico de las sedimentitas terciarias del Grupo Iglesia aflorantes en las cercanías de Tudcum (prov. de San Juan): sus implicancias tectosedimentarias.* MSc thesis, Universidad de Buenos Aires, Argentina.

BALDIS, B. A. & CHEBLI, G. 1969. *Estructura profunda del área central de la Precordillera sanjuanina.* 4° Jornadas Geológicas Argentinas, Mendoza, 1, 47–66.

BALDIS, B. A., BERESI, M., BORDONARO, L. O. & VACA, A. 1982. *Síntesis evolutiva de la Precordillera Argentina.* 5° Congreso Latinoamericano de Geología Argentina, Buenos Aires, 4, 399–445.

BASILE, Y. A. 2004. *Estudio geológico y geofísico del sector sur de las Lomas del Inca, provincia de San Juan.* MSc thesis, Universidad de Buenos Aires, Argentina.

BASTÍAS, H. 1985. *Fallamiento cuaternario en la región sismotectonica de Precordillera.* PhD thesis, Universidad Nacional de San Juan, Argentina.

BASTÍAS, H., WEIDMANN, N. & PÉREZ, M. 1984. *Dos zonas de callamiento plio–cuaternario en la Precordillera de San Juan.* 9° Congreso Geológico Argentino, San Carlos de Bariloche, 2, 329–341.

BECK, M. E., Jr 1989. Análisis of Late Jurassic recent paleomagnetic data from active margins of South America. *Journal of South American Earth Science*, **1**, 39–52.

BEER, J. A. 1990. *Magnetic polarity stratigraphy and depositional environments of the Bermejo Basin and seismic stratigraphy of the Iglesia Basin, Central Andes.* PhD thesis, Cornell University, USA.

BEER, J. A., ALLMENDINGER, R. W., FIGUEROA, D. E. & JORDAN, T. E. 1990. Seismic stratigraphy of a Neogene Piggyback Basin, Argentina. *American Association of Petroleum Geology, Bulletin*, **74**, 1183–1202.

BEVIS, M. & ISACKS, B. 1984. Hypocentral trend surface analysis; probing the geometry of Benioff zones. *Journal of Geophysical Research*, **89**, 6153–6170.

BROOKS, B. A., BEVIS, M. ET AL. 2003. Crustal motion in the southern Andes (26°–36° S): do the Andes behave like a microplate? *Geochemistry Geophysics Geosystems*, **4**, 1085, http://doi.org/10.1029/2003GC000505

CAHILL, T. & ISACKS, B. 1992. Seismicity and shape of the subducted Nazca Plate. *Journal of Geophysical Research*, **97**, 503–529.

CAMINOS, R., NULLO, F. E., PANZA, J. L. & RAMOS, V. A. 1993. *Mapa Geológico de la Provincia de San Juan. Escala 1:500.000.* Servicio Geológico y Minero de Argentina, Buenos Aires.

CARDÓ, R., DÍAZ, I., CEGARRA, M., RODRÍGUEZ, R., HEREDIA, N. & SANTAMARÍA, G. 1998. *Hoja Geológica 3169-I. Rodeo, escala 1: 250.000.* Instituto de Geología y Recursos Minerales, Servicio Geológico Minero Argentino, Boletín, Buenos Aires, **272**.

CASA, A., YAMIN, M., WRIGHT, E., COSTA, C., COPPOLECCHIA, M. & CEGARRA, M. 2011. *Deformaciones Cuaternarias de la República Argentina, Sistema de Información Geográfica.* Instituto de Geología y Recursos Minerales, Servicio Geológico Minero Argentino, digital publication, Buenos Aires, 171.

CEGARRA, M. & RAMOS, V. A. 1996. La faja plegada y corrida del Aconcagua. *In*: RAMOS, V. A. (ed.) *Geología de la región del Aconcagua, provincias de San Juan y Mendoza.* Anales de la Dirección Nacional del Servicio Geológico, Buenos Aires, **24**, 387–422.

CHERNICOFF, C. J. & NASH, C. R. 2002. Geological interpretation of Landsat TM imagery and aeromagnetic survey data, northern Precordillera region, Argentina. *Journal of South American Earth Sciences*, **14**, 813–820.

CICCIOLI, P. L., LIMARINO, C. O., MARENSSI, S. A., TEDESCO, A. & TRIPALDI, A. 2010. Estratigrafía de la cuenca de Vinchina (Terciario), Sierras Pampeanas, provincia de La Rioja. *Revista de la Asociación Geológica Argentina*, **66**, 146–155.

CICCIOLI, P. L., LIMARINO, C. O., MARENSSI, S. A., TEDESCO, A. M. & TRIPALDI, A. 2011. Tectosedimentary evolution of the La Troya and Vinchina depocenters (northern Bermejo Basin, Tertiary), La Rioja, Argentina. *In*: SALFITY, J. A. & MARQUILLAS, R. A. (eds) *Cenozoic Geology of the Central Andes of Argentina.* SCS Publisher, Salta, Argentina, 91–110.

CICCIOLI, P. L., LIMARINO, C. O., FRIEDMAN, R. & MARENSSI, S. A. 2014. New high precision U-Pb ages for the Vinchina Formation: implications for the stratigraphy of the Bermejo Andean foreland basin (La Rioja province, western Argentina). *Journal of South American Earth Sciences*, **56**, 200–213, http://www.sciencedirect.com/science/article/pii/S0895981114001163

CORTÉS, J. M. & CEGARRA, M. 2004. Plegamiento cuaternario transpresivo en el piedemonte suroccidental de la Precordillera sanjuanina. *In*: CORTÉS, J. M., ROSSELLO, E. A. & DALLA SALDA, L. H. (eds) *Avances en Microtectónica.* Asociación Geológica Argentina, Buenos Aires Serie D, Publicación Especial, **7**, 68–75, Buenos Aires.

CORTÉS, J. M., VINCIGUERRA, P., YAMÍN, M. & PASINI, M. M. 1999. Tectónica cuaternaria de la región andina del Nuevo Cuyo (28°–33° LS). *In*: CAMINOS, R. (ed.) *Geología Argentina.* Subsecretaría de Minería de la Nación, Servicio Geológico Minero Argentino, Anales. Buenos Aires, **29**, 760–778.

CORTÉS, J. M., PASINI, M. & YAMÍN, M. 2005. *Paleotectonic controls on the distribution of Quaternary deformation in the southern Precordillera, Central Andes*

*(31°30'-33° SL)*. 6th International Symposium on Andean Geodynamics, Barcelona, 186–189.

CORTÉS, J. M., CASA, A., PASINI, M. M., YAMIN, M. G. & TERRIZZANO, C. 2006. Fajas oblicuas de deformación neotectónica en Precordillera y Cordillera Frontal (31°30'–33°30' LS). Controles paleotectónicos. *Revista de la Asociación Geológica Argentina*, **61**, 639–646.

COSTA, C., MACHETTE, N. M. ET AL. 2000. *Map and database of Quaternary faults and folds in Argentina.* US Geological Survey Open File Report **00–0108**.

COSTA, G. 1994. *Magnetoestratigrafía del Terciario en el Oeste de Mendoza.* MSc thesis, Universidad de Buenos Aires, Argentina.

CRISTALLINI, E. O. 1996. La faja plegada y corrida de La Ramada. *In*: RAMOS, V. A. (ed.) *Geología de la Región del Aconcagua, Provincias de San Juan y Mendoza.* Anales de la Dirección Nacional del Servicio Geológico, Buenos Aires, **24**, 349–385.

CRISTALLINI, E. O. & RAMOS, V. A. 1995. Structural crosssection of Río San Juan. *In*: RAMOS, V. A. (ed.) *Field Guide to the Geology of Precordillera Folded and Thrust Belt (Central Andes)*. ICL-COMTEC-AGA, Buenos Aires, Argentina.

CRISTALLINI, E. O. & RAMOS, V. A. 2000. Thick-skinned and thin-skinned thrusting in La Ramada fold and thrust belt: crustal evolution of the high Andes of San Juan, Argentina (32° SL). *Tectonophysics*, **317**, 205–235.

DALLA SALDA, L. 2005. *El orógeno famatiniano-apalachiano. 16° Congreso Geológico Argentino, CD-ROM*, La Plata.

DEMETS, C., GORDON, R. G., ARGUS, D. F. & STEIN, S. 1990. Current plate motions. *Geophysical Journal*, **101**, 425–478.

DUPONT-NIVET, G., BUTLER, R. F., YIN, A. & CHEN, X. 2003. Paleomagnetism indicates no Neogene vertical axis rotations of the northeastern Tibetan Plateau. *Journal of Geophysical Research*, **108**, 2386, http://doi.org/10.1029/2003JB002399

FAZZITO, S. Y. 2011. *Estudios geofísicos aplicados a la geotectónica de la falla El Tigre, Precordillera de San Juan.* PhD thesis, Universidad de Buenos Aires, Argentina, http://digital.bl.fcen.uba.ar/Download/Tesis/Tesis_4886_Fazzito.pdf

FAZZITO, S. Y., RAPALINI, A. E., CORTÉS, J. M. & TERRIZZANO, C. M. 2011. Kinematic study in the area of the Quaternary oblique-slip El Tigre Fault, western Precordillera, Argentina, on the basis of paleomagnetism and anisotropy of magnetic susceptibility. *Latinmag Letters*, **1**, 1–5. Proceedings Tandil, Argentina.

FISHER, R. A. 1953. Dispersion on a sphere. *Proceedings of the Royal Society of London*, **A217**, 295–305.

FURQUE, G., GONZÁLEZ, P. D. ET AL. 2003. *Hoja Geológica 3169-II San José de Jáchal. Provincias de San Juan y La Rioja.* Instituto de Geología y Recursos Minerales, Servicio Geológico Minero Argentino, Boletín, Buenos Aires, 259.

GANS, C. R., BECK, S. L., ZANDT, G., GILBERT, H., ALVARADO, P., ANDERSON, M. & LINKIMER, L. 2011. Continental and oceanic crustal structure of the Pampean flat slab region, western Argentina, using receiver function analysis: new high-resolution results.

*Geophysical Journal International*, **186**, 45–58, http://doi.org/10.1111/j.1365-246X.2011.05023.x

GIAMBIAGI, L. B., TUNIK, M. A. & GHIGLIONE, M. 2001. Cenozoic tectonic evolution of the Alto Tunuyán foreland bjasin above the transition zone between the flat and the normal subduction segment (33°30' S–34° S), western Argentina. *Journal of South American Earth Sciences*, **14**, 707–724.

GIAMBIAGI, L. B., MESCUA, J., FOLGUERA, A. & MARTÍNEZ, A. 2010. Estructuras y cinemática de las deformaciones pre-andinas del sector sur de la Precordillera, Mendoza. *Revista de la Asociación Geológica Argentina*, **66**, 5–20.

GIAMPAOLI, P. & CEGARRA, M. 2003. Análisis estructural del extremo sur de la Precordillera Central Sanjuanina. *Revista de la Asociación Geológica Argentina*, **58**, 49–60.

GUTSCHER, M. A., SPAKMAN, W., BIJWAARD, H. & ENGDAHL, E. R. 2000. Geodynamics of flat slab subduction: seismicity and tomographic constraints from the Andean margin. *Tectonics*, **19**, 814–833.

HAYWARD, A. B. & GRAHAM, R. H. 1989. Some geometrical characteristics of inversion. *In*: COOPER, M. A. & WILLIAMS, G. D. (eds) *Inversion Tectonics*. Geological Society, London, Special Publications, **44**, 17–39.

HEREDIA, N., RODRIGUEZ FERNANDEZ, L. R., GALLASTEGUI, G., BUSQUETS, P. & COLOMBO, P. 2002. Geological setting of the Argentine Frontal Cordillera in the flat slab segment (30°00'–31°30' S latitude). *Journal of South American Earth Sciences*, **15**, 79–99.

INTROCASO, A. & RUIZ, F. 2001. Geophysical indicators of Neogene strike-slip faulting in the Desaguadero-Bermejo tectonic lineament (northwestern Argentina). *Journal of South American Earth Sciences*, **14**, 655–663.

ISACKS, B. 1988. Uplift of the central Andean plateau and bending of the Bolivian orocline. *Journal of Geophysical Research*, **93**, 3211–3231.

JAPAS, M. S. 1998. Aporte del análisis de fábrica deformacional al estudio de la faja orogénica andina. Homenaje al Dr. Arturo J.AMOS. *Revista de la Asociación Geológica Argentina*, **53**, 15.

JAPAS, M. S. & RÉ, G. H. 2005. *Geodynamic impact of arrival and subduction of oblique aseismic ridges.* 6th International Symposium on Andean Geodynamics, Barcelona, 408–410.

JAPAS, M. S. & RÉ, G. H. 2012a. *Margin curvature at 28–33° SL induced by oblique subduction of the Juan Fernández aseismic ridge: Paleomagnetic constraints.* 8th International Symposium in Andean Geodynamics. Antofagasta, Chile.

JAPAS, M. S. & RÉ, G. H. 2012b. Neogene tectonic block rotations and margin curvature at the Pampean flat slab segment (28°–33° SL, Argentina). *Geoacta*, **37**, 1–4.

JAPAS, M. S., RÉ, G. H. & BARREDO, S. P. 2002. *Lineamientos andinos oblícuos (entre 22° S y 33° S) definidos a partir de fábricas tectónicas. I. Fábricas deformacional y de sismicidad.* 15° Congreso Geológico Argentino, El Calafate, **1**, 326–331.

JAPAS, M. S., URBINA, N. E. & SRUOGA, P. 2010. Control estructural en el emplazamiento del volcanismo y mineralizaciones neógenas, distrito Cañada Honda,

San Luis. *Revista de la Asociación Geológica Argentina*, **67**, 494–506.

JAPAS, M. S., RÉ, G. H., VILAS, J. F. & ORIOLO, S. 2011. *Oblique megashear zones in the Precordillera (Central Andes, Argentina): the Rodeo-Talacasto transpressional belt.* Deformation, Rheology and Tectonics Conference, Oviedo, Spain.

JAPAS, M. S., ORIOLO, S. & SRUOGA, P. 2012. *Análisis cromático de la fábrica de fracturación.* 15° Reunión de Tectónica, Digital abstracts, *San Juan*, 72–73.

JAPAS, M. S., RÉ, G. H., VILAS, J. F. & ORIOLO, S. 2014*a*. *Rotaciones tectónicas neógenas y curvatura del segmento de subducción subhorizontal Pampeano (27°–33° S).* 19° Congreso Geológico Argentino, Córdoba, S20–S12.

JAPAS, M. S., ORIOLO, S., SRUOGA, P. & PERONI, J. 2014*b*. *Control estructural en el emplazamiento de la caldera Diamante, sector norte de la Zona Volcánica Sur.* 19° Congreso Geológico Argentino, Córdoba, S23–S21.

JOHNSON, P., JOHNSON, N., JORDAN, T. & NAESER, C. W. 1984. *Magnetic polarity stratigraphy and age of the Quebrada del Cura, Río Jáchal, and Mogna Formations near Huaco, San Juan province, Argentina.* 9° Congreso Geológico Argentino, Bariloche, **3**, 81–96.

JORDAN, T. E., RUTTY, P., MC RAE, L., BEER, J., TABBUTT, K. & DAMANTI, T. 1990. Magnetic polarity of the Miocene Río Azul section, Precordillera thrust belt, San Juan Province, Argentina. *Journal of Geology*, **98**, 519–539.

JORDAN, T. E., ALLMENDINGER, R. W., DAMANTI, J. F. & DRAKE, R. 1993. Chronology of motion in a complete thrust belt: the Precordillera, 30–31° S, Andes Mountains. *Journal of Geology*, **101**, 135–156.

JORDAN, T. E., SCHLUNEGGER, F. & CARDOZO, N. 1999. *Múltiples hipótesis en la evolución de la cuenca neógena de antepaís de Bermejo, Argentina.* 14° Congreso Geológico Argentino, Salta, **1**, 193–196.

JORDAN, T. E., SCHLUNEGGER, F. & CARDOZO, N. 2001. Unsteady and spatially variable evolution of the Neogene Andean Bermejo foreland basin, Argentina. *Journal of South American Earth Sciences*, **14**, 775–798.

KAY, S. M., ORRELL, S. & ABRUZZI, J. M. 1996. Zircon and whole rock Nd-Pb isotopic evidence for a Grenville age and Laurentia origin for the basement of the Precordilleran terrane in Argentina. *Journal of Geology*, **104**, 637–648.

KIRSCHVINK, J. 1980. The least squares lines and plane and the analysis of paleomagnetic data. *Geophysical Journal of the Royal Astronomy Society*, **62**, 699–718.

KLEIMAN, L. E. & JAPAS, M. S. 2009. The Choiyoi volcanic province at 34–36° S (San Rafael, Mendoza, Argentina): implications for the late Palaeozoic evolution of the southwestern margin of Gondwana. *Tectonophysics*, **473**, 283–299.

LEVERATTO, M. A. 1968. Geología de la zona al oeste de Ullúm-Zonda, borde oriental de la Precordillera de San Juan, eruptividad subvolcánica y estructura. *Revista de la Asociación Geológica Argentina*, **18**, 129–158.

MAFFIONE, M., SPERANZA, F. & FACCENNA, C. 2009. Bending of the Bolivian orocline and growth of the central Andean plateau: paleomagnetic and structural constraints from the Eastern Cordillera (22–24_S, NW Argentina). *Tectonics*, **28**, TC4006, http://doi.org/10.1029/2008TC002402

MARTÍNEZ, M. P., PERUCCA, L. P., GIMÉNEZ, M. E. & RUIZ, F. 2008. Manifestaciones geomorfológicos y geofísicas de una estructura geológica profunda al sur de la Sierra de Pie de Palo, Sierras Pampeanas. *Revista de la Asociación Geológica Argentina*, **63**, 264–271.

MILANA, J. P. 1991. *Sedimentología y Magnetoestratigrafía de Formaciones Cenozoicas en el área de Mogna, y su inserción en el marco tectosedimentario de la Precordillera Oriental.* PhD thesis, Universidad Nacional de San Juan.

ORIOLO, S. 2012. *Análisis de la deformación en la región de Hualilán, Precordillera, San Juan.* MSc thesis, Universidad de Buenos Aires, Argentina.

ORIOLO, S., JAPAS, M. S. & CRISTALLINI, E. O. 2011. *Transtensional megashear zones in the Central Andes: the Hualilán Belt (Precordillera, Argentina).* Deformation, Rheology and Tectonics Conference, Oviedo, Spain.

ORIOLO, S., JAPAS, M. S. & CRISTALLINI, E. O. 2012. *The Hualilan transtensional belt (Precordillera, Argentina).* 8th International Symposium on Andean Geodynamics. Antofagasta, Chile.

ORIOLO, S., JAPAS, M. S., CRISTALLINI, E. O. & GIMÉNEZ, M. 2014. Cross-strike structures controlling magmatism emplacement in a flat-slab setting (Precordillera, Central Andes of Argentina). *In*: LLANA-FÚNEZ, S., MARCOS, A. & BASTIDA, F. (eds) *Deformation Structures and Processes within the Continental Crust.* Geological Society, London, Special Publications, **394**, 113–127, http://doi.org/10.1144/SP394.6

ORIOLO, S., CRISTALLINI, E. O., JAPAS, M. S. & YAGUPSKY, D. 2015. Neogene structure of the Andean Precordillera, Argentina: insights from analogue models. *Andean Geology*, **42**(1), 20–35, http://doi.org/10.5027/andeoV42n1-a02

ORTIZ, A. & ZAMBRANO, J. 1981. *La provincia geológica de Precordillera Oriental.* 8° Congreso Geológico Argentino, San Luis, **3**, 59–74.

ORTIZ, G., FOSDICK, J. & ALVARADO, P. M. 2014. *Edades termocronológicas en la Sierra de Valle Fértil-La Huerta: resultados preliminares de dataciones (U-Th)/He en apatitos.* 19° Congreso Geológico Argentino, Córdoba, S20–S22.

OVIEDO, E. 1989. *Mag88: Un sistema de computación para análisis de datos paleomagnéticos.* PhD thesis, Universidad de Buenos Aires, Argentina.

PÉREZ, D. 2001. Tectonic and unroofing history of Neogene Manantiales foreland basin deposits, Cordillera Frontal (32°30′ S), San Juan Province, Argentina. *Journal of South American Earth Sciences*, **14**, 693–705.

PÉREZ, I. & COSTA, C. 2011. *El braquianticlinal del cerro Negro de Iglesia y su relación con el sistema de fallamiento El Tigre, provincia de San Juan–Argentina.* 18° Congreso Geológico Argentino, *CD*, 823–824.

PÉREZ, M. A. & BAGUR DELPIANO, V. 2012. *Shale Gas & Shale Oil Potential of the Middle to Upper Ordovician in the Precordillera Basin, Argentina Analogies from the Utica Formation (Appalachian Basin from U.S.A. and Canada), to the Gualcamayo, Las Vacas and Las*

*Plantas Formations (Precordillera Basin, Argentina)*. Electronic Report, http://www.artemap.com.ar/index_ htm_files/ English_Shale_G_y_O_Pc.pdf

PÉREZ, M. A., BAGUR DELPIANO, V., GRANEROS, D., BREIER, K. & LAURÍA, M. 2012. Cuenca Precordillera; claves exploratorias para el Paleozoico. Áreas Jáchal y Niquivil, San Juan. Nota técnica. *Petrotecnia*, 76–103.

PERUCCA, L. P. & MARTOS, L. M. 2009. Análisis preliminar de la evolución del paisaje cuaternario en el valle de Iglesia, San Juan. *Revista de la Asociación Geológica Argentina*, **65**, 624–637.

PILGER, R. H. 1981. Plate reconstructions, aseismic ridges, and low angle subduction beneath the Andes. *Geological Society of America Bulletin*, **92**, 448–456.

PUENTE, N. 1992. *Magnetoestratigrafía de la secuencia cenozoica aflorante en la localidad Las Tucumanesas, provincia de La Rioja*. MSc thesis, Universidad de Buenos Aires.

RAGONA, D., ANSELMI, G., GONZÁLEZ, P. & VUJOVICH, G. 1995. *Mapa Geológico de la Provincia de San Juan. Escala 1:500 000*. Servicio Geológico y Minero de Argentina, Buenos Aires, Argentina.

RAMOS, V. A. 1970. Estratigrafía y estructura de la sierra de Los Colorados, provincia de La Rioja. *Revista de la Asociación Geológica Argentina*, **25**, 359–382.

RAMOS, V. A. 1999a. Plate tectonic setting of the Andean Cordillera. *Episodes*, **22**, 183–190.

RAMOS, V. A. 1999b. Los depósitos sinorogénicos terciarios de la región andina. *In*: CAMINOS, R. (ed.) *Geología Regional Argentina*. Anales del Instituto de Geología y Recursos Minerales, Buenos Aires, **29**, 651–682.

RAMOS, V. A. & DALLA SALDA, L. 2011. *Occidentalia: ¿Un terreno acrecionado sobre el margen gondwánico?* 18° Congreso Geológico Argentino, CD-ROM, Neuquén.

RAMOS, V. A., JORDAN, T. E., ALLMENDINGER, R. W., MPODOZIS, C., KAY, S. M., CORTÉS, J. M. & PALMA, M. A. 1986. Paleozoic terranes of the Central Argentine-Chilean Andes. *Tectonics*, **5**, 855–880.

RAMOS, V. A., VUJOVICH, G. I. & DALLMEYER, R. D. 1996. *Los klippes y ventanas tectónicas de la estructura preándica de la Sierra de Pie de Palo (San Juan): edad e implicaciones tectónicas*. 18° Congreso Geológico Argentino y 3° Congreso Exploración de Hidrocarburos, Buenos Aires, **5**, 377–392.

RAMOS, V. A., CRISTALLINI, E. O. & PEREZ, D. J. 2002. The Pampean flat-slab of the Central Andes. *Journal of South American Earth Sciences*, **15**, 59–78.

RAMSAY, J. G. & HUBER, M. I. 1987. *The Techniques of Modern Structural Geology: Folds and Fractures*. Academic Press, London.

RÉ, G. H. 2008. *Magnetoestratigrafía del NO argentino (entre 27° y 31° S) aplicadas al análisis de la deformación andina, y su relación con la subducción de la placa de Nazca durante el Cenozoico Tardío*. PhD thesis, Universidad de Buenos Aires, Argentina.

RÉ, G. H. & BARREDO, S. 1993. *Estudio magnetoestratigráfico y tasa de sedimentación del Grupo Iglesia en sus afloramientos aledaños a la localidad de Angualasto (Prov. de San Juan)*. 12° Congreso Geológico Argentino, Mendoza, **2**, 148–155.

RÉ, G. H. & RAPALINI, A. E. 1995. Non-systematic neogene crustal block rotations in the Andean foothills of Central Argentina. *Curved Orogenic Belts*, Buenos Aires, Abstracts, 31–37.

RÉ, G. H., JAPAS, M. S. & BARREDO, S. P. 2001. Análisis de fábrica deformacional (AFD): El concepto fractal cualitativo aplicado a la definición de lineamientos cinemáticos neógenos en el Noroeste Argentino. *In*: CORTÉS, J. M., ROSSELLO, E. A. & DALLA SALDA, L. H. (eds) *Avances en Microtectónica*. Asociación Geológica Argentina, Buenos Aires Serie D: Publicación Especial, **5**, 75–82.

REYNOLDS, J. H. 1987. *Chronology of Neogene tectonics in the Central Andes (27°–33° S) of western Argentina based on the magnetic polarity stratigraphy of foreland basins sediments*. PhD thesis, Darmouth College, USA.

RIVERA, O. & YÁÑEZ, G. 2009. *Naturaleza y rol de estructuras translitosféricas en la evolución del arco oligomioceno de Chile central*. 12° Congreso Geológico Chileno, Santiago, S9–092.

RODRÍGUEZ BRIZUELA, R. & TAUBER, A. 2006. Estratigrafía y mamíferos fósiles de la Formación Toro Negro (Neógeno), Departamento Vinchina, noroeste de la provincia de La Rioja, Argentina. *Ameghiniana*, **43**, 257–272.

RON, H., FREUND, R. & GARFUNKEL, Z. 1984. Block rotation by strike-slip faulting: structural and paleomagnetic evidence. *Journal of Geophysical Research*, **89**, 6256–6270.

RON, H., AYDIN, A. & NUR, A. 1986. Strike-slip and block rotation in the Lake Mead fault system. *Geology*, **14**, 1020–1023.

ROSSELLO, E. A., MOZETIC, M. E., COBBOLD, P. R., URREIZTIETA, M. & GAPAIS, D. 1996. *El espolón Umango-Maz y la conjugación sintaxial de los lineamientos Tucumán y Valle Fértil (La Rioja, Argentina)*. 13° Congreso Geológico Argentino y 3° Congreso de Hidrocarburos, Buenos Aires, 2187–2194.

SERNAGEOMIN 2003. *Mapa Geológico de Chile: versión digital*. Servicio Nacional de Geología y Minería, Publicación Geológica Digital. Santiago, Chile.

SIAME, L. L., SÉBRIER, M., BELLIER, O., BOURLÈS, D. L., CASTANO, J. C. & ARAUJO, M. 1997. Geometry, segmentation and displacement rates of the El Tigre Fault, San Juan Province (Argentina) from SPOT image analysis and 10Be datings. *Annales Tectonicae*, **1/2**, 3–26.

SIAME, L. L., BELLIER, O., SEBRIER, M. & ARAUJO, M. 2005. Deformation partitioning in flan subduction setting: case of the Andean foreland of western Argentina (28° S–33° S). *Tectonics*, **24**, 1–24, http://doi.org/10.1029/2005TC001787

SOMOZA, R., SINGER, S. & COIRA, B. 1996. Paleomagnetism of upper Miocene ignimbrites in the Puna: an analysis of vertical-axis rotations in the Central Andes. *Journal of Geophysical Research*, **101**, 11 387–11 400.

SRUOGA, P., JAPAS, M. S., ORIOLO, S. & FEINEMAN, M. 2012. *The Diamante Caldera–Maipo Volcano Complex: a potential hazard in the Central Andes of Argentina (34° 10′ S)*. IAVCEI Meeting, Bolsena, Italy.

STEIN, J. 1994. *Estudio magnetoestratigráfico de las sedimentitas terciarias del Grupo Iglesia, localidad de Las Flores (provincia de San Juan), y su aporte a la evolución de la cuenca de Rodeo-Iglesia*. MSc thesis, Universidad de Buenos Aires, Argentina.

TERRIZZANO, C. M., CORTÉS, J. M., FAZZITO, S. Y. & RAPALINI, A. E. 2009. Neotectonic transpressive zones in Precordillera Sur, Central Andes of Argentina: a structural and geophysical investigation. *Neues Jahrbuch für Geologie und Paläontologie, Abhandlungen*, **253**, 103–114.

TERRIZZANO, C. M., FAZZITO, S. Y., CORTÉS, J. M. & RAPALINI, A. E. 2010. Studies of Quaternary deformation zones through geomorphic and geophysical evidence. A Case in the Precordillera Sur, Central Andes of Argentina. *Tectonophysics*, **490**, 184–196.

TORSVIK, T. H. 1992. *IAPD, Interactive analysis of Palaeomagnetic Data*. Manual, NGU, Trondheim, N-7002.

TURNER, J. C. M. 1964. *Descripción geológica de la Hoja 15c. Vinchina (provincia de La Rioja)*. Dirección Nacional de Geología y Minería, Boletín **100**, Buenos Aires.

URBINA, N. E. & SRUOGA, P. 2009. La faja metalogenética de San Luis: mineralización y geocronología en el contexto metalogenético regional. *Revista de la Asociación Geológica Argentina*, **64**, 635–645.

URREIZTIETA, M. 1996. *Tectonique Néogène et bassins transpressifs en bordure méridionale de lAltiplano-Puna (27° S), Nord-Ouest argentin*. PhD thesis, Géosciences Rennes, France.

VANDAMME, D. 1994. A new method to determine paleosecular variation. *Physics of the Earth and Planetary Interiors*, **85**, 131–142.

VIZÁN, H., GEUNA, S. *ET AL.* 2013. Geological setting and paleomagnetism of the Eocene red beds of Laguna Brava Formation (Quebrada Santo Domingo, northwestern Argentina). *Tectonophysics*, **583**, 105–123.

VON GOSEN, W. 1992. Structural evolution of the Argentine Precordillera: the Río San Juan section. *Journal of Structural Geology*, **14**, 643–667.

YAGUPSKY, D., CRISTALLINI, E. O., FANTÍN, J., ZAMORA VALCARCE, G., BOTTESI, G. & VARADÉ, R. 2008. Oblique half-graben inversion of the Mesozoic Neuquén Rift in the Malargüe Fold and Thrust Belt, Mendoza, Argentina: new insights from analogue models. *Journal of Structural Geology*, **30**, 839–853.

YÁÑEZ, G., RANERO, C. R., VON HUENE, R. & DÍAZ, J. 2001. Magnetic anomaly interpretation across the southern central Andes (32°–34° S): the role of the Juan Fernández Ridge in the late Tertiary evolution of the margin. *Journal of Geophysical Research*, **106**, 6325–6345.

ZAPATA, T. R. 1998. Crustal structure of the Andean thrust front at 30° S latitude from shallow and deep seismic refection profiles, Argentina. *Journal of South American Earth Sciences*, **11**, 131–151.

ZAPATA, T. R. & ALLMENDINGER, R. W. 1996a. Thrust-front zone of the Precordillera, Argentina: a thick-skinned triangle zone. *American Association of Petroleum Geology Bulletin*, **80**, 359–381.

ZAPATA, T. R. & ALLMENDINGER, R. W. 1996b. Growth stratal records of instantaneous and progressive limb rotation in the Precordillera thrust belt and Bermejo basin, Argentina. *Tectonics*, **15**, 1065–1083.

ZENCICH, S., VILLAR, H. J. & BOGGETTI, D. 2008. Sistema petrolero Cacheuta-Barrancas de la Cuenca Cuyana, provincia de Mendoza, Argentina. *In*: CRUZ, C. E., RODRÍGUEZ, J. F., HECHEM, J. J. & VILLAR, H. J. (eds) *Sistemas Petroleros de las Cuencas Andinas*. Instituto Argentino del Petróleo y del Gas, Buenos Aires, 109–134.

# New magnetostratigraphic dating of the Palaeogene syntectonic sediments of the west-central Pyrenees: tectonostratigraphic implications


BELÉN OLIVA-URCIA[1,2]*, ELISABET BEAMUD[3], MIGUEL GARCÉS[4], CONCHA ARENAS[1], RUTH SOTO[5], EMILIO L. PUEYO[5] & GONZALO PARDO[1]

[1]*Dpto de Ciencias de la Tierra, Universidad de Zaragoza, c/ Pedro Cerbuna 12, 50009 Zaragoza, Spain*

[2]*Dpto de Geología y Geoquímica, Universidad Autónoma de Madrid, Ciudad Universitaria de Cantoblanco, 28049 Madrid, Spain*

[3]*Laboratori de Paleomagnetisme CCiTUB-CSIC, Institut de Ciències de la Terra "Jaume Almera", c/ Solé i Sabarís s/n, 08028 Barcelona, Spain*

[4]*Departament d'Estratigrafia, Paleontologia i Geociències Marines, Universitat de Barcelona, c/Martí i Franquès s/n, 08028 Barcelona, Spain*

[5]*Instituto Geológico y Minero de España, Unidad de Zaragoza, c/ Manuel Lasala 44, 9°B, 50006 Zaragoza, Spain*

**Corresponding author (e-mail: belen.oliva@uam.es)*



**Abstract:** New magnetostratigraphic results from a 3300 m-thick section across the syntectonic fluvial sediments of the Campodarbe Formation (Upper Eocene–Oligocene) in the Ebro foreland basin (NE Spain) are presented. The new data allow the top of the Campodarbe Formation to be correlated to Chron 7r (Chattian), younger than previously stated (C10r), therefore shifting the age of significant palaeogeographical changes in the foreland basin. The deformation in the southern front produces the cannibalization in the piggyback basin of 1300 m of sediments spanning *c.* 3.7 Myr. Average accumulation rates are lower in the Ebro foreland basin than in the piggyback basin and decrease from 35 to 27 cm kyr$^{-1}$ by the time the San Felices thrust sheet activity decelerates (at *c.* 28 Myr). Shifts of accumulation rates result from accommodation space changes, which occur locally and are linked to the activity of the San Felices thrust, while the sediment supply occurs at orogenic scale (source of sediments is *c.* 200 km to the NE). Finally, sequence boundaries previously considered isochronous in the continental record of the Cenozoic Pyrenean basins are revealed to be 1.8–1 Myr older in the piggyback basin than in the Ebro foreland basin.


Magnetostratigraphy of syntectonic deposits is a classical tool used to determine the age of the tectonic activity during orogenic construction. A precise chronology of syntectonic foreland basin deposits is fundamental to understand the coupling among tectonics, exhumation, sediment supply and accommodation space during shortening and orogenic construction in fold–thrust belts (DeCelles & Giles 1996; Allen & Allen 2005). The punctuated sedimentation in foreland basins results from the interaction of sediment influx and accommodation space. These two factors are governed mainly by tectonic activity but also by eustatic changes and denudation rate that, in turn, can be related to climate (Jordan *et al.* 2001; Clevis *et al.* 2004; Naylor & Sinclair 2007; Hoth *et al.* 2008; Molnar 2009; Allen *et al.* 2013). All these factors contribute to the basin fill and its architectural distribution in the foreland basin. The analyses of the magnetic polarity of thick sedimentary sequences provide an independent chronology when related unambiguously with the geomagnetic polarity time scale (GPTS). Thick and continuous magnetostratigraphic sequences distributed throughout the foreland basins are required to constrain the timing of sedimentation and its distribution in the whole area. This is a key starting point to calculate the accumulation rate within the foreland basin that in turn will be related to the uplift and denudation of the orogen.

The South Pyrenean foreland basin was structured through several stages related to different thrusting episodes of the fold–thrust belt evolution that gave rise to the southwards migration through the Cenozoic of the foreland basin (Labaume *et al.* 1985; Teixell & García-Sansegundo 1995;


*From*: PUEYO, E. L., CIFELLI, F., SUSSMAN, A. J. & OLIVA-URCIA, B. (eds) 2016. *Palaeomagnetism in Fold and Thrust Belts: New Perspectives.* Geological Society, London, Special Publications, **425**, 107–128. First published online August 3, 2015, updated August 21, 2015, http://doi.org/10.1144/SP425.5

Millán-Garrido *et al.* 2000). The South Pyrenean foreland basin was divided from Lutetian to early Miocene times into the piggyback Jaca–Pamplona Basin to the north and the present-day Ebro Basin to the south due to the development of the External Sierras thrust system, the South Pyrenean deformation front at this sector (Puigdefàbregas 1975; Labaume *et al.* 1985; McElroy 1990; Pocoví *et al.* 1990; Millán-Garrido *et al.* 1995; Teixell & García-Sansegundo 1995; Millán-Garrido *et al.* 2000). The Ebro Basin therefore represents the latest stage of the South Pyrenean foreland basin.

The syntectonic continental record of the South Pyrenean foreland basin (Upper Eocene–Miocene) has been extensively studied in order to divide it into lithostratigraphic and allostratigraphic units (tectonosedimentary units, depositional sequences). Note that a tectosedimentary unit (TSU) is a stratigraphic unit made up of strata deposited during a given geological time span, under sedimentary and tectonic conditions characterized by a specific tendency. The TSU boundaries are basin-wide sedimentary discontinuities or their correlative conformities (Garrido-Megías, in Pardo *et al.* 1989). In addition, Pardo *et al.* (1989) defined a TSU as a type of allostratigraphic unit whose vertical evolution is genetically linked to changes in the allocyclic factors that controlled the basin fill dynamics. In all cases, the definition of bounding surfaces is focused on the vertical changes in the sedimentary fill mostly related to pulses in the tectonic activity that affect the sediment source areas (Puigdefàbregas 1975; Arenas *et al.* 2001; Montes 2002). However, the distribution and definition of sedimentary units and bounding surfaces of the Pyrenean syntectonic deposits have varied between authors. The earliest works of Soler & Puigdefàbregas (1970) and Puigdefèabregas (1975) separated the Campodarbe Formation (Upper Eocene–Oligocene) into two main units considering the unconformity linked to the activity of the San Felices thrust sheet, which is part of the deformation in the southern Pyrenean front during Bartonian–Rupelian time (Puigdefàbregas 1975; Millán-Garrido *et al.* 1995). Later work by Montes (2002) divided the infill of the South Pyrenean foreland basin into three megasequences and correlated their boundary surfaces after careful revision of the vertical changes in architecture of the sedimentary systems. According to this author, these bounding surfaces within the Campodarbe Formation were mapped as isochronous surfaces over a large area, both in the foreland and in the piggyback basins.

The continental Uncastillo Formation (Oligo-Miocene; Soler & Puigdefàbregas 1970) overlies the Campodarbe Formation. The boundary between these two formations marks a complete architectural and palaeogeographical change in the Ebro foreland basin due to the tectonic activity in the Pyrenean Front (External Sierras). This change represents the transition from overall WNW-flowing fluvial systems (Campodarbe Formation) to overall southwards-flowing alluvial and fluvial systems (Uncastillo Formation), and is associated with the growing of the Pyrenean fold–thrust belt and, in particular, to the tightening of the WNW–ESE Santo Domingo anticline in the External Sierras during Chattian–Aquitanian times (Puigdefàbregas 1975; Millán-Garrido *et al.* 1995; Arenas *et al.* 2001). The External Sierras therefore represent the southern limit of the west-central part of the Pyrenean orogenic belt and separates the present-day Ebro foreland basin to the south from the Jaca–Pamplona piggyback basin to the north (McElroy 1990; Pocoví *et al.* 1990; Millán-Garrido *et al.* 1995). To the west of the External Sierras, the boundary between the Pyrenees and the Ebro Basin is represented by the continuation of the Santo Domingo anticline (Puigdefàbregas 1975; Oliva-Urcia *et al.* 2012). Three tectonosedimentary units have been described in the Uncastillo Formation in the Ebro Basin regarding the tectonic pulses in the External Sierras (Arenas *et al.* 2001).

In the west-central South Pyrenean foreland basin, palaeomagnetic studies have been carried out since the 1990s. From analyses of new and revised magnetostratigraphic data, it has recently been established that the closing and continentalization of the South Pyrenean foreland basin occurred synchronously during the Priabonian at *c.* 36 Myr (Costa *et al.* 2010). Previous magnetostratigraphic investigations in the Campodarbe and Uncastillo formation near the studied area (Hogan 1993; Hogan & Burbank 1996) include a section close to the San Felices thrust sheet outcrop that contains a long hiatus associated with the activity of such thrust sheet (Hogan 1993 calculated a 4 Myr gap). Together with other magnetostratigraphic sections available in the Campodarbe and Uncastillo formations (Salinas, Agüero and Ayerbe sections, Hogan & Burbank 1996), the top of the Campodarbe Formation was established at Chron 10r in the Jaca–Pamplona piggyback basin, whereas the Uncastillo Formation was set to reach Chron 6C in the Ebro foreland basin. The activity of the San Felices thrust sheet was calculated to begin at 34.4–32.6 Myr and continue until 30.5 Myr (ages considering Hogan & Burbank 1996 interpretation, using the updated 2012 GPTS by Gradstein and Ogg).

In this work, a section *c.* 10 km west of the San Felices thrust sheet outcrop was selected for magnetostratigraphic analysis (the Luesia composite section; Fig. 1). The section provides an *a priori* continuous (punctuated) sedimentation record during the activity of the San Felices thrust sheet,

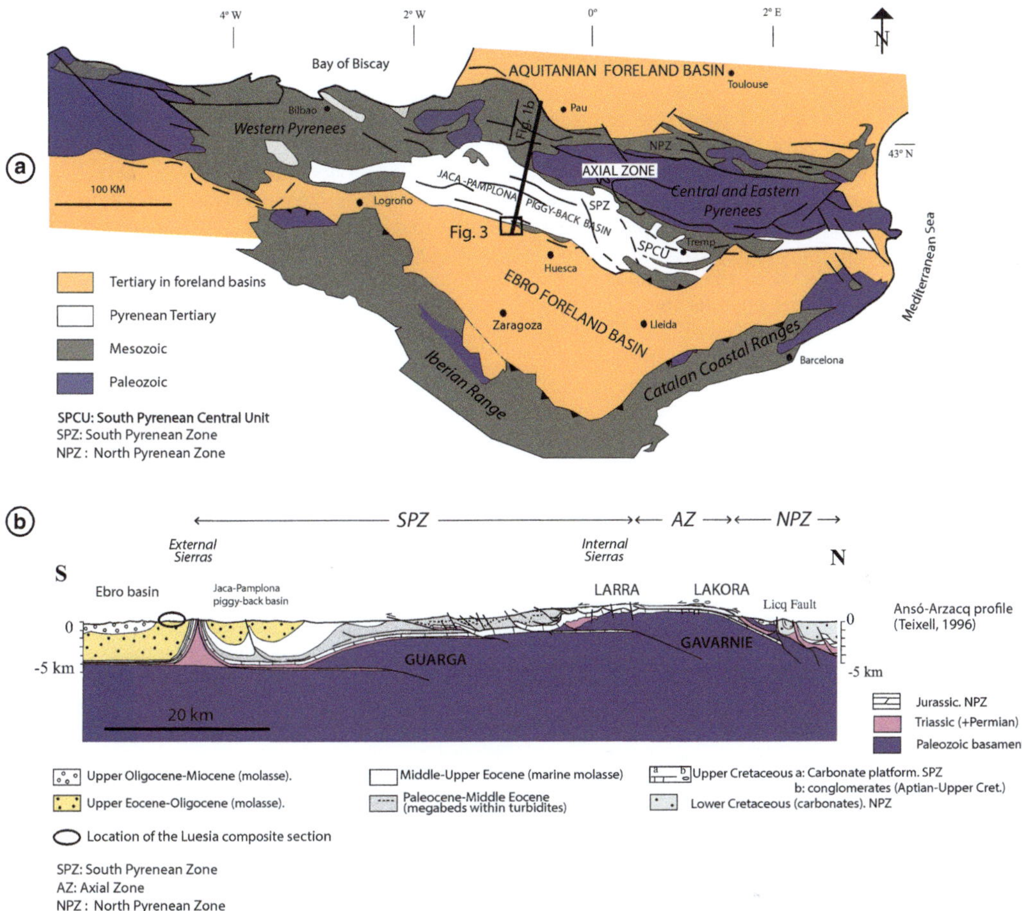

**Fig. 1.** (**a**) Geological map of NE Spain, showing the position of the Ebro Basin surrounded by the Pyrenees, Iberian and Catalan coastal ranges. The square marks the studied area represented in more detail in Figure 3. (**b**) The line is the trace of the geological cross-section represented in (a) which is based on Teixell (1996) and Millán-Garrido (1996) in the southernmost part.

given that the unconformity linked to that thrust sheet evolves to a conformity westwards. With the new magnetostratigraphic data we seek to determine the age of the upper part of the Campodarbe Formation in the selected Luesia composite section, which is crucial to (1) determine the isochronous character of the sequence boundaries in the Pyrenean foreland basin; (2) establish the accumulation rate of the Campodarbe Formation, especially in relation to the activity of the San Felices thrust sheet during its most active period; and (3) constrain unambiguously the age of the changes of the palaeogeography in the foreland basin occurring at the bound of the Campodarbe–Uncastillo formations. Lastly, the integration of all magnetostratigraphic data of the area will contribute to constrain the age of the syntectonic deposits.

## Geological setting

The study area is located at the northern boundary of the Ebro foreland basin, just to the south of the westernmost corner of the External Sierras and to the west of the South Pyrenean Central Unit (Fig. 1a). This is the source area for the Campodarbe Formation, in particular the uplift of the Axial Zone of the South Pyrenean Central Unit (SPCU), as the main uplift of that area (Beamud *et al.* 2010) is roughly coeval with the Campodarbe Formation.

The Pyrenees represent a double-verging asymmetric fold–thrust belt resulting from the collision between the Iberian and the Eurasian plates during late Cretaceous–Miocene times (e.g. Muñoz 1992), which also varies along strike. Close to the study area, in the meridian of the termination of the External Sierras (Santo Domingo anticline), the crustal Ansó–Arzacq cross-section (Teixell 1992, 1996; Fig. 1b) represents the structure as a general foreland-breaking sequence of basement thrusts, namely (from north to south): Lakora (the emplacement of which started during the late Cretaceous); Gavarnie; and Guarga thrusts. In addition, the deformation observed in the northern sector of the South Pyrenean Zone, with thin-skin fold–thrust sheets with south vergence affecting Meso-Cenozoic rocks, is related to the so-called Larra thrust system, which roots towards the north into the Lakora basement thrust (Teixell 1992, 1996). The main deformation activity spans from middle Eocene to early Miocene (Bartonian–Aquitanian) times (Soler & Puigdefàbregas 1970; Labaume *et al.* 1985; Teixell 1992).

The rocks affected by the deformation are Palaeozoic in the Axial Zone and marine platform carbonates, marls and sandstones of Upper Cretaceous–Paleocene age in the Internal Sierras, which unconformably overlie the Palaeozoic rocks. On top there are 3500–4000-m-thick marine turbiditic deposits, and 4500–7000-m-thick Eocene–Miocene transitional-deltaic and continental syntectonic deposits that fill the South Pyrenean foreland basin. In the South Pyrenean Zone, the main tectonic activity started in the north during Bartonian–Priabonian times with folds and thrusts affecting the marine platform deposits and the turbidites (Larra thrust system). The emplacement of the Gavarnie thrust in the north also affected the External Sierras during the deposition of the Arguis–Pamplona Marls Formation (marine deposits of Bartonian age; Gavarnie detachment unit, Choukroune *et al.* 1989). Later, the Guarga basement thrust affected the former thrust systems in a foreland-breaking sequence (Soler & Puigdefàbregas 1970; Labaume *et al.* 1985; Teixell 1992; Millán-Garrido 1996), emerging in the External Sierras and dividing the South Pyrenean foreland basin in the Jaca–Pamplona piggyback basin and the Ebro foreland basin.

## The External Sierras

*Stratigraphy.* The oldest sediments that crop out at the core of the External Sierras are middle–Upper Triassic Muschelkalk and Keuper facies (Fig. 2). Because of the role of the Keuper facies as the main detachment level (Almela & Ríos 1951; Séguret 1972), the thickness of this unit is difficult to determine; it is considered to be *c.* 500 m (Millán-Garrido 1996). The marine deposits on top of the Triassic sediments are represented by a Santonian–Maastrichtian carbonate marine platform (Mey *et al.* 1968) that is overlain by the Garumnian continental facies (Cuevas *et al.* 1989). Neither units reach more than 150 m in thickness. On top of them, carbonate marine limestones with alveolines (Guara Formation) developed during middle Eocene time with differential thickness along strike of the External Sierras (thinner to the west, varying from 1000 to 60 m; Puigdefàbregas 1975). External platform

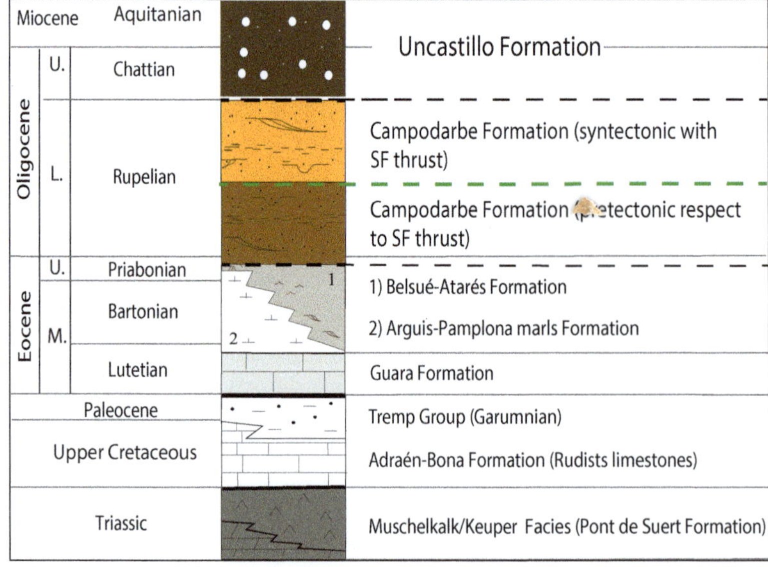



**Fig. 2.** General stratigraphy of the External Sierras and Ebro foreland basin in the studied area. Dashed lines: unconformities

and prodelta deposits of Bartonian age form the Arguis–Pamplona Marls Formation. Prograding deltaic sediments are represented by the Belsué–Atarés Formation of late Eocene age which reaches 2600 m in thickness at its maximum, but has variable thickness along the strike of the External Sierras (thinner to the west) (Teixell 1996).

On top, the youngest sediments are represented by the Campodarbe Formation and the overlying Uncastillo Formation, which filled the Ebro foreland basin with continental sediments between the latest Eocene and Miocene times. In the Jaca–Pamplona piggyback basin, Soler & Puigdefàbregas (1970) and Puigdefàbregas (1975) defined the Bernués Formation on top of the Campodarbe Formation as a lithostratigraphic unit that might be time-equivalent to the Uncastillo Formation (Chattian and maybe part of the Aquitanian for the Bernués Formation and Chattian–Burdigalian for the Uncastillo Formation). However, the Salinas magnetostratigraphic section (Hogan & Burbank 1996) ruled out this correlation. Montes (2002) also proposed an older deposition for the Bernués Formation, starting in the Rupelian (see section 'Previous stratigraphic and magnetostratigraphic framework' below). The Upper Eocene–Miocene detrital deposits are described in more detail in the 'Previous stratigraphic and magnetostratigraphic framework' section, since refining their stratigraphy, age and correlation in the foreland/piggyback basins are the main objectives of our study.

*Structure.* The subsurface structure of the western termination of the External Sierras was interpreted by Pocoví *et al.* (1990) and Millán-Garrido *et al.* (1992, 1995, 2000) as a large WNW–ESE-trending detachment fold (Santo Domingo anticline) which deforms the Meso-Cenozoic cover (cross-sections in Fig. 3). The core of this tight (almost isoclinal) anticline hosts a thrust sheet along the hinge with a hanging-wall movement to the south (South Pyrenean Frontal Thrust). The main detachment level is found within the middle–Upper Triassic sediments (Muschelkalk and Keuper Germanic facies). In its westernmost sector the axis of the Santo Domingo anticline plunges 60° westwards (San Marzal pericline). At the southern flank of this anticline, several small thrust sheets were formed imbricated in a unique frontal ramp (Guarga basement thrust). They formed firstly in the west and migrated towards the east in a hanging-wall breaking sequence (Millán-Garrido *et al.* 1995, 2000; Arenas *et al.* 2001). From west to east, these are: San Felices (early Oligocene), Punta Común (late Oligocene) and Riglos (Miocene) thrust sheets (Millán-Garrido *et al.* 1995, 2000; Arenas *et al.* 2001). Their tectonic activity is recorded in the syntectonic deposits of the Campodarbe Formation

and three tectonosedimentary units described in the area for the Uncastillo Formation (Arenas *et al.* 2001). Deformation continued with the later amplification and tightening of the Santo Domingo isoclinal fold, which partitioned the South Pyrenean foreland basin and tilted the whole thin-skinned thrust system and the Campodarbe Formation adjacent to the Santo Domingo anticline. In addition, the tightening of the Santo Domingo anticline produced a reactivation of the San Felices and Riglos thrust sheets (Millán-Garrido *et al.* 1995; Arenas *et al.* 2001; Fig. 4).

## Previous stratigraphic and magnetostratigraphic framework

The Campodarbe Formation (Upper Eocene–Lower Oligocene) is a succession of mostly alternating sandstones and siltstones of red and brownish colours (Soler & Puigdefàbregas 1970; Puigdefàbregas 1975). The studied rocks of the Campodarbe Formation belong to tectonosedimentary unit T3 (Rupelian; Pardo *et al.* 2004); these continental deposits formed in fluvial systems with a general WNW-flowing direction, mostly sourced on the South Pyrenean Central Unit (SPCU, Fig. 1a). In the Ebro Basin, the Uncastillo Formation corresponds to tectonosedimentary local units $U_1$, $U_2$ (late Oligoene–early Miocene) and $U_3$ (early Miocene–middle Miocene) and represents the alluvial and fluvial deposits derived from the north as a result of erosion of the South Pyrenean Zone (External Sierras, Jaca piggyback basin and northern zone of the Garvarnie unit) from the late Oligocene onwards (Arenas 1993; Arenas *et al.* 2001). Later, units U1 and U2 were correlated to TSU T4 and unit U3 to TSU T5 in the whole Ebro foreland basin (Pardo *et al.* 2004).

As already stated, the mainly fluvial sediments of the Campodarbe Formation show overall WNW-flowing palaeocurrents with a longitudinal distribution with respect to the Pyrenean direction (N110° E). To the west, these fluvial deposits change laterally to marshy and lacustrine sediments (Yeste-Arrés and Guendulain formations; Puigdefàbregas 1975). The Campodarbe Formation increases its thickness westwards (from *c.* 3500 m to *c.* 7000 m; Puigdefàbregas 1975). Since the first sedimentological studies, two lithostratigraphic units have been separated throughout the basins (piggyback and foreland) (Soler & Puigdefàbregas 1970; Puigdefàbregas 1975) and this division has also been used in later structural investigations (Millán-Garrido 1996; Oliva-Urcia *et al.* 2012). The division has its origin in the syntectonic unconformity affecting the Campodarbe Formation related to the San Felices thrust sheet in the External Sierras. The strata involved

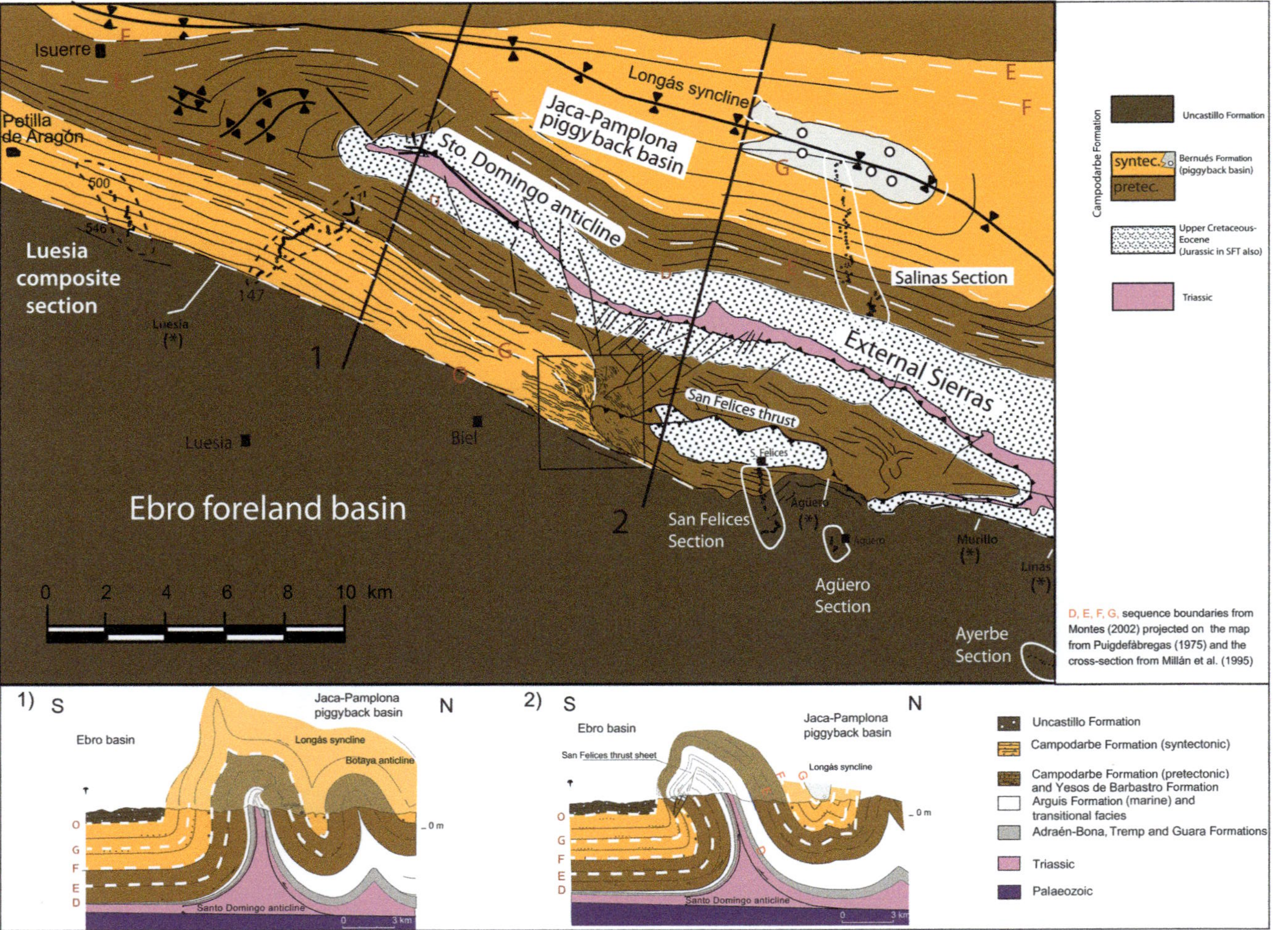



**Fig. 3.** Geological map showing the position of the Luesia composite section; the numbers indicate the position of samples 1–80, 100–147 and 500–546 (samples 78–103 overlap stratigraphically). The previous sections from Hogan & Burbank (1996) are Salinas, San Felices and Agüero sections. The square in the map marks the unconformity due to the San Felices thust sheet (Puigdefàbregas 1975) studied in detail in Millán-Garrido *et al.* (1995). Asterisks indicate the stratigraphical profiles from Arenas *et al.* (2001). Lower: two cross-sections from Millán-Garrido *et al.* (1995). The sequence boundaries defined by Montes (2002) are projected on the Puigdefàbregas (1975) cartography and Millán-Garrido *et al.* (1995) cross-sections.

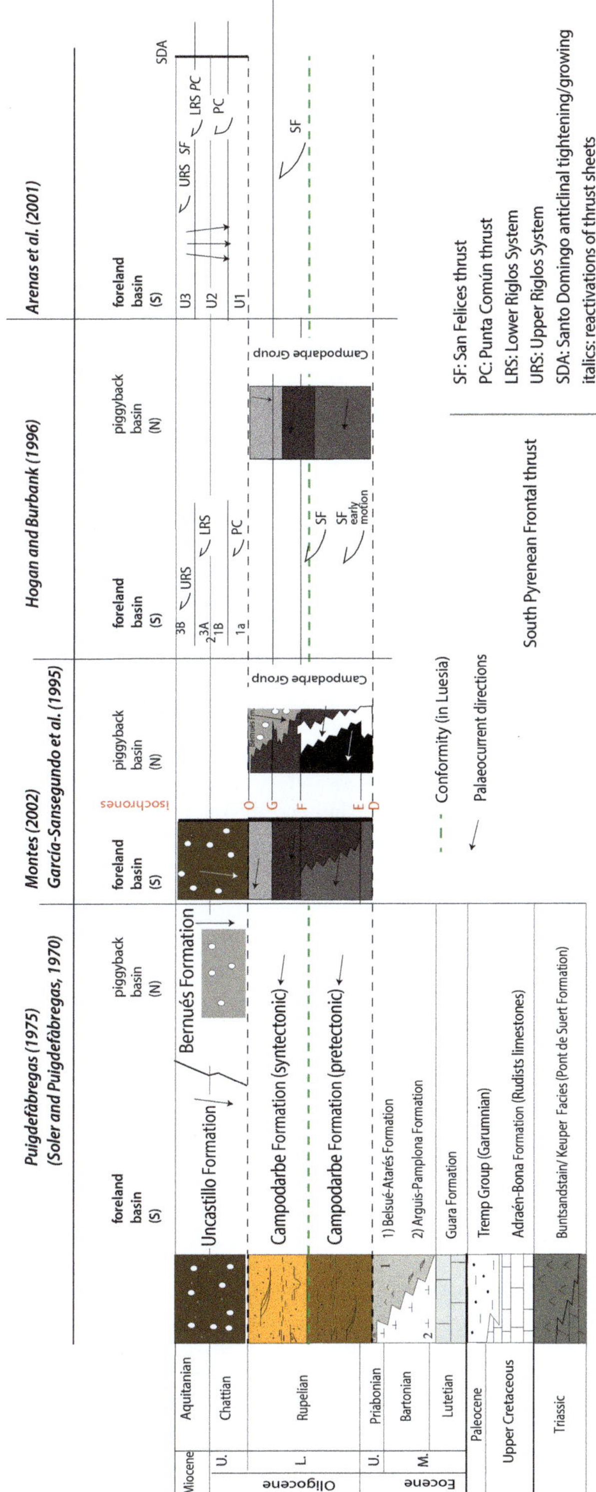


**Fig. 4.** Stratigraphic chart of the External Sierras area, showing the equivalence of the Campodarbe, Uncastillo and Bernués Formation divisions for different authors and the tectonic activity recorded by the Campodarbe and Uncastillo formations.

in such unconformity (offlap-onlap, 'fan-shaped' geometry) and the correlative conformable strata represent the upper part of the Campodarbe Formation (syntectonic Campodarbe unit). The underlying Campodarbe strata (lower part of the Campodarbe Formation) are pretectonic, that is, pre-San Felices thrust (Figs 3 & 4). The top of the Campodarbe Formation in the Ebro foreland basin was dated as being of Rupelian–Chattian age (Fig. 4; Puigdefàbregas 1975; Millán-Garrido *et al.* 1995).

In later investigations on the Campodarbe Formation, Montes (2002) established three depositional sequences in the foreland basin (south of the External Sierras) and in the piggyback basin (north of the External Sierras; Fig. 4), in the same area of the Salinas magnetostratigraphic section (Hogan & Burbank 1996). These depositional sequences are groups of lithofacies with a common vertical sedimentary polarity; the boundaries of these sequences (surfaces D, E, F and G in Montes 2002) represent vertical changes in the architecture of the depositional systems.

Surface E was defined by Montes (2002) to the north of the External Sierras, in the southern part of the piggyback basin, as the conformable superposition of meandering fluvial facies on gravel braided fluvial deposits. The surface represented by this lithological change can be followed in map and aerial photographs.

Surface F was also defined by Montes (2002) in the northern part of the piggyback basin, at the base of a thick conglomerate succession (San Juan de la Peña alluvial fan apex facies, 10 km to the northeast of the Longás syncline) over sandstone and siltstone facies, as part of a progressive unconformity. Despite the disconnection with the Ebro foreland basin, the author interpreted that the beginning of the syntectonic Campodarbe unit in the foreland basin was equivalent to surface F in the piggyback basin. The author considered surfaces E and F as isochronous in both basins as their stratigraphic position in the Salinas and Luesia sections was similar, taking the top of the Eocene transitional (deltaic) facies of Belsué–Atarés Formation as the starting point (surface D).

Surface G was also defined in the north of the piggyback basin, coinciding with a progradational stage of the San Juan de la Peña alluvial fan conglomerates, within the progressive unconformity starting on surface F (Montes 2002). In the foreland basin, the author suggested that surface G is a conformity that to the east corresponds to the offlap–onlap change related to the increasing–decreasing displacement of the San Felices thrust sheet in the foreland basin.

Except for the surface of the syntectonic unconformity linked to the San Felices thrust, the boundaries between those sequences are conformities and therefore isochronous surfaces. In addition, Montes (2002) also postulated the isochronous character of such boundaries within the Campodarbe Formation through the entire South Pyrenean basin (Jaca–Pamplona and Ebro basins).

The Uncastillo Formation (in the foreland Ebro basin) and the Bernués Formation (in the piggyback Jaca–Pamplona Basin) are similar in representing alluvial fan deposits with an overall provenance from the north (Mutti *et al.* 1972; Puigdefàbregas 1975). The Bernués Formation overlies the Campodarbe Formation as part of the San Juan de la Peña progressive unconformity. The Uncastillo Formation adjacent to the Luesia composite section in the Ebro Basin represents the deposition of two coalescent terminal fluvial–alluvial systems with a northern provenance and quasi-radial palaeocurrent distribution (Arenas *et al.* 2001); the thickness is *c.* 1600 m (Arenas 1993). Hirst & Nichols (1986) named this system of quasi-radial palaeocurrent distribution the 'Luna fluvial system'. The Uncastillo Formation overlies the External Sierras in angular unconformity (Riglos–Agüero unconformity), but to the west it forms a large cumulative wedge-out system with the underlying Campodarbe Formation (progressive unconformity of Biel-Gallipienzo, after Soler & Puigdefàbregas 1970; surface O, Figs 3 & 4). The three tectonosedimentary units described in the Uncastillo Formation (Fig. 4) record the effect of the cannibalization of the piggyback Jaca–Pamplona Basin and erosion of areas further north (Internal Sierras and Axial Zone), and show a cyclic retrogradational–progradational stage of the Luna fluvial system. Their boundaries are therefore recognized as the changes from coarsening- to fining-upwards trends in the vertical evolution, and in the foreland basin margin correlate with syntectonic unconformities in the Uncastillo Formation. These unconformities are related to the tectonic pulses affecting the External Sierras (Arenas *et al.* 2001).

## Available magnetostratigraphic ages

Previous magnetostratigraphic sections in the Campodarbe Formation have tried to constrain the age of these units. Hogan (1993) and Hogan & Burbank (1996) (see Fig. 3 for location) bracketed the Campodarbe Formation between Chron 17n (Priabonian) and Chron 10r (Rupelian). Two long sections performed in the Campodarbe Formation by these authors started in the marine deposits of the Arguis Formation, where planktonic foraminifera zoning was available (Canudo *et al.* 1988). However, considering the quality of the local magnetozones, and with new data from the eastern part of the Ebro foreland basin, Costa *et al.* (2010) reinterpreted the continental deposition (Campodarbe

Formation) in the Pyrenean foreland basin as having began in Chron 16n (*c.* 36 Myr). Together with magnetostratigraphic sections in the Uncastillo Formation, the previous magnetostratigraphic sections in the Campodarbe Formation constitute the magnetostratigraphic framework where the new Luesia composite section fits into place. The Salinas section covers the longest sequence of the Campodarbe Formation (193 samples, 25 magnetozones; Hogan 1993). It is located to the north of the Santo Domingo anticline in the piggyback Jaca–Pamplona Basin (Fig. 3). The other reviewed sections are located to the east of the Luesia section in the Ebro foreland basin, just to the south of the San Felices thrust sheet. These three sections are San Felices, Agüero and Ayerbe sections. The Agüero and Ayerbe sections were sampled from the Uncastillo Formation only, whereas the San Felices section was sampled from both Campodarbe and Uncastillo formations (although there is a shortage of sediments from the Campodarbe Formation as mentioned earlier; *c.* 4 Myr). The youngest age for the Uncastillo Formation is Chron 6 (Hogan 1993; Hogan & Burbank 1996). Based on the magnetostratigraphic data of Hogan & Burbank (1996), Arenas *et al.* (2001) proposed the approximate ages of the three tectonosedimentary Uncastillo units as follows: U1: Chattian; U2: upper Chattian–lower Aquitanian; and U3: Aquitanian.

Magnetostratigraphy in the foreland basin, far from unconformities, will help to reduce the uncertainties in the time of deposition of all these units. In combination with previously acquired data, the new magnetostratigraphic section will constrain the timing of those surfaces that were considered as isochrones across the basin by Montes (2002). Clarifying timing of deposition will in turn shed light on the areal distribution of sediment accumulation and its relation to the orogenic activity of the deformation front of the fold–thrust belt.

# Magnetostratigraphy of the Luesia composite section

## Methodology

The Luesia composite section spans the upper part of the Campodarbe Formation to the south of the External Sierras (Fig. 3). The end of the Campodarbe Formation in this section is recognized as a sharp lithological shift between alternating mudstones and sandstones of the Campodarbe Formation (late Eocene–early Oligocene) and conglomerates and sandstones of the Uncastillo Formation (late Oligocene–early Miocene). In the Luesia section 163 magnetostratigraphic sites have been sampled along *c.* 3300 m covering the Campodarbe Formation and a few metres of the Uncastillo Formation. Samples were taken in the field with an electric portable drill machine refrigerated with water, and oriented *in situ* with a magnetic compass coupled to a core orienting device with clinometer. The attitudes of the bedding planes were also measured with a compass in the field, and stratigraphic thickness was measured with a Jacob's staff. Siltstones and claystones were chosen for drilling after cleaning the outcrop by manual digging (40–70 cm deep). The occurrence of suitable lithology determined the sampling spacing with an average of 25 m which, according to previous age constrains (Costa *et al.* 2010), gives a time resolution of 50 kyr, sufficient to allow a complete identification of Oligocene and Lower Miocene geomagnetic polarity reversals. Two to three samples were taken at each site. One sample per site was thermally demagnetized at the Paleomagnetic Laboratory of Barcelona (CCiTUB-CSIC) using furnaces TSD-1 (Schonsted) and MMT80 (Magnetic Measurements). Temperature steps ranged from 25°C to 100°C, and were applied from initial room temperature up to 680°C. The remanence was measured in a cryogenic rock magnetometer (2G Enterprises). The maximum unblocking temperatures range from 550°C to 680°C, which suggest magnetite and hematite as the main carriers of the magnetization (Fig. 5). Characteristic components were obtained by principal component analyses (PCA; Kirschvink 1980) after visual inspection of the demagnetization diagrams. Good-quality characteristic components (class 1 and 2), defined by more than 5 steps and/or MADs (maximum angular deviation) less than 10°, were obtained in 76% of the samples (Table 1).

## Correlation to the GPTS

The local magnetic stratigraphy of the Luesia Section is produced after computing the latitude of the virtual geomagnetic poles (VGP). Positive VGP latitudes are interpreted as normal polarities and negative latitudes are interpreted as reverse polarities. A total of 15 magnetozones, 7 normal (N1–N7) and 8 reverse (R1–R8) were defined by at least two adjacent palaeomagnetic sites with the same polarity (Fig. 6). Short single-site magnetozones were not considered for correlation purposes and were represented as a half-width bar in the local magnetostratigraphy. A correlation with the GPTS (Gradstein & Ogg 2012) is put forward based on the characteristic reversal pattern of the Luesia section and the chronostratigraphic constraints provided by earlier magnetostratigraphic studies in the Jaca–Pamplona Basin (Hogan & Burbank 1996). The unique 1000-m-thick reverse magnetozone (R1) identified in the lower Luesia section is equivalent to the reverse magnetozone of similar thickness and stratigraphic position

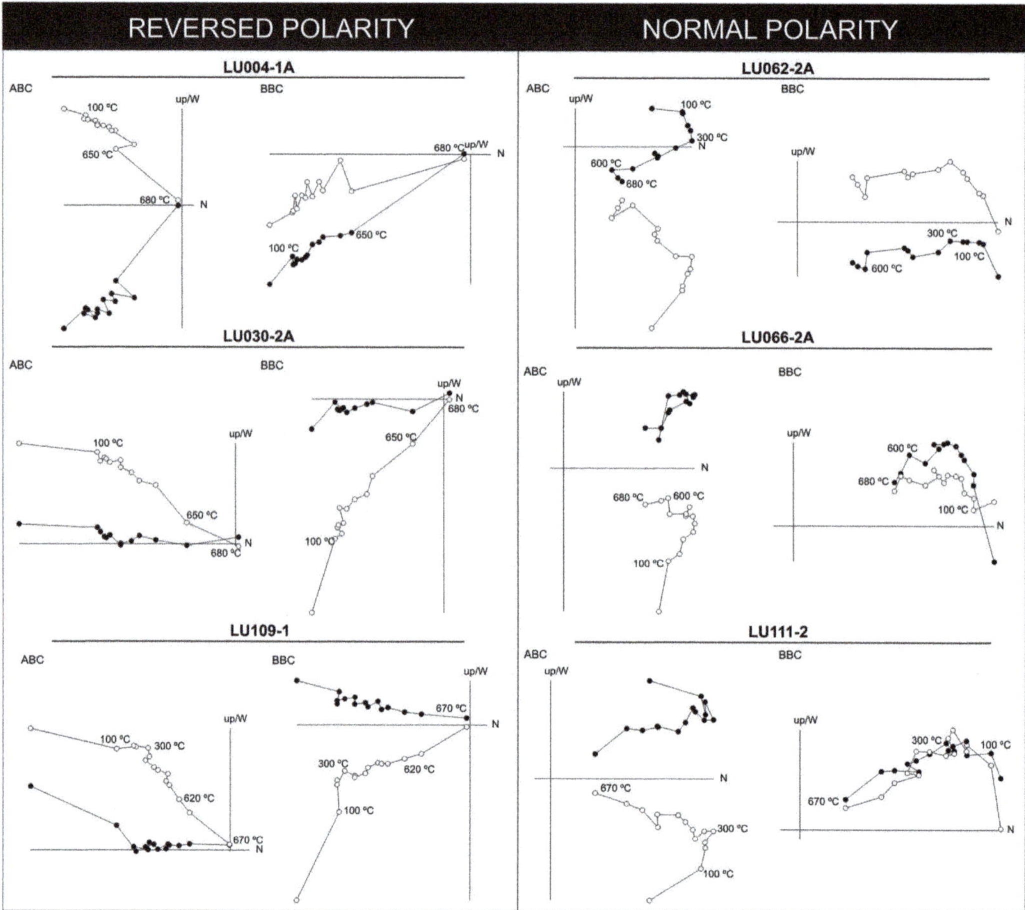



**Fig. 5.** Orthogonal diagrams of selected samples, showing normal and reverse polarities. Number indicates the sample (see Table 1). Black dots represent projection of the palaeomagnetic vector on the horizontal plane and white dots represent projection on the vertical plane. ABC: after bedding correction; BBC: before bedding correction.

found in the Salinas section, on the northern flank of the Santo Domingo anticline (Figs 1 & 7). A magnetostratigraphic correlation between the Salinas and Luesia sections yields a positive match of reversals, and the combined long reversal sequence supports a unique correlation with the GPTS in agreement with earlier work of Hogan & Burbank (1996). Only the correlation of the normal magnetozone N7 at the top of Luesia to either C7n or C7An remains uncertain (Fig. 6). Missing a short (100 kyr) reversed interval such as Chron C7Ar is considered plausible as a result of either the discontinuous nature of alluvial sedimentation or a sampling gap. With the proposed correlation the sedimentation in the Luesia section would range from 33 to 24.5 Myr (option a; Fig. 6) or *c.* 23.5 Myr (option b), thus recording the most of the Oligocene. The top of the Campodarbe Formation yields an age of 24 Myr (base of Chron

C6r), much younger than previously stated (Chron 10r after Hogan & Burbank 1996). The boundary between the Campodarbe and Uncastillo formations therefore occurs in the late Chattian, and not at the Rupelian–Chattian transition as previously considered (as represented in Fig. 2).

## Discussion: new stratigraphic and magnetostratigraphic framework

*Are the bounding surfaces (sequence boundaries) E, F and G in the Campodarbe Formation isochronous?*

The new magnetostratigraphic data from the Luesia section to the south of the External Sierras allows a direct correlation with the previous

**Table 1.** *Palaeomagnetic data*

| Site | Level | Dg | Ig | Ds | Is | Palaeo lat. | Int | Err | Q | T steps | T range (°C) |
|---|---|---|---|---|---|---|---|---|---|---|---|
| LU001-1A | 0 | 227.0 | 64.2 | 216.8 | −24.3 | −46.53 | 333.41 | 4.07 | 2 | 5 | 250–550 |
| LU002-2A | 23 | 182.1 | 48.0 | 186.5 | −37.0 | −67.7 | 776.42 | 4.60 | 2 | 6 | 250–600 |
| LU003-2A | 58 | 172.0 | 60.7 | 189.6 | −23.4 | −58.77 | 257.59 | 3.55 | 2 | 6 | 400–600 |
| LU004-1A | 91 | 148.9 | 10.3 | 126.8 | −31.5 | −38.35 | 3117.82 | 3.32 | 2 | 6 | 300–600 |
| LU005-2B | 117 | 136.5 | 42.2 | 160.6 | −17.2 | −52.39 | 39.38 | 10.84 | 2 | 4 | 250–400 |
| LU006-1B | 142 | 152.6 | 71.7 | 192.3 | −15.2 | −53.76 | 188.05 | 8.83 | 2 | 5 | 400–550 |
| LU007-1A | 195 | 248.4 | 72.9 | 220.9 | −18.2 | −41.35 | 1067.47 | 3.46 | 2 | 6 | 200–600 |
| LU008-1A | 214.5 | 105.9 | 72.7 | 193.0 | −0.7 | −46.44 | 792.35 | 10.07 | 2 | 6 | 250–650* |
| LU009-1A | 239.5 | 261.3 | 55.1 | 239.5 | −25.2 | −31.33 | 1106.89 | 6.17 | 2 | 3 | 250–350 |
| LU010-1A | 259.5 | 176.4 | 44.1 | 181.0 | −36.1 | −67.74 | 496.93 | 6.33 | 2 | 5 | 150–500* |
| LU011-1A | 287.5 | 197.1 | 54.3 | 201.7 | −35.2 | −60.71 | 1809.52 | 2.61 | 2 | 6 | 250–600 |
| LU013-1A | 330.4 | 197.6 | 31.8 | 189.3 | −56.0 | −80.8 | 401.63 | 3.67 | 2 | 5 | 200–500 |
| LU014-1B | 338.4 | 203.7 | 34.3 | 199.5 | −55.6 | −73.75 | 71.24 | 7.50 | 2 | 3 | 200–350 |
| LU015-2B | 375.4 | 206.3 | 30.9 | 201.7 | −59.5 | −73.6 | 112.72 | 10.34 | 2 | 5 | 200–500 |
| LU016-1a | 394.4 | 177.0 | −39.0 | 72.0 | −37.0 | −1.34 | 231.00 | 15.00 | 3 | 4 | 350–610 |
| LU017-1A | 414.4 | 196.6 | 53.9 | 198.5 | −45.2 | −68.33 | 232.51 | 6.98 | 2 | 5 | 150–550* |
| LU018-1A | 430.9 | 201.2 | 32.1 | 193.2 | −67.1 | −78.16 | 1604.05 | 1.25 | 1 | 5 | 150–550 |
| LU019-2a | 451.9 | 221.0 | −22.0 | 21.0 | −66.0 | −2.49 | 151.00 | 8.00 | 3 | 5 | 300–600 |
| LU020-2A | 475.9 | 197.5 | 46.6 | 196.1 | −52.4 | −74.23 | 250.98 | 7.23 | 2 | 4 | 200–400 |
| LU021-1A | 504.9 | 205.8 | 26.7 | 201.6 | −73.2 | −68.7 | 438.12 | 2.26 | 2 | 5 | 400–550 |
| LU022-1A | 516.4 | 191.9 | 35.8 | 181.1 | −60.4 | −88.78 | 216.10 | 2.46 | 2 | 3 | 200–300 |
| LU023-2A | 533.9 | 194.4 | 55.4 | 199.4 | −37.7 | −63.32 | 792.28 | 3.36 | 1 | 6 | 200–680 |
| LU024-3A | 533.9 | 215.5 | 49.2 | 215.3 | −45.6 | −57.58 | 386.75 | 3.67 | 2 | 4 | 350–450 |
| LU025-2A | 596.3 | 284.5 | 43.7 | 257.7 | −16.0 | −14.54 | 583.03 | 2.79 | 2 | 6 | 250–600 |
| LU026-1A | 618.9 | 143.1 | 70.1 | 193.0 | −25.3 | −58.89 | 240.67 | 3.45 | 2 | 5 | 200–550 |
| LU027-2A | 645.3 | 235.1 | 60.3 | 229.1 | −47.0 | −48.14 | 637.23 | 1.97 | 2 | 5 | 200–550 |
| LU028-2A | 667.6 | 142.6 | 25.9 | 143.8 | −24.9 | −47.22 | 687.59 | 3.22 | 2 | 6 | 150–650 |
| LU029-1A | 700.9 | 120.1 | 64.9 | 187.6 | −16.5 | −55.47 | 2047.95 | 3.00 | 2 | 6 | 350–650 |
| LU030-2A | 706.6 | 174.3 | 49.5 | 182.3 | −33.7 | −66.08 | 620.62 | 1.84 | 2 | 6 | 200–600 |
| LU031-2A | 730.6 | 129.8 | 57.7 | 176.0 | −7.2 | −51.16 | 654.67 | 1.21 | 2 | 5 | 400–550 |
| LU032-1A | 749.3 | 247.7 | 76.1 | 216.9 | −11.6 | −41.1 | 160.28 | 12.12 | 2 | 4 | 200–450* |
| LU033-1A | 774.7 | 245.7 | 55.1 | 231.2 | −27.7 | −38.3 | 852.57 | 2.33 | 2 | 1 | 150–500 |
| LU034-1A | 805.1 | 113.7 | 66.3 | 184.2 | 0.7 | −47.18 | 208.32 | 4.59 | 2 | 4 | 200–400 |
| LU035-1A | 837.9 | 247.9 | 39.0 | 239.8 | −27.0 | −31.75 | 856.55 | 2.56 | 2 | 5 | 200–550 |
| LU036-3A | 863.5 | 163.9 | 47.1 | 171.5 | −30.8 | −63.31 | 888.01 | 1.32 | 2 | 4 | 300–400 |
| LU037-1A | 891.1 | 150.2 | 34.0 | 152.1 | −31.2 | −55.15 | 235.91 | 4.37 | 2 | 5 | 350–550 |
| LU038-1A | 904.3 | 186.1 | 58.2 | 190.9 | −27.1 | −60.49 | 372.19 | 2.59 | 2 | 5 | 200–500 |
| LU039-2A | 945.7 | 190.6 | 63.3 | 194.3 | −22.5 | −56.93 | 812.28 | 5.43 | 2 | 6 | 350–680 |
| LU040-1a | 972.2 | 337.0 | −49.0 | 232.0 | −32.0 | −39.45 | 120.00 | 12.00 | 3 | 4 | 350–650 |
| LU041-2A | 983.9 | 153.0 | 60.3 | 186.6 | −18.5 | −56.66 | 1071.33 | 1.85 | 2 | 5 | 200–500 |
| LU042-1A | 1003.1 | 280.4 | 75.1 | 227.7 | −10.7 | −34.01 | 505.80 | 2.82 | 1 | 6 | 200–680 |
| LU043-1A | 1042.4 | 316.7 | −19.9 | 323.1 | 13.8 | 41.99 | 1093.72 | 2.25 | 2 | 6 | 300–680 |
| LU045-2A | 1077.4 | 284.6 | 52.0 | 251.0 | −15.7 | −19.42 | 562.79 | 12.83 | 2 | 5 | 150–500* |
| LU044-2A | 1103.4 | 349.9 | −13.6 | 318.3 | 45.9 | 53.02 | 28.95 | 31.70 | 2 | 4 | 200–400* |
| LU046-2A | 1120.4 | 314.6 | −22.3 | 325.5 | 11.9 | 42.6 | 108.65 | 20.80 | 2 | 6 | 300–600* |
| LU047-2A | 1157.9 | 7.4 | −8.7 | 313.2 | 63.5 | 56.56 | 232.29 | 16.97 | 2 | 6 | 300–680* |
| LU048-4A | 1210.4 | 326.0 | 39.2 | 261.4 | 13.7 | −1.61 | 292.83 | 7.00 | 3 | 5 | 250–680 |
| LU049-1A | 1231.4 | 178.9 | 58.0 | 193.2 | −30.5 | −61.75 | 549.70 | 4.37 | 2 | 6 | 250–680 |
| LU051-1A | 1251.4 | 194.3 | 74.4 | 205.7 | −15.0 | −48.62 | 292.31 | 3.45 | 1 | 6 | 300–680 |
| LU052-1A | 1266 | 209.5 | 52.9 | 209.6 | −37.1 | −57.07 | 128.55 | 6.91 | 2 | 4 | 250–450* |
| LU050-1B | 1297 | 162.2 | 37.4 | 164.8 | −33.8 | −62.92 | 75.14 | 10.19 | 2 | 4 | 300–450 |
| LU053-2A | 1297 | 164.0 | 70.6 | 195.8 | −13.3 | −51.82 | 68.02 | 3.83 | 3 | 3 | 400–550 |
| LU054-2A | 1315 | 157.2 | 68.2 | 192.3 | −13.0 | −52.64 | 135.30 | 8.33 | 3 | 3 | 250–350 |
| LU055-2A | 1340 | 195.5 | 37.2 | 191.5 | −49.8 | −75.1 | 175.79 | 10.67 | 2 | 6 | 300–680* |
| LU056-2B | 1357.5 | 186.1 | 14.9 | 153.4 | −60.4 | −70.21 | 255.93 | 3.37 | 1 | 6 | 400–650 |
| LU057-1A | 1371.2 | 35.3 | 33.4 | 207.0 | 56.4 | −6.97 | 356.02 | 4.56 | 1 | 6 | 250–600 |
| LU058-2A | 1386.5 | 265.7 | 45.8 | 250.1 | −24.4 | −23.22 | 233.59 | 2.29 | 2 | 5 | 400–500 |
| LU059-1A | 1410.5 | 346.3 | −28.2 | 337.7 | 39.7 | 62.9 | 225.85 | 4.33 | 2 | 5 | 300–550 |

*(Continued)*

**Table 1.** *Palaeomagnetic data* (*Continued*)

| Site | Level | Dg | Ig | Ds | Is | Palaeo lat. | Int | Err | Q | *T* steps | *T* range (°C) |
|---|---|---|---|---|---|---|---|---|---|---|---|
| LU060-1b | 1428 | 90.0 | −34.0 | 160.0 | −26.0 | −56.58 | 68.00 | 26.00 | 3 | 3 | 250–350 |
| LU061-1A | 1454 | 154.4 | −74.6 | 42.6 | −8.7 | 29.48 | 349.17 | 6.30 | 3 | 6 | 450–600 |
| LU062-2A | 1474 | 354.6 | −10.0 | 334.7 | 40.9 | 61.74 | 222.15 | 10.67 | 2 | 5 | 300–550* |
| LU063-1a | 1493 | 204.0 | 7.0 | 81.0 | 60.0 | 31.87 | 17.00 | 6.00 | 3 | 3 | 550–680 |
| LU064-1A | 1498 | 351.4 | −27.6 | 350.7 | 28.7 | 61.88 | 19.63 | 10.28 | 2 | 4 | 400–450 |
| LU065-2A | 1514.2 | 352.9 | −46.0 | 6.8 | 17.6 | 56.18 | 430.64 | 14.98 | 2 | 6 | 300–650* |
| LU066-2A | 1527.3 | 331.3 | −16.7 | 330.7 | 17.5 | 47.92 | 1005.86 | 2.34 | 2 | 6 | 350–600 |
| LU068-2B | 1553.5 | 167.2 | 36.9 | 176.5 | −20.3 | −58.05 | 240.50 | 5.74 | 2 | 5 | 350–500 |
| LU069-1A | 1568 | 261.4 | 82.4 | 223.1 | 8.4 | −29.26 | 87.77 | 20.62 | 3 | 3 | 300–400 |
| LU070-1A | 1592.3 | 14.6 | −27.7 | 9.5 | 42.6 | 70.7 | 803.03 | 1.69 | 2 | 4 | 250–400 |
| LU071-1A | 1625.7 | 25.0 | −24.0 | 19.4 | 50.0 | 70.69 | 292.96 | 6.95 | 2 | 6 | 250–600 |
| LU072-2A | 1649.5 | 10.8 | −44.7 | 16.7 | 26.4 | 58.14 | 1186.19 | 2.34 | 2 | 6 | 350–680 |
| LU073-1A | 1673.2 | 22.9 | −26.8 | 18.2 | 46.7 | 69.44 | 606.92 | 3.89 | 2 | 4 | 350–450 |
| LU074-2B | 1707.4 | 196.0 | 50.9 | 199.4 | −28.3 | −58.05 | 678.95 | 1.95 | 2 | 6 | 400–680 |
| LU075-1B | 1720.8 | 224.8 | 55.1 | 218.3 | −23.3 | −45.2 | 760.50 | 2.86 | 2 | 3 | 200–350 |
| LU076-1A | 1740.5 | 220.4 | 58.3 | 214.8 | −20.9 | −46.26 | 94.02 | 10.07 | 2 | 5 | 250–550 |
| LU077-2A | 1761.8 | 161.6 | 31.5 | 162.9 | −29.4 | −59.65 | 185.49 | 5.63 | 2 | 6 | 300–650 |
| LU078-2A | 1779.8 | 200.1 | 56.3 | 203.2 | −23.3 | −53.77 | 158.80 | 3.30 | 2 | 5 | 400–500 |
| LU101-1 | 1793.3 | 188.4 | 48.3 | 195.2 | −25.0 | −58 | 859.71 | 1.41 | 2 | 5 | 300–560 |
| LU102-1A | 1809.8 | 129.5 | 43.6 | 166.6 | 4.1 | −44 | 275.00 | 8.16 | 2 | 5 | 200–560* |
| LU079-1A | 1816.9 | 173.4 | 65.4 | 193.9 | −10.4 | −50.93 | 846.97 | 3.09 | 2 | 6 | 350–650 |
| LU080-1A | 1835.3 | 155.1 | 32.2 | 160.2 | −24.3 | −55.82 | 272.30 | 4.29 | 2 | 5 | 350–500 |
| LU103-1 | 1840.3 | 267.9 | 50.4 | 244.9 | −10.5 | −21.99 | 189.92 | 9.92 | 2 | 5 | 200–480 |
| LU104-2 | 1855.3 | 236.3 | 43.3 | 232.4 | −29.7 | −38.25 | 1196.06 | 1.45 | 1 | 5 | 200–560 |
| LU105-1 | 1894.3 | 176.6 | 11.2 | 155.2 | −46.7 | −65.31 | 148.91 | 5.89 | 2 | 5 | 250–480 |
| LU106-1B | 1930.3 | 195.8 | 45.7 | 199.2 | −29.3 | −58.69 | 658.13 | 11.78 | 2 | 6 | 250–590* |
| LU107-2A | 1948.3 | 178.2 | 28.5 | 174.3 | −36.7 | −67.59 | 149.00 | 27.29 | 2 | 5 | 250–520* |
| LU108-2 | 1970.05 | 212.0 | 32.4 | 211.8 | −44.6 | −59.46 | 423.62 | 10.55 | 2 | 5 | 250–520* |
| LU109-1 | 2000.05 | 190.8 | 18.3 | 177.5 | −51.9 | −80.04 | 434.31 | 8.74 | 1 | 6 | 250–650* |
| LU110-1 | 2022.55 | 221.8 | 54.9 | 218.3 | −21.7 | −44.44 | 495.63 | 5.91 | 1 | 6 | 250–670* |
| LU111-2 | 2040.05 | 338.6 | −35.9 | 348.7 | 19.2 | 56.13 | 335.76 | 14.87 | 2 | 6 | 300–650* |
| LU113-1 | 2061.05 | 356.4 | −56.5 | 13.0 | 14.2 | 53.09 | 433.74 | 5.71 | 2 | 5 | 350–560 |
| LU114-1 | 2071.55 | 22.0 | −31.5 | 19.9 | 44.4 | 66.97 | 498.47 | 11.36 | 1 | 6 | 250–650* |
| LU115-1 | 2133.05 | 248.5 | 32.0 | 249.3 | −33.5 | −27.39 | 621.98 | 3.27 | 2 | 5 | 480–560 |
| LU116-1A | 2151.05 | 347.7 | −50.1 | 4.6 | 15.6 | 55.41 | 437.54 | 3.19 | 2 | 6 | 300–590 |
| LU117-2 | 2167.55 | 358.9 | 60.0 | 233.2 | 36.8 | −10.34 | 190.69 | 4.18 | 2 | 5 | 200–480 |
| LU118-1 | 2205.05 | 237.1 | 9.1 | 261.6 | −57.3 | −29.87 | 273.44 | 13.04 | 2 | 5 | 200–560* |
| LU119-1A | 2224.55 | 218.5 | 10.1 | 226.8 | −66.2 | −57.12 | 304.48 | 5.13 | 2 | 4 | 250–400 |
| LU120-1 | 2244.05 | 165.1 | 38.1 | 173.9 | −22.1 | −58.74 | 100.85 | 26.81 | 2 | 5 | 350–560* |
| LU121-1A | 2259.75 | 174.6 | 38.0 | 179.5 | −27.7 | −62.4 | 622.35 | 7.18 | 1 | 6 | 200–590* |
| LU122-2 | 2283.75 | 189.4 | 57.3 | 199.8 | −17.1 | −52.19 | 256.96 | 3.95 | 1 | 5 | 250–560 |
| LU123-1 | 2306.25 | 260.8 | −32.6 | 335.2 | −42.1 | 19.54 | 429.32 | 12.18 | 2 | 6 | 400–590* |
| LU124-1 | 2336.25 | 339.4 | −1.3 | 313.8 | 35.1 | 44.97 | 475.33 | 3.46 | 2 | 6 | 560–590 |
| LU126-1 | 2367.75 | 267.6 | 74.8 | 225.2 | 4.0 | −29.82 | 345.60 | 3.48 | 2 | 6 | 480–590 |
| LU128-1 | 2397.75 | 7.0 | −15.5 | 350.1 | 51.7 | 77.37 | 316.59 | 9.03 | 1 | 6 | 200–590* |
| LU129-1A | 2410.25 | 135.4 | 4.0 | 129.7 | −11.3 | −32.44 | 414.92 | 6.63 | 1 | 5 | 250–520* |
| LU130-2 | 2475.25 | 7.5 | −38.9 | 9.5 | 33.0 | 64.38 | 493.62 | 7.93 | 1 | 5 | 250–560* |
| LU131-1 | 2525.3 | 9.5 | 3.7 | 322.8 | 65.2 | 63.39 | 193.59 | 10.62 | 2 | 5 | 200–560* |
| LU500-1 | 2535 | 18.6 | −55.5 | 17.7 | 34.4 | 62.19 | 105.61 | 9.99 | 2 | 5 | 400–520 |
| LU501-1 | 2539.5 | 298.1 | −30.7 | 317.1 | 10.6 | 36.98 | 200.95 | 5.21 | 2 | 5 | 250–480 |
| LU502-1 | 2574 | 168.7 | 72.4 | 183.4 | −10.4 | −52.82 | 228.90 | 8.13 | 2 | 6 | 560–620 |
| LU503-1 | 2601 | 198.5 | 44.6 | 198.7 | −33.4 | −61.18 | 120.04 | 10.10 | 2 | 5 | 400–560* |
| LU504-2 | 2614.5 | 142.7 | 43.0 | 160.4 | −16.8 | −52.09 | 160.08 | 7.37 | 2 | 5 | 250–520* |
| LU505-1 | 2650.5 | 230.1 | 51.0 | 216.7 | −7.3 | −39.39 | 500.94 | 2.84 | 2 | 5 | 400–560 |
| LU506-2 | 2659.5 | 145.4 | 31.2 | 155.6 | −13.4 | −48.44 | 645.66 | 2.41 | 1 | 5 | 300–560 |
| LU132-1A | 2670 | 186.6 | 34.4 | 185.9 | −36.5 | −67.43 | 412.10 | 4.93 | 1 | 6 | 200–590 |
| LU133-1 | 2675 | 190.2 | 49.7 | 197.0 | −24.2 | −56.93 | 185.20 | 9.73 | 1 | 6 | 300–620* |
| LU508-2 | 2685.5 | 337.3 | −24.7 | 336.8 | 25.7 | 54.92 | 337.08 | 5.83 | 2 | 6 | 400–590 |

(*Continued*)

**Table 1.** *Palaeomagnetic data (Continued)*

| Site | Level | Dg | Ig | Ds | Is | Palaeo lat. | Int | Err | Q | $T$ steps | $T$ range (°C) |
|---|---|---|---|---|---|---|---|---|---|---|---|
| LU511-3 | 2774 | 33.3 | −22.2 | 31.4 | 57.7 | 65.86 | 237.08 | 5.31 | 2 | 5 | 250–480 |
| LU512-2 | 2787.5 | 144.8 | 26.5 | 155.9 | −12.1 | −47.99 | 304.64 | 3.07 | 2 | 5 | 300–480 |
| LU513-1 | 2823.5 | 343.8 | −24.7 | 336.9 | 43.9 | 64.83 | 230.31 | 10.28 | 2 | 6 | 480–650 |
| LU134-1 | 2873 | 18.4 | −0.3 | 344.6 | 70.5 | 74.06 | 586.30 | 6.36 | 1 | 6 | 250–650* |
| LU135-2 | 2879 | 27.2 | −51.7 | 29.0 | 25.1 | 51.58 | 280.41 | 3.93 | 1 | 5 | 400–560 |
| LU136-1a | 2885 | 358.8 | −15.5 | 342.1 | 45.9 | 69.09 | 228.20 | 19.17 | 1 | 6 | 250–590* |
| LU137-1 | 2889.5 | 331.6 | −31.8 | 341.9 | 16.3 | 52.48 | 74.22 | 18.12 | 2 | 5 | 250–520* |
| LU138-1a | 2895.5 | 7.9 | −57.7 | 19.2 | 16.4 | 52.08 | 633.14 | 3.37 | 2 | 6 | 250–590 |
| LU517 | 2897 | 359.6 | −11.6 | 327.1 | 60.5 | 65.61 | 1097.76 | 1.96 | 1 | 5 | 250–560 |
| LU139-1b | 2901.5 | 12.5 | −39.3 | 13.8 | 34.4 | 63.84 | 213.00 | 18.62 | 2 | 6 | 250–620* |
| LU516-1 | 2904.5 | 355.9 | −32.8 | 350.4 | 44.1 | 71.77 | 391.22 | 2.56 | 2 | 5 | 300–520 |
| LU140-1 | 2907.5 | 209.7 | 64.1 | 211.4 | −12.9 | −44.72 | 42.01 | 21.90 | 3 | 5 | 150–350* |
| LU518-1 | 2946.5 | 292.0 | 50.2 | 244.6 | 1.3 | −18.06 | 149.15 | 4.86 | 2 | 5 | 200–480 |
| LU141-1 | 2957 | 185.8 | 30.9 | 182.7 | −39.0 | −69.62 | 364.20 | 4.00 | 2 | 5 | 250–520 |
| LU142-2a | 2967.5 | 270.6 | 35.9 | 258.5 | −16.7 | −14.25 | 151.45 | 7.36 | 2 | 5 | 250–480 |
| LU143-2 | 2972 | 180.1 | 60.9 | 197.2 | −11.7 | −50.52 | 303.76 | 4.18 | 1 | 6 | 250–590 |
| LU519-1 | 2975 | 218.1 | 52.3 | 214.1 | −7.2 | −40.82 | 229.50 | 6.30 | 2 | 4 | 250–440 |
| LU144-1b | 2976.5 | 175.6 | 28.4 | 172.2 | −35.2 | −66.17 | 70.32 | 16.16 | 3 | 5 | 520–560 |
| LU146-2a | 2991.5 | 151.6 | 58.9 | 185.8 | −2.9 | −48.81 | 100.95 | 2.29 | 3 | 4 | 350–480* |
| LU147-1a | 2995.25 | 157.5 | 61.6 | 189.8 | −3.8 | −48.65 | 323.95 | 3.24 | 3 | 4 | 350–440 |
| LU521-1 | 3030 | 173.8 | 50.6 | 184.9 | −27.3 | −61.85 | 145.40 | 31.64 | 2 | 6 | 250–590* |
| LU523-1 | 3095.5 | 259.5 | 28.3 | 252.6 | −16.8 | −18.59 | 184.52 | 8.43 | 2 | 3 | 200–350 |
| LU524-3 | 3122.5 | 272.7 | 36.1 | 251.1 | −3.3 | −14.98 | 262.35 | 3.60 | 2 | 5 | 350–520 |
| LU525-1 | 3154 | 6.6 | −16.6 | 359.8 | 50.3 | 78.72 | 177.08 | 7.89 | 2 | 6 | 400–590 |
| LU527-2 | 3173.5 | 162.2 | 21.0 | 148.7 | −40.0 | −57.46 | 569.78 | 3.26 | 2 | 6 | 350–640 |
| LU528-3 | 3181 | 334.3 | −48.9 | 344.7 | 24.6 | 57.75 | 489.49 | 3.68 | 2 | 5 | 400–480 |
| LU529-1 | 3199 | 331.6 | −0.6 | 291.4 | 51.7 | 36.01 | 399.05 | 4.17 | 2 | 6 | 400–580 |
| LU530-1b | 3218.5 | 222.3 | 48.9 | 216.3 | −22.6 | −46.13 | 289.86 | 2.06 | 2 | 4 | 250–440 |
| LU531-2a | 3247 | 223.4 | 11.1 | 235.9 | −49.8 | −44.43 | 372.15 | 4.42 | 2 | 3 | 200–300 |
| LU532-3 | 3259 | 231.8 | 39.4 | 228.0 | −24.6 | −39.24 | 264.64 | 3.68 | 2 | 2 | 200–250 |
| LU533-1 | 3272.5 | 242.3 | 7.6 | 259.8 | −41.3 | −22.95 | 226.67 | 2.52 | 2 | 2 | 200–250 |
| LU534-2 | 3273.5 | 238.0 | 52.9 | 223.4 | −9.1 | −36.11 | 836.66 | 2.48 | 2 | 5 | 250–520 |
| LU535-3 | 3287 | 227.1 | 53.3 | 221.2 | −14.5 | −39.69 | 96.97 | 7.43 | 3 | 2 | 300–400 |
| LU536-1 | 3306.5 | 226.7 | 47.5 | 220.1 | −13.2 | −39.78 | 567.59 | 4.90 | 2 | 6 | 150–250* |

Site: name of the sample; Level: thickness (m) from the bottom sample; Dg, Ig: declination and inclination in sample reference system; Ds, Int, intensity of the component; Is: declination and inclination *in situ*; Err, error of the calculated component (which is MAD: maximum angular deviation); Palaeolat.: palaeolatitude; Int. err, MAD; Q: quality of the component (1: good to 3: bad); T range, temperature range over which component was calculated; T steps, number of temperature steps used for the calculation of the component; *Characteristic component is calculated without including the origin.

magnetostratigraphic data from the Salinas section (north of the External Sierras; Hogan & Burbank 1996; Fig. 7). The new magnetostratigraphic framework allows the age of surfaces E, F and G from Montes (2002) across the Santo Domingo anticline to be checked. It is revealed that such surfaces are not time-equivalent in the piggyback and foreland basin; there is a heterochrony of *c.* 1.7 Myr for surface E, *c.* 1 Myr for surface F and *c.* 2 Myr for surface G, being younger in the foreland basin than in the piggyback basin.

### Combined Campodarbe and Uncastillo Formation magnetostratigraphic sections

The integration of the new Luesia section with the previous magnetostratigraphic sections (Hogan 1993; Hogan & Burbank 1996) allows us to construct a composite magnetostratigraphy for the Campodarbe and Uncastillo formations in the Ebro foreland basin (Fig. 8). To do so, we consider not only the local magnetozones of the previous studies but also the sedimentological studies that relocate the Ayerbe section from Hogan & Burbank (1996) in a higher stratigraphic position (younger). The Ayerbe section was constrained by Hogan (1993) from Chron 10n to Chron 6Bn. However, the age of that section is reinterpreted to be the tectonosedimentary units $U_3^2$ and $U_3^3$, the youngest of the area, spanning Chron 6Cn to 6n (Arenas *et al.* 2001; see Fig. 3 for location of the stratigraphic profiles and previous magnetostratigraphic sections). In the Agüero and San Felices previous magnetostratigraphic sections the conglomeratic pulse 1B

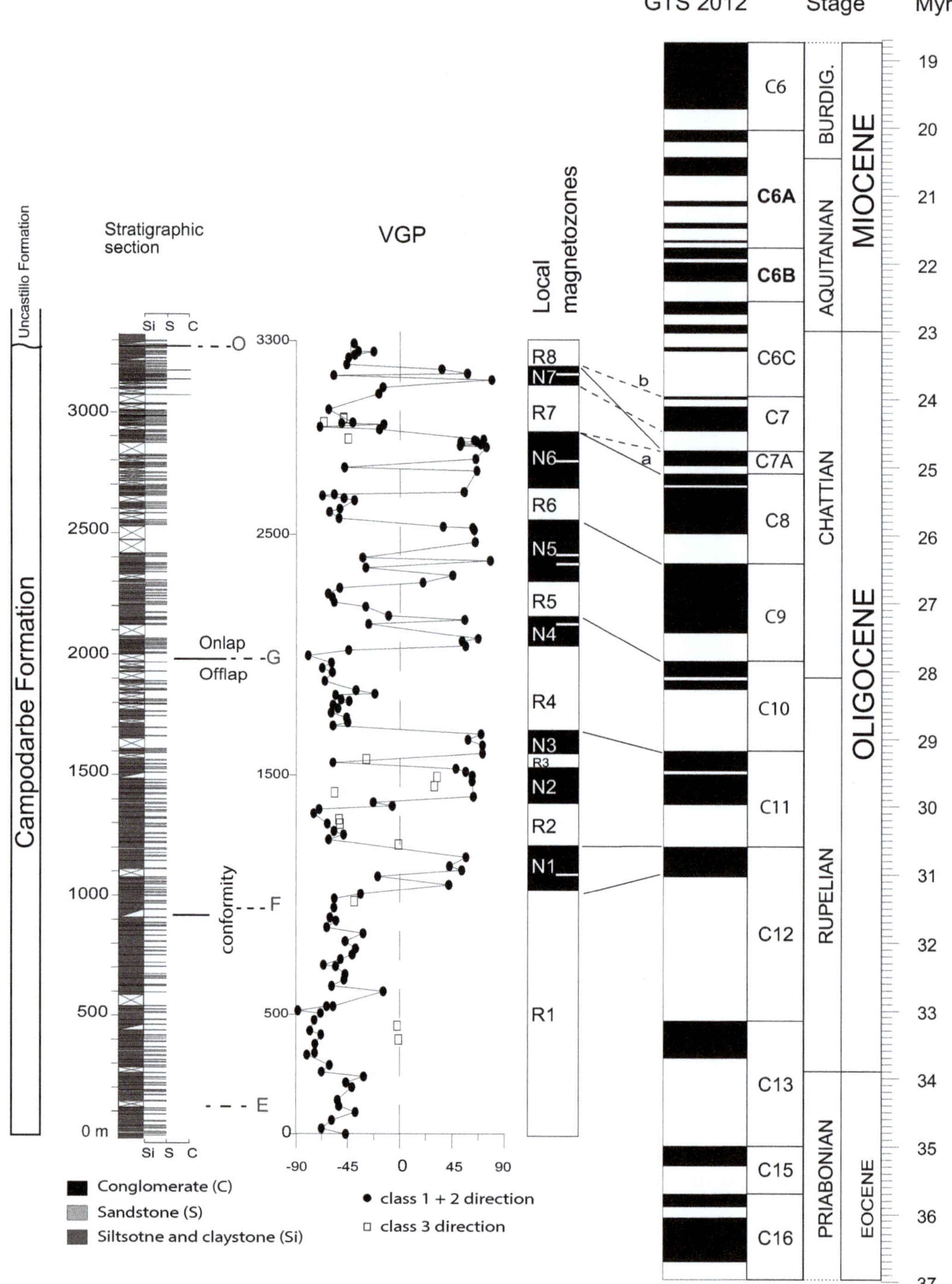



**Fig. 6.** Composite Luesia stratigraphic section, VGP latitudes calculated from the palaeomagnetic data, magnetostratigraphic zonation and correlation with the GPTS (Gradstein & Ogg 2012). Solid lines represent correlation a (without missing a magnetozone), dashed lines correlation b. The isochrone surfaces from Montes (2002) are also highlighted by dashed lines and letters E, F, G, O.

(Hogan & Burbank 1996) is correlated with the beginning of tectonosedimentary Unit $U_2$ in the Uncastillo Formation (C6C) in Arenas *et al.* (2001), which is Chron 6Cr in Figure 8.

Additionally, data from the San Felices section (Hogan & Burbank 1996) are used to extend ~900 m the Luesia section at its lower part which allows completing the Campodarbe sequence. A stratigraphic gap of 200 m between the Lower San Felices and Luesia sections is calculated from geological map and bedding average. The proposed correlation of the local magnetozones with the GPTS2012 misses Chron C13n, which is assumed to be in the gap between the lower San Felices and Luesia.

The interpreted connection of the upper part of the San Felices magnetostratigraphic section (pulse 1B, Hogan & Burbank 1996) with the GPTS (Gradstein & Ogg 2012) leaves open the possibility of missing one normal-polarity magnetozone in the sedimentary record when relating the first normal magnetozone with Chron 6Cn.3n (Fig. 8).

The significant lack of sedimentation (and/or erosion) in the San Felices section has already been highlighted in the stratigraphic work of Arenas *et al.* (2001). They suggested the lack of the upper part of the Campodarbe Formation and of unit U1 (Uncastillo Formation) in the San Felices section, which means there is an important shortage of sediments in that section.

The combination of all available data provides an age range from Chron 16n to 7r (option a) or to 6Cr (option b) for the Campodarbe Formation and from there to Chron 6n for the Uncastillo Formation in the northern part of the Ebro Basin studied in this work.

### Accumulation rates

This investigation was focused on determining the age of the Campodarbe Formation in an area where (1) sedimentation is *a priori* continuous (punctuated); (2) the transition to the Uncastillo Formation is well established in the field; and (3) the section can be laterally correlated to the San Felices unconformity by orthophoto. The accumulation rates have been calculated (a) considering chron boundaries (Table 2); and (b) considering sequence boundaries (Table 3).

In the Luesia sector, the accumulation rates for the Campodarbe and Uncastillo formations over a period of *c.* 20 Myr are obtained from correlation of the composite magnetostratigraphy that results from the integration of the previous magnetostratigraphic studies (Agüero, San Felices, Ayerbe sections; Hogan & Burbank 1996) and the new Luesia section (Table 2, Fig. 9, red & black) to the GPTS (Gradstein & Ogg 2012). We therefore use

the thickness of sediments calculated in Luesia (new data of Campodarbe Formation and measurements from Arenas *et al.* 2001 for Uncastillo Formation) and the timing of chrons considering the new magnetostratigraphic section of Campodarbe Formation and previous magnetostratigraphic sections of the Uncastillo Formation (Table 2). In addition, the accumulation rates between the sequence boundaries defined by Montes (2002) have also been represented in Figure 9 (in green, with option a and b correlations) and in Table 3.

With the accumulation rates considering chron boundaries, the closing and continentalization of the basin which occurs at 36 Myr (Costa *et al.* 2010, after revision of the Hogan & Burbank 1996 data) and which represents a dramatic increase in the piggyback basin, the inflection point in the accumulation rate for the molassic deposits changes from 25 to 45 cm kyr$^{-1}$ (grey arrow 1 in Fig. 9). However, the new data decrease the accumulation rate in the foreland basin due to the longest time recorded by the Campodarbe Formation; the inflection point represents a change from 25 to 35 cm kyr$^{-1}$ (in our composite section in the Ebro foreland basin). The accumulation rate experiences another increase in the piggyback basin at *c.* 33 Myr from 45–63 cm ka$^{-1}$ in Chron 12r (modified from Costa *et al.* 2010). A little bit later (32.8 Myr), the surface boundary E in the Ebro foreland basin occurs, with a change in the sedimentation rate of sedimentary units from 33 to 50 cm kyr$^{-1}$ (Table 3).

Interestingly, the beginning of the activity of the San Felices thrust sheet (31.3 Myr; marked by the F surface by green in Fig. 9) results in decreasing accumulation rates in both the piggyback and the foreland basins (35 and 36 cm ka$^{-1}$, respectively).

However, when looking at accumulation rates considering chron boundaries, more homogeneous rates are observed in the foreland basin with decreasing accumulation rates at two points. One point is close to the end of C10 (*c.* 28 Myr, when accumulation rate diminishes from 35 to 27 cm kyr$^{-1}$; grey arrow 2 in Fig. 9), a time that coincides with the deceleration of the San Felices thrust sheet activity (G surface in green accumulation rates). This is also the end of the sedimentary record in the piggyback basin. The second reduction in the accumulation rate considering chron boundaries occurs during deposition of the Uncastillo Formation at an average rate of *c.* 23 cm ka$^{-1}$ (grey arrow 3). This is the time at which the Santo Domingo Anticline tightening and reactivation of thrust sheets in the front occur (Millán-Garrido *et al.* 1995; Arenas *et al.* 2001).

Considering the accumulation rates for the Uncastillo Formation units, a clear diminishing accumulation rate is seen in the Uncastillo Formation from unit $U_1$ to $U_3$ (39, 28 and 13 cm ka$^{-1}$ in

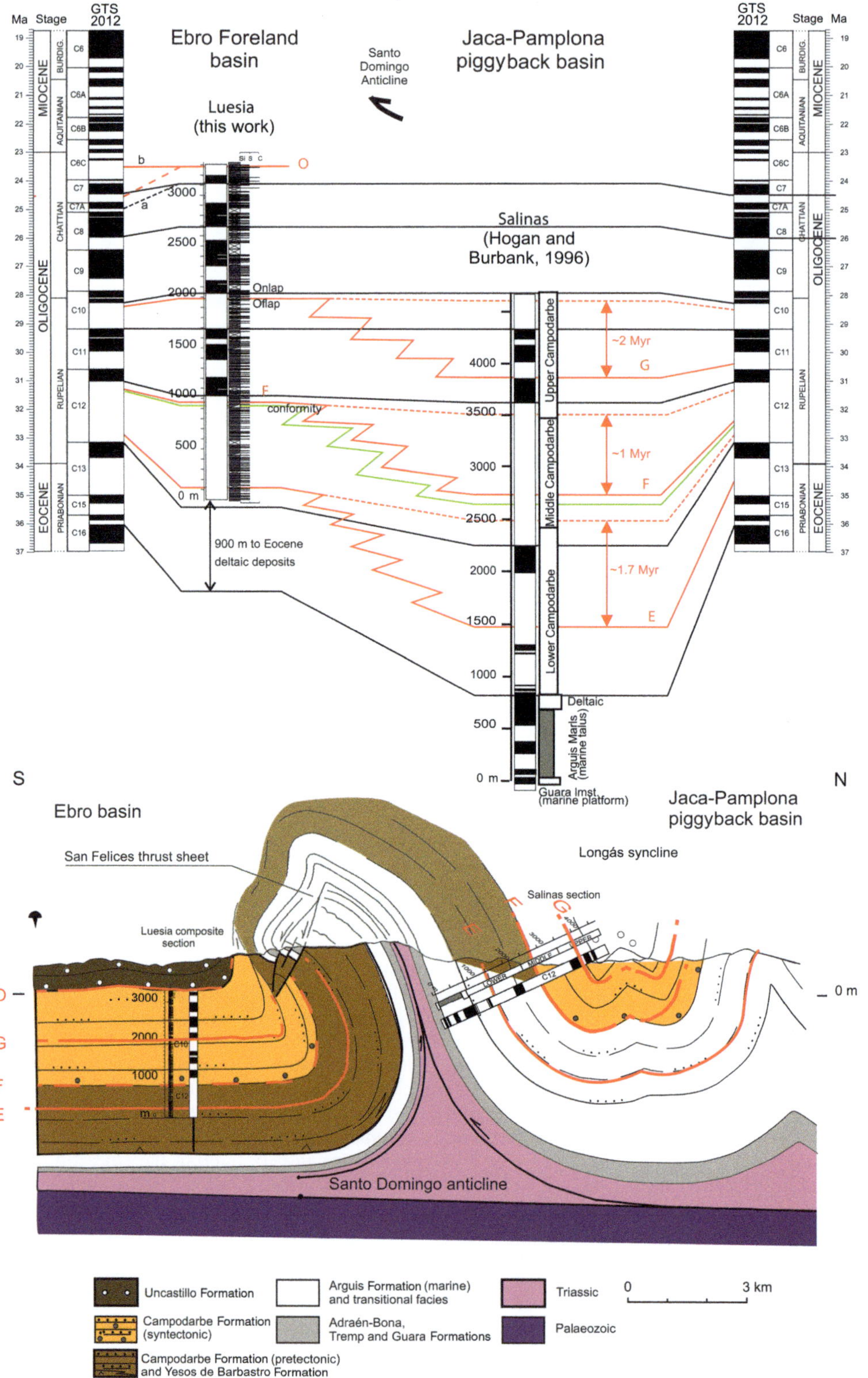
Ma Stage GTS 2012
Ebro Foreland basin
Luesia (this work)
Santo Domingo Anticline
Jaca-Pamplona piggyback basin
GTS 2012 Stage Ma
MIOCENE
BURDIG.
AQUITANIAN
CHATTIAN
OLIGOCENE
RUPELIAN
EOCENE
PRIABONIAN
C6
C6A
C6B
C6C
C7
C7A
C8
C9
C10
C11
C12
C13
C15
C16
Si S C
b
a
O
3000
2500
2000
1500
1000
500
0 m
Onlap
Oflap
F
conformity
900 m to Eocene deltaic deposits
Salinas (Hogan and Burbank, 1996)
Upper Campodarbe
Middle Campodarbe
Lower Campodarbe
4000
3500
3000
2500
2000
1500
1000
500
0 m
~2 Myr
G
~1 Myr
F
~1.7 Myr
E
Deltaic
Arguis Marls (marine talus)
Guara lmst. (marine platform)
S
Ebro basin
San Felices thrust sheet
Luesia composite section
Longás syncline
Salinas section
Jaca-Pamplona piggyback basin
N
O
G
F
E
3000
2000
1000
m
C10
C12
LOWER
MIDDLE E
Santo Domingo anticline
Uncastillo Formation
Campodarbe Formation (syntectonic)
Campodarbe Formation (pretectonic) and Yesos de Barbastro Formation
Arguis Formation (marine) and transitional facies
Adraén-Bona, Tremp and Guara Formations
Triassic
Palaeozoic
0    3 km
0 m

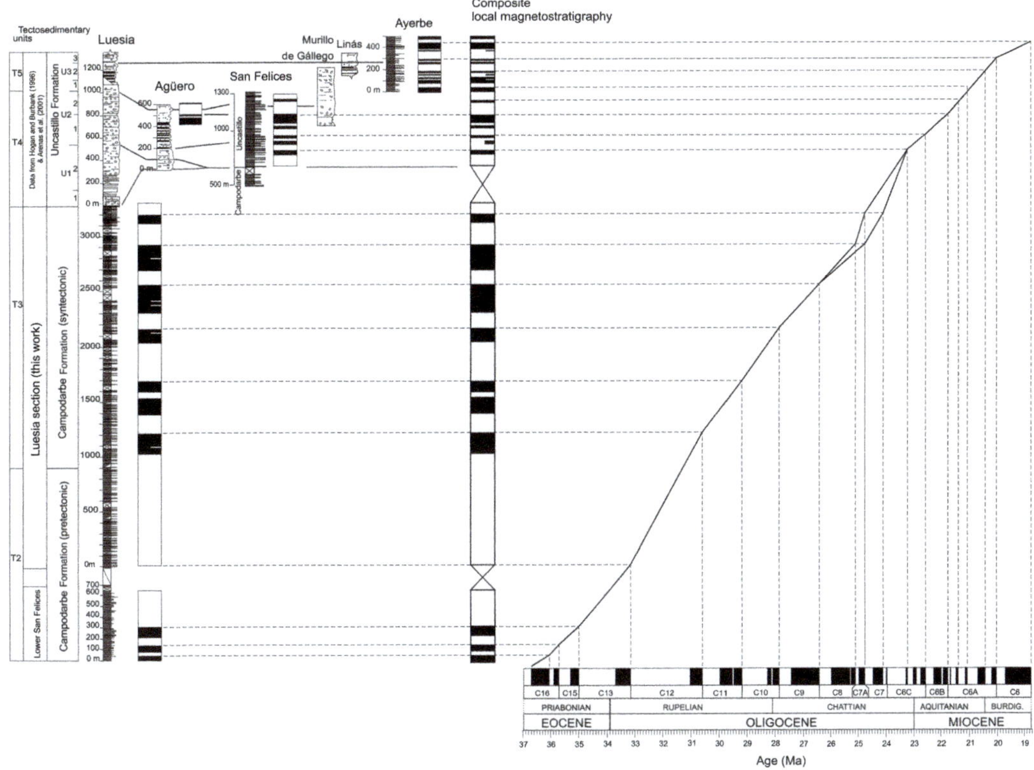

**Fig. 8.** Magnetostratigraphic correlation of sections of the Ebro Basin in the surrounding area of the Luesia section, which includes earlier magnetostratigraphic sections of San Felices, Agüero and Ayerbe (Hogan & Burbank 1996), and the stratigraphic correlations of Arenas *et al.* (2001).

**Table 2.** *Accumulation rates considering chron boundaries*

| Thickness (m) | Age (Myr) | Top of chron | Accumulation rate (cm ka$^{-1}$) |
|---|---|---|---|
| 5106 | 20.4 | C6a | 21.89 |
| 4806 | 21.77 | C6b | 25.31 |
| 4606 | 22.56 | C7a,7,6c | 31.49 |
| 3806 | 25.1 | C8 | 26.51 |
| 3456 | 26.42 | C9 | 27.5 |
| 3060 | 27.86 | C10 | 34.84 |
| 2600 | 29.18 | C11 | 35.46 |
| 2100 | 30.59 | C12,C13 | 40.81 |
| 300 | 35 | C15 | 30 |
| 0 | 36 | | |

**Table 3.** *Accumulation rates considering boundary sequences*

| Thickness (m) | Age (Myr) | Boundary sequence | Accumulation rate (cm ka$^{-1}$) |
|---|---|---|---|
| 5400 | 20 | top U3 | 13.33 |
| 5200 | 21.5 | U2–U3 | 28.12 |
| 4750 | 23.1 | U1–U2 | 39.29 (a), 137 (b) |
| 4200 | 24.5 (a), 23.5 (b) | O | 34.21 (a), 27 (b) |
| 2900 | 28.3 | G | 36.66 |
| 1800 | 31.3 | F | 50 |
| 1050 | 32.8 | E | 32.81 |
| 0 | 36 | D | – |

**Fig. 7.** Magnetostratigraphic correlation of the Campodarbe Formation across the Santo Domingo anticline, between the Salinas and Luesia sections. The position of sequence boundary surfaces mapped by Montes (2002) is indicated by red lines. The green line indicates the limit between the pre-tectonic and syntectonic strata. Errors in the projection (in magnetostratigraphic section or cross-section) can lead to an error of up to 200 kyr. These mapped surfaces are largely heterochronous between the Ebro foreland and the Jaca–Pamplona piggyback.

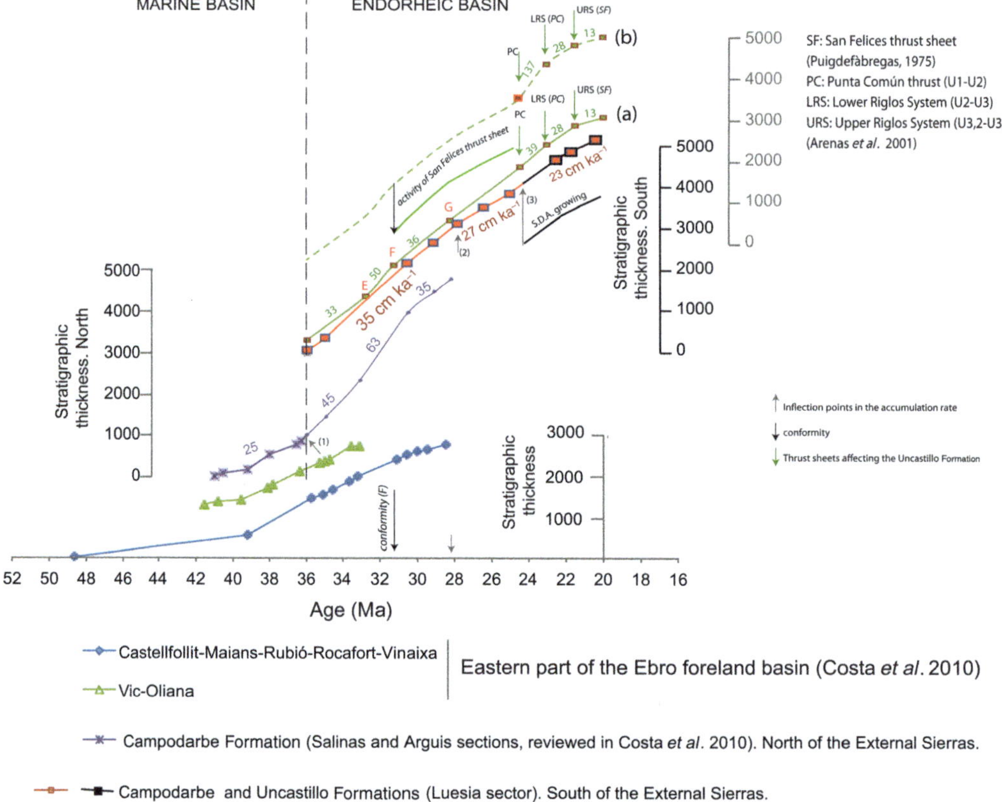



**Fig. 9.** Sedimentation rates for the Luesia composite section compared to the compilation of Costa *et al.* (2010). The sedimentation rates compare chron limits (Table 2; red) and sequence boundaries (Table 3; green with 'a' and 'b' correlation options). Differences in thickness are due to the different stratigraphic sections considered (top of San Felices or top of Luesia).

option a correlation) as expected, which is more dramatic in option b correlation and also unrealistic considering sedimentary unit accumulation rates.

The new data imply a higher sedimentation rate between C16n to C10r in the piggyback basin (*c.* 4000 m) than in the foreland basin (*c.* 2900 m). More importantly, there is no sedimentary record in the piggyback basin from C10r to C7/C6Cr because: (1) either the depocentre of the foreland basin migrated towards the south (accommodation space in the piggyback basin was totally occupied); or, more probably (2) the later cannibalization of the sediments of the piggyback basin during the final building of the External Sierras.

## Sequence boundaries and ages

After comparing the magnetostratigraphic data from the studied area, we propose to continue using the units defined by Puigdefàbregas (1975) for the Campodarbe Formation and that the end of the Campodarbe Formation occurs in C7r at 24.5 Myr (Fig. 10). The boundary between pre- and syntectonic Campodarbe Formation units is therefore not isochronous (1 Myr older in the piggyback basin).

The mean accumulation rate in the Ebro foreland basin seems more constant than the accumulation rate in the piggyback basin, which is higher in average than in the Ebro foreland basin (check the numbers in the colour columns of Fig. 10). A slight decrease in the accumulation rate occurs after the beginning of the activity of the San Felices thrust sheet. In the Salinas section the option with an isochronous F line is also drawn, and the same deceleration in the accumulation rate occurs (even higher).

Additionally, the tectonic activity based on the syntectonic deposits is slightly different from the tectonic activity summarized previously (Hogan & Burbank 1996, fig. 18). We demonstrate that the beginning of the San Felices thrust sheet activity occurred near the end of C12r (*c.* 31.3 Myr). The deceleration of the activity of this thrust sheet

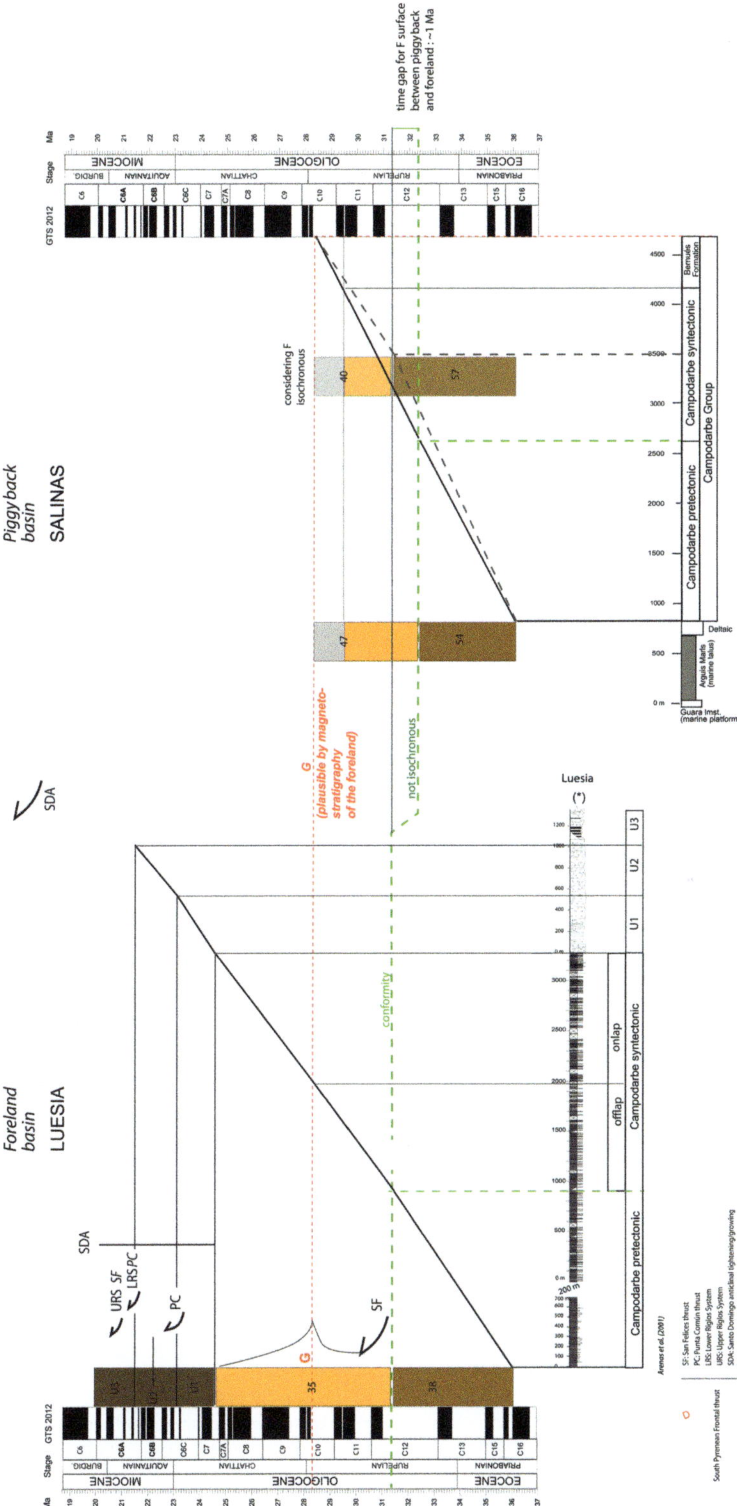


**Fig. 10.** Graphic representation of the accumulation rate of pre- and syntectonic units of the Campodarbe Formation and Bernués and Uncastillo formations (definitions of Puigdefàbregas 1975). Numbers in the shaded columns represent accumulation rates (cm ka$^{-1}$). The tectonic interpretation from Hogan & Burbank (1996) is shaded grey in the foreland basin. In the Salinas section, syntectonic Campodarbe unit and Bernués unit from Puigdefàbregas (1975) are equivalent to Bernués unit from Montes (2002). In this Luesia section two options are represented, the first maintaining the diachronism between pre-tectonic Campodarbe Formation unit as in Puigdefàbregas (1975) and the second considering $F$ as time equivalent. Since the $y$ axis is time and $x$ axis is thickness, the lower the slope, the higher the accumulation rate.

began at the end of Chron 10r (*c.* 28.4 Myr), and ended at the beginning of Chron 6Cr (*c.* 24.5 Myr). The Punta Común thrust sheet activity occurred at *c.* 23 Myr (C6C), and that of the Riglos thrust system at *c.* 21.5 Myr (C6A), taking into account the GPTS 2012.

## Conclusions

In combination with previously acquired data, the magnetostratigraphic study carried out in the composite Luesia section of thickness 3300 m indicates the following.

(1) In the studied area, the Campodarbe Formation spans from C16 (36 Myr, based on data from Costa *et al.* 2010) to C7r in the northern sector of the Ebro foreland basin. Deposition of the Uncastillo Formation began at C7r (*c.* 24.5 Myr) and the boundary between units $U_2$ and $U_3$ is set at *c.* 21.5 Myr.

(2) The tectonic activity of the San Felices thrust sheet spans *c.* 31.5 to *c.* 24 Myr (according to GPTS 2012). In the associated intra-Campodarbe syntectonic unconformity, the change from offlap to onlap geometry is recorded through a change from increasing to decreasing deposition rates ($35-27$ cm ka$^{-1}$ at *c.* 28 Myr; Table 2). This age coincides with the last sedimentary record preserved in the piggyback basin. The change in accumulation rate when the San Felices thrust sheet began to decelerate responds to a local change in the accommodation space, since sediment supply occurs at an orogenic scale (sediment source is in the South Pyrenean Central Zone, 200 km to the NE). A general deceleration rate is also observed in the Campodarbe Formation when San Felices thrust sheet activity starts (Figs 9 & 10), confirming the local change in accommodation space.

(3) The Salinas section is not as complete as the Luesia section, indicating the cannibalization of the piggyback basin during the south Pyrenean front deformation. The cannibalized sediments span *c.* 3.7 Myr and 1300 m.

(4) The Salinas and Luesia magnetostratigraphic sections between surfaces D and G, together with the information of the geological cross-sections, demonstrate that a higher sedimentation rate occurs in the piggyback basin with respect to the Ebro foreland basin. The continental wedge therefore opens to the north and is more abrupt that previously thought. Boundary sequence surfaces are between 1.8 and 1 Myr younger in the Ebro foreland basin than in the Jaca–Pamplona piggyback basin.

(5) The progressive unconformity in the San Juan de la Peña conglomerates (surface F) responds to the movement of a close northern structure (likely a thrust that affects the Eocene South Pyrenean turbiditic basin), *c.* 1 Myr prior to the movement of the San Felices thrust sheet.

The results show that magnetostratigraphy is a very useful tool to check the isochrony of the depositional sequences differentiated in sedimentary basins through stratigraphic analysis. Moreover, such analysis allows quantitative time estimates of hiatuses associated with unconformities.

Financial support to develop this investigation was provided by projects CGL2009-08969, CGL2010-21968-C02-02, CGL2009-14214, CGL2010-17479 and CGL2014-55900-P of the former Spanish Ministry of Science and Innovation (MICINN). We thank the Paleomagnetic Laboratory CCiTUB-ICTJA CSIC where the palaeomagnetic measurements were conducted. This is a contribution of the Geomodels Research Institute and the Research Group of Geodynamics and Basin Analysis 2014SGR467 from the *Agència de Gestió d'Ajuts Universitaris i de Recerca (AGAUR)* and the *Secretaria d'Universitats i Recerca del Departament d'Economia i Coneixement de la Generalitat de Catalunya* GEO-TRANSFER and Continental Sedimentary Basin Analysis Group of the Aragón Government, University of Zaragoza. We are also grateful to an anonymous reviewer and Dr Maodu Yan for their useful comments, which help to improve the first version of the manuscript.

## References

ALLEN, P. A. & ALLEN, J. R. 2005. *Basin analysis. Principles and Applications.* 2nd edn. Blackwell Publishing, Oxford.

ALLEN, P. A., ARMITAGE, J. J. *ET AL.* 2013. The Qs problem: sediment volumetric balance of proximal foreland basin systems. *Sedimentology*, **60**, 102–130.

ALMELA, A. & RÍOS, J. M. 1951. Estudio geológico de la zona subpirenaica aragonesa y de sus sierras marginales. Zaragoza, I Congreso Internacional del Pirineo del Instituto de Estudios Pirenaicos. *Geología*, **3**, 327–350.

ARENAS, C. 1993. *Sedimentología y paleogeografía del Terciario del margen pirenaico y sector central de la Cuenca del Ebro (zona aragonesa occidental).* PhD thesis, Universidad de Zaragoza.

ARENAS, C., MILLÁN, H., PARDO, G. & POCOVÍ, G. 2001. Ebro Basin continental sedimentation associated with late compressional Pyrenean tectonics (north-eastern Iberia): controls on basin margin fans and fluvial systems. *Basin Research*, **13**, 65–89.

BEAMUD, E., MUÑOZ, J. A., FITZGERALD, P. G., BALDWIN, S. L., GARCÉS, M., CABRERA, L. & METCALF, J. R. 2010. Magnetostratigraphy and detrital apatite fission track thermochronology on the exhumation of the South-Central Pyrenees. *Basin Research*, **23**, 309–331, http://doi.org/10.1111/j.1365-2117.2010.00492.x

CANUDO, J. I., MOLINA, E., RIVELINE, J., SERRA-KIEL, J. & SUCUNZA, M. 1988. Les événements biostratigraphiques de la zone prépyreénne d'Aragon (Espagne), de l'Eocène moyen a l'Oligocène inférieur. *Revue de Micropaléontologie*, **31**, 15–29.

CHOUKROUNE, P. & ECORS TEAM. 1989. The ECORS Pyrenean deep seismic profile reflection data and the overall structure of an orogenic belt. *Tectonics*, **8**, 23–39.

CLEVIS, Q., DE JAGER, G., NIJMAN, W. & DE BOER, P. L. 2004. Stratigraphic signatures of translation of thrust-sheet top basins over low-angle detachment faults. *Basin Research*, **145–163**, http://doi.org/10.1111/j.1365-2117.2009.00452.x

COSTA, E., GARCÉS, M., LÓPEZ-BLANCO, M., BEAMUD, E., GÓMEZ-PACCARD, M. & LARRASOAÑA, J. C. 2010. Closing and continentalization of the South Pyrenean foreland basin (NE Spain): magnetochronological constraints. *Basin Research*, **22**, 904–917, http://doi.org/10.1344/105.000001704

CUEVAS, J. L., MARZO, M. & MERCADÉ, L. 1989. *Depósitos de barras de meandro de granulometría gruesa en la Formación Talarn (Tránsito mesozoico-cenozoico de la conca de Tremp, Lérida)*. XII Congreso Español de Sedimentología, Comunicaciones, Bilbao, Spain, 19–22.

DECELLES, P. G. & GILES, K. A. 1996. Foreland basin systems. *Basin Research*, **8**, 105–124.

GRADSTEIN, F. M. & OGG, J. G. 2012. The chronostratigraphic scale. *In*: GRADSTEIN, F. M., OGG, J. G., SCHMITZ, M. D. & OGG, J. G. (eds) *The Geologic Time Scale*. Elsevier, Amsterdam, The Netherlands, 31–42.

HIRST, J. P. P. & NICHOLS, G. J. 1986. Thrust tectonic controls on Miocene alluvial distribution patterns, southern Pyrenees. *In*: ALLEN, P. A. & HONEYWOOD, P. (eds) *Foreland Basins*. International Association of Sedimentologists, Special Publication, Blackwell Scientific Publications, Oxford, **8**, 247–258.

HOGAN, P. J. 1993. *Geochronology, Tectonic, and stratigraphic evolution of the southwest Pyrenean foreland basin, northern Spain*. PhD thesis, University of Southern California.

HOGAN, P. J. & BURBANK, D. W. 1996. Evolution of the Jaca piggyback basin and emergence of the External Sierra, southern Pyrenees. *In*: FRIEND, P. F. & DABRIO, C. J. (eds) *Tertiary Basins of Spain. The Stratigraphic Record of Crustal Kinematics*. Cambridge University Press, Cambridge, 153–160.

HOTH, S., KUKOWSKI, N. & ONCKEN, O. 2008. Distant effects in bivergentorogenic belts; how retro-wedge erosion triggers resource formation in pro-foreland basins. *Earth Planetary Sciences Letters*, **273**, 28–37, http://doi.org/10.1016/j.epsl.2008.05.033

JORDAN, T. E., SCHLUNEGGER, E. F. & CARDOZO, N. 2001. Unsteady and spatially variable evolution of the Neogene Andean Bermejo Foreland Basin, Argentina. *Journal of South American Earth Sciences*, **14**, 775–798, http://doi.org/10.1016/S0895-9811(01)00072-4

KIRSCHVINK, J. L. 1980. The least-squares line and plane and the analysis of paleomagnetic data. *Geophysical Journal of the Royal Astronomical Society*, **62**, 669–718.

LABAUME, P., SÉGURET, M. & SEYVE, C. 1985. Evolution of a turbiditic foreland basin an analogy with an accretionary prism: example of the Eocene South-Pyrenean basin. *Tectonics*, **4**, 661–685.

MCELROY, R. 1990. *Thrust kinematics and syntectonic sedimentation: the Pyrenean frontal ramp, Huesca, Spain*. Unpublished PhD thesis, University of Cambridge.

MEY, P. H. W., NAGTEGAAL, P. J. L., ROBERTI, K. J. & HARTEVELT, J. J. A. 1968. Lithostratigraphic subdivision of post-hercynian deposits in the south-central Pyrénés, Spain. *Leidse Geologische Mededelingen*, **41**, 221–228.

MILLÁN-GARRIDO, H. 1996. *Estructura y cinemática del frente de cabalgamiento surpirenaico en las Sierras Exteriores Aragonesas*. PhD thesis, Universidad de Zaragoza.

MILLÁN-GARRIDO, H., PARÉS, J. M. & POCOCVÍ, A. 1992. Modelización sencilla de la estructura del sector occidental de las sierras marginales aragonesas (Prepirineo, provincias de Huesca y Zaragoza). *III Congreso Geológico de España*, **2**, 140–149.

MILLÁN-GARRIDO, H., POCOVÍ-JUAN, A. & CASAS-SAINZ, A. M. 1995. El frente de cabalgammiento surpirenaico en el extremo occidental de las Sierras Exteriores. *Revista de la Sociedad Geológica de España*, **8**, 73–90.

MILLÁN-GARRIDO, H., PUEYO-MORER, E. L., AURELL-CARDONA, M., LUZÓN-AGUADO, A., OLIVA-URCIA, B., MARTÍNEZ-PEÑA, B. & POCOVÍ-JUAN, A. 2000. Actividad tectónica registrada en los depósitos terciarios del frente meridional del Pirineo central. *Revista de la Sociedad Geológica de España*, **13**, 279–300.

MOLNAR, P. 2009. The state of interactions among tectonics, erosion, and climate: a polemic. *GSA Today*, **19**, 44–45.

MONTES, M. J. 2002. *Estratigrafía del Eoceno-Oligoceno de la Cuenca de Jaca (Sinclinorio del Guarga)*. PhD thesis, Universitat de Barcelona.

MUÑOZ, J. A. 1992. Evolution of a continental collision belt: ECORS-Pyrenees crustal balanced section. *In*: MCCLAY, K. R. (ed.) *Thrust Tectonics*. Chapman and Hall, London, 235–246.

MUTTI, E., LUTERBACHER, H., FERRER, J. & ROSELL, J. 1972. Schema stratigrafico e lineamenti di facies del Paleogeno Marino della zona centrale sudpirenaica tra Tremp (Catalogna) e Pamplona (Navarra). *Memorie della Società Geologica Italiana*, **18**, 15–22.

NAYLOR, M. & SINCLAIR, H. D. 2007. Punctuated thrust deformation in the context of doubly vergent thrust wedges: implications for the localization of uplift and exhumation. *Geology*, **35**, 559–562.

OLIVA-URCIA, B., CASAS, A. M., PUEYO, E. L. & POCOVÍ-JUAN, A. 2012. Structural and paleomagnetic evidence for non-rotational kinematiics of the South Pyrenean frontal Thrust at the western termination of the External Sierras (southwestern central Pyrenees). *Geologica Acta*, **10**, 125–144, http://doi.org/10.1344/105.000001704

PARDO, G., VILLENA, J. & GONZÁLEZ, A. 1989. Contribución a los conceptos y a la aplicación del análisis tecto-sedimentario. Rupturas y unidades tectosedimentarias como fundamento de correlaciones estratigráficas. *Revista de la Sociedad Geológica de España*, **2**, 199–221.

PARDO, G., ARENAS, C. ET AL. 2004. La cuenca del Ebro. *In*: VERA, J. A. (ed.) *Geología de España*.

IGME and Sociedad Geológica de España, Madrid, 533–543.

POCOVÍ, A., MILLÁN, H., NAVARRO, J. J. & MARTÍNEZ, B. 1990. Rasgos estructurales de la Sierra de Salinas y zona de los Mallos (Sierras Exteriores, Prepirineo, provincias de Huesca y Zaragoza). *Geogaceta*, **8**, 36–39.

PUIGDEFÀBREGAS, C. 1975. La sedimentación molásica en la cuenca de Jaca. *Pirineos*, **104**, 188.

SÉGURET, M. 1972. *Etude tectonique des nappes de series décollées de la partie centrale du versant sud des Pyrénées. Caractère synsédimentaire, rôle de la compression et de la gravité*. PhD thesis, Publication de l'Universite des Sciences et Techniques du Langedoc (USTELA), Série Géologie Structurale, Montpellier.

SOLER, M. & PUIGDEFÀBREGAS, C. 1970. Líneas generales de la geología del Alto Aragón occidental. *Pirineos*, **96**, 5–19.

TEIXELL, A. 1992. *Estructura alpina en la transversal de la terminación occidental de la Zona Axial pirenaica*. PhD thesis, Universitat de Barcelona.

TEIXELL, A. 1996. The Ansó transect of the southern Pyrenees: basement and cover thrust geometries. *Journal of the Geological Society, London*, **153**, 301–310, http://doi.10.1144/gsjgs.153.2.0301

TEIXELL, A. & GARCÍA-SANSEGUNDO, J. 1995. Estructura del sector central de la Cuenca de Jaca (Pirineos meridionales). *Revista de la Sociedad Geológica de España*, **8**, 215–228.

# Anisotropy of magnetic susceptibility (AMS) records synsedimentary deformation kinematics at Pico del Aguila anticline, Pyrenees, Spain


DAVID ANASTASIO[1]*, JOSEP M. PARÉS[2], KENNETH P. KODAMA[1], JOANNA TROY[1] & EMILIO L. PUEYO[3]

[1]*Department of Earth and Environmental Sciences, Lehigh University, Bethlehem, PA, United States*

[2]*Paleomagnetism Laboratory, Cenieh, Burgos, Spain*

[3]*Instituto Geológico y Minero de España, Unidad de Zaragoza, Spain*

**Corresponding author (e-mail: dja2@lehigh.edu)*



**Abstract:** Pico del Aguila anticline is a transverse décollement fold located at the Pyrenean thrust front. The anticline is a synsedimentary structure buried during growth by delta front mudstones and sands of the Eocene Arguis and Belsué-Atares formations. Both the anisotropy of magnetic susceptibility measured at 77 K and 294 K and the anisotropy of anhysteretic remanence show that susceptibility is dominated by paramagnetic clay minerals and can be used as a proxy for depositional and tectonic fabric orientations. In general, the maximum and intermediate principal susceptibilities ($k_1$ and $k_2$) of the AMS lie in bedding and the minimum principal susceptibility ($k_3$) is oriented nearly normal to bedding. Layer-parallel shortening (LPS) produced a *c.* north–south-trending magnetic intersection lineation in bedding on anticline limbs and in the adjacent Belsué and Arguis synclines by deforming the depositional and diagenetic compaction fabric. The degree of magnetic anisotropy is higher along axial surfaces than on limbs. At the anticline hinge, oblate magnetic ellipsoids with an east–west-aligned lineation and a bedding-parallel magnetic foliation demonstrate the overprinting of the LPS magnetic fabric during the emplacement of the underlying thrust sheet. AMS data record fold kinematics characterized by constant-length limb rotation about pinned hinges and are compatible with kinematics recorded by growth strata geometries. This study emphasizes that AMS is a very sensitive measure of depositional, compaction and tectonic fabrics in marine clastic rocks in the diagenetic realm.

**Supplementary material:** Data tables including specimen locations, orientation, and AMS matrix elements for new samples and AMS data from Pueyo *et al.* (1997). Bulk susceptibility at 77 K and 294 K for representative specimens is available at http://www.geolsoc.org.uk/SUP18842


Studies of deformation allow assessment of rheology, strain history and kinematics that are necessary prerequisites for conducting dynamic studies, incrementally balancing cross-sections or generating palaeogeographical reconstructions. The location of pin lines during folding creates distinct strain patterns that can be used to test predictions from various kinematic models of folding (fixed v. migrating hinges, flexural v. buckle folding; e.g. Ramsay 1967; Dahlstrom 1990; Fisher & Anastasio 1994; Anastasio *et al.* 1997). Numerous studies have used finite or incremental strain data or mesoscopic fabric variation as a function of structural position to elucidate deformation kinematics (e.g. Cloos 1947; Ramsay & Huber 1983; Fischer *et al.* 1992). However, in orogenic forelands where deformation occurs at shallow depths and low temperatures, ductile penetrative deformation features are subtle and brittle structures may be sparse. Anisotropy of

magnetic susceptibility (AMS) has been successfully used as a proxy for rock strain in sedimentary rocks in regions where other deformation markers were not available (Averbuch *et al.* 1992; Borradaile & Henry 1997; Parés 2004). In general, comparative studies from siliciclastic rocks show good agreement between both the relative magnitude and orientation of penetrative rock strain determined by traditional geometric methods of strain analysis and AMS patterns, whereas only the principal directions of strain were comparable in carbonate rocks (e.g. Borradaile & Tarling 1981; Borradaile 1988; Averbuch *et al.* 1992; Mattei *et al.* 1997; Pueyo *et al.* 1997; Sagnotti *et al.* 1998; Parés & van der Pluijm 2003; Borradaile & Jackson 2004; Larrasoaña *et al.* 2004; Latta & Anastasio 2007; Burmeister *et al.* 2009).

AMS is a second-rank tensor that characterizes a material's magnetization response to an applied magnetic field (e.g. Nye 1957; Tarling & Hrouda


*From*: PUEYO, E. L., CIFELLI, F., SUSSMAN, A. J. & OLIVA-URCIA, B. (eds) 2016. *Palaeomagnetism in Fold and Thrust Belts: New Perspectives*. Geological Society, London, Special Publications, **425**, 129–144.
First published online June 16, 2015, http://doi.org/10.1144/SP425.8

1993). The tensor can be represented by an ellipsoid, with greatest principal axis $k_1$, intermediate axis $k_2$ and shortest axis $k_3$, representing the magnetic susceptibility contributions of a rock's constituent magnetic minerals. Crystallography and grain shape control magnetic mineral properties. For oblate paramagnetic clay minerals, the crystallographic axes also correspond to the susceptibility tensor axes, that is, magnetic axes in such crystals conform to the density distributions of mineral lattice planes obtained by x-ray goniometry (Richter *et al.* 1993; Martín-Hernández & Hirt 2003; Schmidt *et al.* 2009). Natural processes such as current deposition, diagenesis and lithification, or tectonic deformation can all contribute to the measured AMS of a sample. Because we know that the individual particle anisotropies of phyliosilicate minerals are oblate, we can use the AMS fabrics to measure their preferred orientation (e.g. March 1932; Richter *et al.* 1993).

In order to test the sensitivity of AMS to establish variations in penetrative deformation in low strain and shallow crustal environmental conditions to determine fold kinematics, we present results from Pico del Aguila anticline, a submarine, synsedimentary fold that developed near the South Pyrenean mountain front. Pico del Aguila was chosen as the location for this study because deformation conditions and fold kinematics are independently known from previous studies. The anticline formed near the marine sediment surface in the Palaeogene wedge-top basin of the southern Pyrenees (Fig. 1). Among the numerous studies of Pico del Aguila anticline, Puigdefábregas (1975) and more recently Castelltort *et al.* (2003) studied the stratigraphy of the growth strata, Hogan & Burbank (1996) reconstructed fold burial, Kodama *et al.* (2010) used rock magnetic cyclostratigraphy to determine the age of the growth strata, Anastasio & Holl (2001) studied deformation patterns and thermal conditions, and Millán *et al.* (1994), Poblet & Hardy (1995) and Castelltort *et al.* (2003) analysed growth strata geometries to explore fold kinematics. By comparing the AMS fabrics as a function of

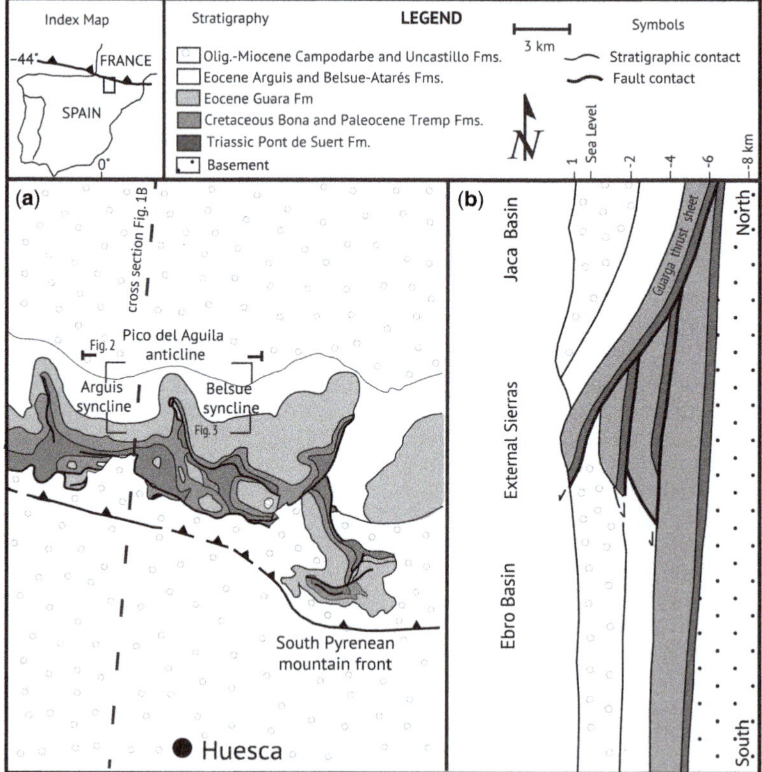



**Fig. 1.** (**a**) Location (inset) and simplified geological map of the central External Sierras and Pico del Aguila anticline (Anastasio 1987). (**b**) Interpretive vertical cross-section of the southern Pyrenees across the Jaca wedge-top basin and External Sierras to the Ebro foreland Basin in the south. Location of the cross-section shown in Figure 2a and area of Figure 3 shown.

structural position and stratigraphic level in the growth strata to predicted strain patterns, we evaluate the ability of AMS to resolve kinematics in the diagenetic realm.

## Geological setting

The Pyrenees Mountains are a dual-vergent mountain belt that formed during the Late Cretaceous–early Miocene as a result of Iberian and Eurasian plate convergence (Roest & Srivastava 1991; Muñoz 1992). The External Sierras are the Spanish foothills that separate the Jaca wedge-top basin from the Ebro foreland basin to the south along the emergent South Pyrenean thrust front (Fig. 1). The Guarga thrust sheet transported the Jaca basin southwards a minimum of 19.4 km during the Middle Tertiary in the west central part of the orogen (Anastasio 1987) with displacement on the Guarga thrust sheet decreasing westwards (DePaor & Anastasio, 1987; Millán 2006). A series of transverse detachment folds, with a décollement level within the Triassic evaporites, developed on top of the thrust sheet during displacement. Pico del Aguila is the best studied of these folds (e.g. Anastasio 1987; Millán et al. 1994, 2000; Pueyo et al. 1997, 2002; Castelltort et al. 2003; Kodama et al. 2010; Vidal-Royo et al. 2013). The Pico del Aguila fold structure was transported southwards within the Guarga thrust sheet on a leading-edge, out-of-sequence imbricate fault (Anastasio 1992) (Fig. 1).

Pico del Aguila is a regional-scale décollement anticline with a thrust-faulted core (Fig. 1). Folding developed above the ductile evaporitic mudstone of the Keuper facies of the Pont de Suert Formation which acts as the décollement and is the base of a c. 1 km thick pre-growth stratigraphy dominated by Triassic–Middle Eocene carbonate rocks. The growth strata are up to 1.5 km thick and shallow in dip to horizontal in the basal Campodarbe Formation. The growth sequence includes deposits of the Arguis and Belsué-Atarés formations, which are of Middle Eocene–Oligocene in age. The Arguis Formation is composed of beds of marly siltstones and sandstones, some of which are bioturbated, and rare interbedded limestones. It represents a large-scale prograding/retrograding flood-dominated delta and silty shelf sequence deposited from east to west, burying the active Pico del Aguila anticline and the other transverse anticlines of the External Sierras. The Belsué-Atarés Formation, which is dominated by coarser sediment deposited on the delta plain, overlies the Arguis Formation and finally the mostly post-folding Campodarbe Formation, which represents Oligocene–Miocene-aged fluvial molasse (Puigdefábregas 1975; Castelltort et al. 2003; Fig. 2). Folding at Pico del

Aguila began in the Middle Eocene (Anastasio 1992; Millán et al. 1994; Kodama et al. 2010; Rodríguez-Pintó et al. 2012) during the deposition of the upper part of the Guara Formation, a foraminiferal carbonate bank that forms the resistant topography of Pico del Aguila. Most of the folding occurred during Arguis Formation deposition. A latest Lutetian–early Priabonian age of the Arguis Formation is supported by multispecies biostratigraphy (Canudo 1990; Canudo et al. 1991) and magnetostratigraphy (Hogan 1993; Hogan & Burbank 1996; Pueyo et al. 2002; Kodama et al. 2010; Rodríguez-Pintó et al. 2012).

The Pico del Aguila anticline, which plunges at 30° to the north, is bounded to the west by the Arguis syncline and to the east by the Belsué syncline. The External Sierras formed under peak burial temperatures of c. 55–60°C based on vitrinite reflectance data (Holl & Anastasio 1995a) and a maximum burial depth of c. 3 km based on subsidence reconstruction by Hogan & Burbank (1996). Deformation at Pico del Aguila included both halotectonic folding and thrust sheet emplacement (Anastasio 1992). The north–south-aligned fold axis was monoclinally folded about an east–west-aligned axis causing an oblique cross-section of the fold to be exposed in map view. The anticline affords excellent access to all structural and stratigraphic positions in the growth strata, which were sampled for AMS analysis.

## Methods

Growth bed attitudes were defined from individual bedding measurements from referenced locations, bed mapping with real-time kinematic GPS positioning and bed tracing from referenced 1:5000 digital elevation models. Figure 2 shows the downplunge projection of these growth strata on a plane perpendicular to the fold axis at the northern end of the structure. The resolution of the spatial position on growth horizons depends on GPS receiver resolution, bed attitude variation, compass precision and projection errors resulting from non-cylindrical folding and fold axis variation. In this study, the traces of growth strata are accurate at the decimetre scale. The traces of fold axial surfaces were determined from deflections in the growth strata. The general thinning and shallowing of the growth beds onto the anticline is consistent with limb rotation about pinned hinges during folding (e.g. Ford et al. 1997). The growth strata suggest limb length has been constant during folding and the position of no interlayer shear (pin line) jumped hinge-wards on the east-dipping limb during fold development. A décollement fold with pinned hinges is expected to experience a change in décollement depth during

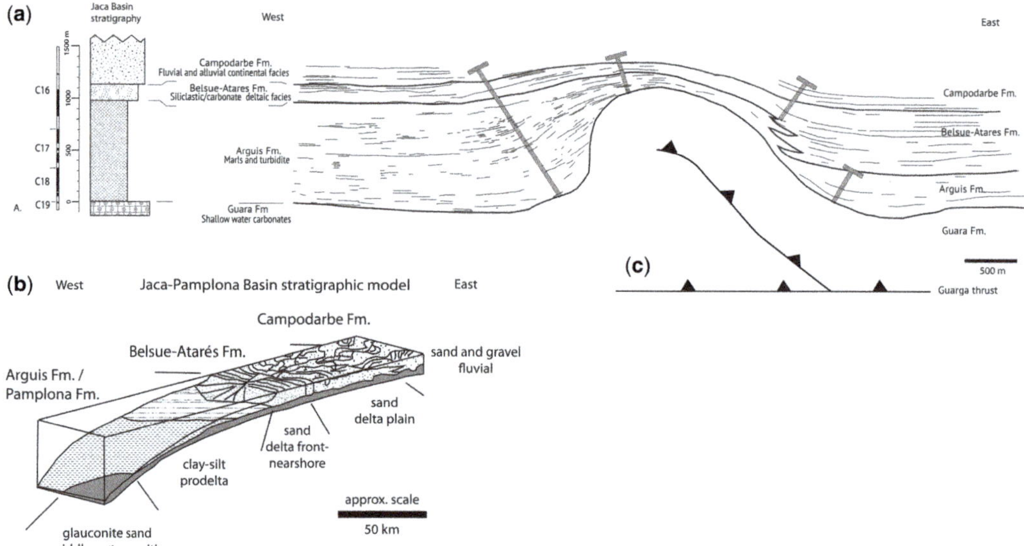



**Fig. 2.** (**a**) Downplunge projection of growth strata on Pico del Aguila anticline. The section is oriented parallel to strike and dips 63° S, perpendicular to the northern anticline fold axis. Magnetostratigraphy is from Kodama *et al.* (2010). Growth strata derived from 1:5000 orthophoto quadrangles IGME, precision GPS surveying and structural measurements located by GPS. Pin lines indicated by grey nails, whose position suggests folding by rotation about fixed hinges. (**b**) Stratigraphic model for the Palaeogene Jaca–Pamplona wedge-top basin. The fluvial deltaic system transgressed from east to west during synsedimentary growth of Pico del Aguila anticline.

fold tightening (Epard & Groshong 1995; Homza & Wallace 1995), easily accommodated within the evaporitic strata coring the fold.

To determine fold kinematics, standard 1 inch oriented cores were collected from 49 sites around Pico del Aguila from the Arguis and Belsué-Atarés formations, which display a growth relationship with the anticline (Fig. 3). Sample sites where chosen from mudstone and fine sand beds at all structural and stratigraphic positions around the anticline, including additional samples from beds sampled by Pueyo *et al.* (1997). At each site, at least four independent cores were drilled and oriented. Multiple cylinders were trimmed from each core to obtain standard 11 cm$^3$ specimens. A total of 6–14 samples from 27 sites were measured for AMS using the KLY-3 s Kappabridge at Lehigh University and applying the 15 directional suscepti-bilities scheme of Jelínek (1978). Ten additional sites (AA, AD, AE, kin1-kin7) were measured in the palaeomagnetic laboratory at the University of Michigan and data from 12 sites reported in Pueyo *et al.* (1997) are also reconsidered in this analysis. Low-field room temperature (294 K) bulk suscepti-bility was measured from all specimens. Additional rock magnetic experiments (see Kodama *et al.* 2010 for further details) were conducted to determine the magnetic mineralogy of representative samples,

including the magnetic mineralogy contributing to the magnetostratigraphy based on thermal de-magnetization of three orthogonal components of isothermal remanent magnetization (IRM; coerciv-ity-unblocking temperature analysis, Lowrie 1990) and the relative contributions to remanence by dif-ferent coercivity components using IRM acquisition modelling (e.g. Kruiver *et al.* 2001).

Fifty specimens from eight sites were selected to evaluate paramagnetic mineral contributions to AMS as a function of grain size and structural position. These samples were measured for bulk suscepti-bility at low temperature (77 K) in liquid nitrogen in a modified susceptibility meter at the palaeo-magnetic laboratory at the University of Michigan, according to the method of Parés & van der Pluijm (2003), in order to enhance the paramagnetic fabric following the Curie–Weiss law. Thirty sam-ples from five sites chosen to represent grain size variations in the Arguis syncline were measured to determine the anisotropy of anhysteretic remanence (AAR). AAR was imparted using a GSD-5 Schon-stedt alternating field demagnetizer and measured with a 2G Enterprises Inc. superconducting magnet-ometer, both at Lehigh University. Each specimen was given an ARM in nine different orientations. A partial ARM was applied with peak alternating field between 0 and 50 mT with 0.097 mT constant

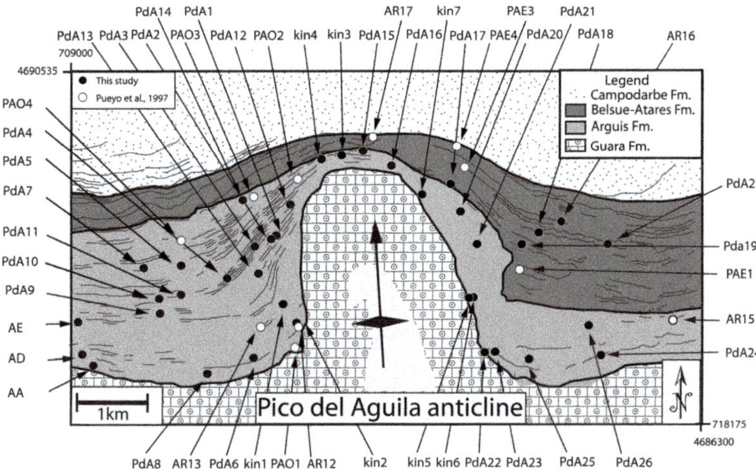


**Fig. 3.** Sample locations for AMS sites ard Pico del Aguila anticline. Solid circles: new sites, this study; open circles: from Pueyo *et al.* (1997). Thin lines: growth strata traces; thick lines: formation contacts. Regional location shown in Figure 1a.

DC field to activate magnetite phases and exclude higher coercivity minerals. Data were analysed using the PMAG software by Tauxe (2002).

## Results

The eigenvalues and eigenvectors of the AMS tensor ($k_1 \geq k_2 \geq k_3$) describe the magnetic fabric that can be related to the mineral fabric. The AMS of sites from the Arguis syncline, west limb, hinge and east limb of the Pico del Aguila anticline and Belsué syncline are shown in Figure 4a–e. Overall, the magnetic fabric is well defined with $k_1$ and $k_2$ in the plane of bedding and $k_3$ parallel to the pole to bedding. At all sites, from all structural positions except the hinge area, $k_1$ plunges shallowly to the north ($c.\ 350 \pm 20°$) nearly parallel to the Pico del Aguila hinge line. In contrast, in the hinge region $k_1$ axes plunge shallowly to the east or west (Fig. 4c).

The shape and anisotropy of the site-averaged AMS data are presented on a Jelínek (1981) plot in Figure 5. Triaxial fabric ellipsoids characterize nearly all sites. The degree of anisotropy ($P_j$) is <1.06 for all samples, consistent with the lack of visible macroscopic tectonic foliation in the field. A stronger oblate AMS fabric characterizes most sites along the axial surfaces and adjacent to the mechanically stiff Guara Formation, which suggests these samples have higher rock strains. The lowest anisotropies ($P_j < 1.03$) are associated with triaxial prolate fabrics at sites in the interior of the Arguis and Belsué synclines.

The mean bulk susceptibility of all samples is $144 \times 10^{-6}$ SI, suggestive of a dominantly paramagnetic susceptibility (e.g. Pueyo-Anchuela *et al.*

2013; Fig. 6a). The distribution of all bulk susceptibility measurements is tightly clustered about the mean with a slight high side tail to the distribution. Low-temperature AMS measurements include specimens from all structural positions and examples of each lithology (Fig. 6b). The average low-temperature bulk susceptibility enhancement for the eight sites was 2.9, despite the inclusion of two coarse-grained Belsué–Atarés Formation sandstones (PdA17 and PdA27) which exhibit the lowest paramagnetic contribution to the bulk susceptibility. AAR measurements from representative samples are shown in Figure 6c. The AAR measurements isolate the anisotropy of remanence-carrying ferromagnetic grains in a sample and can be compared to the orientations of principal susceptibilities from nearby Arguis syncline samples AA, AD, AE, shown in Figures 4 and 7 in geographical and bedding-corrected reference frames, respectively. Comparison of the AMS and AAR from the same sites shows different orientations for $k_1$ axes: WSW for AAR and NNW for AMS. $k_2$ axes are in bedding while $k_3$ axes are normal to bedding for both AAR and AMS measurements.

## Discussion

### Magnetic mineralogy

The use of magnetic fabric to characterize tectonic strain requires knowledge of magnetic mineralogy and pre-tectonic fabric. The AMS of a sample is a measure of the total magnetic fabric due to the preferred orientation of all diamagnetic, ferromagnetic and paramagnetic grains in a sample. The individual

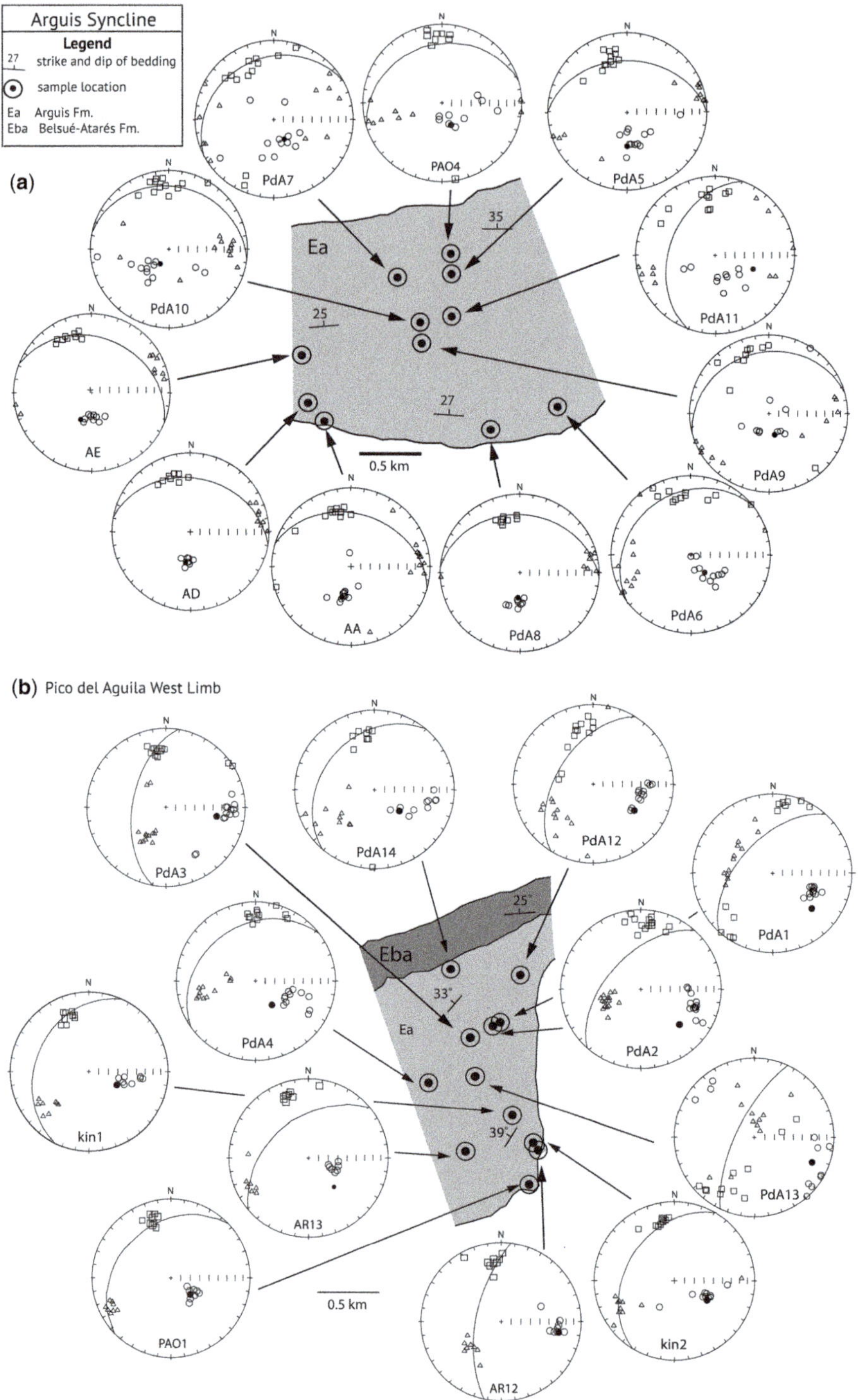



**Fig. 4.** Lower hemisphere stereographic projections of principal magnetic susceptibilities: $k_1$ squares; $k_2$ triangles; $k_3$ circles shown in geographical coordinates. Bedding shown by great circle and the solid dot, which is the pole to bedding. Sample location and bedding attitude shown on map: (**a**) Arguis syncline; (**b**) Pico del Aguila west limb; (**c**) Pico del Aguila hinge; (**d**) Pico del Aguila east limb; (**e**) Belsué syncline.

**(c)** Pico del Aguila Hinge

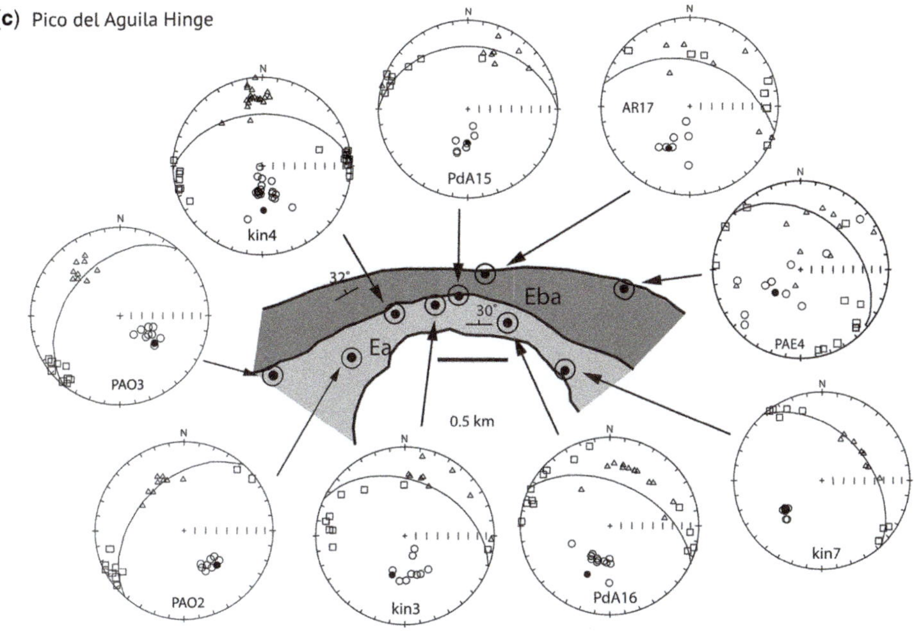


**(d)** Pico del Aguila East Limb

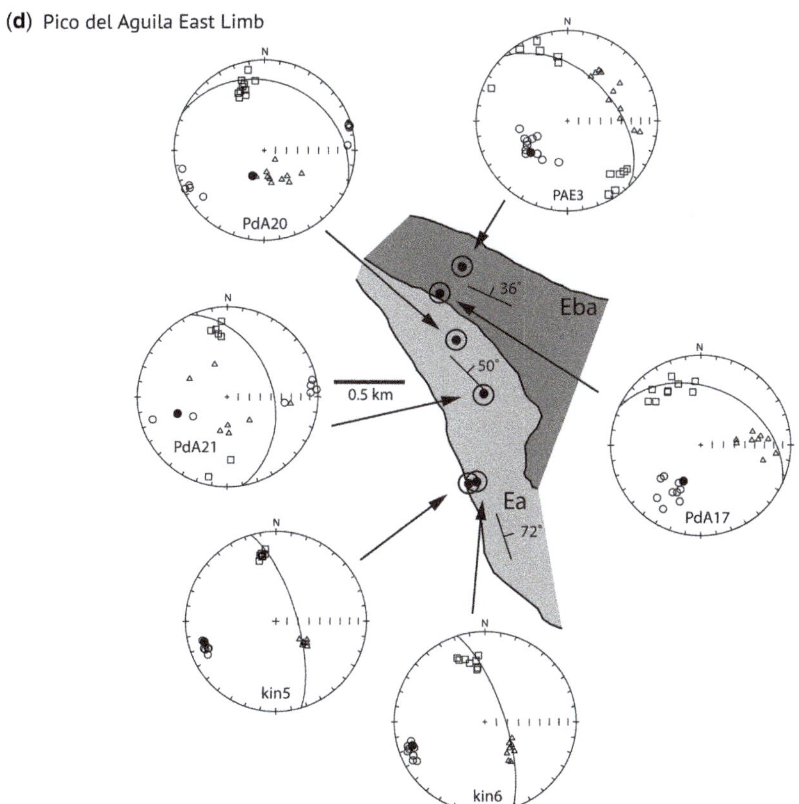


**Fig. 4.** *Continued.*

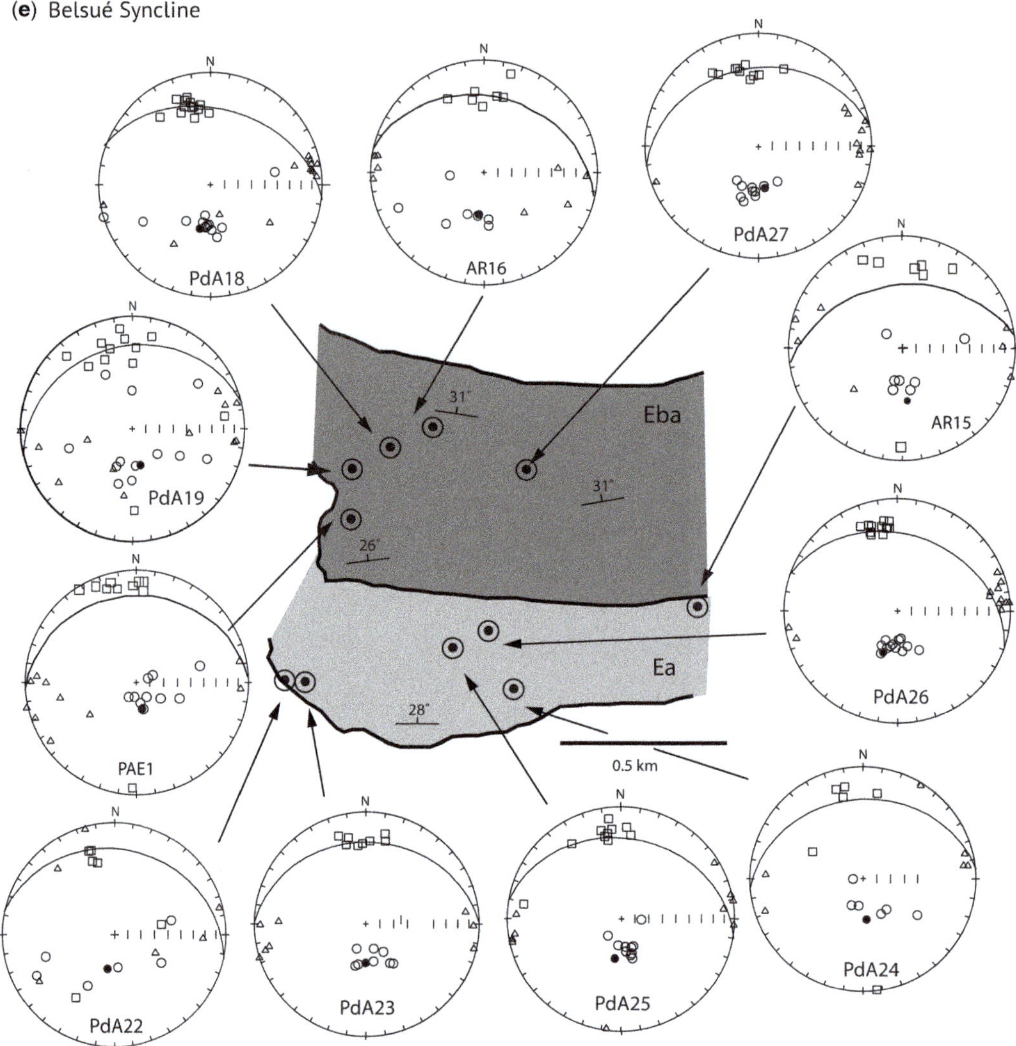



**Fig. 4.** *Continued.*

grain anisotropy of ferromagnetic magnetite grains is typically controlled by grain shape and alignment, whereas paramagnetic phyllosilicate grains have magnetic susceptibility principal axes parallel to the crystallographic axes. The Arguis Formation includes variable amounts of diamagnetic quartz and carbonate grains; however, these minerals will contribute little to the AMS because of their weak negative susceptibilities (Rochette 1987). All magnetic susceptibility measurements in this study were positive.

Kodama *et al.* (2010) report that the magnetic mineralogy of the remanence-carrying grains in the Arguis Formation is mostly detrital magnetite with subsidiary magnetic sulphide, likely pyrrhotite, based on rock magnetic experiments. Because ferromagnetic minerals have high susceptibility, they can dominate the total fabric despite low concentration. The AAR isolates the ferromagnetic fabric and can therefore aid in the understanding of the AMS fabric. Comparison of the AAR (Fig. 6c) and AMS fabric from the same sites (Fig. 4c in geographical and Fig. 7a in stratigraphic reference frames, respectively) show that the principal axes differ in orientation, suggesting the ferromagnetic grains are not dominating the AMS fabric. Both AAR and AMS fabrics from the Arguis syncline samples display a similar compaction fabric with the least principal susceptibility ($k_3$) axes nearly perpendicular to bedding. However, the AAR $k_1$ axes

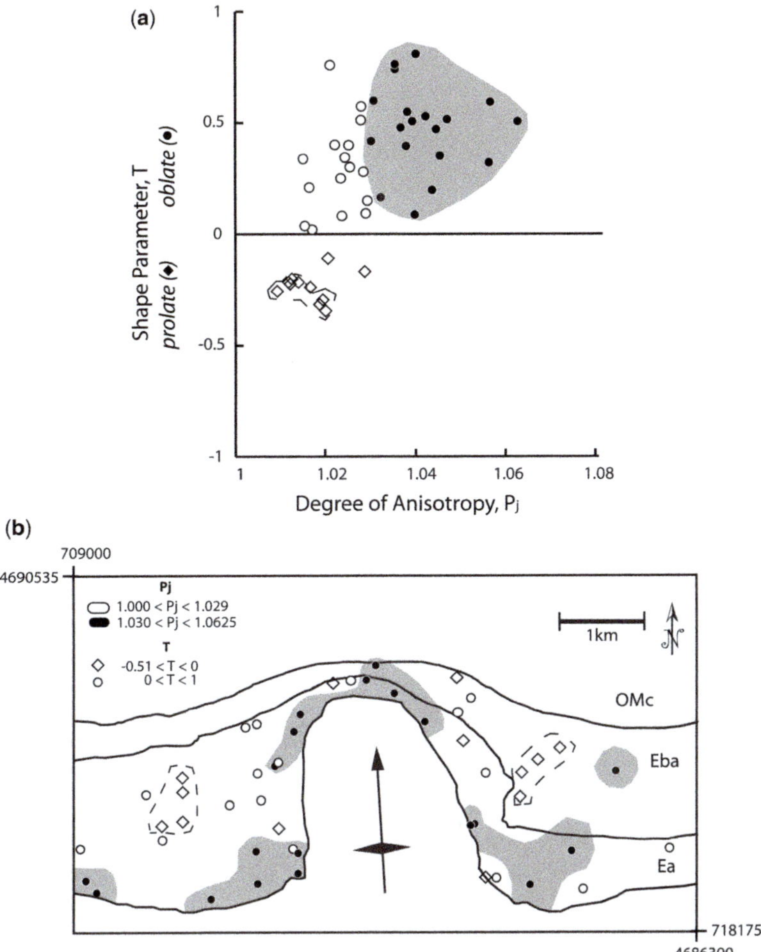



**Fig. 5.** (a) Jelínek (1981) plot of the shape parameter ($T$) and anisotropy degree ($P_j$) of the AMS ellipsoids from the Pico del Aguila structure. The majority of the samples have an oblate shape ($T > 0$) and low anisotropy. (b) Plan view map of $T$ and $P_j$ from Pico del Aguila. Grey shading: highest degree of anisotropy; dashed field: lowest anisotropy. $T$ and $P_j$ are calculated as follows: if $n_1 = \ln(t_1)$, $n_2 = \ln(t_2)$ and $n_3 = \ln(t_3)$, where $t_1$, $t_2$ and $t_3$ are the eigenvalues, then $T = (2n_2 - n_1 - n_3)/(n_1 - n_3)$ and $P' = \exp\{\text{sqrt}(2[(n_1 - n_{\text{mean}})^2 + (n_2 - n_{\text{mean}})^2 + (n_3 - n_{\text{mean}})^2])\}$ and $n_{\text{mean}} = (n_1 + n_2 + n_3)/3$.

orientations are close to the trend of the basin axis, suggesting the magnetite grains could record a depositional fabric, and the AMS $k_1$ axes orientation of the paramagnetic phyllosilicate grains are approximately fold-axis parallel, suggesting they record a tectonic fabric.

Low-temperature AMS measurements were used to enhance the paramagnetic grain contribution to the AMS signal since paramagnetic susceptibility is enhanced as temperature is lowered, according to the Curie–Weiss Law. Susceptibility of a pure paramagnetic material at 77 K is expected to increase by a factor of 3.8 times the room temperature measurement (e.g. Schultz-Krutisch & Heller 1985; Ihmlé *et al.* 1989). Figure 6 shows the comparison of low- and room-temperature susceptibility measurements. The increase of magnetic susceptibility at cold temperature as compared to room temperature measurements suggests the paramagnetic minerals dominate the AMS fabrics of the mudstones and sandstones. This is consistent with expectations of phyllosilicate mineral-rich rocks from analogous rocks elsewhere in the Pyrenean realm (e.g. Parés & van der Pluijm 2002; Pueyo-Anchuela *et al.* 2013). Because the AMS resides mostly on the paramagnetic phyllosilicate fraction,

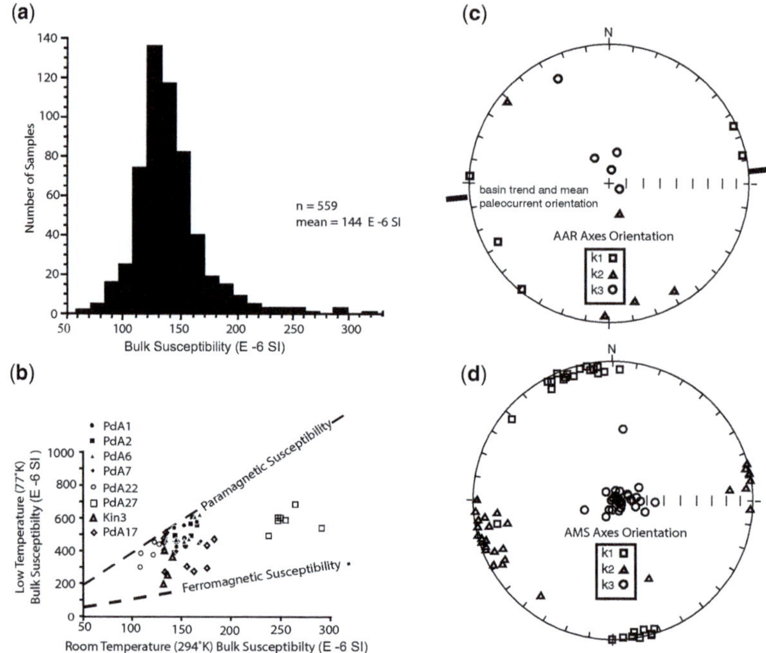



**Fig. 6.** (**a**) Bulk magnetic susceptibility histogram of all specimens sampled. (**b**) Plot of low-temperature (77 K) v. room-temperature (294 K) bulk susceptibility for selected samples. Most samples analysed show a large paramagnetic (clay) enhancement. The highest bulk susceptibilities are from sample PdA27, a sandstone with the lowest magnetic susceptibility enhancement. (**c**) Hand sample (siltstone, very fine sandstone, fine sandstone, medium sandstone, coarse sandstone) means of the anisotropy of anhysteretic remanence (AAR) from Arguis syncline specimens sampled near AMS sites AA, AD and AE plotted in stratigraphic coordinates. The general trend of the Jaca Basin axis determined from palaeocurrent directions from the Arguis Formation are shown (after Puigdefábregas 1975). (**d**) AMS orientations from sites AA, AD and AE in stratigraphic coordinates. Comparison of the principal axes variation in AAR fabric defined by ferromagnetic grain shape and orientation and the AMS fabric recorded by paramagnetic phyllosilicate grain orientations within a site is evident.

we can use the AMS as a proxy for clay grain preferred orientation.

## Magnetic fabric

Sedimentary rocks such as shale, mudstone, and marl, dominated by phyllosilicate minerals, are excellent for AMS studies because of the predominant oblate individual grain anisotropy which adjusts readily during lithification and any subsequent deformation. As grains reorient in response to deposition, compaction or tectonic strain, the magnetic fabric will continuously adjust. During deposition by water currents or simply by settling, the phyllosilicate grains will orient with their basal planes parallel or slightly imbricated to the depositional surface. Because the intermediate and maximum axes are nearly equal in magnitude they will be randomly oriented in bedding, but the minimum axes will tend to be vertical (e.g. Martín-Hernández & Hirt 2003). A weak lineation may form if the imbrication caused by deposition from flowing water

yields a weak intersection lineation parallel to flow (e.g. Aubourg *et al.* 1991).

## Depositional-diagenetic fabric and layer-parallel shortening strain

At all sites $k_3$ is nearly orthogonal to bedding. Some variation in $k_3$ clustering and orientation relative to the pole to bedding does exist, but the dispersion has no preferred bias and is therefore interpreted to record sediment deposition and compaction (Fig. 7a). Compaction during dewatering and lithification will amplify the initial oblate depositional fabric prior to the superposition of any tectonic fabric, and compaction fabrics only should produce oblate fabrics with low anisotropy that are not observed at Pico del Aguila (Fig. 8). The maximum and intermediate susceptibility axes are clustered in bedding rather than widely distributed in the bedding plane, consistent with some post-depositional tectonic strain (e.g. Kligfield *et al.* 1981, 1983; Kodama 2012; Figs 4 & 8).

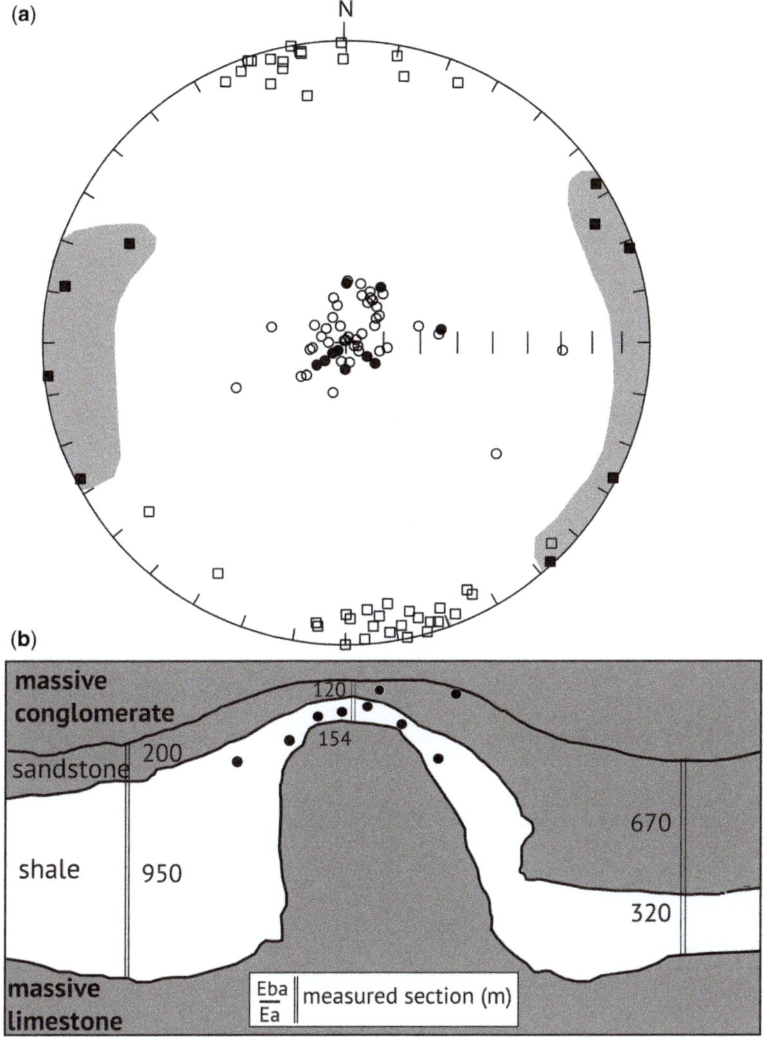


**Fig. 7.** (a) Lower hemisphere stereographic projection of $k_1$ and $k_3$ axes site mean orientations. Plain strain samples are omitted ($|T| \geq 0.01$). Open symbols: samples from synclinal hinges and fold limbs; solid symbols: samples from anticline hinge area identified on Figure 8b. (b) Lithologic map illustrating differences in the thickness of thinner-bedded units in the surrounding synclines as compared to Pico del Aguila anticline. Stratigraphic data derived from Puigdefábregas (1975), Anastasio (1987), Millán *et al.* (1994) and Castelltort *et al.* (2003).

The growth strata on Pico del Aguila have developed bedding fissility but lack a pervasive tectonic foliation including disjunctive or pencil cleavage. The lack of a mesoscopic fabric confirms the low level of finite strain in the rocks (e.g. Ramsay & Huber 1983). At nearly all sites the fabric shows a well-organized magnetic lineation (Fig. 4); however, the lineation magnitude is weak overall. At only a few sites, for example PdA1, PdA7 and PdA16, $k_1$ and $k_2$ have orientations distributed in bedding indicative of pencil cleaved rocks but below the strain threshold to

develop a macroscopic fabric (e.g. Parés 2004). Kligfield *et al.* (1981) describe similar AMS tectonic lineations produced at low strains from siltstones in the Alps Maritime, France, as does Hrouda *et al.* (2009) in samples from the western Carpathians.

The $k_1$ orientation at Pico del Aguila is downdip in the plane of bedding except at the anticline hinge, where it is consistently shallower than bedding dip in plunge. Within the growth strata, $k_1$ is oriented $350 \pm 20°$, parallel to the fold axis, except at anticline hinge area sites (Fig. 7). We interpret all the

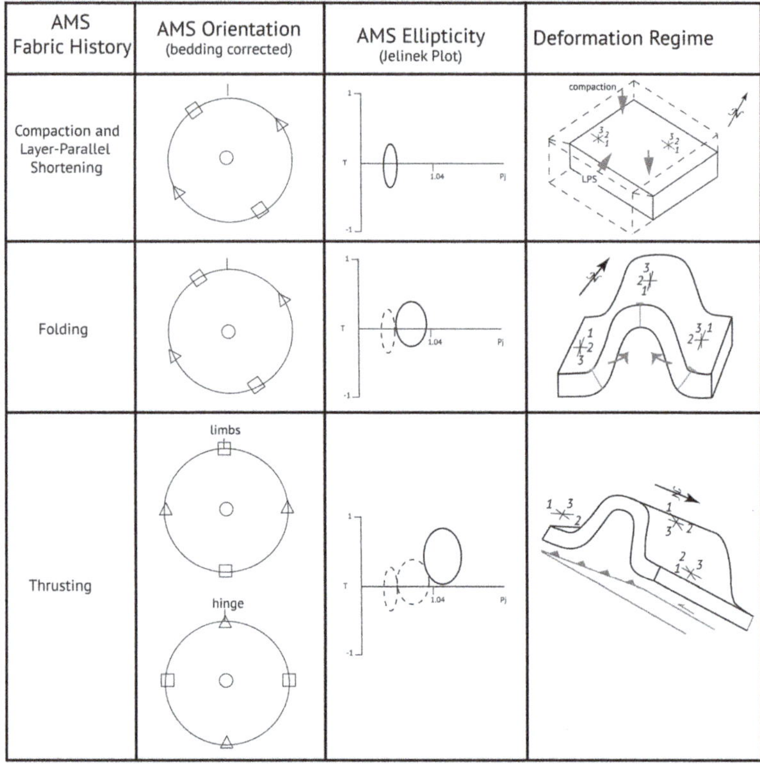

**Fig. 8.** Deformation history of AMS fabric. Top: Depositional, compaction and layer-parallel shortening produced a fabric characterized by bedding-perpendicular $k_3$ axes and NW-trending $k_1$ axes, which were passively rotated clockwise throughout the structure during subsequent emplacement of the Guarga thrust shet. The LPS fabric is closer to orthogonal to plate convergence (e.g. Roest & Srivastava 1991) and orogenic shortening directions (e.g. Holl & Anastasio 1995*b*) before subsequent clockwise vertical axis rotation of the Guarga thrust sheet. Middle: AMS fabric reorientation and anisotropy enhancement during fixed-hinge limb rotation accompanied detachment folding. Lower: Active fabric rotation with superposition of additional flattening strain in anticlinal hinge samples only during Guarga thrust sheet emplacement produced an east–west-trending $k_1$ axes orientation in the hinge area specimens. The evolution of the AMS fabric ellipsoid is shown on the associated Jelínek (1981) plots with ellipses (dashed: previous; solid: current) outlining the data distribution for each phase of the deformation history. The variation in block orientation relative to palaeo-north is to optimize the viewing perspective.

clustered $k_1$ orientations at Pico del Aguila as a tectonic fabric. The dominant palaeocurrent direction along the southern Jaca Basin during the Eocene was to the west (Puigdefábregas 1975), and was recovered by the AAR data (Fig. 6). In rocks dominated by phyllosilicate grains it is difficult to create a well-organized magnetic lineation by aligning grain crystallographic axes; however, an intersection lineation between slightly rotated clay grains orthogonal to a shortening direction has been observed previously (e.g. Henry 1997; Parés *et al.* 1999, 2007; Martín-Hernández & Ferré 2007; Yonkee & Weil 2010).

The *c.* north-trending $k_1$ axes recovered from the growth strata on the limbs of Pico del Aguila are attributed to layer-parallel shortening (LPS) which was subsequently passively reoriented by clockwise vertical-axis rotations during emplacement on the Guarga thrust sheet (Fig. 8b). The evidence for variable rotation magnitude has been recovered from remanence directions throughout the External Sierras (Pueyo *et al.* 2002; Mochales *et al.* 2010; Pueyo-Anchuela *et al.* 2012). Restoration of the rotation shows the LPS fabric is compatible with Pyrenean shortening directions (e.g. Anastasio & Holl 2001). Early LPS fabrics have also been observed in the Appalachians, USA (Nickelson 1979; Geiser & Engelder 1983; Gray & Mitra 1991) and in the Apennines, Italy (Tavernelli & Alverez 2002). Hirt *et al.* (2004) interprets a similar strain history from mineral textures interpreted from an AMS study of the same area of the Appalachian fold–thrust belt where LPS was interpreted in the field.

## Fold kinematics

Growth triangles and constant bedding thickness supports rolling fold hinges during folding, while thinning and shallowing dip of growth strata support pinned fold hinges. Onlap and growth strata thinning towards the Pico del Aguila anticlinal hinge, along with shallowing bedding dip upsection and constant fold limb length in the growth strata, support folding by limb rotation (spin) about a pinned anticlinal hinge (e.g. Riba 1976; Poblet & Hardy 1995; Salvini & Sorti 2002). The overall accumulation of strain as the sediments were lithified, folded and emplaced on the Guarga thrust sheet is faithfully recorded by the AMS (Fig. 8). The distribution of the AMS ellipsoids on a Jelínek (1981) plot and on a geological map of Pico del Aguila clearly show higher magnetic anisotropy ($P_j$) near fold axial surfaces (Fig. 5). With no difference in bulk susceptibility as a function of structural position or significant correlation between bulk susceptibility and the degree of anisotropy $P_j$ (Figs 5 & 6), there is little evidence to support a change in mineral composition or concentration between samples. A higher degree of magnetic fabric anisotropy along fold axial surfaces is compatible with pinned hinge fixed limb-length folding in agreement with growth strata geometries. Fixed limb-length detachment folding requires evacuation of material from the core and décollement depth migration during folding (e.g. Homza & Wallace 1995). The thrust fault and thinning of the Pont de Suert Formation in the fold core are consistent with the rotation of fixed-length limbs during folding (Poblet & McClay 1996; González-Mieres & Suppe 2006).

There is little deflection of the minimum axes ($k_3$) of the AMS ellipsoid from the bedding pole. Penetrative bedding-parallel shear of opposing sense on the fold limbs with clockwise rotation of $k_3$ orientation on the east fold limb and anticlockwise on the west limb would be expected for kinematic dominated by pinned hinge flexural folding (e.g. Ramsay 1967; compare Fig. 4b, d). There is also little field evidence (e.g. shear veins, slickensides) of discrete layer-parallel shear on bedding planes; the AMS fabric is therefore interpreted to support buckle folding about pinned hinges consistent with the growth strata geometry.

## Thrusting fabric

The AMS fabric varies by structural position but not stratigraphic position in the Pico del Aguila samples. Hinge zone sites are the exclusive structural position where $k_1$ orientation is dominantly shallow and east–west plunging rather than northwards plunging, north–south trending and in bedding. The anticlinal hinge and synclinal sites are similar in bedding orientation and strain history. At these locations the beds have not been rotated about a north–south axis during folding, and both of these structural positions were carried in a similar fashion on the Guarga thrust sheet. The orientation of the $k_1$ axis in hinge sites is consistent with c. south-directed shortening during final thrust emplacement rather than recording folding strain. Emplacement of the Guarga thrust sheet continued after transverse fold décollement, which varied in timing along the strike of the External Sierras (DePaor & Anastasio 1987; Anastasio 1992; Millán et al. 2000). Transport up the frontal footwall ramp of the Guarga thrust sheet generated the northwards plunge of the Pico del Aguila fold and further strained sites between the thick stiff mechanical layers of the Guara Formation, a massive bedded carbonate, and the thick bedded sands and gravels of the overlying Campodarbe Formation, where the mudstone and sandstone dominated flysch is thinnest (Fig. 7b). Anticline hinge sites have high anisotropies and a $k_1$ axis oriented east–west rather than the north–south orientation that characterizes synclinal sites at the same stratigraphic level (Figs 4 & 5). Importantly, this means that AMS fabrics in the anticline hinge samples do not preserve the LPS fabric that defined the early intersection lineation in the other samples (e.g. Larrasoaña et al. 2004). The phyllosilicate fabric in the growth strata surrounding Pico del Aguila are only reoriented where the units are thin as additional rock strain was superimposed during thrusting up the frontal footwall ramp of the Guarga thrust sheet.

## Conclusion

The phyllosilicate depositional-compaction fabric surrounding the Pico del Aguila anticline was deformed by layer-parallel shortening oriented east–west (NE–SW in the pre-rotational reference frame), orthogonal to the fold axis orientation, to create a north–south-trending magnetic lineation in bedding on fold limbs and in the adjacent Belsué and Arguis synclines, parallel to the anticlinal fold axes. The magnetic lineation is likely a grain-scale intersection lineation produced by the paramagnetic phyllosilicate grains that dominate the magnetic susceptibility. The magnitude of the $k_1$ lineation varies little across the fold and throughout the growth strata, but varies instead by structural position. The minimum principal susceptibility axes ($k_3$) are perpendicular to bedding across the fold, suggesting that bedding-parallel flexural shear strain did not much deflect the paramagnetic grains. Near the anticlinal hinge, the magnetic lineation is fold axis perpendicular and thought to be the result of later flattening strain related to thrust sheet emplacement.

This study has shown the AMS was useful for resolving fold kinematics and to partition strain history at diagenetic conditions. We showed that the tectonic magnetic fabric developed early in the deformation history and, rather than locking in, the AMS fabric continued to evolve, recording all phases of the progressive deformation history.

Anastasio, Kodama and Parés acknowledge financial support for this work from National Science Foundation grant EAR 0409077. Troy acknowledges a Palmer grant for graduate research administered by the Department of Earth and Environmental Sciences, Lehigh University. Tania Mochales-Lopez, Adriana Rodiguez-Pinto and Carlota Olivan, University of Zaragoza are thanked for field assistance. Christine Regalla assisted with the precision GPS data collection using equipment on loan from the UNAVCO consortium. Anastasio began work on this article during a sabbatical visit to the IGME-Zaragoza and the Dept de Ciencias de la Tierra, University of Zaragoza supported by the Departamento de Ciencia, Tecnología y Universidad (grant FMI046/09), Gobierno de Aragon, Spain. Editorial work by Belen Oliva-Urcia and reviews by Bjarne Almqvist and two anonymous reviewers were helpful in improving the paper.

# References

ANASTASIO, D. J. 1987. *Thrusting, halotectonics, and sedimentation in the External Sierra, southern Pyrenees, Spain.* PhD dissertation, Johns Hopkins University, Baltimore, MD.

ANASTASIO, D. J. 1992. Structural evolution of the External Sierra, Spanish Pyrenees. *In:* MITRA, S. & FISCHER, G. W. (eds) *The Structural Geology of Fold and Thrust Belts.* Johns Hopkins University Press, 239–251.

ANASTASIO, D. J. & HOLL, J. 2001. Transverse fold evolution in the External Sierra, southern Pyrenees, Spain. *Journal of Structural Geology,* **23**, 379–392.

ANASTASIO, D. J., FISHER, D. M., MESSINA, T. A. & HOLL, J. E. 1997. Kinematics of décollement folding in the Lost River Range, Idaho. *Journal of Structural Geology,* **19**, 355–368.

AUBOURG, C., ROCHETTE, P. & VIALON, P. 1991. Subtle stretching lineation revealed by magnetic fabric of Callovian-Oxfordian black shales (French Alps). *Tectonophysics,* **185**, 211–223.

AVERBUCH, O., DELAMOTTE, D. F. & KISSEL, C. 1992. Magnetic fabric as a structural indicator of the deformation path within a fold thrust structure – a test case from the Corbieres (NE Pyrenees, France). *Journal of Structural Geology,* **14**, 461–474.

BORRADAILE, G. 1988. Magnetic susceptibility, petrofabrics and strain. *Tectonophysics,* **156**, 1–20.

BORRADAILE, G. J. & HENRY, B. 1997. Tectonic applications of magnetic susceptibility and its anisotropy. *Earth Science Reviews,* **4**, 49–93.

BORRADAILE, G. J. & JACKSON, M. 2004. Anisotropy of magnetic susceptibility (AMS): magnetic petrofabrics of deformed rocks. *In:* MARTIN-HERNANDEZ, F., LUENEBURG, C. M., AUGOURG, C. & JACKSON, M. (eds) *Magnetic Fabric: Methods and Applications.* Geological Society, London, Special Publications, **238**, 299–360.

BORRADAILE, G. J. & TARLING, D. H. 1981. The influence of deformation mechanisms on magnetic fabrics in weakly deformed rocks. *Tectonophysics,* **77**, 151–168.

BURMEISTER, K. C., HARRISON, M. J., MARSHAK, S., FERRE, E. C. & BANNISTER, R. A. 2009. Comparison of Fry strain ellipse and AMS ellipsoid trends to tectonic fabric trends in very low-strain sandstone of the Appalachian fold-thrust belt. *Journal of Structural Geology,* **9**, 1028–1038.

CANUDO, J. I. 1990. *Los foraminiferos planctonicos del Paleoceno-Eocene del Prepirineo oscense en el sector de Arguis.* Doctoral thesis, University of Zaragoza, Zaragoza, Spain.

CANUDO, J. I., MALAGON, J., MELENDEZ, A., MILLAN, H., MOLINA, E. & NAVARRO, J. J. 1991. Las secuencias deposicionales del Eoceno medio y superior de las Sierras exteriors (Prepirineo meridional aragones). *Geogaceta,* **9**, 81–84.

CASTELLTORT, S., GUILLOCHEAU, F., ROBIN, C., ROUBY, D., NALPAS, T., LAFONT, F. & ESCHARD, R. 2003. Fold control on the stratigraphic record: a quantified sequence stratigraphic study of the Pico del Aguila anticline in the south-western Pyrenees (Spain). *Basin Research,* **15**, 527–551.

CLOOS, E. 1947. Oolite deformation in South Mountain fold, Maryland. *Geological Society of America Bulletin,* **58**, 843–918.

DAHLSTROM, C. D. A. 1990. Geometric constraints derived from the law of conservation of volume and applied to evolutionary models for detachment folding. *American Association of Petroleum Geologists Bulletin,* **74**, 336–344.

DEPAOR, D. G. & ANASTASIO, D. J. 1987. The External Sierra: a case history in the advance and retreat of mountains. *National Geographic Research,* **3**, 199–209.

EPARD, J. L. & GROSHONG, R. H. 1995. Kinematic model of detachment folding including limb rotation, fixed hinges and layer-parallel strain. *Tectonophysics,* **247**, 85–103.

FISCHER, M. P., WOODWARD, N. B. & MITCHEL, M. M. 1992. The kinematics of break-thrust folds. *Journal of Structural Geology,* **14**, 451–460.

FISHER, D. M. & ANASTASIO, D. J. 1994. Kinematic analysis of a large-scale leading edge fold, Lost River Range, Idaho. *Journal of Structural Geology,* **16**, 333–354.

FORD, M., WILLIAMS, E. A., ARTONI, A., VERGÉS, J. & HARDY, S. 1997. Progressive evolution of a fault-related fold pair from growth strata geometries, Sant Llorenç de Morunys, SE Spain. *Journal of Structural Geology,* **19**, 413–441.

GEISER, P. A. & ENGELDER, T. 1983. The distribution of layer parallel shortening fabrics in the Appalachian foreland of New York and Pennsylvania: evidence for two non-coaxial phases of the Alleghanian Orogeny. *In:* HATCHER, R. D. Jr. (ed.) *Contributions to the Tectonics and Geophysics of Mountain Chains.* Geological Society of America, Memoir, **158**, 161–175.

GONZÁLEZ-MIERES, R. & SUPPE, J. 2006. Relief and shortening in detachment folds. *Journal of Structural Geology,* **28**, 1785–1807.

GRAY, M. B. & MITRA, G. 1991. Ramifications of four-dimensional progressive deformation in contractional mountain belts. *Journal of Structural Geology*, **21**, 1151–1160.

HENRY, B. 1997. The magnetic zone axis: a new element of magnetic fabric for the interpretation of magnetic lineation. *Tectonophysics*, **271**, 325–331.

HIRT, A. M., LOWRIE, W., LUENEBURG, C. M., LEBIT, H. & ENGELDER, T. 2004. Magnetic and mineral fabric development in the Ordovician Martinsburg Formation in the Central Appalachian Fold and Thrust Belt, Pennsylvania. *In*: MARTÍN-HERNÁNDEZ, F., LÜNEBURG, C. M., AUBOURG, C. & JACKSON, M. (eds) *Magnetic Fabric: Methods and Application.* Geological Society, London, Special Publications, **238**, 109–126.

HOGAN, P. J. 1993. *Geochronologic, Tectonic, and Stratigraphic Evolution of the Southwest Pyrenean Foreland, Northern Spain.* Unpublished PhD thesis. University of Southern California.

HOGAN, P. J. & BURBANK, D. W. 1996. Evolution of the Jaca piggyback basin and emergence of the External Sierra, southern Pyrenees. *In*: FRIEND, P. F. & DABRIO, C. J. (eds) *Tertiary Basins of Spain: The Stratigraphic Record of Crustal Kinematics.* Cambridge University Press, Cambridge, 153–160.

HOLL, J. E. & ANASTASIO, D. J. 1995*a*. Cleavage development within a foreland fold and thrust belt, southern Pyrenees, Spain. *Journal of Structural Geology*, **17**, 357–369.

HOLL, J. E. & ANASTASIO, D. J. 1995*b*. Kinematics around a large-scale oblique ramp, southern Pyrenees, Spain. *Tectonics*, **14**, 1368–1376.

HOMZA, T. X. & WALLACE, W. K. 1995. Geometric and kinematic models from detachment folds with fixed and variable detachment depths. *Journal of Structural Geology*, **17**, 575–588.

HROUDA, F., KREJCI, O., POTFAJ, M. & STRANIK, Z. 2009. Magnetic fabric and weak deformation in sandstones of accretionary prisms of the Flysch and Klippen Belts of the Western Carpathians: mostly offscraping indicated. *Tectonophysics*, **479**, 254–270.

IHMLÉ, P. F., HIRT, A. M., LOWRIE, W. & DIETRICH, D. 1989. Inverse magnetic fabric in deformed limestones of the Morcles Nappe, Switzerland. *Geophysical Research Letters*, **16**, 1383–1386.

JELÍNEK, V. 1978. Statistical processing of magnetic susceptibility measured in groups of specimens. *Acta Geodaetica et Geophysica Hungarica*, **22**, 52–62.

JELÍNEK, V. 1981. Characterization of the magnetic fabric of rocks. *Tectonophysics*, **79**, 63–67.

KLIGFIELD, R., OWENS, W. H. & LOWRIE, W. 1981. Magnetic susceptibility anisotropy, strain and progressive deformation in Permian sediments from the Maritime Alps (France). *Earth and Planetary Science Letters*, **55**, 181–189.

KLIGFIELD, R., LOWRIE, W., HIRT, A. M. & SIDDANS, A. W. B. 1983. Effect of progressive deformation on remanent magnetization of Permian redbeds from the Alps Maritimes (France), *Tectonophysics*, **97**, 59–85.

KODAMA, K. P. 2012. *Paleomagnetism of Sedimentary Rocks: Process and Interpretation.* Wiley-Blackwell, Oxford.

KODAMA, K. P., ANASTASIO, D. J., NEWTON, M. L., PARÉS, J. M. & HINNOV, L. A. 2010. High-resolution rock magnetic cyclostratigraphy in an Eocene flysch, Spanish Pyrenees. *Geochemistry, Geophysics, Geosystems*, **11**, 1–22.

KRUIVER, P. P., DEKKERS, M. J. & HESLOP, D. 2001. Quantification of magnetic coercivity by the analysis of acquisition curves of isothermal remanent magnetization. *Earth and Planetary Science Letters*, **189**, 269–276.

LARRASOAÑA, J. C., PUEYO, E. L. & PARÉS, J. M. 2004. An integrated AMS, structural, palaeo- and rock-magnetic study of Eocene marine marls from the Jaca-Pamplona Basin (Pyrenees, N Spain): new insights into the timing of magnetic fabric acquisition in weakly deformed mudrocks. *In*: MARTIN-HERNANDEZ, F., LUENEBURG, C. M., AUGOURG, C. & JACKSON, M. (eds) *Magnetic Fabric: Methods and Applications.* Geological Society, London, Special Publications, **238**, 127–143.

LATTA, D. K. & ANASTASIO, D. J. 2007. Multiple scales of mechanical stratification and decollement fold kinematics, Sierra Madre Oriental foreland, northeast Mexico. *Journal of Structural Geology*, **29**, 1241–1255.

LOWRIE, W. 1990. Identification of ferromagnetic minerals in a rock by coercivity and unblocking temperature properties. *Geophysical Research Letters*, **17**, 159–162.

MARCH, A. 1932. Mathematische theorie der regelung nach der korngesalt bei affiner deformation. *Zeitschrift für Kristallographie*, **81**, 285–297.

MARTÍN-HERNÁNDEZ, F. & FERRÉ, E. C. 2007. Separation of paramagnetic and ferromagnetic anisotropies: a review. *Journal of Geophysical Research*, **112**, B03105.

MARTÍN-HERNÁNDEZ, F. & HIRT, A. M. 2003. The anisotropy of magnetic susceptibility in biotite, muscovite and chlorite single crystals. *Tectonophysics*, **367**, 13–28.

MATTEI, M., SAGNOTTI, L., FACCENNA, C. & FUNICIELLO, R. 1997. Magnetic fabric of weakly deformed clay-rich sediments in the Italian peninsula: relationship with compressional and extensional tectonics. *Tectonophysics*, **271**, 107–122.

MILLÁN, H. 2006. *Estructura y cinemática del frente de cabalgamiento Surpirenaico en las Sierras Exteriores Aragonesas.* PhD dissertation, University of Zaragoza, Zaragoza, Spain.

MILLÁN, H., AURELL, M. & MELÉNDEZ, A. 1994. Synchronous detachment folds and coeval sedimentation in the Prepyrenean External Sierras (Spain). A case study for a tectonic origin of sequences and system tracts. *Sedimentology*, **41**, 1001–1024.

MILLÁN, H., PUEYO, E. L., AURELL, M., LUZÓN, A., OLIVA-URCIA, B., MARTÍNEZ PEÑA, M. B. & POCOVÍ, A. 2000. Actividad tectónica registrada en los depósitos terciarios del frente meridional del Pirineo central. *Revista de la Sociedad Geológica de España*, **13**, 117–138.

MOCHALES, T., PUEYO, E. L., CASAS, A. M., BARNOLAS, A. & OLIVA-URCIA, B. 2010. Anisotropic magnetic susceptibility record of the kinematics of the Boltaña Anticline (Southern Pyrenees). *Geological Journal*, **45**, 562–581.

MUÑOZ, J. A. 1992. Evolution of a continental collision belt: ECORS-Pyrenees crustal balanced cross-section. *In*: MCCLAY, K. R. (ed.) *Thrust Tectonics.* Springer, Netherlands, 235–246.

NICKELSON, R. P. 1979. Sequence of structural stages of the Alleghany Orogeny at the Bear Valley Strip Mine, Shamokin. *American Journal of Science*, **279**, 225–271.

NYE, J. F. 1957. *Physical Properties of Crystals: their Representation by Tensors and Matrices*. Oxford University Press, Oxford.

PARÉS, J. M. 2004. How deformed are weakly deformed mudrocks? Insights from magnetic anisotropy. *In*: MARTIN-HERNANDEZ, F., LUENEBURG, C. M., AUGOURG, C. & JACKSON, M. (eds) *Magnetic Fabric: Methods and Applications*. Geological Society, London, Special Publications, **238**, 191–203.

PARÉS, J. M. & VAN DER PLUIJM, B. A. 2002. Evaluating magnetic lineations (AMS) in deformed rocks. *Tectonophysics*, **350**, 283–298.

PARÉS, J. M. & VAN DER PLUIJM, B. A. 2003. Phyllosilicate fabric characterization by low-temperature anisotropy of magnetic susceptibility (LT-AMS). *Geophysical Research Letters*, **29**, 4.

PARÉS, J. M., VAN DER PLUIJM, B. & DINARÈS-TURELL, J. 1999. Evolution of magnetic fabrics during incipient deformation of mudrocks. *Tectonophysics*, **307**, 1–14.

PARÉS, J. M., HASSOLD, N. J. C., REA, D. K. & VAN DER PLUIJM, B. A. 2007. Paleocurrent directions from paleomagnetic reorientation of magnetic fabrics in deep-sea sediments at the Antarctic Peninsula Pacific margin (ODP Sites 1095, 1101). *Marine Geology*, **242**, 261–269.

POBLET, J. & HARDY, S. 1995. Reverse modeling of detachment folds–application to the Pico-del-Aguila Anticline in the south central Pyrenees (Spain). *Journal of Structural Geology*, **17**, 1707–1724.

POBLET, J. & MCCLAY, K. 1996. Geometry and kinematics of single-layer detachment folds. *Bulletin of the American Association of Petroleum Geologists*, **80**, 1085–1109.

PUEYO, E. L., MILLÁN, H., POCOVÍ, A. & PARÉS, J. M. 1997. Determination of the folding mechanism by AMS Data. Study of the relation between shortening and magnetic anisotropy in the Pico del Aguila anticline (Southern Pyrenees). *Physics and Chemistry of the Earth*, **22**, 195–201.

PUEYO, E. L., MILLÁN, H. & POCOVÍ, A. 2002. Rotation velocity of a thrust: a paleomagnetic study in the External Sierras (Southern Pyrenees). *Sedimentary Geology*, **146**, 191–208.

PUEYO-ANCHUELA, O., PUEYO, E. L., POCOVÍ, A. & GIL-IMAZ, A. 2012. Vertical axis rotations in fold and thrust belts: comparison of AMS and paleomagnetic data in the Western External Sierras (Southern Pyrenees). *Tectonophysics*, **532–535**, 119–133.

PUEYO-ANCHUELA, Ó., CASAS-SAINZ, A. M., PUEYO, E. L., POCOVÍ, A. & GIL-IMAZ, A. 2013. Analysis of the ferromagnetic contribution to the susceptibility by low field and high field methods in sedimentary rocks of the Southern Pyrenees and Northern Ebro foreland basin (Spain). *Terra Nova*, **25**, 307–314.

PUIGDEFÁBREGAS, C. 1975. *La Sedimentation Molasica en la Cuenca de Jaca*. Instituto de Estudios Pirenaicos, Jaca, Spain, Monograph 104.

RAMSAY, J. G. 1967. *The Folding and Fracturing of Rocks*. McGraw-Hill, New York.

RAMSAY, J. G. & HUBER, M. I. 1983. *The Techniques of Modern Structural Geology*. Academic Press, San Diego, **1**.

RIBA, O. 1976. Syntectonic unconformities in the Alto Cardener, Spanish Pyrenees: a genetic interpretation. *Sedimentary Geology*, **15**, 213–233.

RICHTER, C., VAN DER PLUIJM, B. & HOUSEN, B. 1993. The quantification of crystallographic preferred orientation using magnetic anisotropy. *Journal of Structural Geology*, **15**, 113–116.

ROCHETTE, P. 1987. Magnetic susceptibility of the rock matrix related to magnetic fabric studies. *Journal of Structural Geology*, **9**, **8**, 1015–1020.

ROEST, W. R. & SRIVASTAVA, S. P. 1991. Kinematics of the plate boundaries between Eurasia, Iberia and Africa in the North Atlantic from late Cretaceous to the present. *Geology*, **19**, 613–616.

RODRÍGUEZ-PINTÓ, A., PUEYO, E. L., SIERRA-KIEL, J., SAMSÓ, J. M., BARNOLAS, A. & POCOVÍ, A. 2012. Lutetian magnetostratigraphic calibration of larger foraminifera zonation (SBZ) in the Southern Pyrenees: the Isuela section. *Paleoecology, Paleoclimatology, Paleoecology*, **333–334**, 107–120.

SAGNOTTI, L., SPERANZA, F., WINKLER, A., MATTEI, M. & FUNICIELLO, R. 1998. Magnetic fabric of clay sediments from the external northern Apennines (Italy). *Physics of the Earth and Planetary Interiors*, **105**, 73–93.

SALVINI, F. & SORTI, F. 2002. Three-dimensional architecture of growth strata associated to fault-bend, fault-propagation, and decollement anticlines in non-erosional environments. *Sedimentary Geology*, **146**, 1–2, 57–73.

SCHMIDT, V., HIRT, A. M., LEISS, B., BURLINI, L. & WALTER, J. 2009. Quantitative correlation of strain, texture and magnetic anisotropy of compacted calcite-muscovite aggregates. *Journal of Structural Geology*, **31**, 1062–1073.

SCHULTZ-KRUTISCH, T. & HELLER, F. 1985. Measurement of magnetic susceptibility anisotropy in Bundsandstein deposits from southern Germany. *Journal of Geophysics*, **56**, 51–58.

TARLING, D. H. & HROUDA, F. 1993. *The Magnetic Anisotropy of Rocks*. Chapman and Hall, London, UK.

TAUXE, L. 2002. *Paleomagnetic Principles and Practices*. Kluwer Academic Publishers, Norwell, MA.

TAVERNELLI, E. & ALVEREZ, W. 2002. The mesoscopic response to positive tectonic inversion processes: an example from the Umbria-Marche Apennines, Italy. *Proceedings of the Scientific Meeting on Geological and Geodynamic Evolution of the Apennines*. Bollettino dela Societa Geologica Italiana, Bologna, Italy, **1**, 715–727.

VIDAL-ROYO, O., MUÑOZ, J. A., HARDY, S., KOYI, H. & CARDOZO, N. 2013. Structural evolution of Pico del Águila anticline (External Sierras, southern Pyrenees) derived from sandbox, numerical and 3D structural modelling techniques. *Geológica Acta*, **11**, 1–26.

YONKEE, A. & WEIL, A. B. 2010. Reconstructing the kinematic evolution of curved mountain belts: internal strain patterns in the Wyoming salient, Sevier thrust belt, U.S.A. *Bulletin of the Geological Society of America*, **122**, 24–49.

# Determining the timing of formation of the Rawil Depression in the Helvetic Alps by palaeomagnetic and structural methods


G. L. CARDELLO[1,2]*, B. S. G. ALMQVIST[1,3], A. M. HIRT[4] & N. S. MANCKTELOW[1]

[1]*Institute of Geology, ETH Zurich, Sonneggstrasse 5, CH-8092 Zürich, Switzerland*

[2]*Present address: Orléans University – ISTO, 1A Rue de la Férrolerie, 45071, Orléans, France*

[3]*Present address: Department of Earth Sciences, Uppsala University, Villavägen 16b, 752 36, Uppsala, Sweden*

[4]*Institute of Geophysics, ETH Zurich, Sonneggstrasse 5, CH-8092 Zürich, Switzerland*

**Corresponding author (e-mail: luca.cardello@univ.orleans.fr)*



**Abstract:** Anisotropy of magnetic susceptibility, palaeomagnetism and structural methods are used in order to test the relative timing of antiform updoming and formation of the Rawil Depression in the Helvetic Alps. Samples were collected from all nappes currently exposed in the study region. The magnetic fabric is consistent with extension oblique and parallel to the regional fold trend and with palaeostress reconstructions from fault planes and veins. Palaeomagnetic analyses show a stable characteristic remanence (ChRM), with samples recording both normal and reverse polarity. A successful fold test performed across the antiformal dome structure suggests that the palaeomagnetic signal was acquired prior to doming. By comparison with thermochronometric data, the ChRM was acquired between 25 and 10 Ma and is pre- to synfolding. A secondary post-doming palaeomagnetic component (A), whose magnetization is likely to have occurred between 10 and 3.5 Ma, appears to be too steep with regards to the inclination of the Earth's field, suggesting recent large-scale tilting has occurred in the region. These combined analyses indicate that widespread orogen-parallel extension occurred prior to the formation of the Rawil Depression, which is finally interpreted as the result of a stepover structure at the curvature between Central and Western Alps.


The Rawil Depression lies in the outer part of the Alpine arc between the domal culminations of two external crystalline massifs, the Aiguilles Rouges–Mont Blanc massifs in the southwest and the Gastern–Aar–Gotthard massifs in the northeast (Lugeon 1914–1918; Argand 1916). As for many similar structural depressions in mountain belts, its formation is not well established and remains controversial in terms of time and mechanism of evolution (e.g. Heim 1920; Burkhard 1988; Dietrich 1989; Levato *et al.* 1994; Pfiffner 2009; Gasser & Mancktelow 2010).

In general, structural depressions such as axial depressions or grabens are places where the highest stratigraphic units have been preserved from erosion (Fig. 1). The axial depression represents a topographic low of a doubly plunging fold hinge and is a product of ductile-related deformational processes, whereas a graben is the result of brittle tectonics. These two structural elements therefore represent the end-members of a structural depression, the first related to folding and the second to faulting. Both processes are active during the progressive exhumation and cooling of the crust,

and may occur at the same time depending on the lithology and the boundary conditions. Different rheological behaviour and structural history may influence the magnetic fabric and the direction of palaeomagnetic components recorded in the rocks. In the case of the Rawil, the resultant structural depression is the combination of an axial depression that is later dissected by transtensive faults (Fig. 1). In the most depressed area, the highest nappes that are preserved by the erosion are shifted *c.* 5 km to the west with respect to the axial depression. Oblique normal faults with displacement up to *c.* 1 km occur in this area (e.g. Iffigensee Fault, Burkhard 1988), suggesting that the more brittle faults dissected the former structure during late formation and progressive cooling (Cardello 2013).

At the beginning of their formation, structural depressions are often associated with stretching parallel to the fold axis where the stretching lineation is horizontal (i.e. 'classical' axial depressions; Dietrich 1989). For completeness and simplicity, we will refer to the Rawil Axial Depression as the Rawil Depression because in this term the brittle part of the structural history is incorporated. In


*From*: Pueyo, E. L., Cifelli, F., Sussman, A. J. & Oliva-Urcia, B. (eds) 2016. *Palaeomagnetism in Fold and Thrust Belts: New Perspectives.* Geological Society, London, Special Publications, **425**, 145–168.
First published online July 22, 2015, updated August 18, 2015, http://doi.org/10.1144/SP425.4

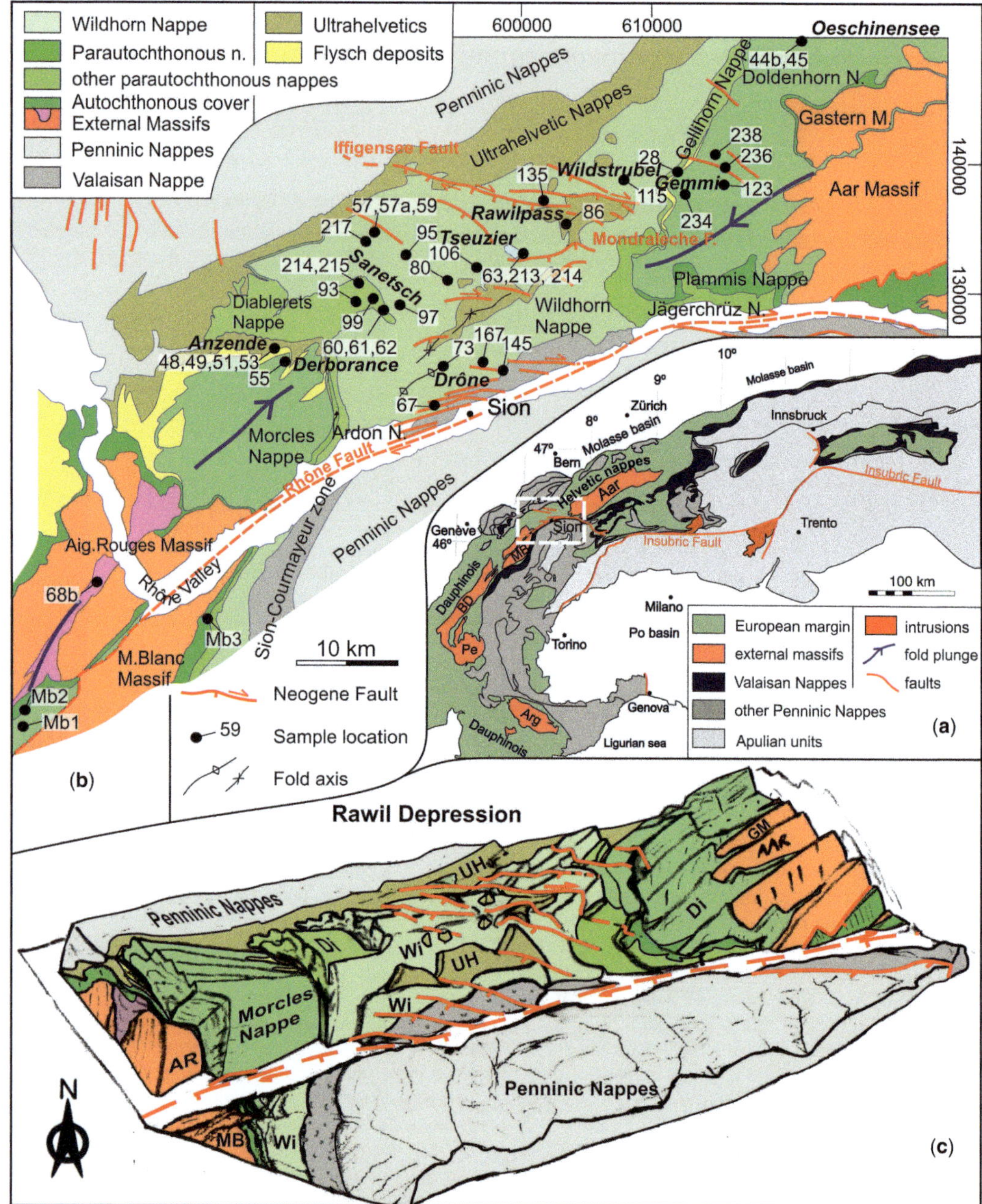



**Fig. 1.** (**a**) Simplified tectonic map of the Alps (modified after Bigi *et al.* 1990). (**b**) Tectonic map of the study area (www.swisstopo.admin.ch, modified). The axis of the Rawil Depression is indicated by the blue line, with the plunge direction indicated by arrows. (**c**) Coloured sketch of the Rawil Depression modified after Heim (1920).

this area, clusters of seismicity show focal mechanisms with similar kinematics to those determined from fault analysis in the field (Maurer *et al.* 1997; Sue *et al.* 2007; Ustaszeweski & Pfiffner 2008; Gasser & Mancktelow 2010; Cardello & Tesei 2013). This suggests that some faults seen in the field were generated by the same stress regime that is responsible for regional earthquakes with $M_w \geq 6$ (Eva *et al.* 1998; Kastrup *et al.* 2004). A comparison between the stress related to this faulting and the incremental strain may help to furnish a more detailed and robust deformational history. Furthermore, a comparison of the anisotropy of magnetic susceptibility (AMS) from rocks in this region with the palaeomagnetic directions can be used to determine possible tectonic rotations, when facilitated by structural observations on both outcrop and large-scale structures. AMS is a valuable tool for investigation of the deformational history of rocks (e.g. Graham 1966; Kligfield *et al.* 1982; Borradaile 1988, 1991; Borradaile & Henry 1997; Parés & van der Pluijm 2002; Borradaile & Jackson 2010) and has been widely used to unravel strain histories, especially in weakly deformed rocks, where conventional strain indicators are either weak or completely absent (e.g. Borradaile & Tarling 1981; Cifelli *et al.* 2004, 2005, 2009). Palaeomagnetism is a powerful tool to infer possible crustal block rotation as shown in numerous studies in fold belts (i.e. Butler 1992; Sussman *et al.* 2004, 2012; Satolli *et al.* 2005; Pueyo *et al.* 2007; Weil & Yonkee 2012).

Magnetic fabrics and palaeomagnetic data will be used together with structural geology in an integrative approach, firstly to better constrain the boundary conditions of the Rawil Structural Depression in the Central Alps and secondly to clarify the processes behind the structural evolution of axial depressions in general. The AMS data will be interpreted in terms of the structural framework of the Alpine rocks, which experienced a complex tectonic and metamorphic history, to understand which deformational phases control the magnetic fabric. The AMS data can then be used to ascertain regional rotations associated with the deformational history. Palaeomagnetic data will further help constrain the timing of magnetization in the rock and local or regional rotation. Comparison of the magnetic results with local structural data will help to establish a regional tectonic evolution of the Rawil Depression and a direct qualitative comparison between magnetic fabric and incremental strain and stress axes from fault stress inversions.

## Geological setting and samples

The Helvetic domain (Fig. 1) represents the sedimentary cover deposited on the European continental crust from the Permo-Carboniferous to the Early Tertiary (Lugeon 1914–1918). The succession was affected more than once during its geological history by synsedimentary normal faulting that is associated with stratigraphic gaps and unconformable contacts (Cardello & Mancktelow 2014). During the subsequent continental collision between Adria and Europe, these units were folded and overthrusted to form the Helvetic nappes; these are

exposed north of the Rhône valley and have been the subject of many studies regarding convergence, crustal thickening, burial and metamorphism (e.g. Lugeon 1914–1918; Masson *et al.* 1980*a, b*; Steck 1980; Ramsay *et al.* 1981, 1983; Steck 1984; Burkhard 1988; Ramsay 1989; Pfiffner 1993; Burkhard 1999; Frey & Ferreiro Mählmann 1999; Herwegh & Pfiffner 2005). After an early stage of overthrusting on the order of 100 km, which juxtaposed the thin Ultrahelvetic units onto the future Helvetic nappes, two major deformation events can be distinguished (Heim 1920; Ramsay *et al.* 1983; Pfiffner 2009). The first event is defined by short-wavelength folding that was associated with thrusting on the order of kilometres at the base of nappe boundaries, and with minor internal displacement on the order of tens of metres within each nappe. On a regional scale the nappe stack in the Rawil Depression is composed of three main units, reflecting the original palaeogeographic domains. From the most internal to most external, these are generally referred to as the Penninic, Ultrahelvetic and Helvetic domains. As a result of Late Tertiary nappe-stacking, the Helvetic sequence is itself repeated three times (Trümpy 1960), comprising the Morcles/Doldernhorn at the bottom, the Diablerets/Gellihorn the centre and the Wildhorn Nappes on top (Fig. 1). Each unit shows internal stratigraphic heterogeneity as well as minor thrusts and associated folds of smaller scale, which indicate thrusting toward the NW. The nappes were emplaced in the time period from *c.* 25 to 15 Ma (Kirschner *et al.* 2003) and metamorphosed under conditions ranging from lower greenschist in the Rhône Valley to anchizone facies a few kilometres further towards the northwest (Bussy & Epard 1984; Burkhard & Kerrich 1988). New data from U/(Th/He) ages on zircons indicate that the Helvetic Nappes in the Rawil Depression were at temperatures $\geq 180–190°C$ for a time span long enough to partially reset the detrital zircons during in the time period *c.* 17–15 Ma (Cardello 2013).

In a second event during the Miocene, the nappe structure started to updome and the older fold and thrust structure was passively folded on a large (10 km) scale, creating an antiformal fold with its axial depression at the Rawil area (Fig. 1). During this late doming of the external massifs the older thrusts were tilted and now dip towards the north in the north-western part, whereas close to the Rhône valley they are tilted to the southeast. The amplitude of the antiformal fold indicates crustal-scale thrusting and folding that occurred during the latest stage of collision (Levato *et al.* 1994; Maurer *et al.* 1997; Burkhard & Sommaruga 1998; Pfiffner 2009). The change in structural style defines a progressive change from thin- to thick-skinned tectonics which, according to the geological profile of Pfiffner (2009), involves the basement and deep thrusting responsible for the ongoing deformation of the outer arc of the Western Alps (Sue *et al.* 2007). The nappe stack has also been overprinted by younger transverse faults. The area has been affected by one of the most rapid recent uplifts in the Alpine chain that went as far as exhuming the underlying para-autochthonous basement in the external crystalline massifs (Gubler *et al.* 1981; Reinecker *et al.* 2008; Glotzbach *et al.* 2011; Egli & Mancktelow 2013).

## Samples

Samples were taken from 29 sites in total from all nappes currently exposed in the Rawil Depression (Fig. 1; Table 1). A site is composed either of two or more block samples (e.g. Tseuzier Lake, samples 63, 213, 214), which were subsequently cored in the lab, or drill cores taken at the site. The blocks and cores were oriented in the field using a magnetic compass. A total of 198 specimens were taken from 37 sample blocks and 5 *in situ* drilling localities.

## Previous magnetic studies in SW Switzerland

Previous palaeomagnetic studies in the area are limited. Heller (1980) conducted an early study in the region and found evidence for a differential anticlockwise rotation of the Lepontine Dome (Central Alps) by *c.* 30° with respect to the extra-Alpine Europe (Aar Massif), documenting the effects of tectonic displacement and rotations especially in the internal part of the Western Alps. In a regional study, Kligfield & Channell (1981)

found that remagnetization occurred post-folding in the Morcles and Diablerets nappes and that the magnetization recorded by the rocks was acquired after the last reversal of the Earth's magnetic field (i.e. Brunhes Chron). However, their sampling did not include the Wildhorn Nappe or the Rawil area. Other unpublished studies from student projects in the Helvetic Alps showed rocks are often remagnetized during the Brunhes Chron magnetization (Hirt, pers. comm., 2013).

Earlier AMS studies in the Helvetic Alps have focused on shear zones (Dick & Burkhard 2001) or single nappe structures (Ihmlé *et al.* 1989; Almqvist *et al.* 2009, 2011), establishing the dependence of the magnetic fabric from single or multiple structural events. Qualitative correspondence between the directions of the principal axes of the strain ellipsoids and the directions of the principal axes of susceptibility anisotropy ellipsoids have been recognized in the limestones of the autochthonous cover of the Aar Massif by Baker (1964; Heller *et al.* 1989) and Kligfield *et al.* (1982). The latter confirmed the spatial relationship of maximum axes of the AMS ellipsoid with the stretching parallel to the fold axes.

# Methods

## Structural analysis criteria

The structural data represent an area of 1.5–3 km² that includes the sampling localities. Mean stress is reconstructed using fault orientation and slickenlines. The directions of the principal axes of stress are calculated using the P-B-T method after Turner (1953) with the Tectonics FP program (Ortner *et al.*

**Table 1.** *Orientation of the PBT axes calculated according to Sperner* et al. *1993 with the program Tectonics FP (Ortner* et al. *2002)*

| Locality | N | P axis | | R | B axis | | R | T axis | | R |
|---|---|---|---|---|---|---|---|---|---|---|
| | | Dip.dir. | Inclination | | Dip.dir. | Inclination | | Dip.dir. | Inclination | |
| Gemmi Fault region | 21 | 141 | 23 | 42 | 327 | 56 | 41 | 43 | 1 | 84 |
| Wildstrubel north | 17 | 1 | 56 | 28 | 161 | 49 | 57 | 358 | 21 | 22 |
| Rhône Valley (St-Léonard North) | 37 | 310 | 53 | 40 | 133 | 58 | 66 | 19 | 8 | 67 |
| Rawil | 99 | 292 | 63 | 53 | 136 | 18 | 28 | 43 | 9 | 63 |
| Tzeusier Lake | 13 | 359 | 71 | 55 | 96 | 7 | 40 | 177 | 1 | 65 |
| Wildhorn SW | 15 | 274 | 55 | 64 | 154 | 47 | 33 | 44 | 22 | 61 |
| Sublage | 3 | 172 | 31 | 94 | 302 | 48 | 94 | 66 | 27 | 96 |
| Rhône Valley (Sion-Mont d'Orge) | 25 | 269 | 50 | 48 | 117 | 30 | 46 | 17 | 9 | 89 |
| Sanetsch North | 5 | 89 | 32 | 38 | 233 | 70 | 83 | 97 | 5 | 21 |
| Col du Sanetsch | 4 | 267 | 64 | 94 | 161 | 13 | 85 | 66 | 27 | 81 |

Dip.dir, Dip direction of axis; *N*, Number of measured faults with slickensides; *R*, stress ratio (expression of orientation and the relative magnitude of the the the axes of the stress ellipsoid, Ortner *et al.* 2002).

2002), where P, B and T represent the principal minimum-, intermediate- and maximum-stress axes, respectively. The eigenvectors of folded bedding (or foliation) provide the fold axis. The direction of fold axes and cross-cutting relationship between veins indicating the stretching directions, as collected in the field, are compared here. Further details are provided in Cardello (2013).

*Rock magnetic properties and AMS*

Isothermal remanent magnetization (IRM) acquisition was used to help identify the ferromagnetic mineralogy. The IRM was acquired with an ASC Model IM-10-30 impulse magnetizer. The first applied field was 2 T along the core long-axis ($-Z$). Subsequent magnetizations were applied in the opposite direction ($+Z$), in increments up to maximum of 2 T. The IRM curves were normalized with regards to the remanence acquired at the highest positive applied field (i.e. $+Z = 2$ T).

Measurements of the AMS were carried out with an AGICO KLY-2 susceptibility bridge, and data were processed with the Anisoft (Agico) software (Chadima & Jelínek 2008). Magnetic susceptibility ($k$) is defined mathematically as a symmetric second-order tensor in which $k_{ij} = M_i/H_j$, where $M_i$ is magnetization and $H_j$ the applied field. The anisotropy of magnetic susceptibility is usually represented by an ellipsoid with three principal axes $k_1 \geq k_2 \geq k_3$. The degree of anisotropy is expressed by the parameter $P_j$ and the shape of the ellipsoid by $T_j$ (Jelínek 1981). The magnetic lineation and foliation are defined as $k_1/k_2$ and $k_2/k_3$, respectively. $T_j > 0$ is an oblate shape ellipsoid, whereas $Tj < 0$ is a prolate ellipsoid. Hext statistics were used to determine whether the AMS in the sample was statistically significant, by comparing confidence angles ($E$) for the principal axes (Jelínek 1981). The limiting value for $E_{ij} > 26°$ has been used to distinguish between isotropic and anisotropic samples, in which at least two of three principal axes are required to have $E \leq 26°$ in order to be considered statistically anisotropic (Hext 1963; Jelínek & Kropáček 1978).

High-field AMS was measured with a home-built torque magnetometer on selected samples to help identify the carriers of the magnetic fabric (Bergmüller *et al.* 1994). Samples were measured in six fields in three planes using a 30° interval of rotation. Data were processed using the method of Martín-Hernández & Hirt (2001).

*Palaeomagnetic measurements and vector discrimination*

Palaeomagnetic measurements were carried out using a 2G Enterprises, three-axis cryogenic magnetometer (Model 755). The natural remanent magnetization (NRM) was measured initially for all specimens. Thermal demagnetization (TDM) was performed with at least 11 steps, to a maximum temperature of 650°C. Additionally, alternating field (AF) demagnetization was performed using up to 15 steps to a maximum field of 140 mT. Principal component analysis of the palaeomagnetic data was performed using Remasoft software (Kirschvink 1980; Chadima & Hrouda 2006). A stable component was defined by a vector component consisting of at least three data points, with a mean angular deviation (MAD) $\leq 15°$. The quality ($Q$) of the palaeomagnetic component was assessed with the criteria of Van der Voo (2005), which generally consisted of $\alpha_{95} \leq 16°$ of site mean, adequate demagnetization, field and reversal tests as well as age control. A site was accepted with $Q \geq 3$.

Incremental fold tests were performed for both the large and small fold tests at Sanetsch and Tseuzier, respectively, in order to get a better understanding of the relative timing between magnetization and folding. A Python script available online (http://earthref.org/PmagPy/cookbook/), based on the orientation matrix method of Tauxe & Watson (1994), was used for the fold tests. This method is suitable when palaeomagnetic directions occur with both normal and reversed polarity.

# Results

*Fold and thrust structure and associated deformation phases*

Depending on geographic location, up to five deformation phases can be recognized in the southeastern part of the study area (Moser 1985; Cardello 2013). From the oldest to youngest, they are: (1) compositional spaced cleavage associated with isoclinal folds; (2) spaced main crenulation cleavage related to asymmetric folds striking NNE–SSW; (3) out-of-sequence overthrusting of the Penninic; (4) box-folding of the previous structure associated in places with crenulational cleavage and veining; and (5) brittle-ductile shear deformation faulting related to dextral transtension on the Rhône fault. In the southern area stretching lineation is related to NW-directed movement, while nappe emplacement progressively turns to more SW-directed stretching (cf. Dietrich 1989). The latter is accompanied by transtension, possibly associated with the activity of the Rhône Fault (Gasser & Mancktelow 2010). In the more Alpine part and along the northern slope of the SW Helvetics, we observe at least two crenulation cleavages in the more shaly units, more open folds and clear patterns of oblique normal faults and veins. The brittle-ductile

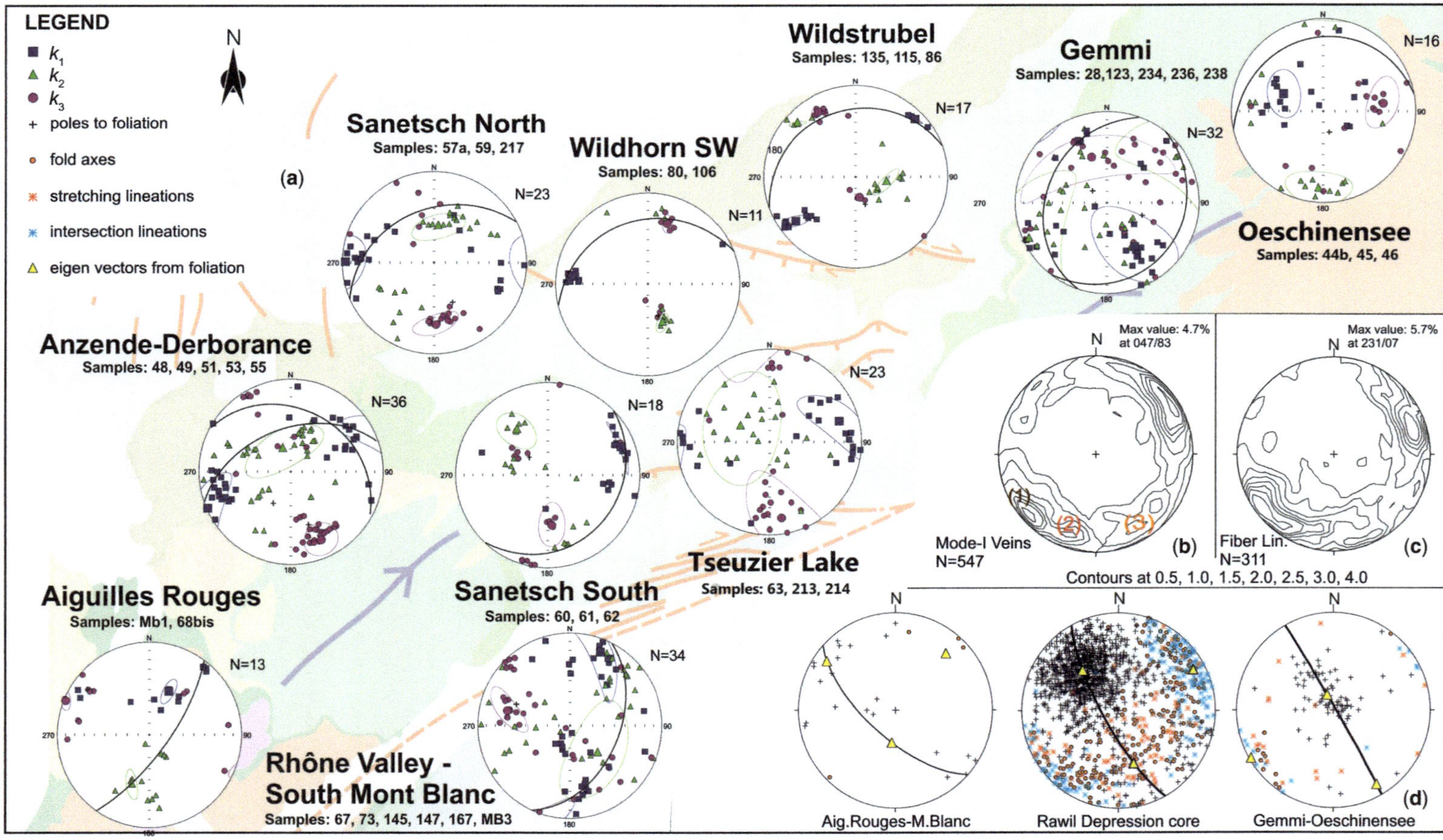


**Fig. 2.** AMS and structural data comaparison. (a) AMS for different areas, shown on a sketched geographical representation (not to scale). Great circles indicate $S_0$ (bedding). (**b–d**) Lower-hemisphere equal-area projections of structural data that show (b) orientation of Mode-I veins (mineralized cracks), having three peaks along the NE–SW, NNE–SSW and NW–SE axes corresponding to progressively younger stretching directions; (c) orientation of fibre lineations in the same cracks; and (d) structural data representative of the SW central and eastern part of the study area. Poles to bedding, fold axes, intersection and stretching lineations are plotted. Great circles represent best fit to poles to bedding calculated with eigenvectors (yellow triangles).

to brittle structures are described in the following sections.

*Veins.* Veins are very common in the Rawil area and constrain the stretching direction at the time of their formation. They developed throughout the deformation history: pre-folding, syn-folding, and coeval with later transtensive faulting. Far from the faults and their associated stress field, the shape of the veins can reflect the regional direction of stretching during veining. In particular, the tips of the veins diverge from the strike of the central part of the veins, indicating that the latest stages of stretching were recorded by the progressive opening of the veins themselves. This implies progressive rotation of the stretching direction, at least locally (Fig. 2b, c). Veins usually occur as mode-I type and are oriented mainly perpendicular to the local fold axes. In Figure 2d, poles to bedding are distributed on a great circle whose pole is roughly parallel to both stretching and intersection lineations and measured fold axes. The calcite fibres of the veins in Figure 2c are mostly parallel to the fold axes (Fig. 2d), indicating that the dominant stretching was parallel to the fold axes and overall trend of the orogen (Fig. 2b–d). Cross-cutting relationships, especially of mode-I veins, and the divergent orientation of different vein sets allow us to establish the relative age. Veins are a few millimetres to few centimetres across. Overall, a progressive anticlockwise relative rotation of the stretching direction was observed. Stretching progressively rotates from WSW (cluster (1) in Fig. 2b) toward SSW (cluster (2) in Fig. 2b) and then towards the SSE (cluster (3) in Fig. 2b).

A gradual increase in the number and size of veins is observed in the more competent lithologies approaching the main damage zone of oblique fault sets, which implies increased fluid circulation toward the fault zones and a common origin of veins and faults (cf. Cardello 2013).

*Faults.* Folding and the initial stage of normal to oblique faulting developed under very low grade metamorphic conditions. Locally there is a transition from an initial ductile mylonitic fabric to cataclasite, possibly indicating that faulting took place with progressive exhumation and cooling of the entire nappe stack. The post-nappe faults can be separated into three distinct sets on the basis of their strike orientation (Fig. 1): (1) NNW/NW-striking; (2) WNW/W-striking; and (3) WSW-striking. The generally transtensional faulting largely post-dates folding, because faults of sets (1) and (2) obliquely cross-cut the fold system and the fold geometry can be matched on either side of the faults. Sets (1) and (2) form together an ENE–WSW-aligned dextral wrench zone coherent with the Rhône fault

kinematics (Fig. 1). Fault set (3) occurs mainly in the Rhône valley, where the fault planes are steep and show a dominant dextral strike-slip component.

Fault kinematic analyses were conducted in the Rawil Depression. The density of faults with respect to the entire structure is much higher between the Rawil Pass and the Wildhorn as well as in the Rhône Valley. Fault orientation and slip direction, which were determined from slickenfibres on the fault planes, show the different orientation of the relative stress-principal axes within the study area. PBT-axes distributions are shown in Table 2. In cases where reconstruction is possible, the position of the minimum stress (T axis) is shallowly dipping to the NE throughout the Rawil Depression. In contrast, the relative positions of the maximum (P axis) and intermediate (B axis) principal stress components are distributed in a girdle corresponding to a steep plane dipping to the southwest. This spread indicates that the difference between $\sigma_2$ and $\sigma_3$ is small and that $T_2$ and $T_3$ axes can switch, with corresponding normal, oblique-slip or strike-slip kinematics. Locally, this is reflected by the interchange in orientation between $\sigma_3$ and $\sigma_2$. For example, in the Sublage and Gemmi areas the intermediate axis is steeply dipping to the northeast and the faults are mainly characterized by strike-slip kinematics; at the Tseuzier Lake however, the measurements indicate extension to the SSW which can be related to the nearby Mondraleche Fault (Fig. 1).

## Magnetic investigations

*Rock magnetic properties and ferromagnetic mineral composition.* Isothermal remanent magnetization (IRM) measurements were performed to help identify carriers of the magnetic remanence in the rocks. Samples show a range of acquisition behaviour, which varies according to lithology as well as with structural position in the nappe stack (Fig. 3). The IRM of samples from the northern Mont Blanc area, which represents the structurally lower (i.e. lower greenschist facies) conditions, are not saturated by 2 T and indicate the presence of a high-coercivity mineral such as hematite or goethite (samples MB2 and MB3). The initial steep increase of remanence in samples MB3, however, also indicates the presence of a low-coercivity mineral, such as magnetite or maghemite. Samples from the structurally highest Wildhorn Nappe contain only a single, low-coercivity phase (e.g. sample 63). Representative samples from intermediate nappes are largely dominated by low-coercivity phases but also contain intermediate- to high-coercivity minerals (e.g. hematite, pyrrhotite). Previous studies in limestones from the Helvetic nappes in southwest Switzerland show similar IRM acquisition

**Table 2.** *Samples description*

| Sample | Latitude | Longitude | Lithology | Nappe | Formation | Depositional age | N |
|---|---|---|---|---|---|---|---|
| **106** | 596 367 | 132 608 | Limestone | Wildhorn | Amden Formation | Late Cretaceous | 4 |
| **80** | 594 824 | 131 080 | Pelagic limestone | Wildhorn | Seewen Formation | Turonian | 7 |
| **115** | 605 343 | 139 367 | Marls | Wildhorn | Amden Formation | Late Cretaceous | 5 |
| **135** | 603 011 | 137 000 | Limestone | Wildhorn | Seewerkalk | Late Cretaceous | 5 |
| **145** | 598 665 | 124 056 | Black slate | South Wildhorn | Schistous Lias | Late Lias | 5 |
| **167** | 598 100 | 124 168 | Limestone | South Wildhorn | Calcareous Lias | Early Lias | 8 |
| **63** | 599 921 | 132 945 | Limestone | Wildhorn | Quinten Formation | Malm | 9 |
| **213** | 599 884 | 132 891 | Limestone | Wildhorn | Quinten Formation | Malm | 8 |
| **214** | 600 332 | 133 006 | Limestone | Wildhorn | Quinten Formation | Malm | 6 |
| **215** | 587 018 | 130 637 | Limestone | Diableretes | Schrattenkalk | Early Cretaceous | 5 |
| **216** | 587 018 | 130 637 | Sandy limestones | Diableretes | Wildstrubel | Eocene | 6 |
| **44B** | 622 023 | 149 191 | Limestone | Dolderhorn | Quinten Formation | Malm | 6 |
| **45** | 622 445 | 148 874 | Limestone | Dolderhorn | Quinten Formation | Malm | 5 |
| **46** | 622 769 | 148 735 | Limestone | Dolderhorn | Quinten Formation | Malm | 5 |
| **48** | 581 633 | 126 138 | Sandstone | Infrahelvetic | Siderolitique | Eocene | 10 |
| **49** | 581 633 | 126 138 | Fine-grained sandstone | Infrahelvetic | flysch | Oligocene (?) | 7 |
| **51** | 579 799 | 126 341 | Fine-grained sandstone | Infrahelvetic | flysch | Oligocene (?) | 5 |
| **53** | 581 499 | 126 133 | Limestone | Morcles | Schrattenkalk | Early Cretaceous | 8 |
| **55** | 581 530 | 125 760 | Pelagic limestone | Morcles | Seewen Formation | Late Cretaceous | 6 |
| **217** | 589 019 | 134 343 | Dark limestones | Wildhorn | Tierwis | Early Cretaceous | 8 |
| **57A** | 589 071 | 134 673 | Sandstone | Wildhorn | Wildstrubel | Eocene | 8 |
| **57** | 589 071 | 134 673 | Sandstone | Wildhorn | Wildstrubel | Eocene | 5 |
| **59** | 589 071 | 134 673 | Limestone | Wildhorn | Schrattenkalk | Early Cretaceous | 7 |
| **60** | 589 041 | 130 217 | Calc-sandstone | Diablerets | Siderolitique (?) | Eocene | 6 |
| **61** | 589 041 | 130 217 | Limestone | Diablerets | Schrattenkalk | Late Cretaceous | 6 |
| **62** | 589 041 | 130 217 | Limestone | Diablerets | Hoghant | Eocene | 6 |
| **67** | 592 333 | 120 468 | Dolomite | South Wildhorn | Dolomite | Late Trias | 7 |
| **73** | 594 842 | 123 486 | Cherty limestone | South Wildhorn | Calcareous Lias | Early Lias | 8 |
| **86** | 603 618 | 135 581 | Calcarenite | UltraHelvetics | Flysch | Oligocene (?) | 7 |
| **95** | 590 396 | 131 762 | Cherty limestone | Wildhorn | Kieselkalk | Early Cretaceous | 5 |
| **97** | 589 861 | 130 173 | Limestone | Wildhorn | Quinten Formation | Malm | 4 |
| **MB1** | 563 789 | 97 594 | Limestone | M. Blanc Cover | Quinten Formation | Malm | 8 |
| **MB2** | 564 199 | 98 417 | Limestone | Aiguilles Rouges | Quinten Formation | Malm | 7 |
| **MB3** | 575 338 | 104 413 | Calcarenite | M. Blanc Cover | Dogger | Dogger | 7 |
| **68** | 567 810 | 107 530 | Mica-rich sandstone | Aiguilles Rouges | Carboniferous | Carboniferous | 5 |
| **123** | 613 565 | 138 659 | Limestone | Dolderhorn | Quinten Formation | Malm | 5 |
| **234** | 611 276 | 138 646 | Dark limestones | Dolderhorn | Hoghant | late Eocene | 10 |
| **236** | 613 325 | 139 140 | Limestone | Dolderhorn | Tithon | Malm | 5 |
| **238** | 612 461 | 139 578 | Limestone | Dolderhorn | Schrattenkalk | Early Cretaceous | 7 |
| **28** | 613 471 | 141 201 | Limestone | Gellihorn | Quinten Formation | Malm | 5 |

Latitude and longitude are given in Swiss coordinates; *N*, number of specimens at a site.

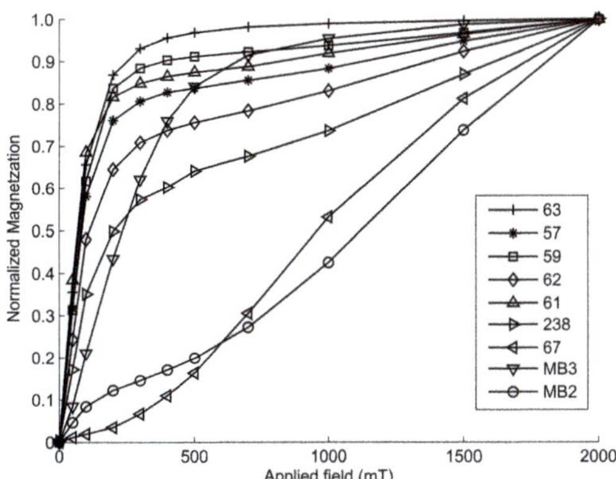

**Fig. 3.** Isothermal remanent magnetization plots for representative samples. Samples listed in the legend illustrate, from bottom to top, the general tectonic position in the nappe stack.

behaviour (Kligfield & Channell 1981; Ihmlé *et al.* 1989; Almqvist *et al.* 2009). Thermal demagnetization of IRM and cross-component IRM have shown that magnetite, pyrrhotite, hematite and goethite can be found in the limestone lithologies, independent of stratigraphic unit.

*Bulk susceptibility and AMS.* Figure 4 shows the mean values of susceptibility for different sample locations. The bulk susceptibility of the anisotropic rocks does not exceed $200 \times 10^{-6}$ SI in general. A few samples (63, 67, 68b, 145) have susceptibilities $>250–350 \times 10^{-6}$ SI. The degree of anisotropy ($P_j$) is typically between 1.0 and 1.2, but a few samples have higher values. In general, $P_j$ is independent of bulk susceptibility (Fig. 4a). The degree of the magnetic foliation ($F$) is variable and larger $F$-values tend to have higher susceptibility (i.e. 63, 67, 68b, 145). In contrast, there is only a small variation in the magnetic lineation ($L$) with values $\leq 1.10$. Oblate to triaxial susceptibility ellipsoid shapes ($F > L$) are the most common, and only a few samples have a more prolate ellipsoid shape (Fig. 4b, c; Table 3). High-field AMS showed that the magnetic fabric was carried by a combination of ferromagnetic and paramagnetic minerals. The low-field AMS could not be specifically related to either component, but reflected the combination of both the paramagnetic and ferromagnetic fabrics.

Regional variation in the orientation of the principal susceptibility axes can be used to examine whether the magnetic fabric records the complete or only part of the stretching history during Neogene exhumation of the Rawil Structural Depression. Recrystallized and veined rocks are weakly paramagnetic or diamagnetic and are nearly

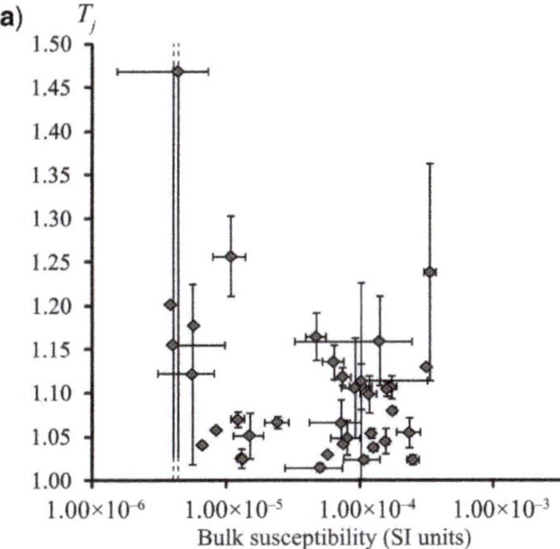

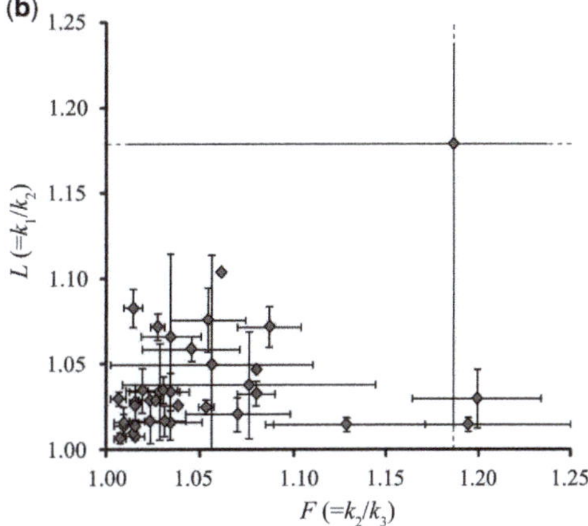

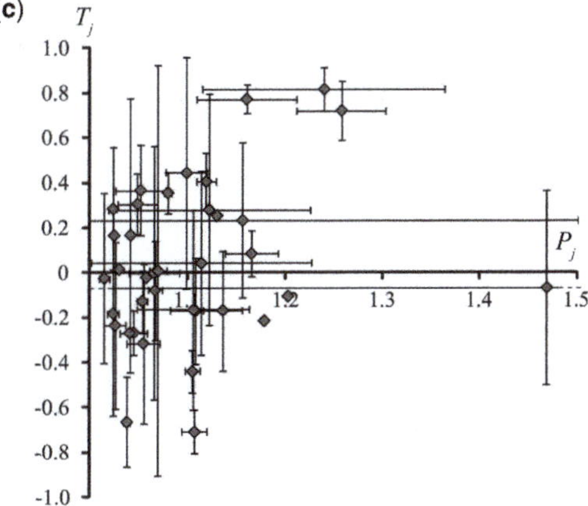

**Fig. 4.** (**a**) Degree of anisotropy ($P_j$) as a function of bulk susceptibility ($k_m$); (**b**) Flinn-type diagram showing magnetic lineation ($L$) v. foliation (F); (**c**) $P_j$ v. the shape factor $T_j$.

magnetically isotropic; only their structural data are considered (samples 46, 57, 60, 216). Nevertheless, a preferred orientation of the principal susceptibility

**Table 3.** *Summary of anisotropy results computed for each site*

| Sample | N | $k_m$ | $\sigma$ | L | F | $P_j$ | $T_j$ | $S_0$ | $D/I\,(k_1)$ | $\alpha_{95}$ | $D/I\,(k_2)$ | $\alpha_{95}$ | $D/I\,(k_3)$ | $\alpha_{95}$ |
|---|---|---|---|---|---|---|---|---|---|---|---|---|---|---|
| **106** | 4 | $1.54 \times 10^{-4}$ | $9.84 \times 10^{-6}$ | 1.028 | 1.015 | 1.045 | $-0.268$ | 343/36 | | | | | | |
| **80** | 7 | $1.01 \times 10^{-4}$ | $1.99 \times 10^{-5}$ | 1.059 | 1.045 | 1.108 | $-0.171$ | 347/32 | 264.1/27.6 | 4.0/3.0 | 158.7/26.9 | 5.5/2.8 | 32.0/49.6 | 4.7/2.9 |
| **115** | 5 | $1.20 \times 10^{-4}$ | $3.74 \times 10^{-6}$ | 1.029 | 1.023 | 1.054 | $-0.128$ | 340/26 | 45.6/10.3 | 6.2/1.1 | 312.8/15.1 | 7.9/1.5 | 168.5/71.6 | 5.3/1.0 |
| **135** | 5 | $1.70 \times 10^{-4}$ | $1.68 \times 10^{-5}$ | 1.083 | 1.014 | 1.107 | $-0.708$ | 052/20 | 229.6/34.7 | 5.1/1.8 | 84.3/49.9 | 18.6/1.5 | 332.2/17.6 | 18.6/2.0 |
| **145** | 5 | $3.32 \times 10^{-4}$ | $3.56 \times 10^{-5}$ | 1.015 | 1.194 | 1.239 | 0.817 | 102/52 | 196.4/3.3 | 29.6/8.5 | 106.0/6.0 | 30.8/11.0 | 314.8/83.1 | 15.3/8.3 |
| **167** | 8 | $1.48 \times 10^{-5}$ | $3.70 \times 10^{-6}$ | 1.016 | 1.034 | 1.052 | 0.367 | 128/46 | 352.1/21.2 | 38.7/15.7 | 260.1/5.2 | 38.7/15.7 | 157.1/68.1 | 16.0/15.4 |
| **63** | 9 | $2.48 \times 10^{-4}$ | $2.42 \times 10^{-5}$ | 1.014 | 1.009 | 1.024 | $-0.182$ | 341/42 | 95.8/1.0 | 12.3/8.8 | 359.0/81.6 | 22.6/11.6 | 186.0/8.4 | 23.1/7.7 |
| **213** | 8 | $7.09 \times 10^{-5}$ | $3.01 \times 10^{-5}$ | 1.034 | 1.028 | 1.066 | $-0.001$ | 140/80 | 59.9/28.9 | 17.1/11.7 | 295.4/45.7 | 17.9/16.8 | 168.6/30.3 | 18.5/10.5 |
| **214** | 6 | $7.90 \times 10^{-5}$ | $1.99 \times 10^{-5}$ | 1.017 | 1.031 | 1.049 | 0.306 | 145/21 | 88.5/1.8 | 15.5/8.7 | 199.1/84.9 | 35.8/11.4 | 358.4/4.8 | 35.3/11.2 |
| **215** | 5 | $5.53 \times 10^{-6}$ | $2.49 \times 10^{-6}$ | 1.038 | 1.076 | 1.122 | 0.28 | 342/27 | 161.6/75.9 | 55.4/37.1 | 313.7/12.5 | 66.3/31.8 | 45.1/6.4 | 68.1/23.2 |
| **216** | 6 | $4.36 \times 10^{-6}$ | $2.84 \times 10^{-6}$ | 1.179 | 1.186 | 1.469 | $-0.068$ | 348/20 | 290.3/63.7 | 35.5/4.7 | 152.1/20.2 | 19.5/9.8 | 56.0/16.1 | 32.5/11.2 |
| **44B** | 6 | $1.21 \times 10^{-5}$ | $1.36 \times 10^{-6}$ | 1.034 | 1.034 | 1.07 | 0.009 | 360/24 | 286.8/47.4 | 14.5/1.8 | 182.4/12.9 | 14.3/5.6 | 81.4/39.7 | 6.4/3.3 |
| **45** | 5 | $4.93 \times 10^{-5}$ | $2.23 \times 10^{-5}$ | 1.007 | 1.007 | 1.015 | $-0.024$ | 333/16 | 298.1/55.7 | 34.1/30.4 | 189.1/12.5 | 39.1/30.5 | 91.3/31.3 | 42.5/26.3 |
| **46** | 5 | $1.29 \times 10^{-5}$ | $8.95 \times 10^{-7}$ | 1.016 | 1.009 | 1.026 | $-0.235$ | 274/14 | 67.3/59.6 | 17.5/7.5 | 303.8/17.9 | 26.9/8.0 | 205.6/23.7 | 28.1/14.6 |
| **48** | 10 | $1.38 \times 10^{-4}$ | $1.06 \times 10^{-4}$ | 1.015 | 1.128 | 1.16 | 0.773 | 151/30 | 247.3/19.8 | 21.6/5.2 | 10.0/56.4 | 21.6/2.0 | 147.2/26.0 | 5.5/1.0 |
| **49** | 7 | $1.74 \times 10^{-4}$ | $8.23 \times 10^{-6}$ | 1.025 | 1.053 | 1.08 | 0.356 | 077/27 | 54.8/21.9 | 7.3/2.0 | 296.1/50.1 | 6.3/2.6 | 159.1/31.5 | 5.6/2.9 |
| **51** | 5 | $7.38 \times 10^{-5}$ | $2.83 \times 10^{-6}$ | 1.017 | 1.023 | 1.042 | 0.167 | 029/32 | 285.8/8.9 | 23.2/15.7 | 21.4/32.2 | 23.2/13.1 | 182.3/56.3 | 19.8/4.4 |
| **53** | 8 | $1.15 \times 10^{-4}$ | $1.61 \times 10^{-5}$ | 1.021 | 1.07 | 1.099 | 0.444 | 176/68 | 67.0/33.6 | 25.6/3.7 | 227.3/54.8 | 26.1/6.4 | 330.7/9.3 | 9.8/3.0 |
| **55** | 6 | $4.59 \times 10^{-5}$ | $7.93 \times 10^{-6}$ | 1.072 | 1.087 | 1.165 | 0.085 | 341/51 | 263.8/17.8 | 7.3/3.5 | 32.7/62.9 | 5.7/5.1 | 167.2/19.8 | 6.7/3.7 |
| **217** | 8 | $9.00 \times 10^{-5}$ | $1.89 \times 10^{-5}$ | 1.066 | 1.034 | 1.106 | $-0.166$ | 338/30 | 104.5/28.7 | 15.0/4.2 | 220.8/39.4 | 22.6/5.7 | 349.5/37.2 | 23.5/11.4 |
| **57A** | 8 | $6.21 \times 10^{-5}$ | $1.14 \times 10^{-5}$ | 1.076 | 1.054 | 1.136 | $-0.169$ | 328/44 | 269.8/14.4 | 14.0/3.7 | 11.1/37.4 | 14.0/4.6 | 162.6/49.0 | 4.8/3.5 |
| **57** | 5 | $3.98 \times 10^{-6}$ | $5.71 \times 10^{-6}$ | 1.212 | 0.461 | 1.156 | 0.233 | 328/44 | 147.1/32.9 | 11.4/2.9 | 304.2/54.9 | 13.6/1.9 | 50.0/10.8 | 8.3/3.5 |
| **59** | 7 | $1.08 \times 10^{-5}$ | $2.95 \times 10^{-6}$ | 1.03 | 1.199 | 1.257 | 0.722 | 328/44 | 287.2/19.6 | 30.8/2.2 | 40.7/48.2 | 30.8/1.3 | 182.6/35.2 | 2.6/1.6 |
| **60** | 6 | $1.06 \times 10^{-4}$ | $3.25 \times 10^{-5}$ | 1.008 | 1.015 | 1.024 | 0.285 | 130/30 | 61.8/16.1 | 11.2/3.9 | 160.3/26.9 | 11.7/2.4 | 304.4/57.9 | 4.9/3.3 |
| **61** | 6 | $1.88 \times 10^{-6}$ | $1.75 \times 10^{-6}$ | 1.873 | $-1.867$ | 0.742 | 0.104 | 120/29 | 65.5/16.4 | 12.2/1.9 | 322.1/38.3 | 10.9/9.2 | 173.9/47.1 | 11.0/8.0 |
| **62** | 6 | $7.30 \times 10^{-5}$ | $1.06 \times 10^{-5}$ | 1.033 | 1.08 | 1.119 | 0.406 | 150/17 | 101.1/24.2 | 20.2/2.6 | 289.0/65.5 | 20.2/3.0 | 192.4/3.0 | 4.6/1.3 |
| **67** | 7 | $2.33 \times 10^{-4}$ | $4.63 \times 10^{-5}$ | 1.035 | 1.019 | 1.055 | $-0.315$ | 105/81 | 30.8/25.1 | 15.7/7.5 | 201.6/64.6 | 36.9/12.1 | 299.1/3.6 | 36.4/9.3 |
| **73** | 8 | $1.25 \times 10^{-4}$ | $9.05 \times 10^{-6}$ | 1.03 | 1.006 | 1.038 | $-0.664$ | 147/24 | 200.1/72.7 | 6.9/3.8 | 46.4/15.6 | 6.5/2.5 | 314.4/7.3 | 4.5/2.6 |
| **86** | 7 | $1.58 \times 10^{-4}$ | $1.06 \times 10^{-5}$ | 1.072 | 1.027 | 1.105 | $-0.44$ | 155/38 | 236.2/8.2 | 4.0/1.8 | 126.1/67.1 | 4.9/2.8 | 329.4/21.2 | 4.9/2.6 |
| **95** | 5 | $1.01 \times 10^{-4}$ | $2.16 \times 10^{-4}$ | 1.05 | 1.056 | 1.114 | 0.043 | 325/42 | 88.8/0.6 | 35.1/5.6 | 179.6/54.9 | 43.9/9.7 | 358.4/35.1 | 35.1/3.5 |
| **97** | 4 | $2.38 \times 10^{-5}$ | $4.66 \times 10^{-6}$ | 1.035 | 1.03 | 1.067 | $-0.078$ | 152/28 | | | | | | |
| **MB1** | 8 | $3.84 \times 10^{-6}$ | $3.15 \times 10^{-6}$ | 1.26 | 0.92 | 1.202 | $-0.106$ | 082/82 | 228.3/5.7 | 12.4/3.3 | 134.9/31.0 | 14.0/3.0 | 327.7/58.4 | 8.1/3.9 |
| **MB2** | 7 | $1.32 \times 10^{-5}$ | $1.26 \times 10^{-6}$ | 1.01 | 1.014 | 1.025 | 0.166 | 113/36 | 257.5/16.0 | 254.3/32.4 | 350.4/10.0 | 55.0/30.0 | 111.3/71.0 | 41.2/28.5 |
| **MB3** | 7 | $5.62 \times 10^{-5}$ | $4.95 \times 10^{-5}$ | 1.014 | 1.015 | 1.03 | 0.016 | 095/54 | 165.6/32.6 | 15.5/6.3 | 46.7/37.0 | 15.5/13.3 | 283.4/36.1 | 13.4/6.3 |
| **68bis** | 5 | $3.12 \times 10^{-4}$ | $6.57 \times 10^{-5}$ | 1.047 | 1.08 | 1.13 | 0.254 | 126/75 | 39.6/7.3 | 1.7/1.0 | 149.6/69.5 | 7.0/0.2 | 307.1/19.0 | 6.9/1.2 |
| **123** | 5 | $6.61 \times 10^{-6}$ | $4.69 \times 10^{-7}$ | 1.026 | 1.015 | 1.041 | $-0.269$ | 316/37 | 268.1/29.0 | 53.9/39.8 | 135.0/50.9 | 54.1/33.9 | 12.2/23.8 | 44.6/32.5 |
| **234** | 10 | $5.69 \times 10^{-6}$ | $1.48 \times 10^{-6}$ | 1.104 | 1.061 | 1.178 | $-0.215$ | 315/36 | 147.2/15.1 | 8.7/7.1 | 257.6/52.2 | 21.9/7.8 | 46.8/33.7 | 21.9/7.9 |
| **236** | 5 | $8.41 \times 10^{-6}$ | $3.23 \times 10^{-7}$ | 1.029 | 1.026 | 1.058 | $-0.02$ | 302/35 | 225.3/27.9 | 28.6/8.2 | 349.0/46.4 | 27.0/13.1 | 117.2/30.5 | 24.2/13.4 |
| **238** | 7 | $1.81 \times 10^{-6}$ | $1.33 \times 10^{-6}$ | 1.191 | 1.561 | 2.059 | 0.301 | 298/23 | 119.6/48.3 | 54.2/19.7 | 224.0/12.6 | 50.2/9.0 | 324.5/39.0 | 53.5/13.8 |
| **28** | 5 | $2.29 \times 10^{-5}$ | $6.95 \times 10^{-6}$ | 1.026 | 1.038 | 1.065 | 0.177 | 290/32 | 338.1/19.3 | 12.1/4.5 | 183.7/68.8 | 14.7/7.3 | 71.1/8.5 | 12.3/4.1 |

$N$: number of specimens from one hand block/drill core; $k_m$: bulk susceptibility; $\sigma$: standard deviation of $k_m$; $L$: magnetic lineation; $F$: magnetic foliation; $P_j$: degree of anisotropy; $T_j$: shape parameter; $S_0$: bedding; $D/I$: declination/inclination in geographic coordinates for respective principal axes, and their associated angles of 95% confidence ($\alpha_{95}$).

axes is still evident, revealing that the former tectonic fabric is partially masked by a more recent fabric, as seen at the Col du Sanetsch (samples 215, 216). Samples of limestone, marly limestone and sandstone are the most strongly anisotropic and are described here with comparison to the structural data, considering sample sites from southwest to northeast (Table 2; Fig. 2).

In the Mont Blanc–Aiguilles Rouges area (samples MB1, 68b), $k_1$ axes are parallel to the stretching direction which is parallel to the trend of the fold axes. In the nappes exposed in the southwestern part of the study area, $k_2$ and $k_3$ axes form a girdle perpendicular to the observed foliation and $k_3$ is always perpendicular to the main structural foliation, which indicates strong flattening. This feature is most likely related to the nappe emplacement stage, indicating that the magnetic foliation developed perpendicular to the main cleavage. Immediately north of the Rhône Valley in the core of the Drône anticline (Fig. 1), the fabric of sample 73 is nearly rotational prolate and is characterized by principal axes lying out of the bedding plane. The steep inclination of $k_1$ may reflect elongation in the axial plane zone of the Drône anticline (cf. Pfiffner 1993), which suggests a tectonized fabric modified from the original stratigraphic fabric.

In the Wildhorn Nappe, along the axial zone of the western and central part of the Rawil Depression, the magnetic fabric has a consistent orientation. The principal axes are well clustered within all samples from the Wildhorn Nappe, having a triaxial shape ellipsoid. The $k_1$ axes are parallel to the fold axis in most cases and coincide with intersection lineations, or in some cases stretching lineations, which are oriented WSW–ENE. In some places (e.g. Sanetsch south, Samples 60, 61, 62 in Figs 1 & 2), the $k_2$ and $k_3$ axes exchange position on the equal area net and $k_2$ lies close to the pole to bedding. The principal axes, however, are controlled by the tectonic lineation and in some samples also by foliation indicating flattening. Different specimens collected from the same sites display an interchange in the position of $k_2$ and $k_3$.

Two types of magnetic fabric are found in the lower nappes: either a disperse grouping of axes, or clustering of axes in more than one direction. In the Gemmi area on the eastern side of the Rawil Structural Depression, $k_1$ clusters predominantly in the southeast and in the southwest, revealing a composite magnetic fabric. Because the AMS ellipsoid is highly prolate, $k_2$ and $k_3$ axes are interchangeable or girdle along a great circle. In the north-easternmost sampling locality at Oeschinensee, the principal susceptibility axes are well grouped but plot systematically 20–30° away from both bedding and foliation, which are both subhorizontal. Principal axes do not coincide with the fold axes, which plunge gently to the northeast, suggesting that the original sedimentary magnetic fabric was overprinted.

*Palaeomagnetic directions and fold tests.* Of the 29 investigated sample localities, 20 were not suitable for palaeomagnetic evaluation either because the initial magnetization was too weak to allow for demagnetization or because the samples did not carry a stable remanent magnetization. Seven sample sites were deemed suitable for calculating site-mean palaeomagnetic directions and had $Q \geq 3$. Two sites have $Q < 3$, and their palaeomagnetic directions are not considered further. Representative vector diagrams from different lithologies show two–three components of magnetization in general (Fig. 5). An unstable viscous remanence is typically removed at fields $<10$ mT; it is therefore excluded from further palaeomagnetic analysis. An intermediate palaeomagnetic component (A), with coercivity between 10 and 50 mT and unblocking temperature $\leq 275°C$, has been recorded in 146 specimens. Component A displays a palaeomagnetic direction that is generally subparallel to the present-day field direction. In the northern area (Oeschinensee, Fig. 1) and at Tseuzier, component A is however subvertical (Fig. 6). A second stable component of magnetization (B) is identified at fields $\geq 50$ mT and demagnetization temperatures $>275°C$ in 28 specimens. This component decays to the origin upon complete demagnetization, and is referred to as the characteristic remanence (ChRM). Both A and B components have been used for palaeomagnetic interpretation, and the results are summarized in Table 4 and Figure 7. Palaeomagnetic directions with reversed polarity are found both in A and B components, indicating that at least some samples acquired a remanence prior to the Brunhes geomagnetic reversal ($>0.78$ Ma).

In fold and thrust belts, the relative age of palaeomagnetic directions with respect to tectonic deformation are based mainly on fold tests (e.g. Graham 1949; Van der Voo 2005). We performed fold tests on two structures in order to establish their relative age of deformation. The first test was performed on a large-scale fold that outcrops at the Sanetsch, which is related to buckling of the nappe-stack structure due to regional-scale thick-skinned tectonics (Fig. 6a, b). The second fold test was made on an older, smaller-scale fold, located at the dam of the Tseuzier Lake (Fig. 6c, d).

At the Sanetsch site (Figs 1 & 6a, b), where the Diablerets and Wildhorn nappes are folded together, the fold test was performed with 39 specimens from 4 sampling localities (Fig. 8; Table 4). The directions of component A are better grouped before tilt correction with $D = 300°$ and $I = 71°$ ($k = 3.2$, $\alpha_{95} = 15.6°$), although the grouping is not

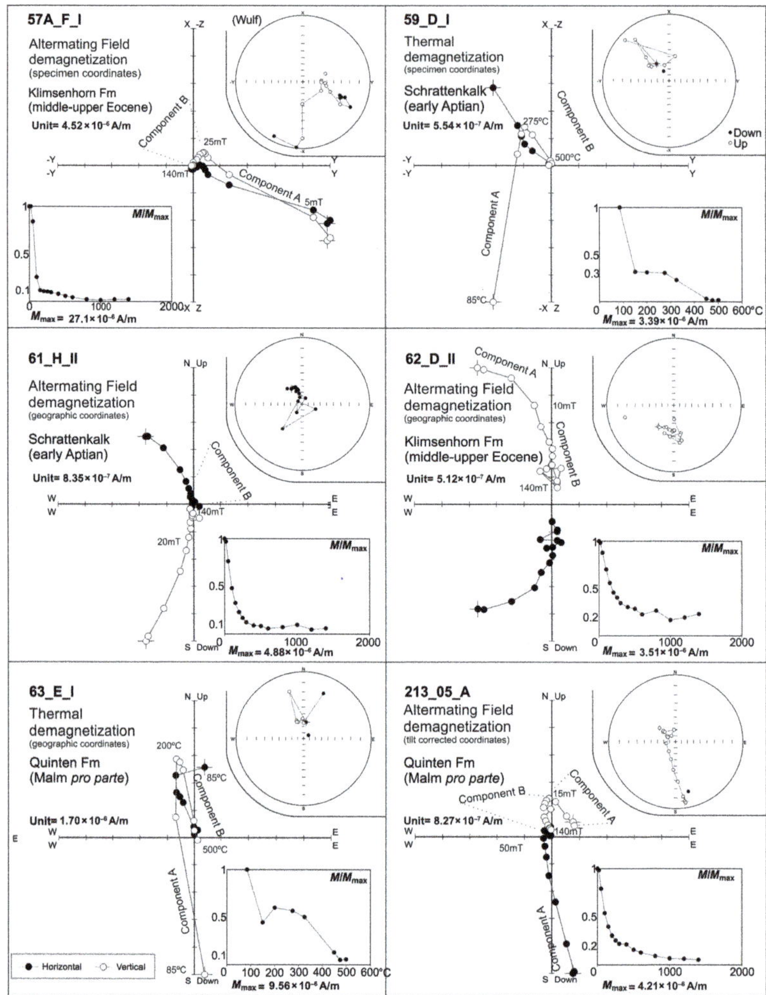

**Fig. 5.** Zijderveld-type diagrams with palaeomagnetic directions from representative samples. The two stable palaeomagnetic components (A and B) that were identified are indicated. Filled (empty) symbols indicate projections in the horizontal (vertical) plane. Decay diagrams and spherical projections are shown.

statistically different from the tilt-corrected direction (Figs 8b & 9a; Table 5). After tilt correction, the ChRM directions are more tightly clustered ($D = 36°$, $I = 70°$, $k = 34.0$, $\alpha_{95} = 4.7°$) than in geographic coordinates ($D = 315°$, $I = -13°$, $k = 1.82$, $\alpha_{95} = 29.8°$) so that they pass the fold test with 100% unfolding (Fig. 9b; Table 5). The ChRM directions show both normal and reversed polarities that are nearly antipodal (Fig. 7), which suggests that the magnetization was acquired prior to folding.

The small fold test at Tseuzier Lake (Fig. 6c, d) was performed with 15 specimens sampled from three sites within the core of the Wildhorn Nappe. The A component only shows normal polarity and

groups significantly better before applying a tilt test with a mean direction $D = 92.9°$ and $I = 84.6°$ ($k = 13.37$, $\alpha_{95} = 10.9°$) (Fig. 9c; Table 5). The ChRM is mainly reversed and was acquired during the folding, as seen from the incremental fold test which is significant from 33% to 76% unfolding (Fig. 9d). At 100% unfolding, the palaeomagnetic directions cluster on a plane orthogonal to the axial plane of the fold (Fig. 6d).

## Discussion

The integration of structural geology, AMS and palaeomagnetism provides new and independent

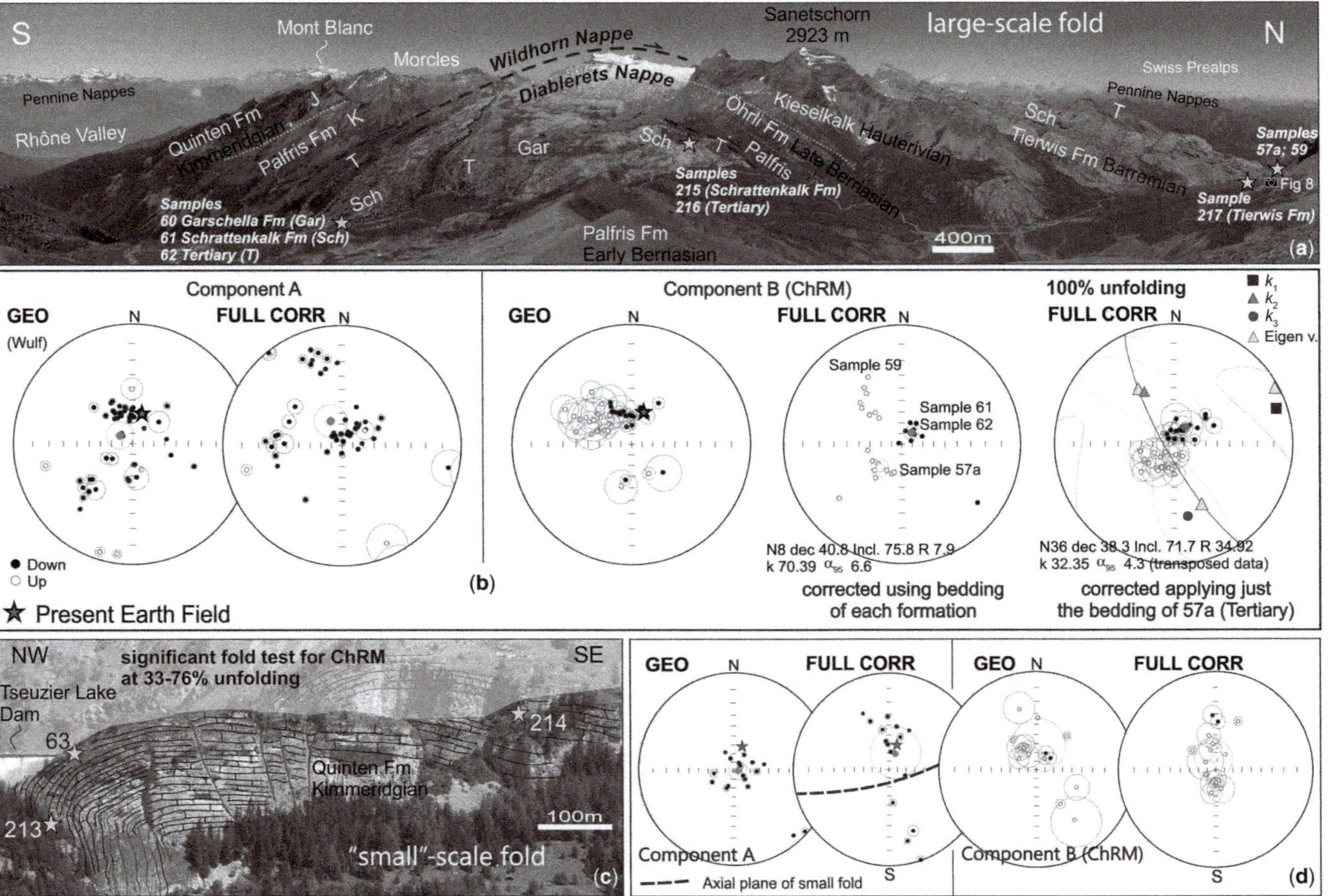



**Fig. 6.** Results from fold tests performed in the Rawil Depression. (**a**) Geological overview of the Sanetsch fold. Coloured lines refer to the age of the sedimentary rocks exposed. (**b**) Equal-angle projections showing mean palaeomagnetic directions of different clusters, the precision parameter ($k$) and the angular standard deviation ($\alpha_{95}$). (**c**) Geological overview of the small-scale fold at Lac Tseuzier. Red lines indicate faults. (**d**) Equal-angle projections showing mean palaeomagnetic directions of different clusters, the precision parameter ($k$) and the angular standard deviation ($\alpha_{95}$; purple circle). After tilt correction the data are still not clustered.

**Table 4.** *Palaeomagnetic directions from the Rawil Depression*

| Site name | Bedding strike (°) | Dip (°) | Component | $N/N_0$ | Geo- corrected | | Tilt-corrected | | $k$ | $\alpha_{95}$ (°) |
|---|---|---|---|---|---|---|---|---|---|---|
| | | | | | Decl | Incl | Decl | Incl | | |
| 63 | 341 | 42 | A | 10/12 | 73 | 76 | 359 | 46 | 19.6 | 11.2 |
| | | | B (ChRM) | 10/12 | 335 | −55 | 183 | −82 | 22.2 | 10.5 |
| 213 | 140 | 80 | A | 3/3 | 238 | 68 | 159 | 12 | 41.2 | 19.5 |
| | | | B (ChRM) | 3/3 | 134 | −38 | 327 | −62 | 21.5 | 27.3 |
| 214 | 145 | 21 | A | 2/2 | 188 | 84 | 151 | 65 | 12.0 | 80.2 |
| | | | B (ChRM) | 2/2 | 39 | −58 | 11 | −48 | 13.5 | 74.4 |
| Tzeusier Lake area | | | A | 15/17 | 93 | 85 | 22 | 70 | 2.7 | 28.8 |
| | | | B (ChRM) | 15/17 | 359 | −70 | 330 | −85 | 9.5 | 13.1 |
| 234 | 315 | 36 | A | 8/8 | 42 | 70 | 342 | 49 | 137.0 | 4.7 |
| | | | B (ChRM) | 8/8 | 236 | −73 | 159 | −54 | 148.7 | 4.6 |
| 236 | 302 | 35 | A | 4/4 | 34 | 70 | 330 | 51 | 197.1 | 6.6 |
| | | | B (ChRM) | 4/4 | 98 | 65 | 338 | 75 | 8.6 | 33.3 |
| 238 | 298 | 23 | A | 6/6 | 352 | 59 | 330 | 42 | 112.9 | 6.3 |
| | | | B (ChRM) | 5/6 | 349 | 17 | 344 | 2 | 5.5 | 35.8 |
| Gemmi area | | | A | 18/18 | 19 | 68 | 335 | 47 | 92.2 | 3.6 |
| | | | B (ChRM) | 8/18 | 236 | −73 | 159 | −53 | 148.7 | 4.6 |
| 57a | 333 | 54 | A | 11/11 | 224 | 36 | 272 | 34 | 13.8 | 13.5 |
| | | | B (ChRM) | 7/10 | 294 | −43 | 220 | −61 | 40.6 | 9.6 |
| 59 | 328 | 44 | A | 9/9 | 349 | 60 | 345 | 18 | 99.9 | 5.2 |
| | | | B (ChRM) | 8/8 | 306 | −52 | 212 | −75 | 37.2 | 9.2 |
| 217 | 338 | 30 | A | 2/2 | 182 | 63 | 276 | 78 | 55.6 | 34.2 |
| | | | B (ChRM) | 2/2 | 160 | 33 | 164 | 63 | 7.3 | 113.3 |
| Sanetsch north | | | A | 22/23 | 257 | 68 | 314 | 34 | 5.5 | 15 |
| | | | B (ChRM) | 14/19 | 300 | −48 | 221 | −69 | 29.1 | 7.5 |
| 61 | 120 | 29 | A | 9/9 | 344 | 61 | 46 | 69 | 37.4 | 8.5 |
| | | | B (ChRM) | 6/6 | 344 | 59 | 41 | 69 | 47.7 | 9.8 |
| 62 | 150 | 17 | A | 8/9 | 6 | 66 | 45 | 76 | 26.0 | 11.1 |
| | | | B (ChRM) | 7/8 | 359 | 60 | 21 | 73 | 32.2 | 10.8 |
| Sanetsch south | | | A | 17/19 | 353 | 64 | 46 | 72 | 31.2 | 6.5 |
| | | | B (ChRM) | 11/14 | 57 | 60 | 36 | 71 | 48.8 | 6.6 |
| MB1 | 82 | 82 | A | 3/3 | 350 | 55 | 31 | 8 | 52.6 | 17.2 |
| MB3 | 92 | 54 | A | 6/6 | 336 | 66 | 62 | 43 | 68.1 | 8.2 |
| | | | B (ChRM) | 5/6 | 106 | −68 | 244 | −57 | 163.3 | 7.2 |
| Mont Blanc | | | A | 11/12 | 351 | 62 | 45 | 30 | 11.3 | 14.2 |
| | | | B (ChRM) | 5/6 | 106 | −68 | 244 | −57 | 163.3 | 7.2 |
| 44b | 360 | 24 | A | 6/6 | 346 | 78 | 355 | 55 | 94.9 | 6.9 |
| | | | B (ChRM) | 4/4 | 57 | 69 | 28 | 51 | 23.4 | 19.4 |
| 45b | 333 | 16 | A | 4/4 | 130 | 84 | 347 | 79 | 9.6 | 31.2 |
| | | | B (ChRM) | 4/4 | 195 | 84 | 314 | 78 | 10.1 | 30.4 |
| Oeschinensee | | | A | 10/10 | 360 | 85 | 353 | 64 | 15.6 | 12.6 |
| | | | B (ChRM) | 8/8 | 71 | 81 | 12 | 67 | 8.5 | 20.2 |
| 48 | 151 | 30 | A | 8/8 | 359 | 17 | 350 | 43 | 14.6 | 15 |
| 49 | 77 | 27 | A | 4/4 | 311 | 73 | 34 | 69 | 51.5 | 12.9 |
| | | | B (ChRM) | 2/2 | 352 | 73 | 41 | 57 | 671.8 | 9.7 |
| 51 | 29 | 32 | A | 4/4 | 84 | 40 | 74 | 18 | 10.8 | 29.3 |
| | | | B (ChRM) | 4/4 | 66 | 43 | 60 | 15 | 2.7 | 70.2 |
| 53 | 176 | 68 | A | 5/5 | 360 | 59 | 116 | 53 | 62.6 | 9.7 |
| | | | B (ChRM) | 2/2 | 4 | 57 | 111 | 55 | 33.3 | 44.8 |
| 55 | 341 | 51 | A | 5/5 | 74 | 78 | 13 | 39 | 48.7 | 11.1 |
| Derborance | | | B (ChRM) | 3/4 | 110 | −44 | 143 | −5 | 8.8 | 44.4 |
| | | | A | 18/24 | 38 | 72 | 59 | 53 | 5.4 | 16.4 |
| | | | B (ChRM) | 6/13 | 0 | 70 | 56 | 60 | 10.7 | 21.4 |
| 86 | 155 | 38 | A | 4/4 | 344 | 67 | 139 | 74 | 386.9 | 4.7 |
| | | | B (ChRM) | 2/2 | 323 | 36 | 300 | 72 | 3.0 | 0 |
| 115 | 340 | 26 | A | 4/5 | 312 | −20 | 306 | −42 | 70.6 | 11 |
| | | | B (ChRM) | 3/3 | 306 | −26 | 295 | −46 | 20.5 | 27.9 |

$N/N_0$, number of specimens used from total number of samples for site mean; $k$: Fisher dispersion parameter; $\alpha_{95}$: angle of 95% confidence about the mean direction.

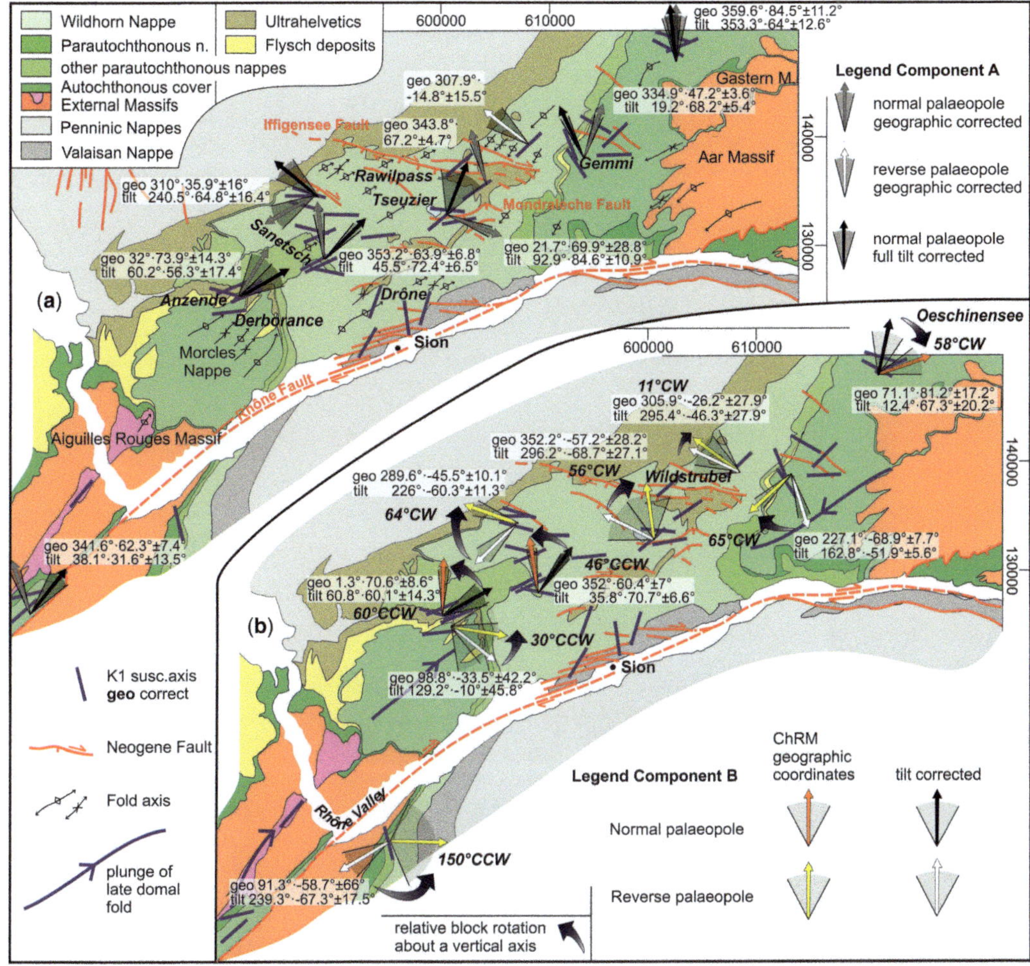


**Fig. 7.** Geographical distribution of ChRM directions and AMS, the latter according to their field orientation. (**a**) Regional distribution of Component A site mean directions in different nappes. (**b**) Regional distribution of Component B (ChRM) site mean directions in different nappes. Relative direction deviation of ChRM are shown in bold italics. CW: clockwise; CCW: anticlockwise.

information to constrain the relative timing and mode of formation of the Rawil Structural Depression.

## Structure and AMS

The anticlockwise rotation of the stretching observed in the cross-cutting vein sets is consistent with the observed change from oblique slip to subsequent dip-slip on faults of sets (1) and (2). Overall, there is an almost 180° anticlockwise change in the stretching direction. Initial orogen-perpendicular stretching of reclined fold limbs during NW-directed thrusting is followed by a gradual transition to orogen-parallel stretching. Orogen-parallel

stretching is reflected in the dominant period of veining and oblique, transtensional faulting that is overprinted during a subsequent (at least locally) period of limited, more orogen-perpendicular extension. Veins also occur in the damage zones within limestone units, where they are sheared into mylonitic shear zones that form the precursors to later brittle faults. They also form a major component of subsequently developed cataclasites, indicating cooler conditions associated with antiformal doming and interaction with fluids. Antiformal doming is not strictly cylindrical because the antiform is somewhat narrower in the core of the Rawil Depression, simulating an embryonic double plunging conic folding. Better examples of conic folding

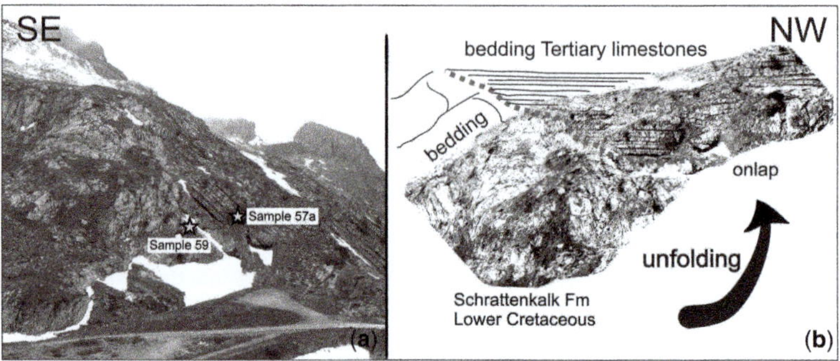



**Fig. 8.** The correction applied for the stratigraphic test at the Sanetsch in the forelimb of the Wildhorn Nappe and its graphical tilt correction. (**a**) Field view of collected samples. (**b**) Unfolded and untilted K/T boundary. Cretaceous limestones were already tilted and eroded before Tertiary sedimentation.

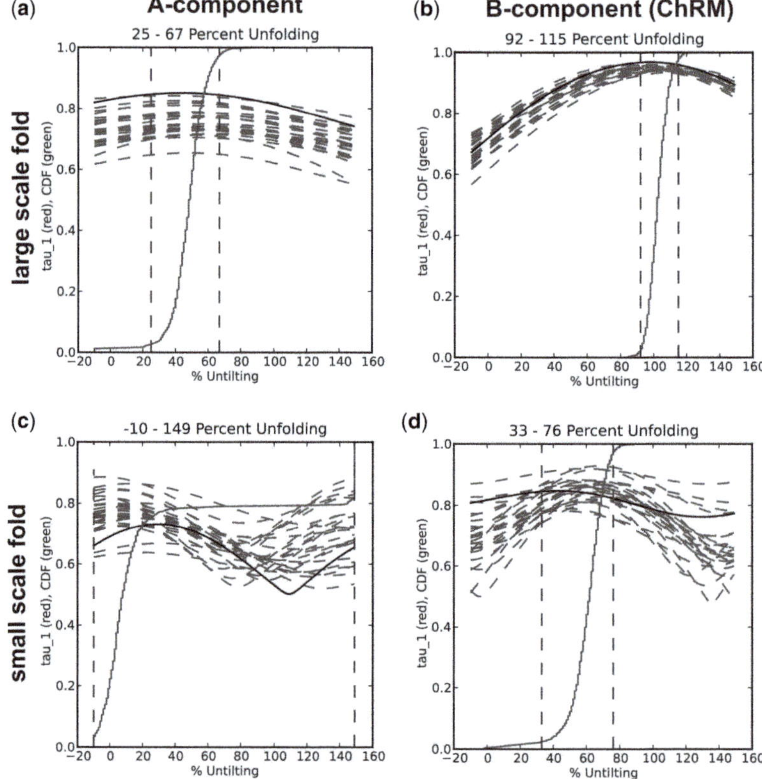



**Fig. 9.** Incremental fold test that shows grouping of A and B palaeomagnetic components as a function of progressive unfolding for (**a, c**) the large Sanetsch fold and (**b, d**) the smaller fold at Lake Tseuzier. Unfolding is expressed in percent, ranging from −20% to 160%. Dashed lines (from left to right) indicate trends of the eigenvalues $\tau_1$ obtained from the orientation matrices. The curves represent the cumulative distribution function of the eigenvalues, and provide a reference for where the grouping of palaeomagnetic direction is probabilistically highest. The 95% confidence interval is indicated by the vertical dashed lines.

**Table 5.** *Summary of results from fold tests*

| Site name | Samples | Component | $N/N_0$ | Geo-corrected | | | | Full-tilt-corrected | | | |
|---|---|---|---|---|---|---|---|---|---|---|---|
| | | | | Decl. | Incl. | $k$ | $\alpha_{95}$ (°) | Decl. | Incl. | $k$ | $\alpha_{95}$ (°) |
| Sanetsch | 57a, 59, 61, 62 | A | 39/51 | 300 | 71 | 3.8 | 13.6 | 324 | 59 | 3.2 | 15.6 |
| | | B (ChRM) | 28/51 | 315 | −13 | 1.8 | 29.8 | 36 | 70 | 34.0 | 4.7 |
| Tzeusier Lake | 63, 213, 214 | A | 15/17 | 93 | 85 | 13.4 | 10.9 | 22 | 70 | 2.7 | 28.8 |
| | | B (ChRM) | 15/17 | 5 | −64 | 2.5 | 29.0 | 347 | −78 | 3.7 | 21.7 |

are found in the Pyrenees (e.g. Pueyo *et al.* 2003; Pastor-Galán *et al.* 2012). The latter-stage formation of cataclasites and the overall cooling of the structure inhibit further development of the magnetic fabric, which reflects mainly ductile processes.

Any original sedimentary magnetic fabric (i.e. $k_3$ perpendicular to bedding and $k_1$ and $k_2$ within the bedding plane) is not preserved, such that the AMS records deformation under peak metamorphic conditions. The original sedimentary fabric has therefore been overprinted as a result of burial, metamorphism, folding and the early stages of transtension, which dissected the original thin-skinned fold and thrust structure. This deformation-related AMS is homogeneous within sites belonging to the same nappe (i.e. Wildhorn Nappe). The orientation of the principal axes is consistent with palaeo-stress reconstructions from fault planes, allowing at least a geometrical comparison between the increment of finite strain recorded in the magnetic fabric and incremental palaeostrain indicators measured using fault analysis (Sperner *et al.* 1993). Similar to that observed in the metamorphic domes outcropping in the Aegean Sea (Jolivet *et al.* 2004), $k_1$ corresponds to the stretching lineations in a general context of crustal thinning related to back-arc extension. In many sedimentary rocks, $k_1$ is oriented quite parallel to the trend in fold axis (e.g. Borradaile & Henry 1997). In the Rawil Depression, the orientation of $k_1$ corresponds to the stretching parallel to the fold axes and with the stretching perpendicular to the most common mode-I veins (Fig. 2; Table 3), but also to the T axes calculated from the faults (Table 1). Coaxial deformation most likely occurred throughout the history of the Rawil Depression, with the extension direction (T axis) fixed and B and P axes being interchangeable or girdling along a great circle perpendicular to the fold axis (Cardello 2013). A similar distribution of $k_2$ and $k_3$ suggests that the magnetic fabric of most samples was imprinted during coaxial extension. The shape of the magnetic ellipsoid suggest that the magnetic fabric was imprinted under true extension parallel to the fold axes (Dietrich 1989; Maurer *et al.* 1997; Kastrup *et al.* 2004; Cardello 2013), and it is not related to

two different deformational styles (shortening later followed by flattening). Our results therefore indicate that the magnetic fabric may only record the initial stage of stretching associated with ductile-brittle extension. Because the magnetic lineation is subparallel to fold axes and stretching, we interpret the resultant magnetic fabric as the result of orogen-parallel stretching (Figs 2 & 7). The progressive rotation of the stretching direction, observed from faults and relative cross-cutting relationships of veins, is not recorded in the magnetic fabric, although the late orogen-perpendicular stretching is clearly recorded in late movements on faults and in veins (Fig. 2). Faulting, fracturing and veining occur at conditions below 185°C in the Wildhorn Nappe (Cardello 2013). Under these conditions, the temperatures were possibly not high enough to overprint the magnetic fabric. The magnetic fabric is most likely acquired under greenschist metamorphic facies only in the south-eastern part of the study area, where it appears to be unaffected by the late buckling, veining and trantensional faulting that pervade the area.

## Palaeomagnetic directions, origin of the signal and tectonic rotations

The $k_1$ axis of the magnetic susceptibility coincides with the ChRM (B-component geographic-corrected palaeomagnetic vector) at several locations. The pattern of rotation of component B is shown in Figure 7 and follows the Alpine arc, being anticlockwise in the southwest and clockwise in the northeast. The curved grey arrows measure the angle between geo- and tilt-corrected directions, being the projection of the directions on the horizontal plane. The angles related to the relative rotation are likely to exceed the true block rotations on either side of the Rawil Depression. Some variations may occur due to minor block rotation related to transtensive faulting across the Rawil region (e.g. samples from the Wildstrubel Nappe). B-component palaeomagnetic vectors for nappes exposed on the south-western side of the Rawil Depression deviated by *c.* 30–60° in an anticlockwise sense with respect to stable Europe (cf. Heller *et al.* 1989).

On the north-eastern side of the Rawil Depression, a clockwise deviation of about the same amount is calculated (30–60°, Fig. 7). This can probably be related to the change in orientation reported in the map of Pfiffner *et al.* (2011), where in the Gemmi area fold axes are 20° more to the south with respect to the fold axes in the same nappe in the Oeschinensee area (Fig. 7). At this locality, ChRM corrected for folding corresponds to the present north pole position which has not changed much since the Tertiary. Comparing ChRM directions from Gemmi, Wildstrubel and Oeschinensee, a clockwise rotation of *c.* 20° around a vertical axis may be possibly related to late refolding around the updoming Aar Massif during the Late Neogene.

The difference between the two sides of the Rawil Depression depends most probably on the degree of deformation expressed by stretching parallel to fold axes, which depends on the amount of deformation imparted on the magnetic fabric and the particular metamorphic conditions at the time of acquisition of both A and B (ChRM) components. Although the amount of palaeomagnetic data is too small to warrant a causal relationship between the AMS and palaeomagnetic signals, this observation is of potential interest for future integrated

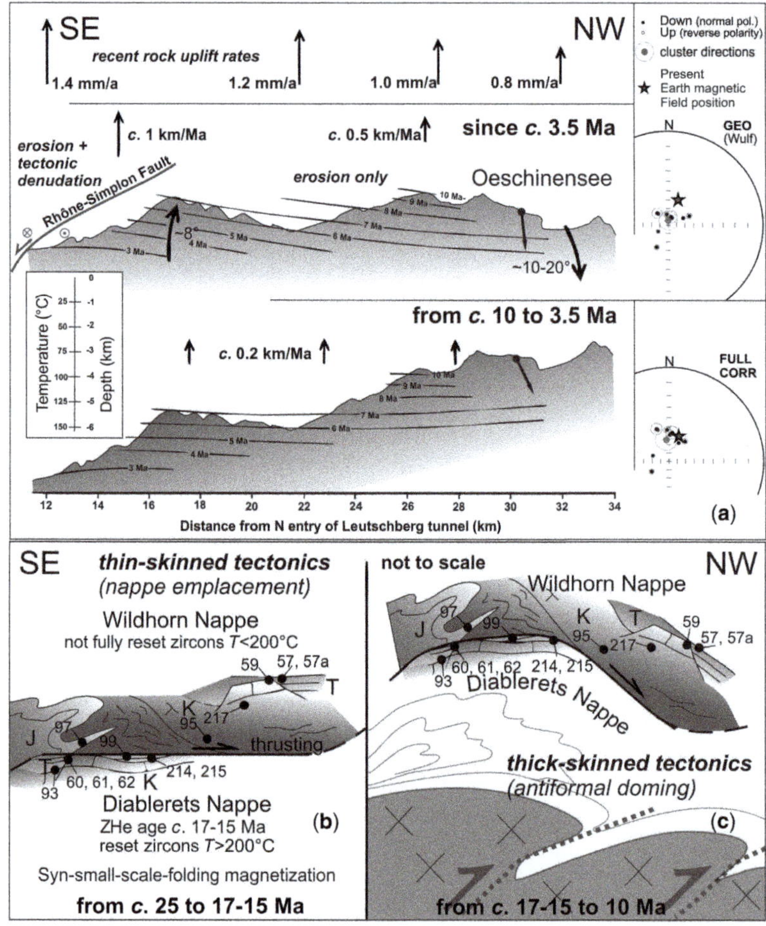


**Fig. 10.** (**a**) Sketch of the recent tectonic evolution of the Aar Massif, modified after Reinecker *et al.* (2008). Arrows with circles at ends indicate the pole position before and after 3.5 Ma. The tilt observed in the south corresponds to an opposite tilt in the north due to thrusting. This explains the oversteepened direction of the Component A registered at Oeschinesee. Equal-area projections are shown with the angular standard deviation ($\alpha_{95}$; circle). (**b, c**) Sketch of the evolution along the section of the Sanetsch. (b) Unfolded situation showing not fully reset zircons in the Wildhorn Nappe and partially to fully reset zircons in the parautochthonous nappes (i.e. Diablerets and Doldenhorn nappes), indicating that these rocks experienced $T > 200°C$ at around 17 Ma before updoming. (c) Antiformal doming with folded thrust at the base of the Wildhorn Nappe (further details in Cardello 2013).

magnetic studies in the study region. Furthermore, detailed structural studies in the region (e.g. Dietrich 1989) indicate intense deformation that might have partially rotated the palaeomagnetic vectors to somewhat parallel to the main stretching direction.

IRM acquisition and thermal demagnetization of NRM indicate that the limestone samples carry several ferromagnetic minerals, including goethite, magnetite and pyrrhotite. This has also been found in other studies examining magnetic properties of the Helvetic limestones (e.g. Kligfield & Channell 1981; Ihmlé *et al.* 1989; Almqvist *et al.* 2009). Because the ferromagnetic minerals that carry the palaeomagnetic signal also contribute to the magnetic fabric, the AMS was also acquired after the metamorphic peak and, more precisely, at the end of the thin-skinned tectonics and before the late buckling of the nappe stack. Most of the strain was likely already recorded by the magnetic fabric when the palaeomagnetic signal was blocked in.

The origin of the palaeomagnetic signal is uncertain, but probably represents a thermo-chemical remanence. Furthermore, the acquisition of ChRM in the samples at the northern limb of the Sanetsch (Figs 1 & 8) suggests that a primary sedimentary palaeomagnetic signal has been overprinted. The polarities are not lithology dependent, but rather nappe dependent. Reversed magnetizations are found in the Tertiary limestones (sample 57a) and Schrattenkalk Formation (sample 59) belonging to the Wildhorn Nappe, whereas normal polarity is found in the Schrattenkalk Formation (sample 61) and Tertiary limestones (sample 62) of the Diablerets Nappe (Fig. 8). Further information about the relative timing of remanence acquisition is provided by ChRM directions in tilted Early Cretaceous limestones (Schrattenkalk Formation) and the overlying Tertiary nummulite-rich limestones, which are separated by an angular unconformity (Fig. 8). The ChRM directions for sample 59 (Schrattenkalk Formation) show affinity with the overlying Tertiary sample 57. This suggests that the ChRM was acquired during regional metamorphism that occurred between the burial and the exhumation. We interpret this ChRM component to be a post-sedimentation overprint recorded during and/or after the metamorphic peak, which preceded antiformal doming and exhumation.

Kligfield & Channell (1981) found that limestones in the Morcles and Diablerets nappes were pervasively remagnetized during the Brunhes Epoch. This can be attributed to a viscous component of magnetization. We also identify either unstable magnetization at many of the sample sites, or a remanence direction parallel to the Earth's field in the majority of our samples (i.e. intermediate

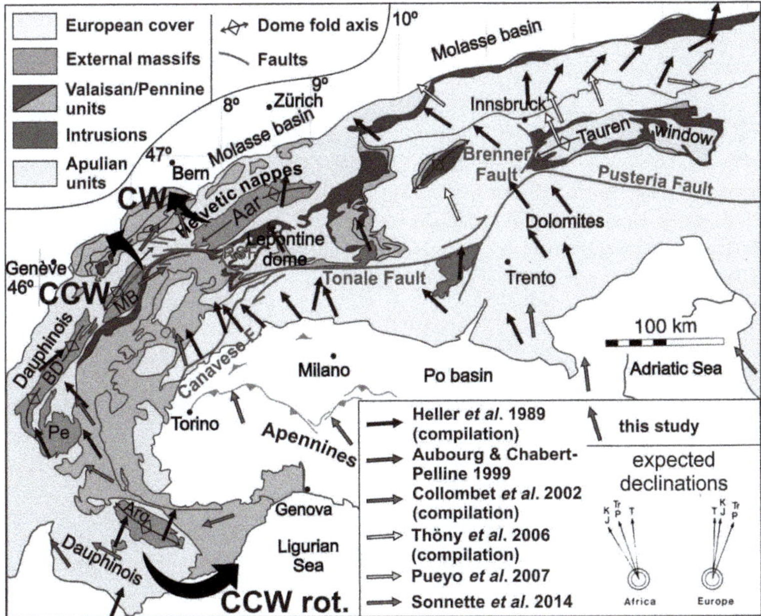


**Fig. 11.** Declination of palaeomagnetic vectors at sites in different tectonic units of the Alps, updated from a previous compilation by several authors. The contribution made by this study is highlighted in the rectangle. CW, clockwise; CCW, anticlockwise; RSF, Rhône-Simplone Fault (Mancktelow 1992); MB, Mont Blanc Massif; BD, Belledonne Massif; Pe, Pelvoux Massif; Arg, Argentera Massif.

component A). However, in contrast to the study of Kligfield & Channell (1981), our measurements indicate that a reversed palaeomagnetic field direction is recorded in several samples collected from nappes currently exposed in the Rawil Depression, in particular where the regional metamorphism is lower.

Incremental fold tests in both folds indicate that the A component is acquired post-folding and may be related to very late tilting (Fig. 9a, c). In contrast, the B component (ChRM) displays clear improvement in grouping during or after unfolding (Fig. 9b, d). In several sampling locations within nappes at different structural levels which have undergone different metamorphic conditions (i.e. Oeschinensee, Tseuzier and Sanetsch), the intermediate remanence component (A) appears to have a nearly vertical inclination (Fig. 10; Table 4). As shown in Figure 10, the steep inclination is significantly different from the inclination of the present Earth field, suggesting that recent regional tilting of the area towards the north has occurred. In this work, we distinguish between tilt and folding; 'tilt' is a block movement that involves localities from the same tilted-structure and 'folding' is the result of bending after continuous deformation that creates double-dipping flanks around a hinge. Similar northwards tilting has also been reported from palaeomagnetic data from the Aar Massif and its autochthonous cover rocks (Heller *et al.* 1989).

Thermochronometric data support the young age of the tilted magnetization. Reinecker *et al.* (2008) demonstrated *c.* 8° of tilt of the Aar Massif and its cover north of the Rhône Valley towards the north, attributing the tilting to normal displacement along the Rhone Fault, which took place during the last 3.5 Ma (Fig. 10). Consequentially, the age of component A acquisition is probably older than 3.5 Ma and younger than *c.* 10 Ma. The isotherms were subhorizontal before 10 Ma and uplift rate homogenous across the region, as calculated across the Aar Massif (Reinecker *et al.* 2008). The metamorphic conditions preceding domal uplift are illustrated in Figure 10. The late tilt of the intermediate component (A) with respect to horizontal axes can be attributed to late orogenic isostastic rebound, the latter being related to erosion as argued by Vernon *et al.* (2008), Champagnac *et al.* (2009), Willet (2010), Sternai *et al.* (2012, 2013) and Fox *et al.* (2015). These authors argue that orogenic isostatic rebound is the main mechanism of exhumation and faulting in the Western Alps during the Late Neogene (last 10 Ma). This northwards tilt is probably synchronous with the rotations that occurred about vertical axes as recorded at different sites by the component A (i.e. Mont Blanc, Sanetsch South, Wildstrubel and Gemmi; Figs 1, 7 & 10).

The palaeomagnetic B component yields a stable remanence (interpreted as ChRM), with samples recording both normal and reverse polarity. The palaeopole calculated from the B component suggests that this magnetization was acquired during stretching parallel to fold axes that possibly occurred during and after nappe emplacement (i.e. 25–15 Ma, Kirschner *et al.* 2003), but before the updoming of the large-scale fold (i.e. 17–15 Ma, Cardello 2013; 12 Ma, Reinecker *et al.* 2008, fig. 12).

The example shown in this work from the Rawil Depression can also be compared with the other structures occurring between the domal culmination of the Alps (e.g. Tauern window, Engadina and the other external crystalline massifs of the western arc; Fig. 11). Their en echelon distribution is coherent with the Neogene dextral strike-slip kinematics (Pleuger *et al.* 2012) and is associated with extension parallel and oblique to the fold axes. The oblique kinematic is largely due to the differential block rotation between culminations, which is an effect of the indentation mechanics of Adria and the curvature of the orogen. Different thrusting directions at the thrust front and the undulated geometry of the basal décollement may also contribute to create local rotation of the domes, axial depressions and wrench zones.

## Conclusions

Results from this structural and magnetic study show how the two datasets can be used to unravel tectonic evolution of the Rawil Depression in the Helvetic Alps. Despite a relatively pervasive recent remagnetization in the Helvetic Alps, deformational events can be tracked in both the magnetic fabric and in the palaeomagnetic directions, providing useful information about the kinematics of deformation and the thermal conditions reached during metamorphism. The AMS data show a composite fabric related to nappe emplacement in the more metamorphosed nappes and to extension parallel to the chain in the higher coldest nappes. The magnetic fabric records only part of the deformational history with a timing around peak metamorphism, which corresponds to orogen-parallel stretching. The later progressive rotation of stretching from parallel to perpendicular to the regional fold axes is not recorded. Component A from the palaeomagnetic data is often steeper than the present Earth's field direction, which supports a more recent tilt of the entire area. This low-temperature/coercivity component can therefore be used for tectonic interpretation. The high-temperature/coercivity palaeomagnetic directions indicate that widespread orogen-parallel extension took place prior to the formation of the Rawil Depression, which subsequently formed by more discrete ductile-brittle transtensional faulting. This interpretation is also supported by structural observations and AMS data. The direction of the palaeomagnetic vectors implies that the north-western side of the Rawil Depression experienced an anticlockwise rotation with respect to stable Europe, whereas the north-eastern side was locally rigidly rotated clockwise. This event possibly occurred partially during and after nappe emplacement and before cooling at *c.* 10 Ma. On either side of the depression we observe rigid block rotation about vertical axes. This appears to be an important mechanism related to the updoming of the Aar Massif and the formation of the Rawil Depression, which is itself a Late Neogene event. This work encourages a more extensive integrated structural and palaeomagnetic study of the Helvetic Alps.

This work was supported by the Swiss National Science Foundation (project grant 2-77178-11). We thank Jara Schnyder, Rolf Brujin, Hans-Peter Hächler and Andrea Biedermann for their help in the field and in the lab and Lionel Sonnette for his suggestions regarding AMS and palaeomagnetism in the Western Alps. We are particularly grateful to Carlo Doglioni and Jan Pleuger for fruitful discussions and Aviva Sussman and Emilio Pueyo for having improved this manuscript with their review. This paper is dedicated to the memory of Roberto Lanza for his inspiring comments.

# References

ALMQVIST, B. S. G., HIRT, A. M., SCHMIDT, V. & DIETRICH, D. 2009. Magnetic fabrics of the Morcles Nappe complex. *Tectonophysics*, **466**, 89–100.

ALMQVIST, B. S. G., HIRT, A. M., HERWEGH, M. & LEISS, B. 2011. Magnetic anisotropy reveals Neogene tectonic overprint in highly strained carbonate mylonites from the Morcles nappe, Switzerland. *Journal of Structural Geology*, **33**, 1010–1022.

ARGAND, E. 1916. Sur l'arc des Alpes Occidentales. *Eclogae Geologicae Helvetiae*, **14**, 145–191.

AUBOURG, C. & CHABERT-PELLINE, C. 1999. Neogene remagnetization of normal polarity in the Late Jurassic black shales from the southern Subalpine Chains (French Alps). Evidence for late anticlockwise rotations. *Tectonophysics*, **308**, 473–486.

BAKER, D. W. 1964. *Structural Petrology of the Windgällen Fold*. Unpublished PhD thesis, University of Zürich.

BERGMÜLLER, F., BÄRLOCHER, C., GEYER, B., GRIEDER, M., HELLER, F. & ZWEIFEL, P. 1994. A torque magnetometer for measurements of the high-field anisotropy of rocks and crystals. *Measurement Science and Technology*, **5**, 1466–1470.

BIGI, G., COSENTINO, D., PAROTTO, M., SARTORI, R. & SCANDONE, P. 1990. Structural model of Italy and gravity map. *Consiglio Nazionale delle Ricerche P.G.F., Quaderni de La Ricerca scientifica*, **114**.

BORRADAILE, G. J. 1988. Magnetic susceptibility, petrofabrics and strain. *Tectonophysics*, **156**, 1–20.

BORRADAILE, G. J. 1991. Correlation of strain with anisotropy of magnetic susceptibility (AMS). *Pure and Applied Geophysics*, **135**, 15–29.

BORRADAILE, G. J. & HENRY, B. 1997. Tectonic applications of magnetic susceptibility and its anisotropy. *Earth Science Reviews*, **42**, 49–93.

BORRADAILE, G. J. & JACKSON, M. 2010. Structural geology, petrofabrics and magnetic fabrics (AMS, AARM, AIRM). *Journal of Structural Geology*, **32**, 1519–1551.

BORRADAILE, G. J. & TARLING, D. H. 1981. The influence of deformation mechanisms on magnetic fabrics in weakly deformed rocks. *Tectonophysics*, **77**, 151–168.

BURKHARD, M. 1988. L'Helvétique de la bordure occidentale du massif de l'Aar (évolution tectonique et métamorfique). *Eclogae Geologicae Helvetiae*, **81**, 63–114.

BURKHARD, M. 1999. Strukturgeologie und Tektonik im Bereich AlpTransit. *In*: LÖW, S. & WYSS, R. (eds), *Vorerkundung und Prognose der Basistunnels am Gotthard und am Lötschberg*. A.A. Balkema, Rotterdam, 45–56.

BURKHARD, M. & KERRICH, R. 1988. Fluid regimes in the deformation of the Helvetic Nappes, Switzerland, as inferred from stable isotope data. *Contributions to Mineralogy and Petrology*, **99**, 416–429.

BURKHARD, M. & SOMMARUGA, A. 1998. Evolution of the western Swiss Molasse basin: structural relations with the Alps and the Jura belt. *Cenozoic Foreland Basins of Western Europe*, **134**, 279–298.

BUSSY, F. & EPARD, J.-L. 1984. Essai de zonéographie métamorphique entre les Diablerets et le massif de l'Aar (Suisse occidentale), basée sur l'étude des Grès de Taveyanne. *Schweizerische Mineralogische und Petrographische Mitteilungen*, **64**, 131–150.

BUTLER, R. F. 1992. *Paleomagnetism: Magnetic Domains to Geologic Terranes*. Blackwell Scientific Publications, Boston.

CARDELLO, G. L. 2013. *The Rawil Depression: its Structural History from Cretaceous to Neogene*. PhD thesis, Swiss Federal Institute of Technology, Zurich (ETH).

CARDELLO, G. L. & MANCKTELOW, N. S. 2014. Cretaceous syn-sedimentary faulting in the Wildhorn Nappe (SW Switzerland). *Swiss Journal of Geosciences*, **107**, 223–250.

CARDELLO, G. L. & TESEI, T. 2013. Transtensive faulting in carbonates at different crustal levels: examples from SW Helvetics and Central Apennines. *Rendiconti online della Società Geologica Italiana*, **29**, 20–23.

CHADIMA, M. & HROUDA, F. 2006. Remasoft 3.0 a user-friendly paleomagnetic data browser and analyzer. *Travaux Géophysiques*, **27**, 20–21.

CHADIMA, M. & JELÍNEK, V. 2008. Anisoft 4.2.–Anisotropy data browser. *In*: *Paleo, Rock and Environmental Magnetism, 11th Castle Meeting, Contribution to Geophysics and Geodesy, Special Issue*. Geophysical Institute of the Slovak Academy of Sciences. *Bojnice Castle, Slovac Republic*.

CHAMPAGNAC, J. D., SCHLUNEGGER, F., NORTON, K., VON BLANCKENBURG, F., ABBÜHL, L. M. & SCHWAB, M. 2009. Erosion-driven uplift of the modern Central Alps. *Tectonophysics*, **474**, 236–249.

CIFELLI, F., ROSSETTI, F., MATTEI, M., HIRT, A. M., FUNICIELLO, R. & TORTORICI, L. 2004. An AMS, structural and paleomagnetic study of quaternary deformation in eastern Sicily. *Journal of Structural Geology*, **26**, 29–46.

CIFELLI, F., MATTEI, M., CHADIMA, M., HIRT, A. M. & HANSEN, A. 2005. The origin of tectonic lineation in extensional basins: combined neutron texture and magnetic analyses on 'undeformed' clays. *Earth and Planetary Science Letters*, **235**, 62–78.

CIFELLI, F., MATTEI, M., CHADIMA, M., LENSER, S. & HIRT, A. M. 2009. The magnetic fabric in 'undeformed clays', AMS and neutron texture analyses from the Rif Chain (Morocco). *Tectonophysics*, **466**, 79–88.

COLLOMBET, M., THOMAS, J. C., CHAUVIN, A., TRICART, P., BOUILLIN, J. P. & GRATIER, J. P. 2002. Counterclockwise rotation of the western Alps since the Oligocene: new insights from paleomagnetic data. *Tectonics*, **21**, 14–21.

DICK, P. & BURKHARD, M. 2001. Magnetic anisotropy and X-ray diffraction study of clay minerals in the decollement horizons of the western Helvetic nappes, SW Switzerland. *Clay Minerals*, **36**, 181–196.

DIETRICH, D. 1989. Fold-axis parallel extension in an arcuate fold-and thrust belt: the case of the Helvetic nappes. *Tectonophysics*, **170**, 183–212.

EGLI, D. & MANCKTELOW, N. 2013. The structural history of the Mont Blanc massif with regard to models for its recent exhumation. *Swiss Journal of Geosciences*, **106**, 469–489.

EVA, E., PASTORE, S. & DEICHMANN, N. 1998. Evidence for ongoing extensional deformation in the western Swiss Alps and thrust-faulting in the southwestern Alpine foreland. *Journal of Geodynamics*, **26**, 27–43.

FOX, M., HERMAN, F., KISSLING, E. & WILLETT, S. D. 2015. Rapid exhumation in the Western Alps driven by slab detachment and glacial erosion. *Geology*, **43**, 379–382.

FREY, M. & FERREIRO MÄHLMANN, R. 1999. Alpine metamorphism of the Central Alps. *Schweizerische Mineralogische und Petrographische Mitteilungen*, **79**, 135–154.

GASSER, D. & MANCKTELOW, N. S. 2010. Brittle faulting in the Rawil depression: field observations from the Rezli fault zones, Helvetic nappes, Western Switzerland. *Swiss Journal of Geosciences*, **103**, 15–32.

GLOTZBACH, C., VAN DER BEEK, P. A. & SPIEGEL, C. 2011. Episodic exhumation and relief growth in the Mont Blanc massif, Western Alps from numerical modelling of thermochronology data. *Earth and Planetary Science Letters*, **304**, 417–430.

GRAHAM, J. F. 1949. The stability and significance of magnetism in sedimentary rocks. *Journal of Geophysical Research*, **54**, 131–168.

GRAHAM, J. W. 1966. Significance of magnetic anisotropy in Appalachian sedimentary rocks. *Geophysical Monograph Series*, **10**, 627–648.

GUBLER, E., KAHLE, H. G., KLINGELE, E., MÜLLER, S. & OLIVIER, R. 1981. Recent crustal movements in Switzerland and their geophysical interpretation. *Tectonophysics*, **38**, 297–315.

HEIM, A. 1920. *Geologie der Schweiz. Band II*. Tauchnitz Verlag Gmbh, Leipzig.

HELLER, F. 1980. Paleomagnetic evidence for Late Alpine rotation of the Lepontine area. *Eclogae Geologicae Helvetiae*, **73/2**, 607–618.

HELLER, F., LOWRIE, W. & HIRT, A. M. 1989. A review of paleomagnetic and magnetic anisotropy results from the Alps. *In*: COWARD, M. P., DIETRICH, D. & PARK, R. G. (eds) *Alpine Tectonics*. Geological Society, London, Special Publications, **45**, 399–420.

HERWEGH, M. & PFIFFNER, O. A. 2005. Tectonometamorphic evolution of a nappe stack: a case study of the Swiss Alps. *Tectonophysics*, **404**, 55–76.

HEXT, G. R. 1963. The estimation of second-order tensors, with related tests and designs. *Biometrika*, **50**, 353–373.

IHMLÉ, P. F., HIRT, A. M., LOWRIE, W. & DIETRICH, D. 1989. Inverse magnetic fabric in deformed limestones of the Morcles Nappe, Switzerland. *Geophysical Research Letters*, **16**, 1383–1386.

JELÍNEK, V. 1981. Characterization of the magnetic fabric of rocks. *Tectonophysics*, **79**, T63–T67.

JELÍNEK, V. & KROPÁČEK, R. V. 1978. Statistical processing of anisotropy of magnetic susceptibility measured on groups of specimens. *Studia Geophysica et Geodaetica*, **22**, 50–62.

JOLIVET, L., FAMIN, V., MEHL, C., PARRA, T., AUBOURG, C., HÉBERT, R. & PHILIPPOT, P. 2004. Strain localization during crustal-scale boudinage to form extensional metamorphic domes in the Aegean Sea. *In*: COWARD, M. P., DIETRICH, D. & PARK, R. G. (eds) *Alpine Tectonics. Geological Society of America, Special Paper*, **380**, 185–210.

KASTRUP, U., ZOBACK, M. L., DEICHMANN, N., EVANS, K. F., GIARDINI, D. & MICHAEL, A. J. 2004. Stress field variations in the Swiss Alps and the northern Alpine foreland derived from inversion of fault plane solutions. *Journal of Geophysical Research*, **109**, B01402, http://doi.org/10.1029/2003JB002550

KIRSCHNER, D. L., MASSON, H. & COSCA, M. A. 2003. An 40Ar/39Ar, Rb/Sr, and stable isotope study of micas in low-grade fold-and-thrust belt: an example from the Swiss Helvetic Alps. *Contributions to Mineralogy and Petrology*, **145**, 460–480.

KIRSCHVINK, J. L. 1980. The least-squares line and plane and the analysis of palaeomagnetic data. *Geophysical Journal International*, **62**, 699–718.

KLIGFIELD, R. & CHANNELL, J. E. T. 1981. Widespread re of Helvetic limestones. *Journal of Geophysical Research*, **86**, 1888–1900.

KLIGFIELD, R., LOWRIE, W. & PFIFFNER, O. A. 1982. Magnetic properties of deformed oolitic limestones from the Swiss Alps: the correlation of magnetic anisotropy and strain. *Eclogae Geologicae Helvetiae*, **75**, 127–157.

LEVATO, L., SELLAMI, S., EPARD, J. L., PRUNIAUX, B., OLIVIER, R., WAGNER, J. J. & MASSON, H. 1994. The cover-basement contact beneath the Rawil axial depression (western Alps): true amplitude seismic processing, petrophysical properties, and modelling. *Tectonophysics*, **232**, 391–409.

LUGEON, M. 1914–1918. *Les Hautes Alpes calcaires entre la Lizerne et la Kander: (Wildhorn, Wildstrubel, Balmhorn et Torrenthorn): (explication de la carte spéciale no. 60). 3 Fascicule. Matériaux pour la carte géologique de la Suisse. Nouvelle série 60.*

MANCKTELOW, N. S. 1992. Neogene lateral extension during convergence in the Central Alps: evidence from interrelated faulting and backfolding around the

Simplonpass (Switzerland). *Tectonophysics*, **215**(3), 295–317.

MARTÍN-HERNÁNDEZ, F. & HIRT, A. M. 2001. Separation of ferrimagnetic and paramagnetic anisotropies using a high-field torsion magnetometer. *Tectonophysics*, **337**, 209–222.

MASSON, H., BAUD, A., ESCHER, A., GABUS, J. & MARTHALER, M. 1980a. Compte rendu de l'excursion del la Société Géologique Suisse du 1 au 3 octobre 1979: coupe Préalpes-Helvétique-Pennique en Suisse occidentale. *Eclogae Geologicae Helvetiae*, **73**, 331–349.

MASSON, H., HERB, R. & STECK, A. 1980b. *Helvetic Alps of Western Switzerland, The Geology of Switzerland, A Guide Book. Part B.* Birkhäuserverlag, Basel.

MAURER, H. R., BURKHARD, M., DEICHMANN, N. & GREEN, A. G. 1997. Active tectonism in the central Alps: contrasting stress regimes north and south of the Rhone Valley. *Terra Nova*, **9**, 91–94.

MOSER, H.-J. 1985. *Strukturgeologische Untersuchungen in der Rawil-Depression*. PhD thesis, Universitaetsdruckierei Bern.

ORTNER, H., REITER, F. & ACS, P. 2002. Easy handling of tectonic data: the programs TectonicVB for Mac and TectonicsFP for Windows™. *Computers & Geosciences*, **28**, 1193–1200.

PARÉS, J. M. & VAN DER PLUIJM, B. 2002. Evaluating magnetic lineations (AMS) in deformed rocks. *Tectonophysics*, **350**, 283–298.

PASTOR-GALÁN, D., GUTIÉRREZ-ALONSO, G., MULCHRONE, K. F. & HUERTA, P. 2012. Conical folding in the core of an orocline. A geometric analysis from the Cantabrian Arc (Variscan Belt of NW Iberia). *Journal of Structural Geology*, **39**, 210–223.

PFIFFNER, A. O. 1993. The structure of the Helvetic nappes and its relation to the mechanical stratigraphy. *Journal of Structural Geology*, **15**, 511–521.

PFIFFNER, O. A. 2009. *Geologie der Alpen*. Haupt Verlag, Bern.

PFIFFNER, O. A., BURKHARD, M. ET AL. 2011. *Structural map of the Helvetic zone of the Swiss Alps, Including Vorarlberg (Austria) and Haute Savoie (France), 7 Map Sheets. Swisstopo/Landesgeologie,* Geological Special Map, **128**, 1–7.

PLEUGER, J., MANCKTELOW, N., ZWINGMANN, H. & MANSER, M. 2012. K–Ar dating of synkinematic clay gouges from Neoalpine faults of the Central, Western and Eastern Alps. *Tectonophysics*, **550**, 1–16.

PUEYO, E. L., PARÉS, J. M., MILLÁN, H. & POCOVI, A. 2003. Conical folds and apparent rotations in paleomagnetism (a case study in the Southern Pyrenees). *Tectonophysics*, **362**, 345–366.

PUEYO, E. L., MAURITSCH, H. J., GAWLICK, H. J., SCHOLGER, R. & FRISCH, W. 2007. New evidence for block and thrust sheet rotations in the central northern Calcareous Alps deduced from two pervasive remagnetization events. *Tectonics*, **26**, TC5011, http://doi.org/10.1029/2006/tC001965

RAMSAY, J. G. 1989. Fold and fault geometry in the western Helvetic nappes of Switzerland and France and its implication for the evolution of the arc of the Western Alps. *In*: COWARD, M. P., DIETRICH, D. & PARK, R. G. (eds) *Alpine Tectonics*. The Geological Society, London, Special Publications, **45**, 33–45.

RAMSAY, J. G., CASEY, M. ET AL. 1981. Strain features of the Helvetic nappes of Switzerland. *In*: *The Geological Society of America, 94th Annual Meeting 13*. Abstracts with Programs, Geological Society of America.

RAMSAY, J. G., CASEY, M. & KLIGFIELD, R. 1983. Role of shear in development of the Helvetic fold-thrust belt of Switzerland. *Geology*, **11**, 439–442.

REINECKER, J., DANIŠÍK, M., SCHMID, C., GLOTZBACH, C., RAHN, M., FRISCH, W. & SPIEGEL, C. 2008. Tectonic control on the late stage exhumation of the Aar Massif (Switzerland): constraints from apatite fission track and (U-Th)/He data. *Tectonics*, **27**, TC6009, http://doi.org/10.1029/2007TC002247

SATOLLI, S., SPERANZA, F. & CALAMITA, F. 2005. Paleomagnetism of the Gran Sasso range salient (central Apennines, Italy): pattern of orogenic rotations due to translation of a massive carbonate indenter. *Tectonics*, **24**, TC4019, http://doi.org/10.1029/2004 TC001771

SONNETTE, L., HUMBERT, F., AUBOURG, C., GATTACCECA, J., LEE, J. C. & ANGELIER, J. 2014. Significant rotations related to cover–substratum decoupling: Example of the Dôme de Barrôt (Southwestern Alps, France). *Tectonophysics*, **629**, 275–289.

SPERNER, B., RATSCHBACHER, L. & OTT, R. 1993. Fault-striae analysis: a Turbo Pascal program package for graphical presentation and reduced stress tensor calculation. *Computers & Geosciences*, **19**, 1361–1388.

STECK, A. 1980. Deux directions principales de flux symmétamorphiques dans les Alpes centrales. *Bulletin de la Société Vaudoise des Sciences Naturelles*, **75**, 141–149.

STECK, A. 1984. Structures de déformation tertiaires dans les Alpes centrales (transversale Aar-Simplon-Ossola). *Eclogae Geologicae Helvetiae*, **77**, 55–100.

STERNAI, P., HERMAN, F., CHAMPAGNAC, J. D., FOX, M., SALCHER, B. & WILLETT, S. D. 2012. Pre-glacial topography of the European Alps. *Geology*, **40**, 1067–1070.

STERNAI, P., HERMAN, F., VALLA, P. G. & CHAMPAGNAC, J. D. 2013. Spatial and temporal variations of glacial erosion in the Rhône valley (Swiss Alps): insights from numerical modeling. *Earth and Planetary Science Letters*, **368**, 119–131.

SUE, C., DELACOU, B., CHAMPAGNAC, J.-D., ALLANIC, C., TRICART, P. & BURKHARD, M. 2007. Extensional neotectonics around the bend of the Western/Central Alps: an overview. *Annual Review of Earth and Planetary Sciences*, **96**, 1001–1029.

SUSSMAN, A. J., BUTLER, R. F., DINARÈS-TURELL, J. & VERGÉS, J. 2004. Vertical-axis rotation of a foreland fold and implications for orogenic curvature: an example from the Southern Pyrenees, Spain. *Earth and Planetary Science Letters*, **218**, 435–444.

SUSSMAN, A. J., PUEYO, E. L., CHASE, C. G., MITRA, G. & WEIL, A. J. 2012. The impact of vertical-axis rotations on shortening estimates. *Lithosphere*, **4**, 383–394, http://doi.org/10.1130/L177.1

TAUXE, L. & WATSON, G. S. 1994. The fold test: an eigen analysis approach. *Earth and Planetary Science Letters*, **122**, 331–341.

THÖNY, W., ORTNER, H. & SCHOLGER, R. 2006. Paleomagnetic evidence for large en-bloc rotations in the Eastern Alps during Neogene orogeny. *Tectonophysics*, **414**, 169–189.

TRÜMPY, R. 1960. Paleotectonic evolution of the central and western Alps. *Geological Society of America Bulletin*, **71**, 843–908.

TURNER, F. J. 1953. Nature and dynamic interpretation of deformation lamellae in calcite of three marbles. *American Journal of Science*, **251**, 276–298.

USTASZEWESKI, M. & PFIFFNER, A. 2008. Composite faults in the Swiss Alps formed by the interplay of tectonics, gravitation and postglacial rebound: an integrated field and modelling study. *Swiss Journal of Geosciences*, **101**, 223–235.

VAN DER VOO, R. 2005. *Paleomagnetism of the Atlantic, Tethys and Iapetus Oceans*. Cambridge University Press, Cambridge.

VERNON, A. J., VAN DER BEEK, P. A., SINCLAIR, H. D. & RAHN, M. K. 2008. Increase in late Neogene denudation of the European Alps confirmed by analysis of a fission-track thermochronology database. *Earth and Planetary Science Letters*, **270**, 316–329.

WEIL, A. B. & YONKEE, W. A. 2012. Layer-parallel shortening across the Sevier fold-thrust belt and Laramide foreland of Wyoming: spatial and temporal evolution of a complex geodynamic system. *Earth and Planetary Science Letters*, **357**, 405–420.

WILLET, S. 2010. Late Neogene erosion of the Alps: a climate driver? *Annual Review of Earth and Planetary Sciences*, **38**, 411–437.

# Integration of palaeomagnetic data, basement-cover relationships and theoretical calculations to characterize the obliquity of the Altomira–Loranca structures (central Spain)

M. VALCÁRCEL[1]*, R. SOTO[2], E. BEAMUD[3], B. OLIVA-URCIA[4], J. A. MUÑOZ[5] & C. BIETE[5]

[1]*Instituto Geológico y Minero de España (IGME), Departamento de Investigación en Recursos Geológicos, Des. 108, c/ La Calera, 1, 28760 Tres Cantos, Spain*

[2]*Instituto Geológico y Minero de España (IGME), Unidad de Zaragoza, c/ Manuel Lasala 44, 9°B, 50006 Zaragoza, Spain*

[3]*Laboratori de Paleomagnetisme (CCiTUB-ICTJA CSIC), Institut de Ciències de la Terra "Jaume Almera", Solé i Sabarís, s/n, 08028 Barcelona, Spain*

[4]*Departamento de Geología y Geoquímica, Facultad de Ciencias, Universidad Autónoma de Madrid, Campus Cantoblanco, 28049 Madrid, Spain*

[5]*Grup Geodinàmica i Anàlisi de Conques, Universitat de Barcelona, Zona Univ, Pedralbes, 08028 Barcelona, Spain*

**Corresponding author (e-mail: manoel.valcarcel@gmail.com)*

**Abstract:** The main objective of this work is to characterize the structures belonging to the Altomira Range and Loranca Basin (SW Iberian Chain, Central Spain) in terms of understanding their present-day orientation, highly oblique with respect to the NW–SE orientation of adjacent structures of the Iberian Chain. The Altomira and Loranca fold and thrust belts present a slightly curved geometry with a general north–south orientation at their central sector, and structures oriented NNE–SSW and north–south to NNW–SSE at their northern and southern sectors, respectively. Palaeomagnetic data from Middle Eocene–Lower Miocene rocks (including clays, fine sandstones and limestones) reveal the absence of vertical-axis rotations in the central sector of the studied area, where structures are oriented north–south, and up to 17° of clockwise and 21° of anticlockwise vertical-axis rotations in the northern and southern sectors, respectively. These data are supported by calculations of the theoretical vertical-axis rotations from shortening estimates and basement-cover structural relationships. This approach highlights the importance of integrating different datasets to characterize the obliquity of fold and thrust belts.

**Supplementary material:** Complete information about the paleomagnetic sites and directions is available at http://www.geolsoc.org.uk/SUP18841

Oblique structures are characterized by their lack of perpendicularity with respect to the regional transport direction. They can be primary, progressive or secondary based on the timing relationships between their formation and the development of their obliquity (Sussman & Weil 2004). The deformation mechanisms and kinematics of structures vary according to the type of obliquity. This has important implications in evaluating the characteristics of potential geological reservoirs or estimation of shortening, for example. Primary oblique structures form already oblique with respect to the regional tectonic transport direction during an initial deformation event, and their trends do not change during subsequent deformation. They can develop as a result of several causes such as lateral rheological and thickness changes of the pre-tectonic series, lateral variations of the detachment level (e.g. Cotton & Koyi 2000; Soto *et al.* 2002, 2003; Vidal-Royo *et al.* 2009) or along-strike variations of syntectonic sedimentation and erosion rates. Secondary oblique structures acquire their obliquity subsequently by means of vertical-axis rotations. In progressive curved structures, the strike of the structure changes progressively during deformation and rotation and thrusting occur simultaneously.

Numerous methods have been used to unravel vertical-axis rotations in fold and thrust belts (see Weil & Sussman 2004): palaeomagnetism (e.g. McCaig & McClelland 1992); changes in strain orientations (e.g. Mitra & Yonkee 1985;

*From*: PUEYO, E. L., CIFELLI, F., SUSSMAN, A. J. & OLIVA-URCIA, B. (eds) 2016. *Palaeomagnetism in Fold and Thrust Belts: New Perspectives*. Geological Society, London, Special Publications, **425**, 169–188. First published online August 21, 2015, updated September 3, 2015, http://doi.org/10.1144/SP425.7

                                                                M. VALCÁRCEL *ET AL.*

Craddock *et al.* 1988; Kollmeier *et al.* 2000); changes in palaeocurrent orientation (e.g. Freund 1970; Puigdefábregas 1975); and along-strike variability in thrust transport directions (e.g. Nur *et al.* 1986; Allerton 1994; Knon & Mitra 2004). Among those, the strongest method of unravelling the kinematics of curved orogens and/or oblique structures is the combination of palaeomagnetism and detailed structural analysis (Weil & Sussman 2004). The use of palaeomagnetism to quantify vertical-axis rotations is not always possible however, because of: (1) rocks with intensities of the natural remanent magnetization (NRM) below the instrumental noise (e.g. *c.* $10^{-12}$ Am$^2$) of current superconducting magnetometers; (2) rocks recording a magnetization acquired after the rotational movement of structures; (3) complex areas (e.g. superimposed folding, non-coaxial folding) without having a reliable control of their evolution in order to correctly restore palaeomagnetic data (see Pueyo 2010); and/or (4) in areas located at very high latitudes showing subvertical inclinations of the palaeomagnetic vector (see Baraldo *et al.* 2003).

Other factors include a lack of suitable rocks from which to obtain palaeomagnetic samples and/or the existence of unstable palaeomagnetic components and remagnetizations.

In the interior of the Iberian plate, despite Cenozoic deformation responding to the relative movements between Africa, Iberia and Europe with a general north–south and NW–SE orientation (e.g. Rosenbaum *et al.* 2002), the resulted contractional structures in the Iberian Chain do not present a homogeneous trending (Fig. 1a). This variety has been interpreted in different ways, for example multiphase evolutionary models (e.g. Liesa & Simón 2007) or the influence of the inversion of previous Variscan and/or Mesozoic faults with different orientation (e.g. Guimerà *et al.* 2004; De Vicente *et al.* 2009). One of the most interesting and attractive structural features of the SW deformation front of the Iberian Chain (NE Spain) is the occurrence of a series of approximately north–south to NNE–SSW and NNW–SSE-trending structures forming a subtle arc, oblique with respect to the general NW–SE trend of the Iberian Chain.

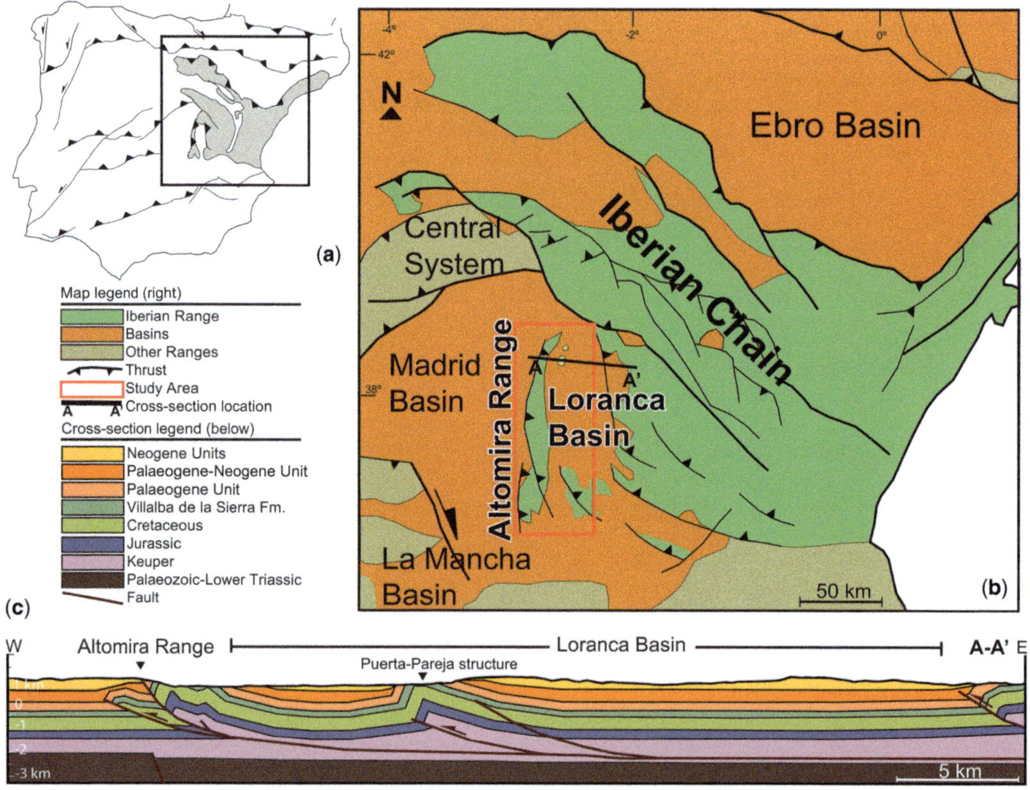


**Fig. 1.** Geographic and geological location of the studied area in (**a**) the Iberian Peninsula and (**b**) the Iberian Chain. (**c**) West–east cross-section of the Altomira–Loranca structures at the northern part (A–A′). Red square represents location of Figure 2.

In this work, we use palaeomagnetic data to detect possible vertical-axis rotations to account for their obliquity together with two other datasets: (1) theoretical vertical-axis rotations calculated using shortening estimates from balanced cross-sections from Muñoz-Martín & De Vicente (1998), following the methodology proposed by Pueyo *et al.* (2004); and (2) basement-cover structural relationships in two areas located inside the Loranca Basin from subsurface and surface data.

## Geological setting

### Structural features

The Altomira Range and Loranca Basin constitute the SW termination of the Iberian Chain (i.e. the deformation front at this latitudinal portion; Fig. 1). The Iberian Chain is an intraplate range formed in a generalized north–south-aligned compressive context between the Paleocene and the Early Miocene (e.g. Casas-Sainz & Faccenna 2001; De Vicente *et al.* 2004) due to the relative movements between Africa, Iberia and Europe with a general north–south and NW–SE orientation (e.g. Rosenbaum *et al.* 2002). It formed by the tectonic inversion of previous Permo-Triassic and Upper Jurassic–Cretaceous basins during the Alpine Orogeny (e.g. De Vicente *et al.* 2009). Structures belonging to the Altomira Range and Loranca Basin are oblique with respect to the general NW–SE trend of the Iberian Chain (Fig. 1). They are oriented north–south at its central portion, NNE–SSW at its northern sector and north–south to NNW–SSE at its southern sector, therefore tracing a subtle arc convex towards the west (Figs 1b & 2). Muñoz-Martín (1997) described several 'transfer zones' (sets of strike-slip and transtensional faults) that would accommodate these differences in the orientation between sectors along the Altomira Range and the Loranca Basin.

Deformation in the Altomira Range and Loranca Basin did not occur simultaneously both across- and along-strike. At its northern and central sector, the Altomira Range structures formed firstly during Eocene–Late Oligocene time (Gómez *et al.* 1996) with a westwards transport direction. Their formation individualized the Loranca Basin, an inward piggyback basin, also containing approximately north–south-aligned west-vergent thrust-related folds formed later during Late Oligocene–Early Miocene time (Díaz Molina *et al.* 1995), in an out-of-sequence thrusting with respect to Altomira's structures (Gómez *et al.* 1996; Fig. 1). At the southern sector, by contrast, the Loranca structures formed firstly during the Late Eocene?–Oligocene, following an in-sequence thrusting that ended there with the formation of the Altomira structures

(Muñoz-Martín 1997; Muñoz-Martín & De Vicente 1998). Although most Loranca structures do not crop out, they are identifiable in seismic profiles (Querol 1989; ITGE 1990). Both the Altomira and Loranca structures affect a Mesozoic–Cenozoic cover detached over a regional décollement level formed by Middle–Upper Triassic evaporites and clays (e.g. Gómez *et al.* 1996; Muñoz-Martín & De Vicente 1998; Fig. 1c).

The obliquity of the Altomira and Loranca thrusts and folds in relation to the NW–SE Iberian orientation and the north–south general compression in Iberia is controversial. It has been classically related to either: (1) their origin as oblique ramps of the Iberian Chain (Guimerà 1988); (2) the influence of the across-strike westwards thinning of the main detachment level, the Upper Triassic rocks (from 800 m thick to the east to less than 400 m thick at the western part of the studied area; Suárez-Alba 2007) and even its disappearance to the west of the Altomira Range (Van Wees & Stephenson 1995); (3) the influence of previous north–south-trending basement faults, which could have controlled both the Pemo-Triassic and Upper Jurassic–Cretaceous rifts, their normal fault pattern and the orientation of posterior structures (Muñoz-Martín & De Vicente 1998); and (4) the superposition of transmitted major palaeostress fields coming from the Betic and Pyrenean active margins added to factors described in (2) and (3) (Muñoz 1997; Muñoz-Martín *et al.* 1998; Andeweg *et al.* 1999).

### Stratigraphy of the Altomira Range–Loranca Basin

The stratigraphic sequence of the SW border of the Iberian Range can be divided into the following units (Muñoz-Martín 1997; Muñoz-Martín & De Vicente 1998): (1) the Palaeozoic and Lower Triassic rocks (Buntsandstein facies) that constitute the basement; (2) the Middle–Upper Triassic evaporites and clays (Muschelkalk and Keuper facies), acting as the regional detachment level; (3) 400-m-thick Jurassic dolomites with interbedded marls and *c.* 500 m thickness of Cretaceous rocks mainly including limestones, dolomites and marls; (4) the so-called Villalba de la Sierra Formation (Campanian–Lutetian), a transition formation between the Mesozoic marine rocks and the fully continental Cenozoic rocks including 200 m thickness of marls, limestones, sandstones, clays and gypsum (Muñoz-Martín 1997; Hernaiz *et al.* 1999*a*; Torres *et al.* 2006; and (5) Bartonian–Pliocene continental rocks, consisting of several levels of clay, sandstone, lacustrian limestones and evaporites grouped into seven different units (Hernaiz *et al.* 1999*a, b*; Gabaldón *et al.* 1999*a, b, c, d*): Palaeogene

Unit (PU) (Bartonian–Rupelian); Palaeogene–Neogene Unit (PNU) (Rupelian–Aquitanian); first Neogene Unit (NU1) (Aquitanian); second Neogene Unit (NU2) (Burdigalian–Langhian); third Neogene Unit (NU3) (Langhian–Tortonian); fourth Neogene Unit (NU4) (Tortonian–Messinian); and Pliocene sediments.

The tectono-stratigraphic relationships between the Bartonian–Pliocene continental units and the Altomira and Loranca structures vary both along and across strike. In the northern and central sectors the out-of-sequence thrusting of the Loranca structures with respect to the Altomira (e.g. Gómez *et al.* 1996) means that the first syntectonic units at Altomira are older than those located at Loranca. The Palaeogene Unit (PU) (Bartonian–Rupelian) has therefore been described as pre-tectonic with respect to the Puerta–Pareja anticline (Hernaiz *et al.* 1999*a*), a structure located in the northern Loranca Basin (Fig. 1c), and syntectonic with respect to the adjacent Altomira structures (Gómez *et al.* 1996). The syntectonic deposits of the Loranca Basin, Rupelian–Langhian in age, are included in the PNU, NU1 and NU2 units (Hernaiz *et al.* 1999*a*, *b*; Gabaldón *et al.* 1999*a*, *b*, *c*, *d*). The post-tectonic deposits (late Langhian–present), display a similar lithology as the syntectonic deposits.

Sampling was focused on Cretaceous to Lower Miocene pre- and syntectonic rocks of the Altomira Range and Loranca Basin in order to detect possible vertical-axis rotations which occurred during the formation of their structures.

## Palaeomagnetic methodology

### Sampling

A total of 45 palaeomagnetic sites were drilled distributed across the Altomira Range and the Loranca Basin, including 12 sites of Cretaceous rocks (containing 1 Villalba de la Sierra Formation site), 20 sites of Bartonian–Rupelian rocks (Palaeogene Unit), 9 sites of Rupelian–early Aquitanian rocks (Palaeogene–Neogene Unit) and 4 sites of Aquitanian–Langhian rocks (first and second Neogene Units) (Fig. 2). Sampling was focused on clays, fine-grain sandstones and limestones in order to obtain maximum success in the palaeomagnetic analysis. Most rocks cropping out at the Altomira Range consist of Mesozoic dolomites that were not sampled to avoid remagnetizations and/or rocks with very low intensities of the NRM. On the other hand, most deposits exposed at the Loranca Basin belong to the post-tectonic units, which do not provide any structural information in relation to our objectives. Pre- and syntectonic rocks only crop out at the western part of the Loranca Basin, close to the Altomira Range. Sampling was

therefore constrained to the very limited non-dolomitic outcrops located at the Altomira Range and to a few suitable outcrops for palaeomagnetic sampling in the pre- and syntectonic deposits located at the westernmost part of the Loranca Basin (Fig. 2). Seven to twelve samples per site were drilled and oriented *in situ* with a magnetic compass coupled to a core-orienting device. Clays and fine-grained sandstones, which represent most of the sampled materials, were drilled with an electrical power drill cooled with water, whereas limestones were sampled with a gas power drill. All cores were sliced in standard palaeomagnetic specimens (cylinders 25 mm in diameter and 22 mm in height) to conduct the palaeomagnetic and rock-magnetic study. A total of 385 samples from our 45 sites were analysed in this study.

### Palaeomagnetic analysis

Palaeomagnetic analyses were carried out in the Paleomagnetic Laboratory of Barcelona (CCiTUB-ICTJA CSIC). These consisted of progressive thermal demagnetization of the natural remanent magnetization (NRM) up to 660°C with a maximum of 20 steps per specimen. Thermal demagnetizers MMTD-80 (Magnetic Measurements) and TSD-1 (Schonstedt) and a superconductor rock magnetometer SRM 755R (2G Enterprises) were used. Magnetic susceptibility was monitored after each demagnetization step using a KLY-2 (Agico) in order to control possible mineralogical changes during heating. Characteristic components were determined by linear regression techniques after visual inspection of the orthogonal demagnetization diagrams. The obtained directions were classified as either Class 1, 2 or 3. Class 1 directions show straight trajectories directed towards the origin and maximum angular deviations (MADs) of the linear regression lower than 10. Directions also well directed towards the origin but more scattered, or with MAD greater than 10, were classified as Class 1 directions. Finally, scattered directions not directed towards the origin were classified as Class 3 directions (see Table 1 for the complete set of directions). The site mean directions and dispersion parameters ($k$ and $\alpha_{95}$) were calculated following Fisher's statistics (Fisher 1953; Table 1). Site mean directions were also classified into three groups: Class 1 sites were defined for $\alpha_{95}$ lower than 10; Class 2 sites for $\alpha_{95}$ between 10 and 20; and Class 3 sites with $\alpha_{95}$ higher than 20. Class 3 sites were not analysed further.

Magnetic mineralogy experiments were performed in representative samples to characterize the ferromagnetic particles of the different sampled lithologies. These experiments were: curves of progressive acquisition of isothermal remanent

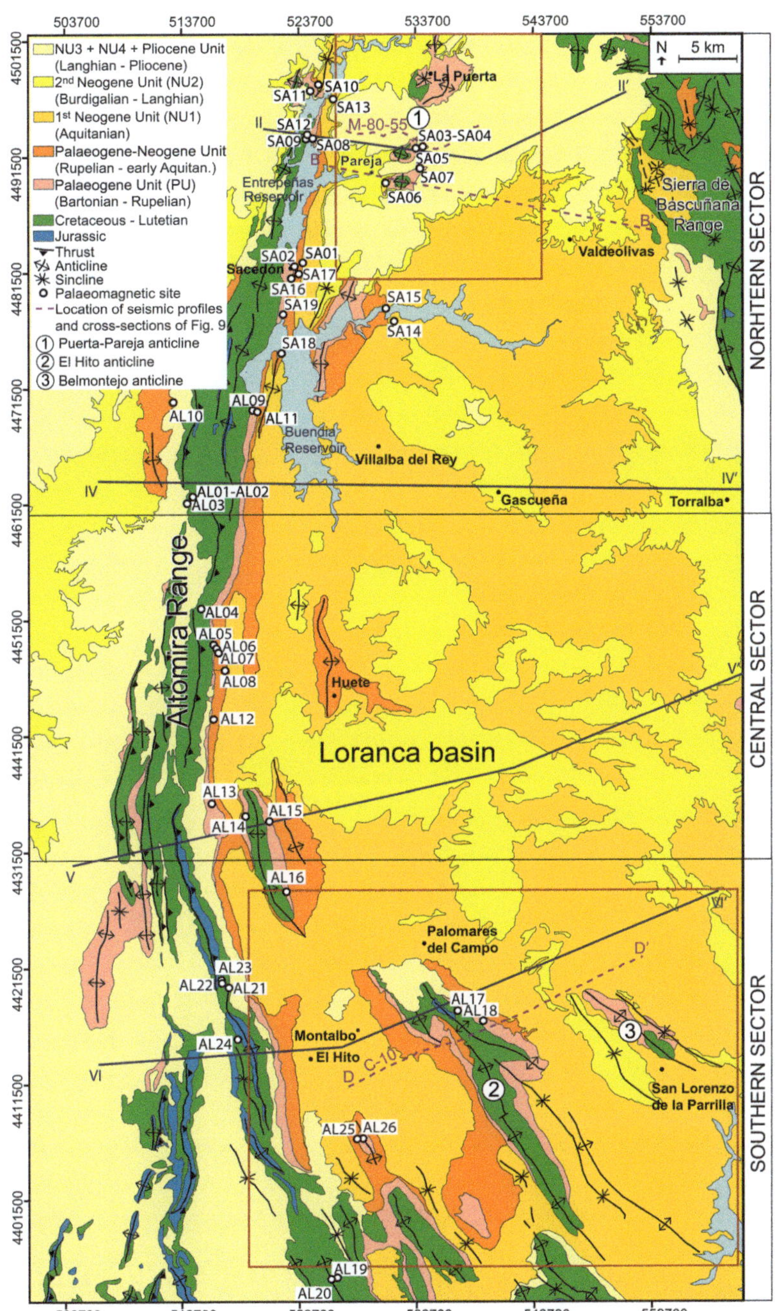


**Fig. 2.** Detailed map of the sampling sites in their geological context. Location of cross-sections II, IV, V and VI (from Muñoz-Martín & De Vicente 1998) selected for theoretical calculation of vertical-axis rotation. Red squares represent locations of Figures 7 and 8.

magnetization (IRM) up to 1.2 T using a pulse magnetizer IM10-30 (ASC Scientific) and demagnetization curves of three-axes IRM (in fields of 1.2, 0.3 and 0.1 T) according to Lowrie (1990) using IM10-30 and a thermal demagnetizer TSD-1 (Shondstedt).

**Table 1.** *Mean palaeomagnetic directions for class 1, 2 and 3 sites*

| Site | Unit | Age | No. specimens $N$ | Bedding attitude $S_0$ (dip direction/dip) | In situ | | | | Bedding-corrected | | | |
|---|---|---|---|---|---|---|---|---|---|---|---|---|
| | | | | | $D$ | $I$ | $\alpha_{95}$ | $k$ | $D$ | $I$ | $\alpha_{95}$ | $k$ |
| Class 1 and 2 ($\alpha_{95} < 20°$) | | | | | | | | | | | | |
| *Northern Sector* | | | | | | | | | | | | |
| SA14 | NU1 | Aq | 7 | 090/10 | 192 | −40 | 14 | 19.6 | 199 | −37 | 14 | 19.6 |
| SA02* | PNU | Rup-Aq | 8 | 083/34 | 0.2 | 17.0 | 13.1 | 18.8 | 6.7 | 13.2 | 12.2 | 21.7 |
| AL11 | PU | Bar-Rup | 6 | 095/49 | 342.8 | 36.4 | 17.4 | 15.8 | 22.7 | 37.2 | 17.4 | 15.8 |
| SA07 | PU | Bar-Rup | 7 | 047/20 | 9 | 54.1 | 10.9 | 31.9 | 199.9 | −36.8 | 10.9 | 31.9 |
| SA11* | PU | Bar-Rup | 9 | 075/40 | 13 | 52 | 8 | 42.2 | 38 | 24 | 8 | 42.2 |
| AL01 | Cr | Tur | 4 | 233/23 | 8.3 | 54.3 | 7.5 | 149.6 | 332.0 | 66.0 | 7.5 | 149.6 |
| *Central Sector* | | | | | | | | | | | | |
| AL07 | PNU | Rup-Aq | 5 | 030/15 | 5.2 | 56.4 | 11.4 | 45.7 | 11.3 | 42.3 | 11.4 | 45.7 |
| AL13* | PU | Bar-Rup | 7 | 030/20 | 158.0 | −23.6 | 6.3 | 91.6 | 161.5 | −14.6 | 6.5 | 88.2 |
| *Southern Sector* | | | | | | | | | | | | |
| AL25 | PNU | Rup-Aq | 4 | 250/30 | 17.5 | 29.8 | 18.2 | 26.4 | 356.9 | 44.7 | 18.2 | 26.4 |
| AL26 | PU | Bar-Rup | 8 | 250/30 | 353.1 | 27.3 | 9.7 | 33.3 | 336 | 30 | 8 | 49 |
| Class 3 ($\alpha_{95} > 20°$) | | | | | | | | | | | | |
| *Northern Sector* | | | | | | | | | | | | |
| AL10 | PNU | Rup-Aq | 4 | 085/39 | 356 | 38.6 | 34.7 | 8 | 21.8 | 27.6 | 34.7 | 8 |
| SA08 | PNU | Rup-Aq | 6 | 106/25 | 13.4 | 37.5 | 35.7 | 4.5 | 31.6 | 33.9 | 35.7 | 4.5 |
| SA15 | PNU | Rup-Aq | 5 | 000/00 | 8 | 39.4 | 52.7 | 3.1 | 7.8 | 39.4 | 52.7 | 3.1 |
| SA06 | PU | Bar-Rup | 7 | 130/74 (overturned) | 18.9 | −6.6 | 84.6 | 1.5 | 50.7 | −16.7 | 84.6 | 1.5 |
| SA12 | PU | Bar-Rup | 7 | 109/40–117/55 | 7 | 26.8 | 48.9 | 2.5 | 38.5 | 26.4 | 51.6 | 2.3 |
| SA16 | PU | Bar-Rup | 7 | 090/25 | 24.8 | 32.6 | 33.7 | 4.2 | 35.4 | 19.3 | 33.7 | 4.2 |
| SA19 | PU | Bar-Rup | 10 | 115/40 | 335.2 | 17.6 | 25 | 4.7 | 354.5 | 43.5 | 25 | 4.7 |
| *Central Sector* | | | | | | | | | | | | |
| AL08 | PNU | Rup-Aq | 7 | 088/40 | 3 | 33 | 39.2 | 3.3 | 23.4 | 20.8 | 39.2 | 3.3 |
| AL06 | PU | Bar-Rup | 7 | 124/14 | 358.7 | 20 | 29.4 | 5.2 | 4 | 27.2 | 29.4 | 5.2 |

NU1: First Neogene Unit (Aq: Aquitanian); PNU: Palaeogene–Neogene Unit (Rup-Aq: Rupelian–Aquitanian); PU: Palaeogene Unity (Bar-Rup: Bartonian–Rupelian); Cr: Cretaceous (Tur: Turonian). $D, I$: declination, incination; $\alpha_{95}$, $k$: statistical parameters (Fisher 1953). *Class 1 and 2 sites with anomalously shallow inclinations ($I < 25°$) after bedding correction, discarded for further interpretation.

## Results

The samples showed two magnetic behaviours corresponding to the two sampled lithologies (i.e. red clays/fine-grained sandstones and limestones). The characteristic remanent magnetization (ChRM) component ranged between 200–450 and 600°C for red clays and fine sandstones and between 200–380°C and 380–470°C for limestones (Fig. 3). These maximum unblocking temperatures point to hematite as the main remanence carrier for the red clays and fine sandstones and to (titano-)magnetite for limestones. This is confirmed by the thermal demagnetization of the three-axes IRM. All red clays and fine sandstones show a similar IRM acquisition behaviour (Fig. 4a, b, left); saturation is not reached in magnetic fields of 1.2 T. Thermal demagnetization of three-axes IRMs also shows the dominance of the high-coercivity phase with magnetization decaying smoothly from 0 to 675°C (Fig. 4a, b, right). This indicates that the magnetic mineralogy of these samples is controlled by hematite. Regarding the limestones, most samples show IRM acquisition curves dominated by low-coercivity minerals saturated at fields of 300–400 mT, suggesting that (titano-)magnetite is the main carrier of the magnetization (Fig. 4c, d, left). The thermal decay of the magnetization also shows the contribution of low- and medium-coercivity minerals decaying at between 200°C and 350°C, probably due to the presence of iron sulphides. However, the final decay of the magnetization occurs at c. 550°C for the low-coercivity minerals, indicating the presence of magnetite (SA03, Fig. 4c, right). Site SA07 also shows a variable contribution of a higher-coercivity phase not saturated at 1.2 T (Fig. 4d). This high-coercivity phase decays between 100 and 200°C during thermal demagnetization, suggesting the presence of goethite at this site (Fig. 4d).

After analysing the demagnetization diagrams of all samples, only samples with unambiguous ChRMs following the quality criteria mentioned in the last section and sites with low $\alpha_{95}$ angles ($<20°$) were selected. From the 12 sites sampled on Cretaceous rocks, only site AL01 shows reliable ChRMs (see Table 1). However, this result has not been considered for further interpretation due to the impossibility of constraining the age of magnetization with data from only one site. Moreover, sites with anomalously shallow inclinations ($I < 25°$) after bedding correction (SA02, SA11 and AL13; Table 1) were not analysed further.

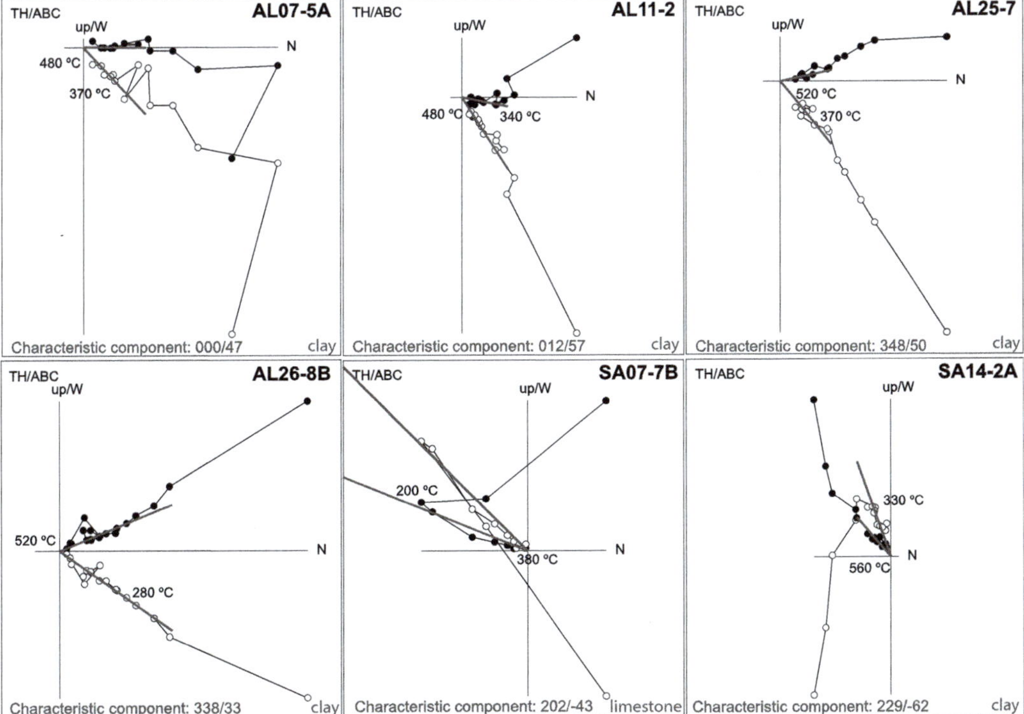

**Fig. 3.** Demagnetization diagrams of representative samples from selected sites. Black (white) dots represent projection onto the horizontal (vertical) plane. TH: thermal demagnetization; ABC: after bedding correction.

                    M. VALCÁRCEL *ET AL.*

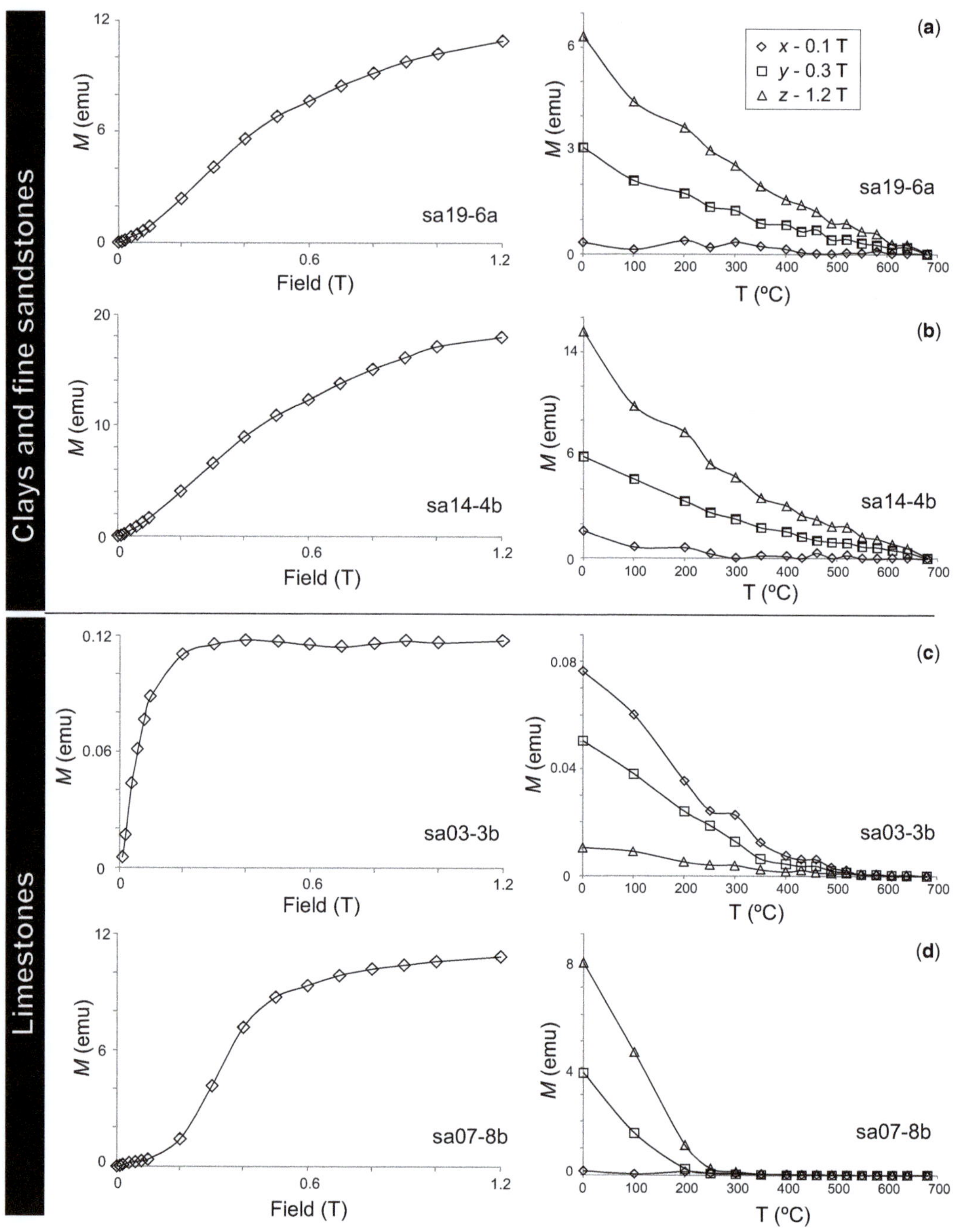



**Fig. 4.** Magnetic mineralogy experiments for two lithological groups of samples (red clays/fine-grained sandstones and limestones/marls) (IRM and thermal demagnetization of three orthogonal IRM components was induced in successive fields of 0.1 T, 0.3 T, and 1.2 T). M, magnetization; T, temperature.

It is worth noting that only 6 sites satisfied all these quality criteria (see Table 1), representing only 13% of the original sites in the dataset. In the studied area, rocks showing unsuccessful palaeomagnetic data, not analysed further because they do not satisfy the quality criteria probably due to unstable and/or very low NRM intensities are: (1) all sampled Cretaceous marls and limestones, despite one sample (AL01) being performed on Turonian marls; (2) Campanian–Lutetian sandstones belonging to the Villalba de la Sierra Formation; and (3) sandstones and sandy clays sampled both on the Palaeogene (PU; Bartonian–Rupelian) and Palaeogene–Neogene units (PNU; Rupelian–Aquitanian), whereas clays and limestones of these units yielded successful lithologies.

Samples of both lithological groups show either normal or reverse polarity of the ChRM (Table 1; Fig. 5). Before bedding correction, site-mean directions are slightly scattered with a north direction and variable inclinations. After tectonic correction, their clustering does not improve significantly but the mean directions acquire similar inclinations (34–43°; Table 1; Fig. 5). We interpret the observed dispersion in declinations after tectonic correction of site-mean directions as being caused by differential vertical-axis rotation of individual sites with respect to the reference direction for the Eocene–Miocene interval for the geographic coordinates of the study area ($D = 004°$, $I = 48°$, $\alpha_{95} = 6°$). This reference direction was calculated from the Eocene, Oligocene and Miocene reference directions from Taberner et al. (1999), Barberà et al. (1996) and Larrasoaña et al. (2006), respectively, relocated to the geographic coordinates of the study area by the pole conversion method (Noël & Batt 1990). Sites AL07 and AL25 therefore show no vertical-axis rotation considering their error intervals and that of the reference direction; sites SA07, SA14 and AL11 (all of them located at the northern sector of the studied area) display an average clockwise rotation of 17°; and site AL26, located at the southern sector, displays an anticlockwise rotation of 21° (Fig. 5c). A comparison of the strike deviations relative to declination deviations (strike test; see Eldredge et al. 1985) does not yield a conclusive result due to the scarcity of reliable data.

## Theoretical vertical-axis rotations from shortening estimates

Most vertical-axis rotations deduced in compressional scenarios are related to the differential displacement of thrust sheets and/or lateral variations of tectonic shortening in fold and thrust belts (e.g. Bates 1989; McCaig & McClelland 1992; Allerton 1998; Wilkerson et al. 2002; Bayona et al. 2003; Sussman et al. 2004, 2012; Soto et al. 2006; Muñoz et al. 2013). This implies that simple map-view models of the rotational movement of thrust sheets can quantify shortening and how it varies along strike by means of trigonometric relations as

proposed by Pueyo et al. (2004) and Oliva-Urcia & Pueyo (2007) (Fig. 6a). In a natural example with a high-resolution palaeomagnetic dataset of vertical-axis rotations, the theoretical shortening will coincide with the measured shortening from restored cross-sections only if it has been correctly estimated taking into account possible out-of-plane motions and internal deformation, which is often underestimated (Sussman et al. 2012).

In order to test this approach in the studied area and reinforce our palaeomagnetic results, a theoretical estimate of the amount of vertical-axis rotations was performed using four balanced cross-sections from Muñoz-Martín & De Vicente (1998), built perpendicular to the present-day orientation of the structures, with approximate WNW–ESE, west–east and WSW–ENE orientations (II-II', IV-IV', V-V' and VI-VI' sections, Fig. 6). From these cross-sections, we select only shortening values (ii, iv, v and vi in Fig. 6b) regarding the Altomira and Loranca structures, avoiding shortening from structures belonging to hinterland-wards Iberian structures (Sierra de Bascuñana Range, located to the east). According to Muñoz-Martín & De Vicente (1998), these shortening values are 7.2, 9.2, 14.3 and 14.3 km for the selected parts (ii, iv, v and vi, Fig. 6d) of cross-sections II, IV, V and VI, respectively. After applying the trigonometric technique by Oliva-Urcia & Pueyo (2007), the theoretical vertical-axis rotations obtained are 3.6° and 9.7° in a clockwise sense for sectors between cross-sections II and IV and IV and V, respectively (Fig. 6c). This would imply that the northern and central sectors of the Altomira–Loranca fold and thrust belt, as outlined in Figure 2, experienced clockwise vertical-axis rotations of 3.6° and 9.7°, respectively. However, the sector located between cross-sections V and VI (see Fig. 2) demonstrates an absence of vertical-axis rotations, as shortening considered from both cross-sections is the same (Fig. 6c).

## Basement-cover structural relationships

The Altomira and Loranca fold and thrust belts display an Alpine contractional thin-skinned structure detached on the Middle–Upper Triassic evaporites and clays that affect the Jurassic–lower Palaeogene sediments (Muñoz-Martín 1997). Below the Mesozoic-Cenozoic cover, available seismic data show quite a flat basement (Palaeozoic and Lower Triassic rocks) top cut by high-angle Mesozoic faults (Muñoz-Martín & De Vicente 1998). Several authors have already postulated the influence of these previous basement faults on the nucleation of the posterior Tertiary structures (Muñoz-Martín 1997; Muñoz-Martín & De Vicente 1998; Biete et al. 2012).

                    M. VALCÁRCEL *ET AL.*

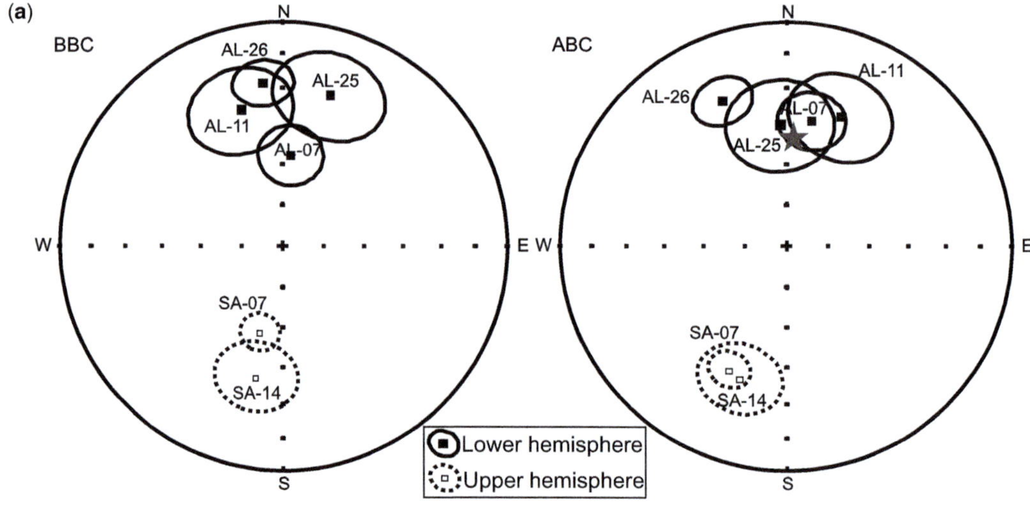


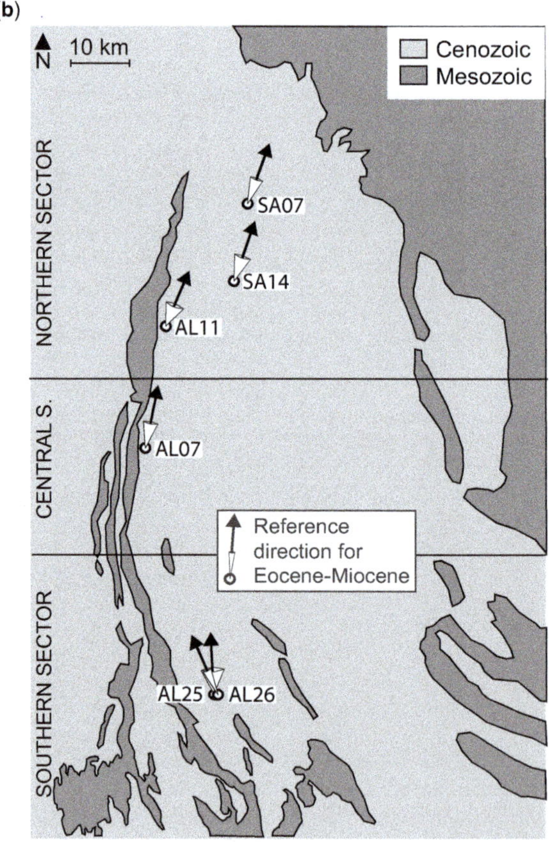


**Fig. 5.** (**a**) Mean palaeomagnetic directions before (BBC) and after (ABC) bedding correction. The star represents the reference direction for the Eocene–Miocene interval for the geographical coordinates of the study area ($D = 004$, $I = 48$, $\alpha_{95} = 6$; see text). (**b**) Geological sketch of the studied area showing palaeomagnetic data considered for further kinematic interpretation. Arrows represent the site mean declination after bedding correction and cones represent the declination error of the mean directions (Dec err $= \alpha_{95}/\cos$ Inc).

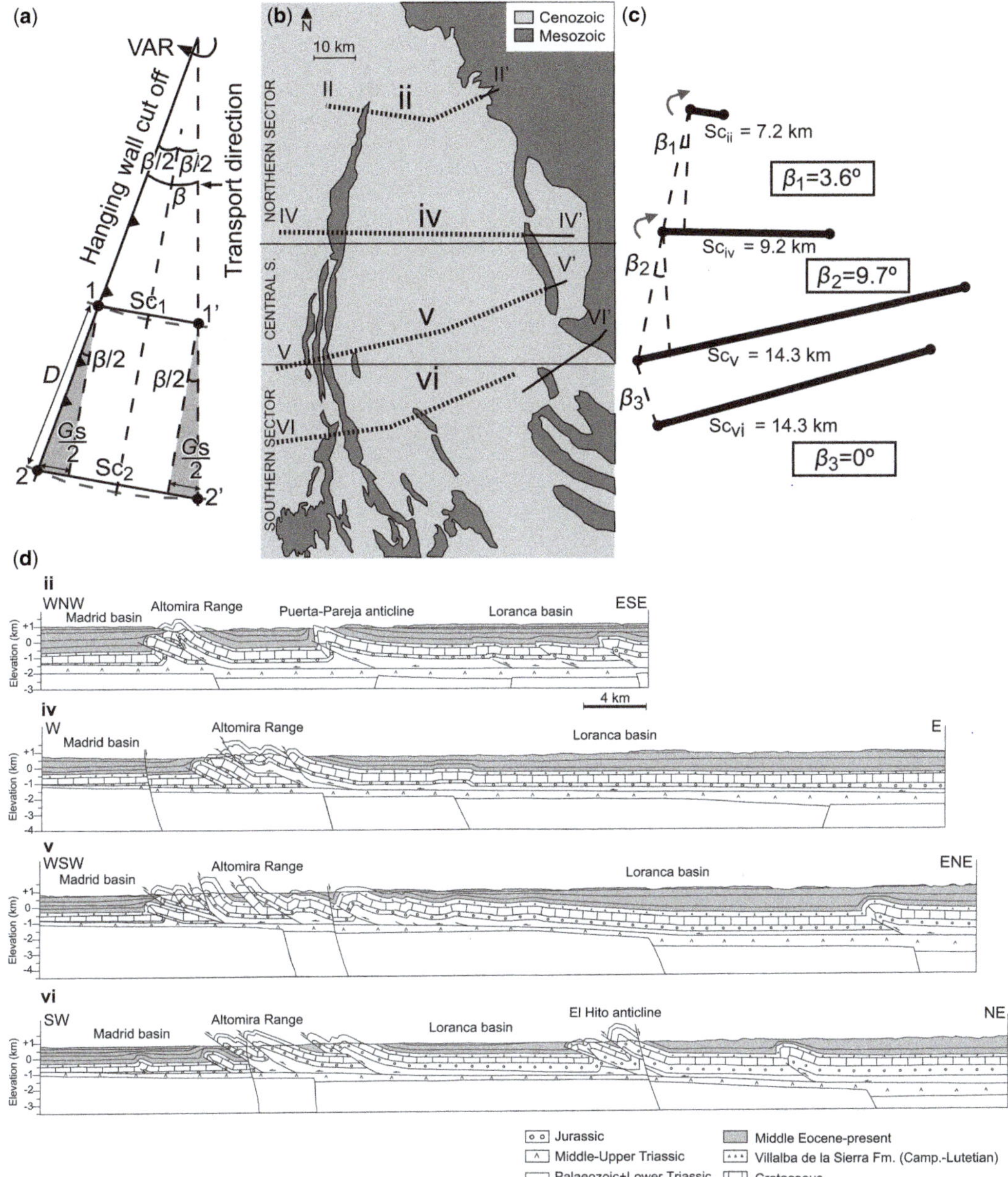



**Fig. 6.** (**a**) Trigonometric model for theoretical estimate of the vertical-axis rotations by means of gradient of shortening calculations. $G_s = (S_{c1} - S_{c2})/D = 2 \sin (\beta/2)$. Sketch modified from Oliva-Urcia & Pueyo (2007). (**b**) Location of balanced cross-sections by Muñoz-Martín & De Vicente (1998) (II-II′, IV-IV′, V-V′, V-VI′) in bold lines and parts of the sections considered in this work (ii, iv, v, vi) in dotted lines. (**c**) Application of the trigonometric technique by Oliva-Urcia & Pueyo (2007) to the Altomira–Loranca case. Estimated angles obtained through the comparison of shortenings of balanced cross-sections II v. IV ($\beta_1$), IV v. V ($\beta_2$) and V v. VI ($\beta_3$). (**d**) Cross-sections considered in this study, modified from Muñoz-Martín & De Vicente (1998).

In this work, we study basement-cover structural relationships in order to deduce possible vertical-axis rotations, assuming the decoupling of deformation produced by the Middle–Upper Triassic décollement between the basement and cover units. This study was performed by means of the detailed analysis of surface and subsurface data of two selected areas located in the northern and southern sectors of the Loranca Basin around the Puerta–Pareja and El Hito anticlines, respectively (see Fig. 2). With respect to the El Hito area in the southern sector, this work benefits from the 3D geological model performed by Biete *et al.* (2012). After seismic interpretation, the conversion of time-domain to depth-domain data for both areas was carried out by generating a velocity model considering the sonic information from available well logs. The depth-converted seismic interpretations and surface data (i.e. construction of digital terrain model and data management) were integrated into a common 3D framework using the software gOcad®. From these 3D models, maps of structural contours were created for the top of the Lower Triassic and top of the Upper Triassic (Keuper facies) for each area (Figs 7 & 8).

In the northern sector, the only major structure deforming sediments filling the Loranca Basin is the Puerta–Pareja anticline (see cross-section in Fig. 1). In this case, surface data – bedding dip measurements, bedding and fault traces (Lendínez *et al.* 1989; Torres *et al.* 1990; Gabaldón *et al.* 1999*a*; Hernaiz *et al.* 1999*a*) – and subsurface data – 14 seismic reflection profiles (GEODE 2011) and three hydrocarbon exploration wells (Torralba-1, Santa Bárbara-1 and Belmontejo-1A; Lanaja 1987; GEODE 2011) – were used to create a 3D model. The detailed seismic interpretation of the Puerta–Pareja structure reveals the presence of a velocity pull-up seismic distortion (e.g. Gadallah & Fisher 2008) under the main Puerta–Pareja thrusts in the basement (Fig. 9a, b), discarding the existence of a possible basement structure controlling the nucleation of the Puerta–Pareja structure. This effect was already suggested by Muñoz-Martín & De Vicente (1998) based on gravimetric data. The 3D model shows how the geometry of the Puerta–Pareja structure varies along-strike from a relatively simple fault-propagation fold with a total fault displacement of *c.* 500 m in the south to a more complex geometry with two main thrusts following a breakthrough style and a total of 1200 m of fault displacement at its northern part (Fig. 7). No basement-cover structural relationships can therefore be inferred from this sector of the Loranca Basin in order to detect possible vertical-axis rotations.

Below the Altomira Range in the northern sector, stratigraphic and aeromagnetic data (Ardizone *et al.* 1989) indicate the presence of a large basement fault (Sacedón fault) which controlled the Triassic and Lower Cretaceous synrift deposition (Sopeña *et al.* 1988; Peropadre & Meléndez 2004) and the nucleation of the posterior Cenozoic compressional structures (Perucha *et al.* 1995; Muñoz-Martín & De Vicente 1998; Muñoz-Martín *et al.*

1998; Van Wees *et al.* 1996). However, the poor quality of available seismic data prevents analysis of its exact geometry and relationships with the cover structures there.

In the southern sector, the 3D geological model of El Hito area was performed using surface data (bedding dip measurements and bedding and fault traces from Albert & Ferrero 1990; Diaz de Neira & Cabra 1990; Diaz Molina & Lendinez 1992; Hernaiz *et al.* 1999*b*) and subsurface data, namely 21 seismic reflection profiles (GEODE 2011) and two hydrocarbon exploration wells (El Hito and Belmontejo-1A; Lanaja 1987; GEODE 2011). The El Hito area is characterized by several NNW–SSE-trending west-verging folds and thrusts that overlie several smooth basement highs (Figs 8 & 9c, d). The origin of these basement highs is related to the inversion, probably latest Cretaceous–Paleocene in age, of Late Permian?–Early Triassic extensional faults (Biete *et al.* 2012). Taking into account the fact that inversion predated the formation of the cover Tertiary structures (i.e. formed mainly during the late Oligocene-early Miocene), these basement highs have been interpreted as mechanical perturbations controlling the nucleation of the posterior Tertiary structures (Biete *et al.* 2012). The orientation of these basement highs and the posterior Tertiary cover structures does not coincide exactly and highlights the existence of a partial decoupling between them (Fig. 8). The orientation of structures affecting the top of the basement (i.e. Lower Triassic) and the top of the Keuper facies shows an angle of 13° comparing the orientation of the main faults (see Fig. 8). Such an angle might be coherent with an anticlockwise vertical-axis rotation of the cover sediments with respect to a fixed basement in this sector. Other possible causes to explain the observed difference in the orientation of the basement and cover structures without implying vertical-axis rotation of the cover sediments could be: variation of thickness in the cover series; and/or the presence of anisotropies in the cover sediments as previous extensional faults. However, available subsurface data from this area do not show significant changes of thickness in the cover series and/or the presence of important previous extensional faults coinciding with these structures (Querol 1989).

## Discussion

### Reliability of palaeomagnetic data

Palaeomagnetism was chosen in this study as it is the only technique able to determine absolute vertical-axis rotations (e.g. Norris & Black 1961) and it is a robust technique to characterize the origin of obliquity (i.e. primary, progressive or secondary) in fold and thrust belts. However,

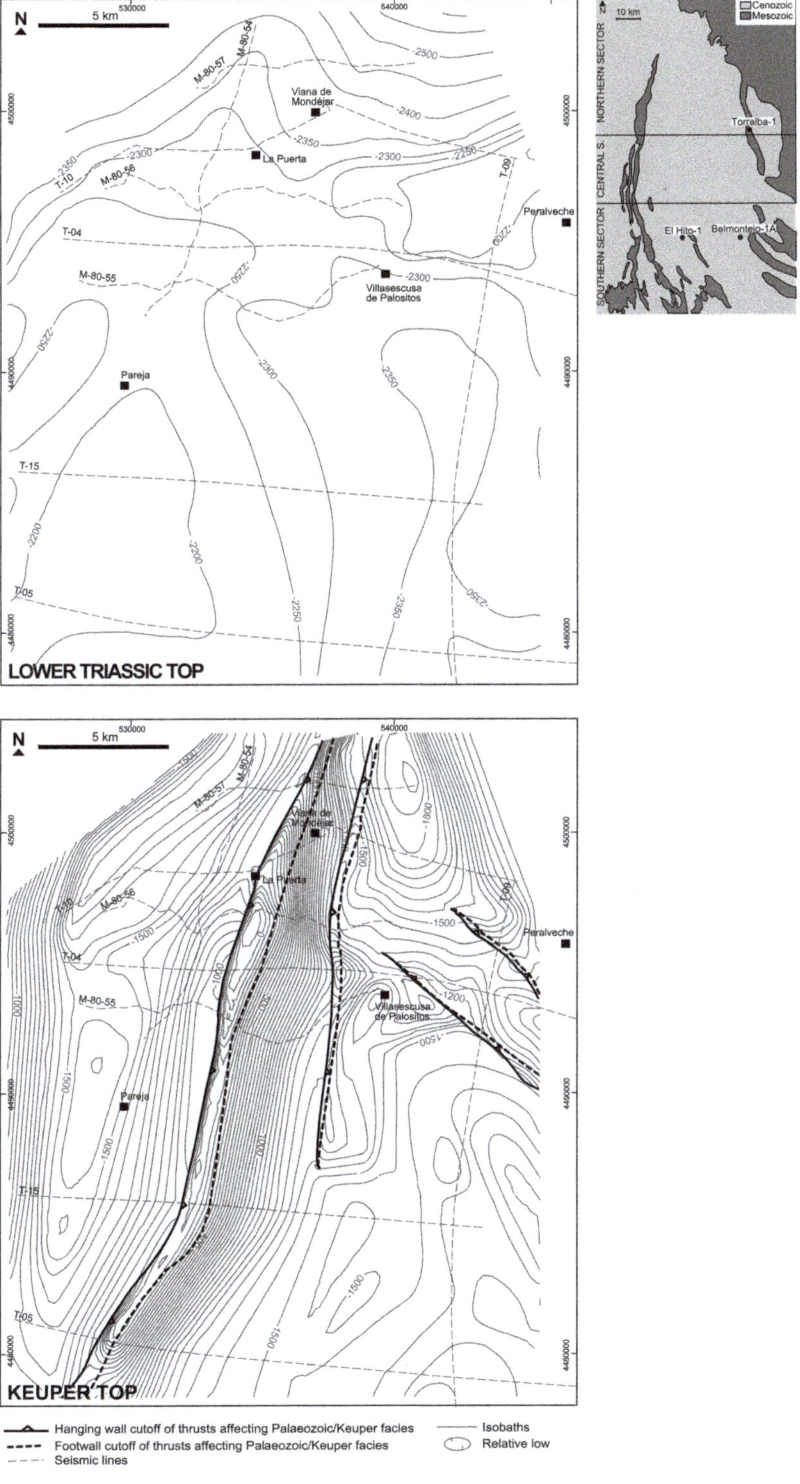



**Fig. 7.** Contour maps of the top of Palaeozoic–Lower Triassic basement and of the top of the Upper Triassic (Keuper facies) rocks in the Puerta–Pareja area. See location on Figure 2.

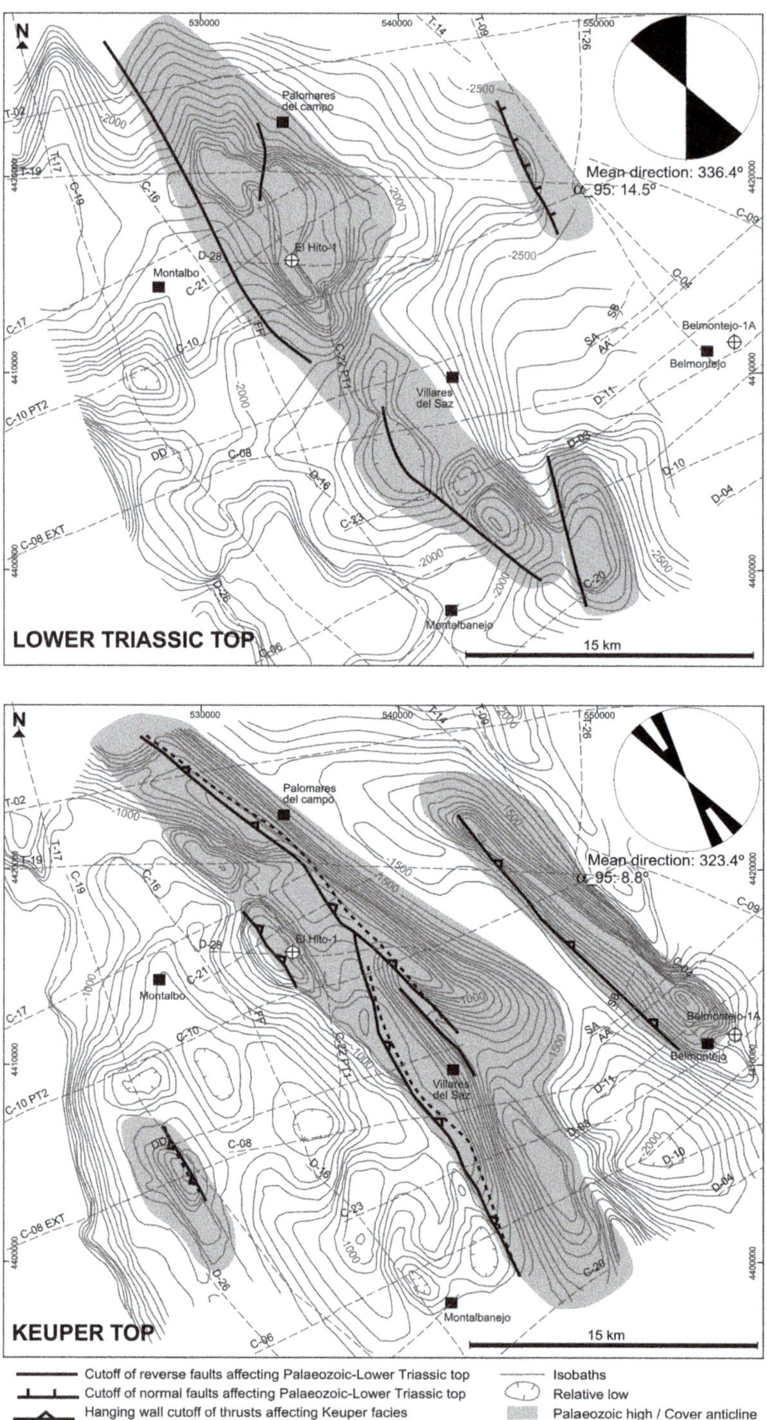



Cutoff of reverse faults affecting Palaeozoic-Lower Triassic top
Cutoff of normal faults affecting Palaeozoic-Lower Triassic top
Hanging wall cutoff of thrusts affecting Keuper facies
Footwall cutoff of thrusts affecting Keuper facies
Seismic lines
Isobaths
Relative low
Palaeozoic high / Cover anticline
Hydrocarbon exploration well

**Fig. 8.** Contour maps of the top of Palaeozoic–Lower Triassic basement and of the top of the Upper Triassic (Keuper facies) rocks in the El Hito area. Modified from Biete *et al.* (2012). See location on Figure 2. Rose diagrams represent the orientation of main faults.

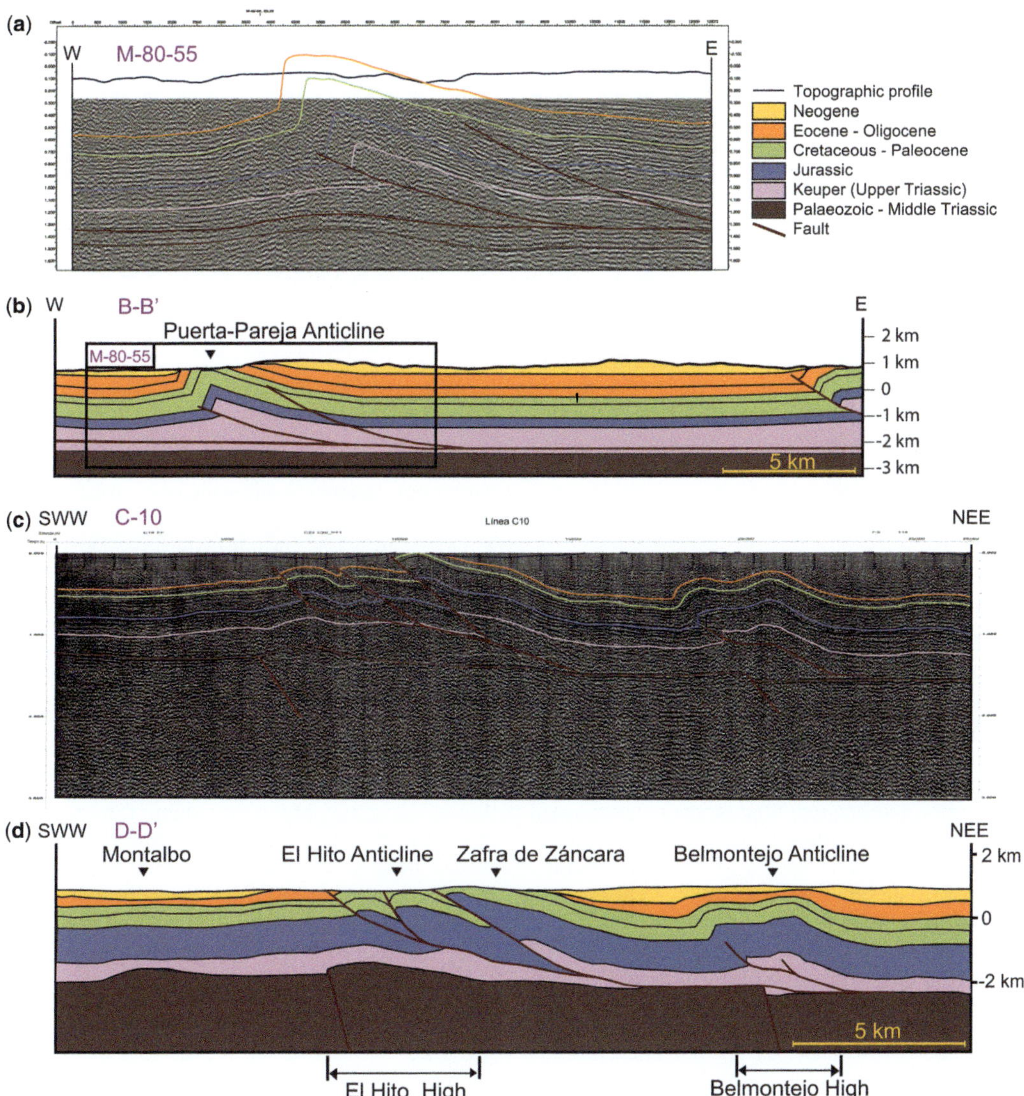


**Fig. 9.** Seismic profiles and cross-sections showing the geometry of (**a, b**) the Puerta–Pareja and (**c, d**) El Hito (modified from Biete *et al.* 2012) structures in the northern and southern sectors, respectively. The apparent antiform at the top of the Middle Triassic in (a) has been interpreted as a velocity pull-up seismic distortion.

palaeomagnetic data have several limitations and sources of errors when applying them to understand fold and thrust belt kinematics (see Van der Voo 1990; Pueyo 2010): difficulties in the correct constraining of the age of magnetization and correct isolation of magnetic components often overlapped; internal deformation affecting the palaeomagnetic vector; and/or incorrect restoration of palaeomagnetic components in non-coaxial and/or complex structures. In this work, we assume that no important deformation affected the sampled rocks and

palaeomagnetic vectors. Also, simple bed restoration of palaeomagnetic data was performed as we sampled in simple structures, avoiding periclinal terminations of folds or complex structures. However, our data still have the following two limitations: (1) a primary magnetization component is assumed; and (2) the scarcity of palaeomagnetic data that satisfy the statistics quality criteria.

A correct constraining of the age of magnetization acquisition is fundamental to obtain reliable structural interpretations. In most works, the primary magnetization components are distinguished based on the results of both the reverse and fold tests (e.g. Graham 1949; Van der Voo 1990; Weil & Van der Voo 2002). In this work we interpret the ChRM as a primary component because of: (1) the presence of normal and reverse polarity of this component; and (2) the mean directions acquire similar inclinations after the tectonic correction, despite declination scattering. On the other hand, we obtained a low number of reliable palaeomagnetic data for the calculation of vertical-axis rotations; only 6 sites from 45 can be considered, due to few suitable outcrops for sampling and the poor magnetic signal of most samples in the laboratory. However, as this work shows, the integration of other types of data with this palaeomagnetic dataset can help to characterize the origin of obliquity of the Altomira and Loranca fold and thrust belts.

## Vertical-axis rotations in the Altomira and Loranca fold and thrust belts deduced from palaeomagnetism

Although scarce, some kinematic information can be deduced from the palaeomagnetic sites. The results of the palaeomagnetic analysis show that the present-day north–south orientation of the Altomira and Loranca structures at its central portion is a primary feature, that is, they were formed with a roughly north–south orientation already oblique with respect to the NW–SE orientation of the Iberian Chain as our results indicate non-vertical-axis rotations there or minor vertical-axis rotations (site AL07, Table 1). Sites located at the northern sector show a clockwise vertical-axis rotation up to 17°. It is worth noting that, despite being sparse, reliable sites in the northern sector (SA14, AL11 and SA07, Table 1) are not clustered and restricted to a particular region but distributed all along this sector, which makes us confident about the consistency of these results. In the southern sector, only site AL26 shows an anticlockwise vertical-axis rotation up to 21° (Fig. 5 & Table 1; location in Fig. 2). Site AL25, close to site AL26 but sampled in younger rocks, does not record any vertical-axis rotation, however (Fig. 5 & Table 1; location in Fig. 2). This might help to constrain the end of the vertical-axis rotation in this sector as it occurred before the deposition of site AL25 (Rupelian–early Aquitanian in age), although more data would be necessary to confirm it as its associated uncertainty is quite high. In the northern sector, site SA14 is the youngest site recording rotation (Fig. 5 & Table 1; location in Fig. 2), which suggests that the vertical-axis rotation would be posterior to the Aquitanian.

The 17° clockwise and 21° anticlockwise vertical-axis rotations inferred for the northern and southern sectors, respectively, would imply a secondary component to account for the subtle curved geometry of the Altomira and Loranca structures in plan-view (see Fig. 2), that is, the NNE–SSW and NNW–SSE orientation of structures in the northern and southern sectors, respectively.

## Integration of basement-cover relationships, theoretical calculations and palaeomagnetic data

Kinematics of the Altomira and Loranca fold and thrust belts, mainly concerning the relative timing of formation of the different structures, has been well constrained by macro- and micro-structural data and tectonic–sedimentation relationships (Díaz Molina *et al.* 1995; Gómez *et al.* 1996; Muñoz-Martín 1997; Muñoz-Martín & De Vicente 1998; Hernaiz *et al.* 1999*a*, *b*). However, the characterization of the primary or secondary origin of its roughly north–south oblique orientation has not been analysed before by means of palaeomagnetism.

Previous works have postulated a primary origin (i.e. absence of vertical-axis rotations to explain their present-day trend) of structures of the Altomira Range and Loranca Basin based on the following features: variations in the basal detachment conditions in a roughly north–south direction; the existence of oblique normal basement faults that would have conditioned the oblique nucleation of the cover structures (Van Wees & Stephenson 1995; Muñoz-Martín & De Vicente 1998; Biete *et al.* 2012); and the influence of the double convergence of the Pyrenean–Iberian and Betic Chains added to these oblique anisotropies (Muñoz-Martín *et al.* 1998; Andeweg *et al.* 1999). Analogue and numerical modelling has proven that the localization of outer deformation fronts, such as the Altomira Range, is highly influenced by the boundary between sectors with different basal detachment conditions (e.g. Cotton & Koyi 2000; Schreurs *et al.* 2001; Bahroudi & Koyi 2003; Storti *et al.* 2007) and the nucleation of single thrusts is conditioned by pre-existing basement highs (e.g. Schedl & Wiltschko 1987). Our palaeomagnetic data confirm the absence of vertical-axis rotation to account for the north–south orientation of structures in the central sector of the Altomira Range and Loranca Basin. At this central sector, theoretical calculations of vertical-axis rotations from shortening based in balanced 2D cross-sections perpendicular to the principal structures from Muñoz-Martín & De Vicente (1998) show 9.7° of clockwise vertical-axis rotation (Fig. 6). These data would match well, as vertical-axis rotations under 10° are in the limit of the resolution of palaeomagnetic measurements. Close to this central sector southwards (between sections V–V' and VI–VI', see Fig. 2), theoretical calculations of vertical-axis rotation indicate the absence of vertical-axis rotation for the Altomira and Loranca fold and thrust belts, supporting the primary origin of the north–south orientation of structures there.

Palaeomagnetic data also indicate the secondary acquisition of the slightly curved geometry of these two fold and thrust belts by clockwise and anticlockwise vertical-axis rotations in the northern and southern sectors, respectively. In the southern sector, basement-cover relationships analysed in the Loranca Basin (i.e. El Hito area) might be coherent with an anticlockwise vertical-axis rotation of the cover sediments with respect to a fixed basement of 13° (Fig. 8). This angle matches fairly well with data of the only available palaeomagnetic site (AL26) located there (Fig. 2; Table 1) that shows 21° ($\pm 8°$) of vertical-axis rotation in the same sense. Data obtained from these two techniques would therefore indicate the same kinematic evolution in this southern sector.

With respect to the northern sector of the Loranca Basin, the absence of basement structures controlling the nucleation of the Puerta–Pareja anticline does not allow to inference of vertical-axis rotation there using this technique. Considering both the Altomira and Loranca fold and thrust belts, theoretical calculations of vertical-axis rotations yield a 3.6° clockwise vertical-axis rotation (Fig. 6). Despite a difference of 13° between the theoretical angle and that obtained from palaeomagnetic data (17°), both methods deduce clockwise vertical-axis rotations, supporting a similar kinematic model. We interpret the along-strike fault displacement differences of the Puerta–Pareja anticline in terms of accommodating the general slight clockwise vertical-axis rotation of the area.

In summary, the origin of the oblique orientation of structures of the Altomira and Loranca fold and thrust belts based on palaeomagnetic data, theoretical calculations and basement-cover relationships is primary in the central sector of the studied area (i.e. in sector with structures oriented north–south), and the origin of the small deviation structures with respect to the north–south direction is secondary at the northern and southern sectors.

## Conclusions

This work deals with the integration of palaeomagnetic data and other two datasets to characterize the primary v. secondary obliquity of structures of the Altomira and Loranca fold and thrust belts (Iberian Chain, Central Spain).

Pre- and syntectonic Cretaceous–Lower Miocene rocks cropping out along the Altomira Range and Loranca Basin were sampled for palaeomagnetism. Scarce palaeomagnetic data (only 13% of sites satisfied the quality criteria) were obtained due to few suitable outcrops for sampling and rocks with unstable and/or very low intensities of the NRM. This makes the comparison of palaeomagnetic results with other data, from (1) theoretical calculations of vertical-axis rotations that must occur to accommodate the along-strike shortening variations and (2) basement-cover relationships, necessary. The integration of these three methodologies supports the conclusion that the origin of the oblique orientation of structures of the Altomira and Loranca fold and thrust belts is primary in the central sector of the studied area (i.e. in sector with structures oriented north–south). In the northern and southern sectors, the subtle curved geometry of the Altomira and Loranca structures in planview (i.e. the NNE–SSW and NNW–SSE orientation of structures in the northern and southern sectors, respectively) is secondary. This approach highlights the importance of integrating different datasets to characterize the obliquity of fold and thrust belts.

This work was funded by the project KINESAL (CGL2010-21968-C02-02) of the Spanish Ministry of Economy and Competitiveness and a PhD-IGME grant for the first author. It is a contribution of the Geomodels Research Institute and the *Grup de Recerca de Geodinàmica i Anàlisi de Conques* (2014SGR467, supported by the *Agència de Gestió d'Ajuts Universitaris i de Recerca* and the *Secretaria d'Universitats i Recerca del Departament d'Economia i Coneixement de la Generalitat de Catalunya*). We thank the Paleomagnetic Laboratory of Barcelona (CCiTUB-ICTJA CSIC) where the palaeomagnetic analyses were carried out. Paradigm generously provided the gOcad® software license used. We are very grateful to the editor, O. Pueyo-Anchuela, A. Maestro and an anonymous reviewer for their comments.

## References

ALBERT, V. & FERRERO, E. 1990. *Hoja Geológica 1:50,000 MAGNA (San Lorenzo de la Parrilla)*. ITGE, Madrid, **634**.

ALLERTON, S. 1994. Vertical-axis rotation associated with folding and thrusting: an example from the Eastern Subbetic Zone of southern Spain. *Geology*, **22**, 1039–1042.

ALLERTON, S. 1998. Geometry and kinematics of vertical-axis rotations in fold and thrust belts. *Tectonophysics*, **299**, 15–30.

ANDEWEG, B., DE VICENTE, G., CLOETINGH, S., GINER, J. & MUÑOZ-MARTÍN, A. 1999. Local stress fields and intraplate deformation of Iberia: variations in spatial and temporal interplay of regional stress sources. *Tectonophysics*, **305**, 153–164.

ARDIZONE, J., MEZCUA, J. & SOCIAS, I. 1989. *Mapa aeromagnético de España Peninsular. Escala 1:1,000,000.* IGN, Madrid.

BATES, M. P. 1989. Palaeomagnetic evidence for rotations and deformation in the Nogueras Zone, Central Southern Pyrenees, Spain. *Journal of the Geological Society, London*, **146**, 459–476.

BAHROUDI, A. & KOYI, H. A. 2003. Effect of spatial distribution of Hormuz salt on deformation style in the Zagros fold and thrust belt: an analogue modelling approach. *Journal of Geological Society, London*, **160**, 719–733.

BARALDO, A., RAPALINI, A. E., BÖHNEL, H. & MENA, M. 2003. Paleomagnetic study of Deception Island, South Shetland Islands, Antarctica. *Geophysical Journal International*, **153**, 333–343.

BARBERÀ, X., CABRERA, L., GOMIS, E. & PARÉS, J. M. 1996. Determinación del polo paleomagnético para el límite Oligoceno-Mioceno en la Cuenca del Ebro. *Geogaceta*, **20**, 1014–1016.

BAYONA, G., THOMAS, W. A. & VAN DER VOO, R. 2003. Kinematics of thrust sheets within transverse zones: a structural and paleomagnetic investigation in the Appalachian thrust belt of Georgia and Alabama. *Journal of Structural Geology*, **25**, 1193–1212.

BIETE, C., ROCA, E. & HERNAIZ-HUERTA, P. P. 2012. The Alpine structure of the basement beneath the southern Loranca Basin and its influence in the thin-skinned contractional deformation of the overlying Mesozoic and Cenozoic cover. *Geo-Temas*, **13**, 173.

CASAS-SAINZ, A. M. & FACCENNA, C. 2001. Tertiary compressional deformation of the Iberian plate. *Terra Nova*, **13**, 281–288.

COTTON, J. T. & KOYI, H. A. 2000. Modeling of thrust fronts above ductile and frictional detachments: application to structures in the Salt Range and Potwar Plateau, Pakistan. *Geological Society American Bulletin*, **112**, 351–363.

CRADDOCK, J. P., KOPANIA, A. A. & WILTSCHKO, D. W. 1988. Interaction between the Northern Idaho-Wyoming thrust belt and bounding basement blocks, Central Western Wyoming. *Geological Society of America Memoirs*, **171**, 333–352.

DE VICENTE, G., VEGAS, R. & CASAS, A. 2004. Estructura y evolución alpina de la Cadena Ibérica. *In*: VERA, J. A. (ed.) *Geología de España.* SGE-IGME, Madrid, 525–527.

DE VICENTE, G., VEGAS, R. *ET AL.* 2009. Oblique strain partitioning and transpression on an inverted rift: the Castilian Branch of the Iberian Chain. *Tectonophysics*, **470**, 224–242.

DIAZ DE NEIRA, A. & CABRA, P. 1990. *Hoja Geológica 1:50,000 MAGNA (Valverde de Júcar).* ITGE, Madrid, **662**.

DIAZ MOLINA, M. & LENDINEZ, A. 1992. *Hoja Geológica 1:50,000 MAGNA (Palomares del Campo).* ITGE, Madrid, **633**.

DÍAZ MOLINA, M., ARRIBAS, J., GÓMEZ, J. J. & TORTOSA, A. 1995. Geological modelling of a reservoir analogue: cenozoic meander belts, Loranca Basin, Spain. *Petroleum Geoscience*, **1**, 43–48.

ELDREDGE, S., BACHTADSE, V. & VAN DER VOO, R. 1985. Paleomagnetism and the orocline hypothesis. *Tectonophysics*, **119**, 153–179.

FISHER, R. A. 1953. Dispersion on a sphere. *Proceedings of the Royal Society of London*, **A217**, 295–305.

FREUND, R. 1970. Rotation of strike slip faults in Sistan, southeast Iran. *Journal of Geology*, **78**, 188–200.

GABALDÓN, V., DÍAZ, J. A., CABRA, P., PORTERO, J. M. & DEL OLMO, P. 1999a. *Hoja Geológica 1:50,000 MAGNA (Sacedón).* ITGE, Madrid, **562**.

GABALDÓN, V., LENDÍNEZ, A., MUÑOZ, J. L., OLMO, P. & PORTERO, J. M. 1999b. *Hoja Geológica 1:50,000 MAGNA (Almonacid de Zorita).* ITGE, Madrid, **585**.

GABALDÓN, V., LENDÍNEZ, A., DÍAZ MOLINA, M., OLMO, P. & PORTERO, J. M. 1999c. *Hoja Geológica 1:50,000 MAGNA (Huete).* ITGE, Madrid, **608**.

GABALDÓN, V., LENDÍNEZ, A. & DÍAZ MOLINA, M. 1999d. *Hoja Geológica 1:50,000 MAGNA (Palomares del Campo).* ITGE, Madrid, **633**.

GADALLAH, M. R. & FISHER, R. 2008. *Exploration Geophysics.* Springer Science & Business Media, Berlin.

GEODE 2011. Mapa Geológico Digital continuo de España [on-line]. *In*: NAVAS, J. (ed) *Sistema de Información Geológica Continua.* SIGECO, IGME, Madrid, http://cuarzo.igme.es/sigeco/default.htm

GÓMEZ, J. J., DÍAZ MOLINA, M. & LENDÍNEZ, A. 1996. Tectonosedimentary analysis of the Loranca Basin (Upper Oligocene-Miocene, central Spain): a 'non sequenced' tforeland basin. *In*: FRIEND, P. F. & DABRIO, C. J. (eds) *Tertiary basins of Spain.* Cambridge University Press, Cambridge, 285–294.

GRAHAM, J. W. 1949. The stability and significance of magnetism in sedimentary rocks. *Journal of Geophysics Research*, **54**, 131–167.

GUIMERÀ, J. 1988. *Estudi estructural de l'enllaç entre la Serralada Ibèrica i la Serralada Costanera Catalana.* PhD thesis, University of Barcelona, Spain.

GUIMERÀ, J., MAS, R. & ALONSO, A. 2004. Intraplate deformation in the NW Iberian Chain: Mesozoic extension and Tertiary contractional inversion. *Journal of the Geological Society, London*, **161**, 291–303.

HERNAIZ, P. P., CABRA, P., SOLÉ, J., PORTERO, J. & OLMO, P. 1999a. *Hoja Geológica 1:50,000 MAGNA (Auñón).* ITGE, Madrid, **537**.

HERNAIZ, P. P., CABRA, P. & SOLÉ, J. 1999b. *Hoja Geológica 1:50,000 MAGNA (Villarejo de Fuentes).* ITGE, Madrid, **661**.

ITGE 1990. *Documentos sobre la Geología del Subsuelo de España. Tomo III (Madrid - Depresión Intermedia).* ITGE, Madrid.

KOLLMEIER, J. M., VAN DER PLUIJM, B. A. & VAN DER VOO, R. 2000. Analysis of Variscan dynamics; early bending of the Cantabrian-Asturias Arc, northern Spain. *Earth and Planetary Science Letters*, **181**, 203–216.

KNON, S. & MITRA, G. 2004. Strain distribution, strain history, and kinematic evolution associated with the formation of arcuate salients in fold-thrust belts: the example of the Provo Salient, Sevier orogen, Utah. *In*: SUSSMAN, A. J. & WEIL, A. B. (eds) *Orogenic Curvature: Integrating Paleomagnetic and Structural Analyses.* Geological Society of America, Boulder, Special Publications, **383**, 205–224.

LANAJA, J. M. 1987. *Contribución de la Exploración Petrolífera al Conocimiento de la Geología de España.* Instituto Geológico y Minero de España, Madrid.

LARRASOAÑA, J. C., MURELAGA, X. & GARCÉS, M. 2006. Magnetobiochronology of Lower Miocene (Ramblian) continental sediments from the Tudela Formation (western Ebro basin, Spain). *Earth and Planetary Science Letters*, **243**, 3–4.

LENDÍNEZ, A., TENA, M. & GABALDÓN, V. 1989. *Hoja Geológica 1:50,000 MAGNA (Valdeolivas)*. ITGE, Madrid, **538**.

LIESA, C. L. & SIMÓN, J. L. 2007. A probabilistic approach for identifying independent remote compressions in an intraplate region: the Iberian Chain (Spain). *Mathematical Geology*, **39**, 337–348.

LOWRIE, W. 1990. Identification of ferromagnetic minerals in a rock by coercivity and unblocking temperature properties. *Geophysical Research Letters*, **17**, 159–162.

McCAIG, A. M. & McCLELLAND, E. 1992. Palaeomagnetic techniques applied to thrust belts. *In*: McCLAY, K. R. (ed.) *Thrust Tectonics*. Chapman & Hall, London, 209–216.

MITRA, G. & YONKEE, W. A. 1985. Spaced cleavage and its relationship to folds and thrusts in the Idaho-Utah-Wyoming thrust belt of the Rocky Mountain Cordilleras. *Journal of Structural Geology*, **7**, 361–373.

MUÑOZ, J. A., BEAMUD, E., FERNÁNDEZ, O., ARBUÉS, P., DINARÈS-TURELL, J. & POBLET, J. 2013. The Ainsa Fold and thrust oblique zone of the central Pyrenees: kinematics of a curved contractional system from paleomagnetic and structural data. *Tectonics*, **32**, 1142–1175.

MUÑOZ-MARTÍN, A. 1997. *Evolución geodinámica del borde oriental de la cuenca del Tajo desde el Oligoceno hasta la actualidad*. PhD thesis, Universidad Complutense de Madrid, Spain.

MUÑOZ-MARTÍN, A. & DE VICENTE, G. 1998. Cuantificación del acortamiento alpino y estructura en profundidad del extremo sur-occidental de la Cordillera Ibérica (Sierras de Altomira y Bascuñana). *Revista de la Sociedad Geológica de España*, **11**, 39–58.

MUÑOZ-MARTÍN, A., CLOETINGH, S., DE VICENTE, G. & ANDEWEG, B. 1998. Finite-element modelling of Tertiary paleostress fields in the Eastern part of the Tajo Basin (central Spain). *Tectonophysics*, **300**, 47–62.

NOËL, M. & BATT, C. M. 1990. A method for correcting geographically separated remanence directions for the purpose of archaeomagnetic dating. *Geophysics Journal International*, **102**, 753–756.

NORRIS, D. K. & BLACK, R. F. 1961. Application of palaeomagnetism to thrust mechanics. *Nature*, **192**, 933–935.

NUR, A., RON, H. & SCOTTI, O. 1986. Fault mechanics and the kinematics of block rotations. *Geology*, **14**, 746–749.

OLIVA-URCIA, B. & PUEYO, E. L. 2007. Gradient of shortening and vertical-axis rotations in the Southern Pyrenees (Spain), insights from a synthesis of paleomagnetic data. *Revista de la Sociedad Geológica de España*, **20**, 105–118.

PEROPADRE, C. & MELÉNDEZ, N. 2004. Las facies continentales del Cretácico Inferior en la zona meridional de la Sierra de Altomira (Cordillera Ibérica): estratigrafía, sedimentología y discusión sobre su correlación regional. *Revista de la Sociedad Geológica de España*, **17**, 1–2.

PERUCHA, M. A., MUÑOZ-MARTÍN, A., TEJERO, R. & BERGAMÍN, J. F. 1995. Estudio de una transversal entre la Cuenca de Madrid y la Cordillera Ibérica a partir de datos estructurales, sísmicos y gravimétricos. *Geogaceta*, **18**, 15–18.

PUEYO, E. L. 2010. Evaluating the paleomagnetic reliability in fold and thrust belt studies. *Trabajos de Geología*, **30**, 145–154.

PUEYO, E. L., POCOVÍ, A., MILLÁN, H. & SUSSMAN, A. J. 2004. Map-view models for correcting and calculating shortening estimates in rotated thrust fronts using paleomagnetic data. *In*: SUSSMAN, A. J. & WEIL, A. B. (eds) *Orogenic Curvature: Integrating Paleomagnetic and Structural Analyses*. Geological Society of America, Boulder, Special Publications, **383**, 57–71.

PUIGDEFÁBREGAS, C. 1975. La sedimentación molásica en la cuenca de Jaca. *Pirineos*, **104**, 188.

QUEROL, R. 1989. *Geología del subsuelo de la Cuenca del Tajo*. Escuela Técnica Superior de Ingenieros de Minas de Madrid, Madrid, **48**.

ROSENBAUM, G., LISTER, G. S. & DUBOZ, C. 2002. Relative motions of Africa, Iberia and Europe during alpine orogeny. *Tectonophysics*, **359**, 117–129.

SCHEDL, A. & WILTSCHKO, D. V. 1987. Possible effects of pre-existing basement topography on thrust fault ramping. *Journal of Structural Geology*, **9**, 1029–1037.

SCHREURS, G., HÄNNI, R. & VOCK, P. 2001. Four-dimensional analysis of analog models: experiments on transfer zones in fold and thrust belts. *In*: RAMBERG, H., KOYI, H. A. & MANCKTELOW, N. S. (eds) *Tectonic Modeling: A Volume in Honor of Hans Ramberg*. Geological Society of America, Memoir, **193**, 179–190.

SOPEÑA, A., LÓPEZ, J., ARCHE, A., PEREZ-ARLUCEA, M., RAMOS, A., VIRGILI, C. & HERNANDO, S. 1988. Permian and Triassic rift basins of the Iberian Peninsula. *In*: MANSPEIZER, W. (ed.) *Triassic-Jurassic rifting; Continental Breakup and the Origin of the Atlantic Ocean and Passive Margins*. Elsevier, Amsterdam, Developments in Geotectonics, **22**, 757–786.

SOTO, R., CASAS, A. M., STORTI, F. & FACCENNA, C. 2002. Role of lateral thickness variation on the development of oblique structures at the Western end of the South Pyrenean Central Unit. *Tectonophysics*, **350**, 215–235.

SOTO, R., STORTI, F., CASAS, A. M. & FACCENNA, C. 2003. Influence of along-strike pre-orogenic sedimentary tapering on the internal architecture of experimental thrust wedges. *Geological Magazine*, **140**, 253–264.

SOTO, R., CASAS-SAINZ, A. M. & PUEYO, E. L. 2006. Along-strike variation of orogenic wedges associated with vertical-axis rotations. *Journal of Geophysical Research (Solid Earth)*, **111**, 1–22, http://doi.org/10.1029/2005JB004201

STORTI, F., SOTO MARÍN, R., ROSSETTI, F. & CASAS SAINZ, A. M. 2007. Evolution of experimental thrust wedges accreted from along-strike tapered, silicone-floored multilayers. *Journal of the Geological Society, London*, **164**, 73–85.

SUÁREZ-ALBA, J. 2007. La Mancha Triassic and Lower Lias Stratigraphy, a well log interpretation. *Journal of Iberian Geology*, **33**, 55–78.

SUSSMAN, A. J. & WEIL, A. B. (eds) 2004. *Orogenic Curvature: Integrating Paleomagnetic and Structural Analyses.* Geological Society of America, Special Paper, **383**.

SUSSMAN, A. J., BUTLER, R. F., DINARÈS-TURELL, J. & VERGÉS, J. 2004. Vertical-axis rotation of a foreland fold and implications for orogenic curvature: an example from the Southern Pyrenees, Spain. *Earth and Planetary Science Letters*, **218**, 435–449.

SUSSMAN, A. J., PUEYO, E. L., CHASE, C. G., MITRA, G. & WEIL, A. B. 2012. The impact of vertical-axis rotations on shortening estimates. *Lithosphere*, **4**, 383–394.

TABERNER, C., DINARÈS-TURELL, J., GIMÈNEZ, J. & DOCHERTY, C. 1999. Basin infill architecture and evolution from magnetostratigraphic cross-basin correlations in the southeastern Pyrenean foreland basin. *Bulletin of the Geological Society of America*, **111**, 1155–1174.

TORRES, T., PORTERO, J., DEL OLMO, P. & ELIZAGA, E. 1990. *Hoja Geológica 1:50,000 MAGNA (Priego).* ITGE, Madrid, **563**.

TORRES, T., ORTIZ, J. E. & ARRIBAS, I. 2006. El anticlinal y las discordancias de Pareja (Guadalajara): definición de las unidades cenozoicas de la Depresión Intermedia (provs. Cuenca y Guadalajara, España). *Estudios Geológicos*, **62**, 89–102.

VAN DER VOO, R. 1990. The reliability of paleomagnetic data. *In*: VAN DER VOO, R. & SCHMIDT, P. W. (eds) *Reliability of Paleomagnetic Data.* Tectonophysics, **184**, 1–9.

VAN WEES, J. D. & STEPHENSON, R. A. 1995. Quantitative modelling of basin and rheological evolution of the Iberian Basin (Central Spain): implications for lithospheric dynamics of intraplate extension and inversion. *Tectonophysics*, **252**, 163–178.

VAN WEES, J. D., CLOETINGH, S. & DE VICENTE, G. 1996. The role of pre-existing faults in basin evolution: constraints from 2D finite element and 3D flexure models. *In*: BUCHANAN, P. G. & NIEUWLAND, D. A. (eds) *Modern Developments in Structural Interpretaion, Validation and Modelling.* Geological Society, London, Special Publications, **99**, 297–320.

VIDAL-ROYO, O., KOYI, H. A. & MUÑOZ, J. A. 2009. Formation of orogen-perpendicular thrusts due to mechanical contrasts in the basal décollement in the Central External Sierras (Southern Pyrenees, Spain). *Journal of Structural Geology*, **31**, 523–539.

WEIL, A. B. & SUSSMAN, A. J. 2004. Classifying curved orogens based on timing relationships between structural development and vertical-axis rotations. *In*: SUSSMAN, A. J. & WEIL, A. B. (eds) *Orogenic Curvature: Integrating Paleomagnetic and Structural Analyses.* Geological Society of America, Boulder, Special Publications, **383**, 1–15.

WEIL, A. B. & VAN DER VOO, R. 2002. The evolution of the paleomagnetic fold test as applied to complex geologic situations, illustrated by a case study from northern Spain. *Physics and Chemistry of the Earth*, **27**, 1223–1235.

WILKERSON, M. S., APOTRIA, T. & FARID, T. 2002. Interpreting the geologic map expression of contractional fault-related folds terminations: lateral/oblique ramps v. displacement gradients. *Journal of Structural Geology*, **24**, 593–607.

# Evidence of Late Cretaceous oroclinal bending in north-central Anatolia: palaeomagnetic results from Mesozoic and Cenozoic rocks along the İzmir–Ankara–Erzincan Suture Zone

MUALLA CENGIZ CINKU[1]*, MUMTAZ HISARLI[1], ANN M. HIRT[2], FRIEDRICH HELLER[2], TIMUR USTAÖMER[3], NURCAN KAYA[1], ERDINÇ ÖKSÜM[1] & NACI ORBAY[1]

[1]Istanbul University, Faculty of Engineering, Department of Geophysical Engineering, 34320 Avcılar Istanbul, Turkey

[2]Institute of Geophysics, Department of Earth Sciences, Sonneggstrasse 5, ETH Zurich, 8092 Zurich, Switzerland

[3]Istanbul University, Faculty of Engineering, Department of Geological Engineering, 34320 Avcılar Istanbul, Turkey

*Corresponding author (e-mail: mualla@istanbul.edu.tr)

**Abstract:** The Sakarya Zone and the Kırşehir Block of northern Turkey are separated by the İzmir–Ankara–Erzincan Suture (IAES) Zone which is the remnant of the northern branch of the Neotethys Ocean. During the closure of the IAES in the Late Cretaceous, northwards drift of the Kırşehir Block and its eventual indentation into the Sakarya Zone produced crustal deformation defined by thrusts and reverse faults, mainly between the indenting Kırşehir Block and the Sakarya Zone. Previous palaeomagnetic studies in the eastern part of the Pontides and the Sakarya Zone showed that palaeomagnetic declinations could record the deformation that resulted in the curvature of the IAES. In order to define the tectonic deformation of the northern part of the Kırşehir Block, we present new palaeomagnetic data from 57 different sites that include Mesozoic–Cenozoic sedimentary and volcanic rocks. The results from Late Cretaceous rocks (40 sites) indicate that large clockwise rotations of *c.* 140–165° occurred in the eastern limb of the bend, while anticlockwise rotations progressively decreased from *c.* 80° to 55° from SW to NW in the western limb of the bend. In contrast, small clockwise and anticlockwise rotations are observed in the flat-lying segment of the suture zone. These rotation patterns are consistent with the geometrical trends of the IAES in northern Turkey. Declinations of seven different Middle Eocene sites within the Kırşehir Block are rotated anticlockwise by *c.* 30–10°. This indicates that the deformation in the Sakarya Zone and the Kırşehir Block continued in the Middle Eocene.

The main tectonic domains of Turkey consist of several assemblages of microcontinents represented by the Pontides (i.e. the Istranca Massif, Istanbul Zone and Sakarya Zone), the Niğde–Kırşehir Massif or Kırşehir Block (Robertson & Dixon 1984) (synonymous with the Central Anatolian Crystalline Complex or CACC; Akıman *et al.* 1993), the Anatolide–Tauride Block and the Arabian Platform. They are confined by different suture zones which resulted from the closure of different branches of the Neotethys Ocean during Late Cretaceous– Early Cenozoic time (Şengör & Yılmaz 1981; Şengör *et al.* 1982; Görür *et al.* 1984; Okay 1989; Okay *et al.* 1994; Yılmaz *et al.* 1997; Robertson 2002; Fig. 1). The Pontides are assumed to have been part of Eurasia since the Triassic, while the other continental pieces were detached from Gondwana in the Triassic and carried northwards during rifting and subsequent spreading of the different branches of the Neotethyan Ocean.

The northern Neotethys oceanic basin was progressively closed by northwards subduction during the Late Mesozoic and Cenozoic. The remnants of the northern branch of the Neotethys are collectively known as the İzmir–Ankara–Erzincan Suture Zone. This zone separates the Pontides in the north from the many units of the Anatolide–Tauride blocks and the Kırşehir Block in the south. It is sharply folded into an 'omega' shape between the Kırşehir Block and the Pontides (Kaymakçı *et al.* 2003*a*). This shape has been explained by the northwards drift of the Kırşehir Block and its eventual indentation into the Pontides continent (Kaymakçı *et al.* 2003*a, b*; Çinku *et al.* 2011). As a result of this collision ophiolitic slivers, melanges and ensimatic volcanic arc rocks, which formed the İzmir–Ankara–Erzincan Suture between these continental fragments, were imbricated with metamorphic rocks of the Kırşehir Block basement (Şengör & Yılmaz 1981; Tüysüz

*From*: Pueyo, E. L., Cifelli, F., Sussman, A. J. & Oliva-Urcia, B. (eds) 2016. *Palaeomagnetism in Fold and Thrust Belts: New Perspectives*. Geological Society, London, Special Publications, **425**, 189–212.
First published online August 3, 2015*, updated August 14, 2015, http://doi.org/10.1144/SP425.2

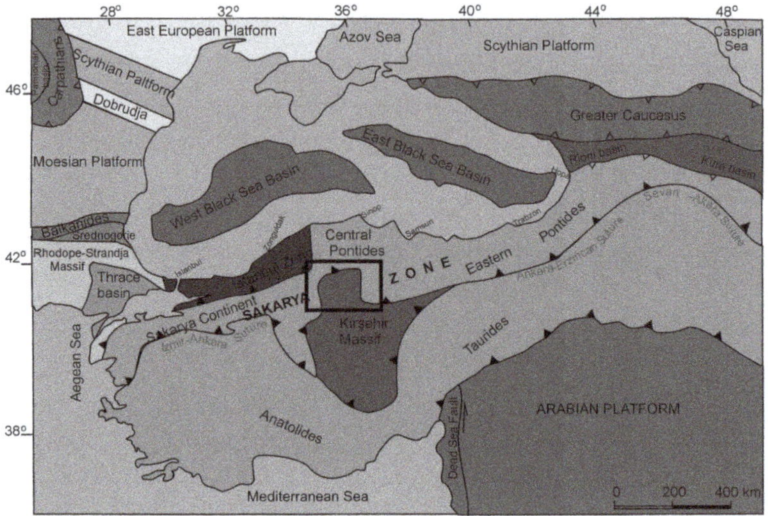

**Fig. 1.** Main tectonic units of Anatolia and surroundings (after Okay & Tüysüz 1999). The outlined area shows the location of the sampling sites.

& Dellaloğlu 1992; Yılmaz *et al.* 1993; Tüysüz *et al.* 1995; Kaymakçı *et al.* 2000, 2003*a*, *b*; Rice *et al.* 2006, 2009).

Recent systematic palaeomagnetic studies of Late Cretaceous–Middle Eocene rocks in the Sakarya Zone of the Pontides showed that palaeomagnetic declinations could evaluate the rotational component of deformation of the curvature in the Central Pontides (Meijers *et al.* 2010). Çinku *et al.* (2011) showed that the palaeomagnetically derived senses of rotations of Middle Eocene rocks in the eastern part of the North Central Anatolian region are compatible with the thrust sheet displacements that developed as a result of the indentation of the Kırşehir Block into the Sakarya Zone. Palaeomagnetic declinations in the eastern limb of the Kırşehir Block are rotated clockwise while in the area between the southern part of the Kırşehir Block and the northern part of the ophiolites, anticlockwise rotations occur depending on the orientation of the thrusts. In the western limb and inside the Kırşehir Block, however, reliable palaeomagnetic studies are missing and results from rocks of Upper Cretaceous age are lacking.

In this study we present new palaeomagnetic data from Upper Cretaceous, Middle Eocene and Miocene volcanic–volcanoclastic rocks and sedimentary rocks from the northern borders of the Kırşehir Block. The new results provide insights into the indentation process of the Kırşehir Block into the Sakarya Zone by constraining the time of deformation.

## Regional geology and palaeomagnetic sampling sites

The geological features framing our study area include major fragments of the Kırşehir Block, the Central Pontides and the Sakarya Zone. Several subunits have been recognized, namely the Karakaya Unit (Triassic), the Sakarya Unit (Liassic–Cretaceous), the Kalecik Unit (Upper Cretaceous), the İskilip Unit (Eocene–Oligocene) and the Çankırı Unit (Miocene) (Tüysüz *et al.* 1995).

The Kırşehir Block (Robertson & Dixon 1984), also called Niğde Kırşehir Massif (Robertson *et al.* 2009), forms a triangular continental fragment that occupies a large part of central Turkey. It is composed of a Mesozoic–Palaeozoic metamorphic basement overlying MORB and supra-subduction zone ophiolitic fragments which are interpreted to have been obducted southwards onto the Kırşehir Block from the north. This assemblage was then intruded by a series of Upper Cretaceous intrusive granitoid rocks (Robertson & Dixon 1984; Göncüoğlu *et al.* 1991; Boztuğ 2000; Kadıoğlu *et al.* 2003, 2006; İlbeyli *et al.* 2004; Robertson *et al.* 2009; Fig. 2). A Late Cretaceous age ($<100$ Ma) is commonly assigned to the metamorphism and intrusion of the Kırşehir Block (Göncüoğlu *et al.* 1993; Whitney & Hamilton 2004; Peters, 2010). At the northern and western margin of the Kırşehir Block, several basins developed during regional plate convergence during Upper Cretaceous–Middle Eocene time (the Central Anatolian Basins; Görür *et al.* 1984). The basin sequences include Upper

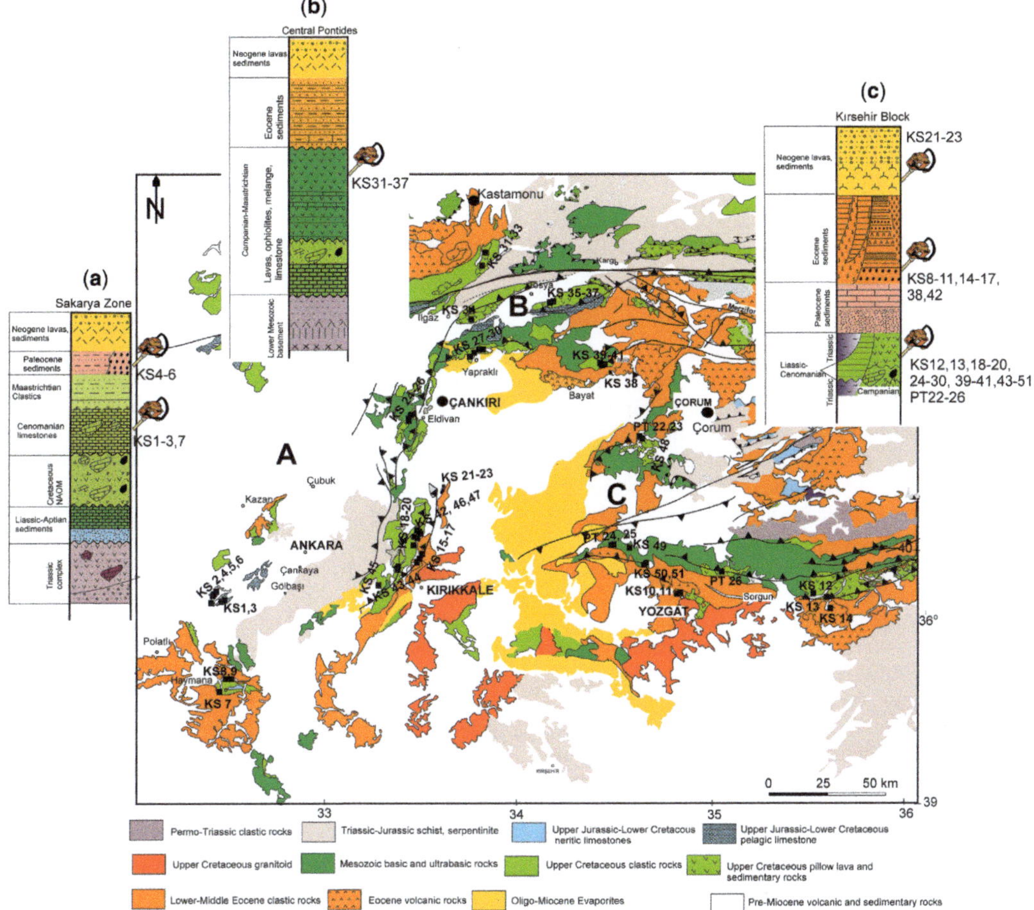



**Fig. 2.** Geological map and modifed stratigraphic section with numbered sampling site locations for (**a**) Sakarya Zone (after Rojay & Süzen 1997); (**b**) Central Pontides (after Rice *et al.* 2006); and (**c**) Kırşehir Block (after Tüysüz *et al.* 1995). Geological map modified after MTA, 1:500 000 geological map.

Cretaceous–Pliocene sedimentary as well as volcanic rocks (e.g. Koçyiğit 1991; Poisson *et al.* 1996; Görür *et al.* 1998; Clark & Robertson 2002, 2005).

The central Pontides, situated between the Kırşehir Block to the south and the Black Sea to the north, are made up of Palaeotethyan and Neotethyan accretionary subduction complexes, which are unconformably overlain by Upper Jurassic–Oligocene sedimentary and volcanic rocks. A magmatic arc and a mélange complex were formed during the Late Cretaceous as a result of subduction of the northern Neotethys Ocean (Yoldaş 1982; Tüysüz *et al.* 1995; Fig. 2).

The strongly deformed and locally metamorphosed Permo-Triassic basement of the Sakarya Zone is named the Karakaya Complex (Bingöl *et al.* 1975; Ustaömer & Robertson 1997; Okay & Göncüoğlu 2004). Liassic–Lower Cretaceous passive

margin sediments and Late Cretaceous volcanic-arc rocks unconformably overlie the Permo-Triassic basement. An accretionary complex indicating the former position of the İzmir–Ankara–Erzincan ocean is known as the North Anatolian Ophiolitic Melange (NAOM; Rojay 1995). It is composed of an Upper Triassic–Lower Jurassic metamorphic melange (Pickett & Robertson 1996), a Permo-Triassic limestone melange (Pickett & Robertson 1996) and Upper Cretaceous accretionary complex (Okay *et al.* 2006) in the margins of the Çankırı Basin (Fig. 2), and separates the Kırşehir Block from the Sakarya Zone. The Upper Cretaceous ophiolitic melange is composed of thrust sheets of mantle peridotites, basalts, dykes, cherts, sandstones and limestones (Norman 1984; Dilek & Thy 2006) and are either intruded by Late Cretaceous–Early Cenozoic granitic rocks in the south of the Çankırı

Basin (İlbeyli *et al.* 2004; Boztuğ & Jonckheere 2007; Kaymakçı *et al.* 2009) or covered by Upper Cretaceous–Palaeogene epi-ophiolitic associated forearc sequences (Görür *et al.* 1984; Rice *et al.* 2006). Along the ophiolitic belt, slices of Campanian–Maastrichtian volcanic and volcanoclastic rocks are found, which are defined as the Yaylaçayı Formation (Yoldaş 1982). Sedimentary cover units consist of a 500-m-thick sequence (Yapraklı Formation; Birgili *et al.* 1974) that unconformably overlies the volcanic-arc rocks, whereas clastics of the Oligocene–Eocene İskilip Unit and evaporites of the Oligo- Late Miocene Çankırı Basin rest on the older units with an angular unconformity (Fig. 2).

In the area between the Kırsehir Block and the Sakarya Zone along the NAOM, Late Cretaceous pelagic limestones and volcanoclastic rocks were sampled at four sites around Ankara (KS 1–3, 7, Fig. 2, stratigraphic section A). In the same area three sites were sampled from Paleocene sandstones and clastic rocks (KS 4–6), whereas basaltic clasts embedded in the sandstones were sampled together with their matrix for conglomerate tests (KS 5b, Fig. 2, stratigraphic section A).

Volcanoclastic sandstones and lavas of the Campanian–Maastrichtian Yapraklı Formation were sampled at seven sites (KS 31–37, Fig. 2, stratigraphic section B). Arc-type volcanic rocks, pelagic sediments and andesitic lavas of Campanian–Maastrichtian age were sampled at six sites from the Yaylaçayı Formation (KS 12, 13, 18–20, 46, Fig. 2, stratigraphic section C). Cherts and lavas were sampled at 23 sites along the Upper Cretaceous ophiolitic belt (KS 24–30, 39–41, 43–45, 47–51, PT 22–26, Fig. 2, stratigraphic section C).

Middle Eocene andesites and sandstones were sampled at three localities around Yozgat (KS 10, 11, 14) and seven sites from the Middle Eocene İskilip Unit (KS 8, 9, 15–17, 38, 42), whereas three sampling sites were established in Miocene evaporites in the Çankırı Basin (KS 21–23, Fig. 2, stratigraphic section B).

## Previous palaeomagnetic data from north-central Anatolia (NCA)

The first study in Central Pontides was carried out by Sarıbudak (1989), who evaluated the palaeotectonic evolution of the Pontides in the Eocene and Upper Cretaceous. Later studies focused on detecting the deformation that occurred close to the North Anatolian Fault (NAF) (Morris and Robertson 1993; Gürsoy *et al.* 1998; Kaymakçı *et al.* 2007; Çinku & Orbay 2010; Piper *et al.* 2010). Platzman *et al.* (1994) provided palaeomagnetic data from Niksar along a transect across the NAF that indicated *c.* 30° of anticlockwise rotation. In the Pontides, Piper *et al.* (1996, 2010) detected

anticlockwise rotations in Middle Eocene lavas on both sides of the NAF. They interpreted their results in terms of block rotations on second-order faults that splayed westwards from the NAF during the neotectonic period. İşseven & Tüysüz (2006) assumed that the eastern Central Pontides consisted of individual blocks that represent segments of a dextral shear zone. They found clockwise rotations based on palaeomagnetic data from 46 sites in Eocene volcanic rocks, suggesting evidence for fault-bounded blocks rotated about vertical axes within the right-lateral NAF zone.

Kaymakçı *et al.* (2001) first proposed that the omega-shaped curvature of the Çankırı Basin could be due to the indentation of the Kırşehir Block into the Pontides during the Middle Eocene. The authors proposed that the western and southeastern margins of the Çankırı Basin experienced anticlockwise rotations and the eastern part was exposed to clockwise rotations, whereas no rotations have occurred in the Kırşehir Block since the Oligocene (Kaymakçı *et al.* 2003a). The study of Çinku *et al.* (2011) supported this hypothesis through palaeomagnetic data compiled from 72 sites in Middle Eocene rocks in the eastern part of north-central Anatolia.

## Palaeomagnetic methods

Stepwise thermal and alternating field (AF) demagnetization was applied to isolate the characteristic remanent magnetization (ChRM) component of the natural remanent magnetization (NRM) of the samples. They were demagnetized both thermally, in 14 steps between 75°C and 680°C using a Magnetic Measurements MTD 80 oven, and with alternating fields (AF), in steps between 1 and 100 mT using a tumbling Molspin AF demagnetizer. Remanence measurements were performed using a JR-6 spinner magnetometer (AGICO). Magnetic components were identified and calculated using orthogonal projections (Zijderveld 1967) and principal component analysis (Kirschvink 1980). Fisher's (1953) statistics was used to calculate site mean directions and the palaeomagnetic pole. Magnetic mineralogy was investigated by measuring hysteresis curves of crushed samples in fields up to 1 T deploying a Molspin vibrating sample magnetometer (VSM) at the Natural Magnetism Laboratory of ETH Zürich. Low field thermomagnetic measurements were carried out on all lithologies using a MS2 Bartington susceptibility meter in the University of Istanbul Yılmaz İspir palaeomagnetic laboratory. Normalized isothermal remanent magnetization (IRM) acquisition curves were obtained for one specimen per site up to maximum fields of 1 T along the sample $z$-axis with an ASC pulse magnetizer. The Lowrie test (Lowrie 1990) was

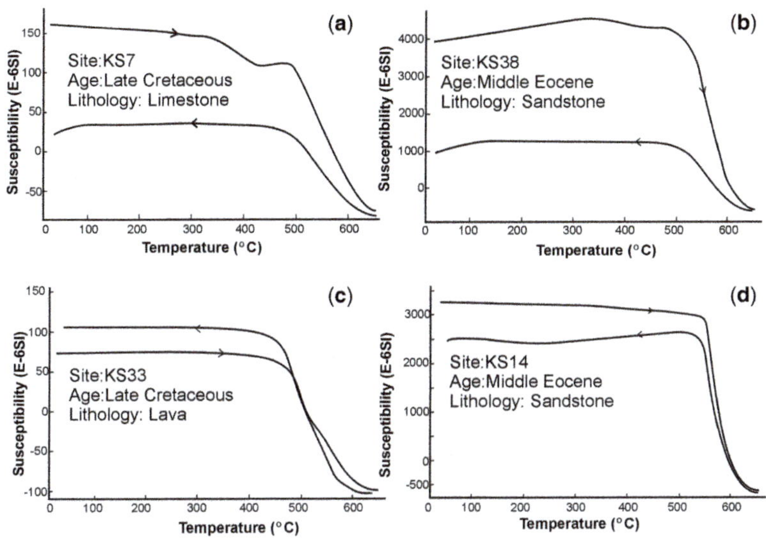

**Fig. 3.** Thermomagnetic curves for representative samples. (**a**) Limestone sample of site KS 7; (**b**) sandstone sample of site KS 38 with alteration upon heating and intensity drop at *c.* 450°C showing the presence of titanium-rich magnetite; (**c, d**) nearly reversible thermomagnetic curves of a lava sample from site KS 33 and a sandstone sample from site KS 14 show only one magnetic phase with a Curie temperature of *c.* 580°C.

performed by applying fields of 1 T along the sample *z*-axis (hard component), 0.4 T (medium component) along the *y*-axis and 0.12 T (soft component) along the sample *x*-axis. Subsequently, samples were thermally demagnetized to identify the magnetic carriers based on their coercivity and laboratory unblocking behaviour. Anisotropy of low-field magnetic susceptibility (AMS) was measured in 18 different directions to determine the magnetic fabric of the rocks using the MS2 and AGICO KLY2 kappabridge. Eigenvalues and principal directions of the AMS ellisoid were calculated with the AGICO Anisoft package (Chadima & Jelínek 2008).

In order to determine the acquisition time and stability of the remanence, a conglomerate test was performed in the volcanic clasts as well as the remanence study of the matrix. The statistical test of Watson (1956) was performed to identify the randomness of directions in the clasts. Incremental palaeomagnetic fold tests (McElhinny 1964; McFadden 1990; Enkin & Watson 1996) associated with progressive unfolding were applied in suitable tectonic situations.

## Magnetic mineralogy

Representative results of high-temperature thermomagnetic analyses are shown in Figure 3. Most samples show irreversible behaviour and Curie temperatures of *c.* 580°C, independent of rock type. The irreversible heating and cooling curves suggest

the transformation of Ti-rich to Ti-poor magnetite (Fig. 3a, b; Dunlop & Özdemir 1997). The slight decrease during heating between 400°C and 450°C may indicate the production of a new magnetic phase, likely the transformation of magnetite into maghemite or titanomagnetite. The nearly reversible thermomagnetic behaviour with Curie temperatures between 500°C and 580°C in other cases (Fig. 3c, d) is typical for Ti-poor magnetite (Dunlop & Özdemir 1997). The slightly smeared Curie temperatures, which exceed even 580°C in these samples, indicate further oxidation of the maghemite-hematite phases (Dunlop & Özdemir 1997).

Normalized IRM acquisition curves are illustrated in Figure 4a. The remanence is usually dominated by a phase with coercivity <300 mT, as the saturation of IRM curves suggests. The thermal demagnetization of the cross-component IRM shows that the soft coercivity component (0.12 T) is gradually unblocked up to 600°C (Fig. 4b–d), suggesting magnetite as carrier. However, a second inflexion of the thermal demagnetization curves in 60% of the samples between 350°C and 400°C (Fig. 4b, d) indicates the unblocking of a second magnetic phase in this coercivity range, which is typical for titanomagnetite. A predominating middle–high coercivity component unblocks at *c.* 650°C in some rock types (e.g. volcanic sample of site KS 37, Fig. 4e), and represents the occurrence of hematite (Fig. 4a). The lower-coercivity component in these samples also persists beyond 600°C, which may indicate the presence of stable

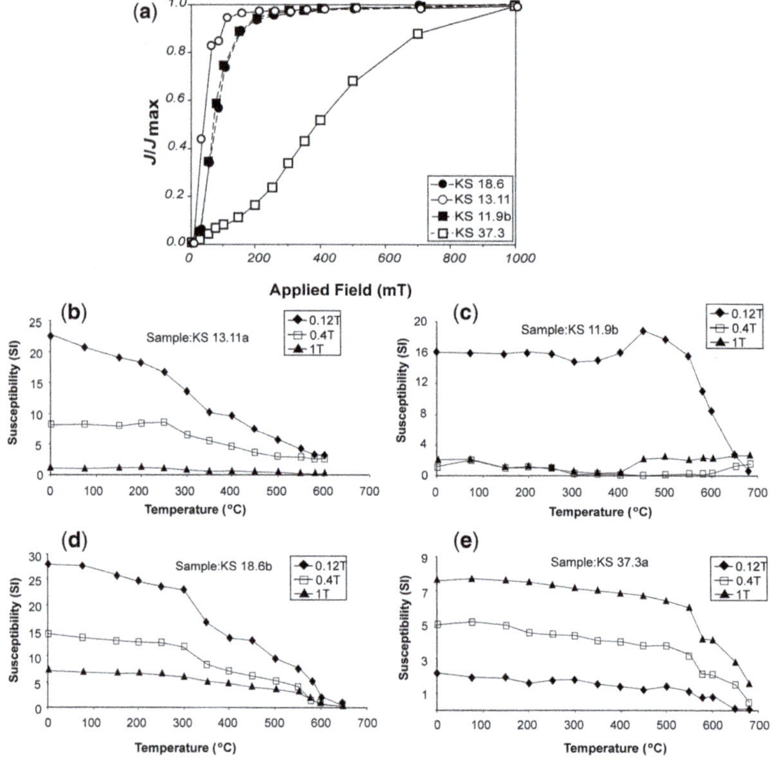

**Fig. 4.** (**a**) Normalized IRM acquisition curves showing the predominance of low-coercivity minerals in almost all samples and an additional high-coercivity mineral in sample 37.3. (**b–e**) Thermal demagnetization of three-axial IRM: hard (1 T along the sample *z*-axis), medium (0.4 along the sample *y*-axis) and soft (0.12 T along the sample *x*-axis).

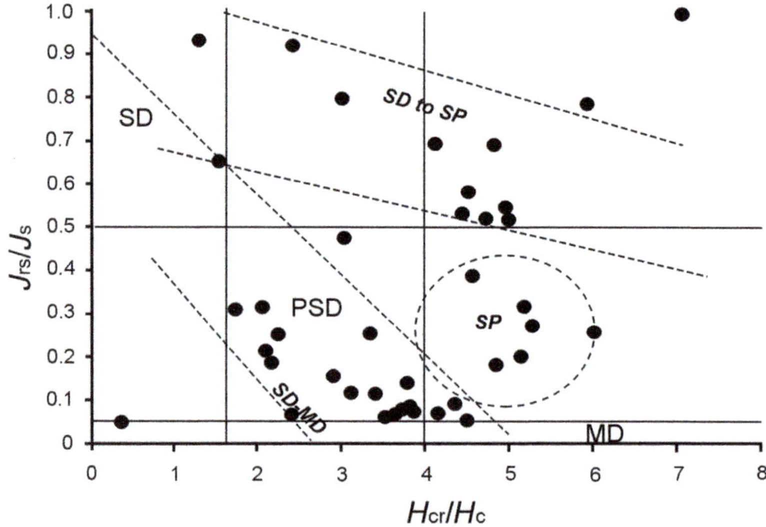

**Fig. 5.** Hysteresis parameters on the Day plot (Day *et al.* 1977). SD: single domain; PSD: pseudosingle domain; MD: multidomain; SP: superparamagnetic. Dashed lines refer to the grain size fields as defined by Dunlop (2002).

maghemite or coarse-grained hematite (Fig. 4e; see also Fig. 3c, d).

The magnetic grain size is analysed in Figure 5 using the Day diagram (Day *et al.* 1977; Dunlop 2002). Most samples fall into the pseudosingle domain region (PSD) of the Day plot, along the single-domain–multidomain (SD-MD) mixing line. Some samples have high magnetization ratios that suggest a mixture of SD and superparamagnetic (SP) grain sizes.

## Palaeomagnetism

The demagnetization behaviour of the volcanic rocks, volcanoclastic sandstones and limestones collected at different sites is dependent on lithology, but stable magnetization components are observed in most samples after removal of a secondary NRM component (Fig. 6). The secondary component, which shows mostly random orientation, was removed either by AF demagnetization between 0 and 5 mT (Fig. 6; KS 10.10b, 13.10a, 30.1, 38.1,

47.2b) or thermally by 350°C (Fig. 6; KS 11.4b, 26.2, 27.7, 33.1). In some cases, for example KS 40.1, this component is consistent with the present-day field $(D/I = 359.6°/61.6°)$. A ChRM is defined after removal of the secondary component by temperatures $\leq 300$°C or alternating fields generally around 10 mT. The ChRM vector decays linearly to the origin with maximum unblocking temperatures between 500°C and 580°C or destructive alternating fields up to 40 mT, consistent with the presence of SD-PSD (titano)magnetite (Fig. 6, KS 4.10). Several volcanic samples unblock between 600°C and 650°C however, which is ascribed to the presence of titanohematite as remanence carrier (Fig. 6, KS 26.2). Volcanic sandstone samples are thermally unstable above 400°C, so a ChRM could not be isolated; AF treatment was more effective in defining the ChRM in this case.

Seven (KS18, 20, 21, 23, 32, 36 and 45) out of 57 sites had to be excluded from the palaeomagnetic analysis because of unstable demagnetization behaviour. Only a secondary component could

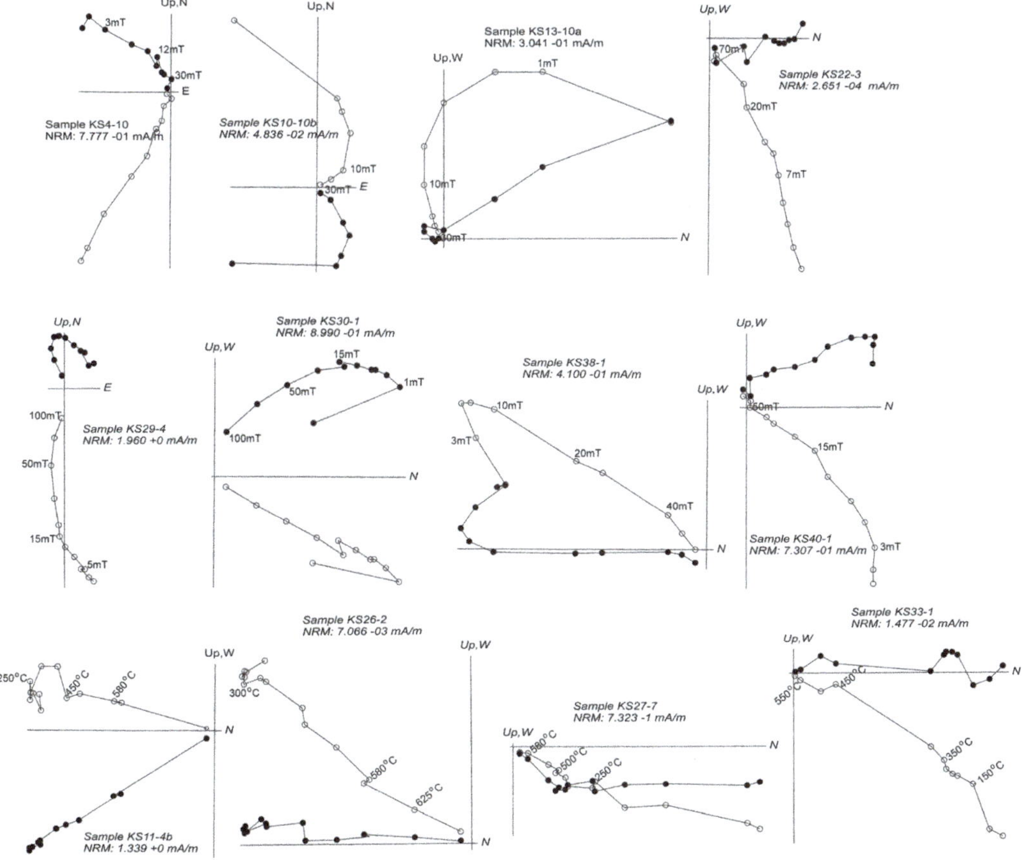

**Fig. 6.** Zijderveld diagrams of representative samples during stepwise thermal and alternating field demagnetization. Solid (open) symbols for horizontal (vertical), respectively.

**Table 1.** *Palaeomagnetic results for the Late Cretaceous, Paleocene, Middle Eocene and Miocene formations in the study region*

| Site no. | Lithology | Latitude (°N) | Longitude (°E) | Strike/Dip | N/n | $D_g$ | $I_g$ | $D_s$ | $I_s$ | $\alpha_{95}$ | $k$ |
|---|---|---|---|---|---|---|---|---|---|---|---|
| *Late Cretaceous Group C-1* | | | | | | | | | | | |
| KS1 | Pelagic limestone | 39.46.787 | 31.30.211 | 219/66 | 10/8 | 93.8 | 61.4 | 331.9 | 44.7 | 13.3 | 18.4 |
| KS2 | Volcano-clastic | 39.47.991 | 32.27.823 | 210/19 | 14/14 | 334.8 | 43.7 | 327.9 | 28.2 | 7.4 | 29.6 |
| KS3 | Pelagic limestone | 39.47.080 | 32.30.769 | 218/23 | 8/8 | 279.2 | 46.3 | 286.4 | 25.4 | 14.6 | 18.1 |
| KS7 | Limestone | 39.27.403 | 32.31.086 | 234/32 | 16/15 | 295.9 | 48.5 | 310.2 | 26.2 | 13.2 | 9.31 |
| **KS1–3, 7** | | | | | **48/45** | **319.2** | **68.3** | | | **10.2** | **5.35** |
| **=C-1A** | | | | | | | | **316.1** | **34.9** | **6.9** | **10.63** |
| **C-1B** | **Group Mean from 4 sites** | | | | | **316.6** | **68.6** | | | **43.3** | **5.5** |
| | | | | | | | | **313.4** | **35.0** | **23.1** | **16.8** |
| *KS18** | *Sandstone* | *39.57.054* | *33.27.347* | *6/88* | *10/7* | *94.4* | *20.1* | *2.2* | *−18.6* | *17.3* | *7.9* |
| *KS19** | *Sandstone* | *39.58.499* | *33.27.785* | *193/63* | *11/11* | *357.6* | *65.3* | *–* | *–* | *10.4* | *42.7* |
| *KS20** | *Sandstone* | *39.59.420* | *33.27.137* | *182/70* | *10/9* | *304.9* | *29.0* | *305.9* | *−31.6* | *32.8* | *4.4* |
| *Group C-2* | | | | | | | | | | | |
| KS24 | Lava | 40.30.811 | 33.25.632 | 132/37 | 8/5 | 348.5 | 23.4 | 326.6 | 40.1 | 10.3 | 27.7 |
| KS25 | Pelagic limestone | 40.30.811 | 33.25.632 | 132/37 | 8/7 | 157.5 | −27.3 | 134.2 | −36.6 | 12.1 | 27.2 |
| KS26 | Pelagic limestone | 40.31.301 | 33.26.814 | 202/51 | 11/11 | 173.5 | −85.3 | 117.2 | −36.7 | 6.1 | 57.1 |
| **KS24–26** | | | | | **27/23** | **342.9** | **58.1** | | | **15.6** | **6.2** |
| **=C-2A** | | | | | | | | **307.1** | **37.9** | **7.1** | **26.2** |
| **C-2B** | **Group Mean from 3 sites** | | | | | **343.5** | **44.6** | | | **58.4** | **5.5** |
| | | | | | | | | **337.1** | **39.7** | **18.2** | **46.9** |
| *Group C-3* | | | | | | | | | | | |
| KS27 | Sandstone | 40.46.103 | 33.45.682 | 230/27 | 10/10 | 41.2 | 41.4 | 21.5 | 32.5 | 12.8 | 15.1 |
| KS28 | Sandstone | 40.46.485 | 33.46.309 | 55/52 | 7/5 | 356.2 | 25.8 | 39.3 | 61.1 | 12.3 | 20.3 |
| KS29 | Sandstone | 40.47.030 | 33.46.977 | 55/52 | 8/6 | 339.9 | 18.0 | 2.3 | 66.1 | 10.3 | 81.2 |
| KS30 | Sandstone | 40.47.275 | 33.47.112 | 315/52 | 7/7 | 305.4 | 29.3 | 334.1 | 24.2 | 13.4 | 21.3 |
| KS31 | Lava | 41.08.482 | 33.46.942 | 210/36 | 8/7 | 233.0 | −46.5 | 191.8 | −48.1 | 5.4 | 124.6 |
| *KS32** | *Sandstone* | *41.08.589* | *33.46.738* | *57/29* | *7/7* | *100.5* | *−16.7* | *152.3* | *−27.2* | *47.7* | *2.6* |
| KS33 | Lava | 41.06.523 | 33.45.080 | 305/39 | 8/7 | 305.6 | 43.6 | 336.3 | 32.1 | 13.5 | 209.3 |
| KS34 | Sandstone | 40.54.709 | 33.47.437 | 155/30 | 8/7 | 2.5 | 15.4 | 351.4 | 26.9 | 11.0 | 38.2 |
| **KS27–31, 33, 34** | | | | | **56/49** | **357.0** | **35.9** | | | **10.5** | **4.7** |
| **=C-3A** | | | | | | | | **7.9** | **38.0** | **6.6** | **10.6** |
| **C-3B** | **Group Mean 7 sites** | | | | | **352.7** | **37.2** | | | **28.4** | **5.5** |
| | | | | | | | | **358.7** | **43.6** | **18.4** | **11.7** |
| *Group C-4* | | | | | | | | | | | |
| KS35 | Lava | 40.59.093 | 34.06.038 | 79/27 | 8/8 | 153.7 | −21.8 | 147.7 | −47.5 | 10.0 | 31.8 |
| *KS36** | *Lava* | *40.59.678* | *34.09.869* | *222/29* | *10/6* | *139.4* | *26.6* | *339.6* | *40.2* | *46.0* | *3.1* |
| KS37 | Lava | 40.59.631 | 34.10.060 | 230/30 | 10/9 | 31.6 | 66.9 | 353.3 | 47.1 | 8.9 | 38.5 |
| KS39 | Lava | 40.45.123 | 34.25.507 | 65/55 | 8/8 | 331.0 | −13.7 | 329.8 | 41.2 | 8.7 | 41.2 |
| KS40 | Sandstone | 40.45.279 | 34.24.989 | 205/18 | 15/12 | 346.5 | 44.8 | 336.1 | 32.3 | 10.0 | 59.4 |
| KS41 | Sandstone | 40.44.831 | 34.26.162 | 90/50 | 7/7 | 328.3 | 10.4 | 307.7 | 49.2 | 10.8 | 32.3 |

| | | | | | | | | | | | |
|---|---|---|---|---|---|---|---|---|---|---|---|
| **KS35, 37, 39–41** | | | | | **48/44** | **153.3** | **− 22.6** | | | **9.4** | **8.9** |
| **=C-4A** | | | | | | | | **325.6** | **43.0** | **5.8** | **21.2** |
| **C-4B** | **Group Mean 5 sites** | | | | | **339.2** | **27.1** | | | **36.2** | **5.4** |
| | | | | | | | | **331.5** | **44.4** | **12.7** | **37.2** |
| *Group C-5* | | | | | | | | | | | |
| KS43 | Lava | 39.54.036 | 33.21.545 | 105/80 | 13/11 | 322.2 | 8.0 | 282.3 | 37.9 | 6.4 | 58.1 |
| KS44 | Limestone | 39.54.036 | 33.21.480 | 105/80 | 15/10 | 326.1 | 25.9 | 259.2 | 41.1 | 9.9 | 26.7 |
| *KS45** | *Limestone* | *39.46.761* | *33.13.716* | *236/62* | *8/6* | *240.8* | *23.0* | *258.6* | *6.7* | *28.1* | *6.6* |
| KS46 | Sandstone | 40.01.268 | 33.27.404 | 215/42 | 15/13 | 314.6 | 71.5 | 309.2 | 33.7 | 6.6 | 48.5 |
| KS47 | Flysh | 40.01.271 | 33.27.170 | 35/20 | 8/6 | 275.3 | 12.0 | 271.3 | 29.1 | 7.3 | 23.7 |
| **KS43, 44, 46, 47** | | | | | 51/40 | **312.8** | **34.1** | | | **10.5** | **5.6** |
| **=C-5A** | | | | | | | | **285.0** | **36.3** | **6.0** | **15.2** |
| **C-5B** | **Group Mean 4 sites** | | | | | **308.6** | **28.3** | | | **47.5** | **4.7** |
| | | | | | | | | **283.5** | **35.6** | **21.9** | **18.6** |
| *Group C-6* | | | | | | | | | | | |
| PT22 | Lava | 40.25.910 | 34.37.307 | 205/21 | 17/17 | 156.5 | 28.8 | 167.8 | 43.2 | 4.2 | 33.7 |
| PT23 | chert | 40.25.910 | 34.37.307 | 205/21 | 10/10 | 125.8 | 35.3 | 130.8 | 55.8 | 4.3 | 53.2 |
| KS48 | chert | 40.21.087 | 34.36.787 | 148/75 | 10/9 | 284.5 | 14.9 | 117.1 | 35.2 | 6.2 | 71.0 |
| **PT22, 23, KS48** | | | | | **37/36** | **134.9** | **22.7** | | | **10.2** | **6.5** |
| **=C-6A** | | | | | | | | **145.9** | **48.1** | **6.9** | **12.9** |
| **C-6B** | **Group Mean 3 sites** | | | | | **128.0** | **18.0** | | | **62.3** | **5.9** |
| | | | | | | | | **138.0** | **46.9** | **34.0** | **14.2** |
| *Group C-7* | | | | | | | | | | | |
| KS49 | Sandstone | 40.00.630 | 34.33.964 | 115/71 | 17/15 | 58.6 | 55.3 | 178.7 | 45.9 | 6.7 | 39.1 |
| PT24 | lava | 40.00.393 | 34.32.162 | 85/69 | 10/10 | 4.8 | 48.4 | 161.4 | 61.5 | 7.8 | 45.4 |
| PT25 | chert | 40.00.393 | 34.32.162 | 115/75 | 10/6 | 59.1 | 28.7 | 144.3 | 55.7 | 8.5 | 45.5 |
| **KS49, PT24, 25** | | | | | **37/31** | **57.5** | **52.0** | | | **7.5** | **12.8** |
| **=C-7A** | | | | | | | | **167.1** | **47.6** | **6.0** | **19.4** |
| **C-7B** | **Group Mean 3 sites** | | | | | **42.6** | **47.0** | | | **40.8** | **10.2** |
| | | | | | | | | **162.6** | **55.2** | **19.9** | **39.5** |
| *Group C-8* | | | | | | | | | | | |
| PT26 | Sandstone | 39.54.791 | 34.56.277 | 230/23 | 8/8 | 4.7 | 60.0 | 347.9 | 41.1 | 7.8 | 25.5 |
| KS12 | Limestone | 39.50.383 | 35.27.049 | 224/36 | 10/7 | 49.1 | 69.7 | 346.9 | 50.4 | 9.4 | 42.0 |
| KS13 | Sandstone | 39.49.370 | 35.30.801 | 268/18 | 13/13 | 178.2 | −70.2 | 178.1 | −52.2 | 10.5 | 16.5 |
| KS50 | Lava | 39.57.516 | 34.36.890 | 258/70 | 12/7 | 151.9 | 64.4 | 357.6 | 44.3 | 8.4 | 27.9 |
| KS51 | Chert | 39.57.520 | 34.36.881 | 258/70 | 19/16 | 38.8 | 87.3 | 350.2 | 18.3 | 7.3 | 22.3 |
| **PT26, KS12, 13, 50, 51** | | | | | **62/58** | **23.5** | **80.2** | | | **6.1** | **10.3** |
| **=C-8A** | | | | | | | | **351.4** | **34.8** | **5.0** | **14.8** |
| **C-8B** | **Group Mean 5 sites** | | | | | **26.3** | **75.7** | | | **24.8** | **10.4** |
| | | | | | | | | **352.0** | **41.6** | **13.6** | **32.8** |

(*Continued*)

**Table 1.** *Palaeomagnetic results for the Late Cretaceous, Paleocene, Middle Eocene and Miocene formations in the study region* (*Continued*)

| Site no. | Lithology | Latitude (°N) | Longitude (°E) | Strike/Dip | N/n | $D_g$ | $I_g$ | $D_s$ | $I_s$ | $\alpha_{95}$ | $k$ |
|---|---|---|---|---|---|---|---|---|---|---|---|
| *Paleocene* | | | | | | | | | | | |
| KS4 | Sandstone | 39.47.100 | 32.27.679 | 210/20 | 14/14 | 342.3 | 52.9 | 330.4 | 36.7 | 8.2 | 24.5 |
| KS5a | Clastic | 39.48.225 | 32.27.934 | 180/10 | 10/10 | 338.3 | 47.7 | 329.1 | 43.2 | 5.7 | 73.2 |
| KS5b | Volcanic pebbles | 39.48.225 | 32.27.934 | 180/10 | 10/10 | 354.6 | −81.7 | 47.8 | −77.6 | 52.1 | 1.8 |
| KS6 | Sandstone | 39.48.390 | 32.27.827 | 222/15 | 10/10 | 334.8 | 51.5 | 329.7 | 37.4 | 10.3 | 23.0 |
| **KS4, 5a, 6** | | | | | **34/34** | **337.8** | **50.8** | | | **4.2** | **32.8** |
| **=PaleoceneA** | | | | | | **329.2** | **38.5** | **4.2** | **33.1** | | |
| **PaleoceneB Group Mean 3 sites** | | | | | | **338.5** | **50.7** | | | **5.5** | **505.2** |
| | | | | | | | **329.8** | **39.1** | **5.5** | **495.0** | |
| *Middle Eocene Group E-1* | | | | | | | | | | | |
| KS8 | Volcano-clastic | 39.28.206 | 32.31.788 | 294/42 | 8/8 | 346.7 | 75.2 | 12.8 | 37.2 | 12.2 | 13.6 |
| KS9 | Volcano-clastic | 39.28.365 | 32.31.713 | 300/48 | 17/17 | 96.8 | −37.7 | 138.7 | −39.8 | 4.5 | 63.9 |
| **KS8, 9** | | | | | **25/25** | **339.5** | **67.1** | | | **7.5** | **20.9** |
| **=E-1A** | | | | | | | | **8.9** | **27.3** | **7.3** | **21.9** |
| *Group E-2* | | | | | | | | | | | |
| KS10 | Lava | 39.49.437 | 34.47.121 | 210/12 | 10/10 | 156.9 | −44.3 | 151.3 | −34.4 | 7.7 | 35.1 |
| KS11 | Sandstone | 39.48.827 | 34.46.884 | 215/20 | 22/22 | 154.4 | −51.5 | 149.2 | −32.3 | 3.7 | 72.6 |
| KS14 | Sandstone | 39.46.995 | 35.32.604 | 252/16 | 12/12 | 151.4 | −47.0 | 153.2 | −31.2 | 8.6 | 32.4 |
| **KS10, 11, 14** | | | | | **44/42** | **159.7** | **−46.4** | | | **5.0** | **19.3** |
| **=E-2A** | | | | | | | | **155.6** | **−31.7** | **5.2** | **18.1** |
| **E-2B** | **Group Mean 3 sites** | | | | | **154.3** | **−47.6** | | | **6.2** | **391.0** |
| | | | | | | | | **151.4** | **−32.6** | **3.7** | **1096** |
| *Group E-3* | | | | | | | | | | | |
| KS15 | Sandstone | 39.53.260 | 33.25.938 | 232/45 | 15/15 | 357.9 | 59.4 | 340.3 | 18.5 | 5.1 | 226.0 |
| KS16 | Sandstone | 39.53.430 | 33.25.860 | 215/50 | 10/10 | 347.5 | 62.6 | 324.1 | 18.1 | 11.2 | 19.4 |
| KS17 | Sandstone | 39.54.010 | 33.25.899 | 331/22 | 13/13 | 334.0 | 58.0 | 349.9 | 33.7 | 11.8 | 13.4 |
| KS42 | Sandstone | 39.54.343 | 33.21.545 | 330/40 | 10/10 | 325.9 | 26.9 | 345.2 | 22.9 | 12.7 | 15.4 |
| **KS15-17, 42** | | | | | **48/48** | **340.1** | **49.8** | | | **7.3** | **9.0** |
| **=E-3A** | | | | | | | | **340.0** | **25.0** | **6.9** | **9.9** |
| **E-3B** | **Group Mean 4 sites** | | | | | **338.2** | **47.4** | | | **22.7** | **17.3** |
| | | | | | | | | **339.5** | **23.6** | **14.4** | **41.5** |
| KS38 | Sandstone | 40.42.292 | 34.29.352 | 210/37 | 7/7 | 203.3 | −50.4 | 170.4 | −34.8 | 12.2 | 25.2 |
| *Middle Miocene* | | | | | | | | | | | |
| *KS21** | *Evaporite* | *40.18.450* | *33.36.833* | *125/28* | *8/4* | *4.6* | *31.8* | *352.4* | *50.5* | *68.2* | *2.8* |
| KS22 | Evaporite | 40.19.873 | 33.39.222 | 175/12 | 8/8 | 18.5 | 59.4 | 357.4 | 62.2 | 12.4 | 24.7 |
| *KS23** | *Evaporite* | *40.20.159* | *33.39.069* | *175/12* | *10/10* | *209.1* | *−43.4* | *198.0* | *−49.2* | *36.2* | *2.7* |

N denotes number of samples per locality, n the number of samples used for site mean calculation. Declination $D_{g(s)}$ and inclination $I_{g(s)}$ describe the mean directions in geographic (before tilt correction) and stratigraphic coordinates (after tilt correction), respectively. $\alpha_{95}$ is the 95% confidence circle; $k$ is the precision parameter (Fisher 1953). Group means are calculated first using the individual samples (group number + index A), and secondly using the site mean directions (group number + index B). *Sites in italics have been discarded (see main text). Bold lines show the group mean directions.

be distinguished at site KS 19, but the remaining sites gave stable ChRM site means (Table 1; Figs 7 & 8).

Data groups have been chosen according to their geographic distribution and tectonic situation in order to differentiate the influence of thrust slices (Table 1; Fig. 2).

Group mean directions have been analysed statistically, both at the sample (index A in Table 1) as well the site (index B in Table 1) hierarchical level because of the low number of sample sites per group (three to seven) and the highly variable local deformation (Figs 7 & 8). The Late Cretaceous volcanic and sedimentary rocks show well-defined group mean directions on both statistical levels (Table 1).

The group mean declinations after tilt correction of the Late Cretaceous sites are largely scattered throughout the study area. The mean ChRM direction of group C-1A (individual samples from KS

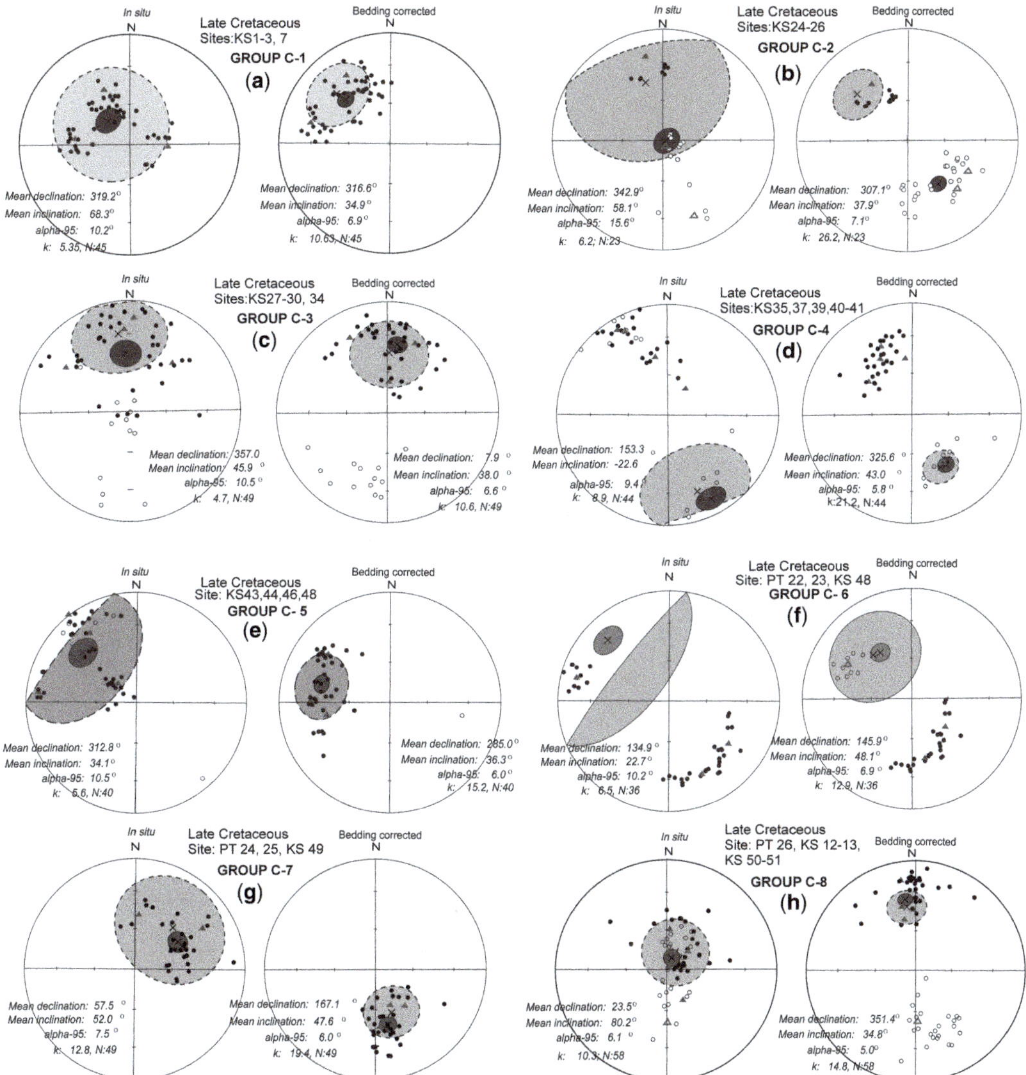

**Fig. 7.** Palaeomagnetic directions of Late Cretaceous sites before and after tilt correction (**a**) Group C-1; (**b**) Group C-2; (**c**) Group C-3; (**d**) Group C-4; (**e**) Group C-5; (**f**) Group C-6; (**g**) Group C-7; and (**h**) Group C-8. Group mean directions with 95% circle of confidence have been calculated from individual samples (dark grey confidence circle and black dots, index A) and from site mean directions (light grey confidence circle and red triangles, index B). Solid (open) symbols on lower (upper) hemisphere of equal area stereographic projection.

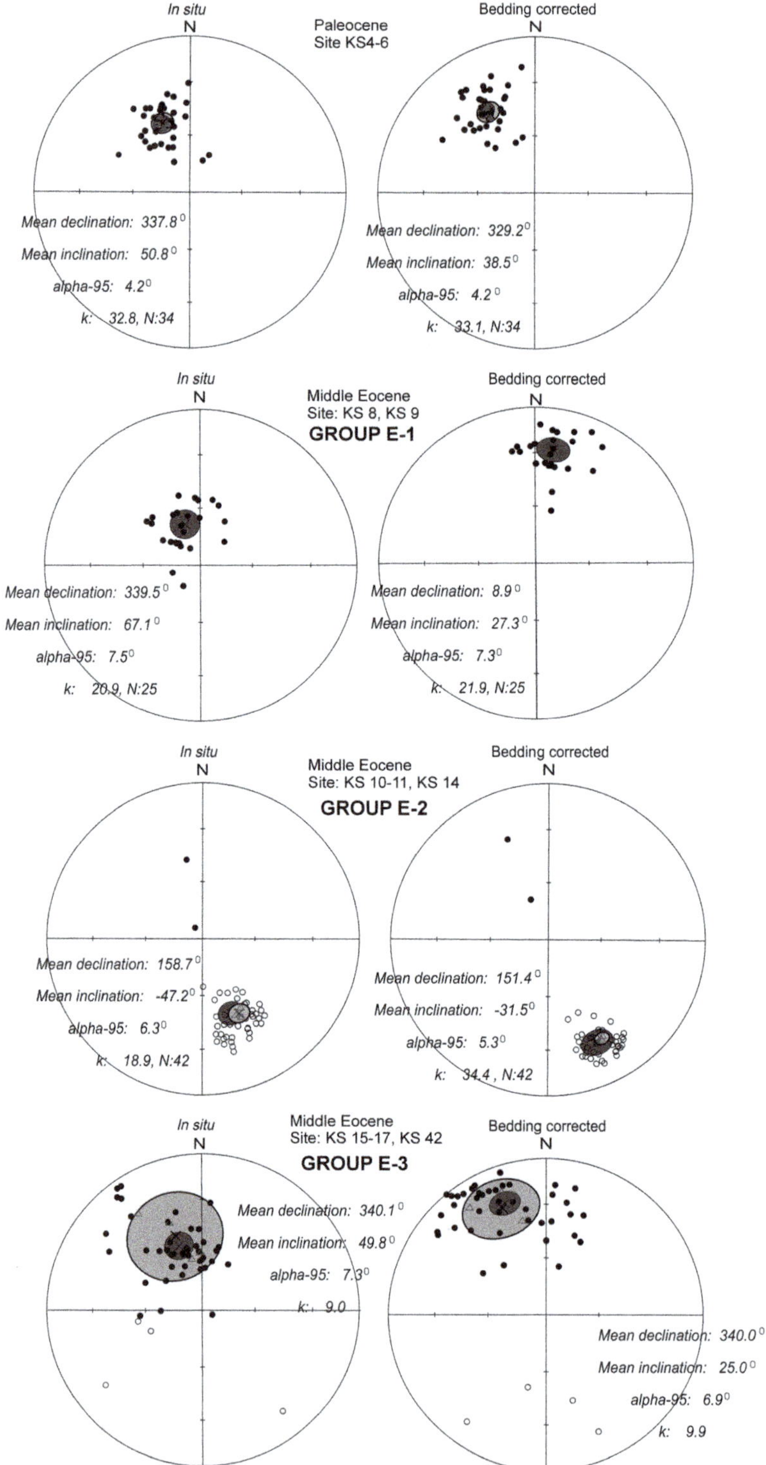

**Fig. 8.** Palaeomagnetic directions of Paleocene and Middle Eocene sites before and after tilt correction. Data significance as in Figure 7.

1–3 and 7) is $D = 316.1°$ and $I = 34.9°$ ($k = 6.99$, $\alpha_{95} = 10.63°$) in stratigraphic coordinates and has only normal polarity. The site mean direction of this group (sites of group C-1B) is also well defined with a nearly identical mean direction and significantly reduced directional scatter after tilt correction (Table 1; Fig. 7a).

We argue that the geomagnetic secular variations are averaged in the data for several reasons. Firstly, both the lavas and the sedimentary rocks are distributed over a long time interval within the geological formations, which should be sufficient to average out the geomagnetic secular variation. Secondly, in almost all groups the sample sites are independent and widely distributed to again average out secular variation sufficiently (Fig. 2).

Group C-2A (KS 24–26) includes lavas and pelagic limestones. The mean direction is $D = 327.1°$ and $I = 37.9°$ ($k = 26.2$, $\alpha_{95} = 7.1°$) after tilt correction, without systematic difference in direction between lithologies (Fig. 7b; Table 1). In addition, the site mean declinations vary between 297° and 327°; when using only the site mean directions the dispersion is therefore extremely high (group C-2B). The mean ChRM direction for Group C-3A (KS 27–31, 33, 34) is $D = 7.9°$, $I = 38.0°$ ($k = 10.6$ and $\alpha_{95} = 6.6°$) in stratigraphic coordinates and is based on six normal-polarity and only one reverse-polarity sites (Fig. 7c). The site mean direction in group C-3B yields a steeper inclination with north to northeast declination. Group C-4A (KS 35–41) has a mixed polarity mean direction of $D = 325.6°$, $I = 43.0°$ ($k = 21.2$ and $\alpha_{95} = 5.8°$), whereas Group C-5 (KS 43–46) has a mean ChRM direction with $D = 285.4°$, $I = 36.3°$ ($k = 15.27$ and $\alpha_{95} = 6.0°$) in stratigraphic coordinates. The mean direction for both group C-4B and C-5B is again similar to the mean direction based on individual samples, showing well-defined directions after tilt correction (Fig. 7d, e). We exclude site KS 45 in the mean direction even though the inclination is shallow, because all the inclination values are considered for further inclination-only analysis. Group C-6 (PT 22–23, KS 48) has a mean direction of $D = 145.9°$, $I = 48.1°$ ($k = 12.9$, $\alpha_{95} = 6.9°$) in stratigraphic coordinates and Group C-7 (PT 24, 25, KS 49) has a mean ChRM direction of $D = 150.3°$, $I = 50.6°$ ($k = 9.1$, $\alpha_{95} = 10.2°$) in stratigraphic coordinates (Fig. 7f, g). The palaeomagnetic mean declinations of groups C-6 and C-7 vary largely between 117° and 179°. For group C-8A (PT 26, KS 12, 13, 50–51) the mean direction is $D = 351.4°$, $I = 34.8°$ ($k = 14.8$ and $\alpha_{95} = 5.0°$) in stratigraphic coordinates; here the mean inclination is slightly shallower (but not significantly different) than at other sites if chert site KS 51 is included (Fig. 7h). However, when using the site mean directions (group C-8B),

an inclination similar to the rest of the groups is found, with an average confidence circle $<14°$ (Table 1).

Three sites (KS 4–6) were obtained from Paleocene sandstones in the southwest part of the suture zone near Ankara. Both the site mean and the sample mean group directions are similar for this group of sites with the mean direction based on samples being $D = 329.3°$, $I = 38.5°$ ($k = 33.1$ and $\alpha_{95} = 4.2°$) in stratigraphic coordinates (Fig. 8a). Nine out of ten Middle Eocene sites in volcanoclastic rocks and sandstones fall into one of three groups: group E-1A (KS 8, 9) shows a mean direction of $D = 8.9°$, $I = 27.3°$ ($k = 21.9$, $\alpha_{95} = 7.3°$) obtained from only two sites (Fig. 8b); group E-2A (KS 10, 11, 14) has reverse polarity with $D = 155.6°$, $I = -28.7°$ ($k = 18.1$, $\alpha_{95} = 5.2°$; Fig. 8c); and the mean direction of group E-3A (KS 15–17, 42) was obtained from four normal-polarity sites with $D = 340.0°$, $I = 25.0°$ ($k = 9.9$, $\alpha_{95} = 6.9°$; Fig. 8d). Both group means E-2B and E-3B have small confidence circles (3.7° and 14.4°, respectively) after tilt correction. A single site (KS 38) was not included because the location is too far away from the other sites. Three sites were sampled from Middle Miocene evaporates strata inside the Çankırı Basin. Only one site carried a stable remanent magnetization with a mean direction of $D = 357.4°$, $I = 62.2°$ ($k = 24.7$, $\alpha_{95} = 12.4°$) after tilt correction (Table 1), which is similar to the present magnetic field duration.

## Age of magnetization

The Upper Cretaceous strata showed a wide dispersion of declinations due to inferred rotation (Fig. 9a). Because the Upper Cretaceous rocks consist of both lavas and sedimentary rocks, we applied the inclination-only fold test for each rock type separately. Progressive tilt correction for the lavas at eight sites indicates that the maximum precision parameter ($k = 27.85$) is obtained at 90% unfolding (Fig. 9b), whereas the precision parameter is maximum in the sediments ($k = 24.10$) at 80% untilting. In both cases, the maximum precision parameter is not significantly different from the value at 100% untilting, which we interpret to indicate that the lavas and sediments acquired their ChRM before tectonic deformation.

A conglomerate test was performed using Paleocene basalt cobbles at site KS 5b (Fig. 9c). The NRM generally consists of a single component that is unblocked over a temperature range 400–585°C or in alternating fields of 15–40 mT (Fig. 9d). The ChRM directions are strongly scattered (Fig. 9c). The resultant vector $R$ (5.5) of 13 sample directions is slightly smaller than the 99% critical value $R_0 = 5.59$ in Watson's (1956) randomness

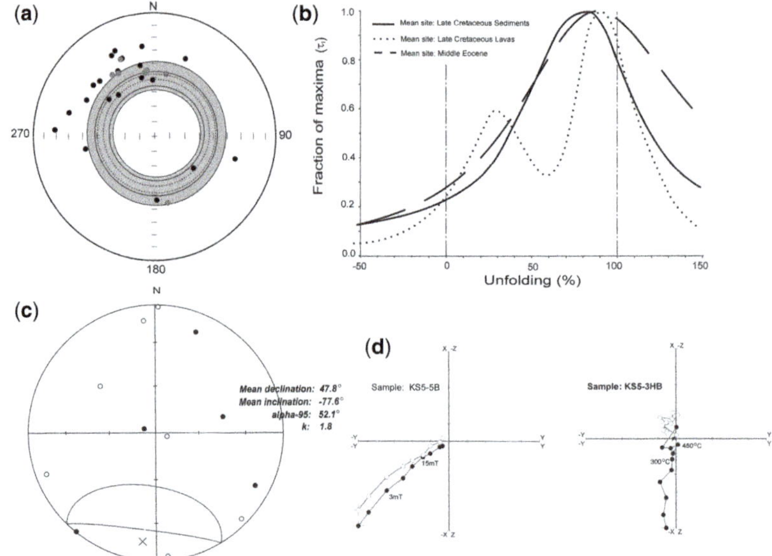



**Fig. 9.** (**a**) Inclination distribution of Late Cretaceous sediments (black) and lavas (grey) with their 95% confidence circles on the stereonet. (**b**) Enkin & Watson (1996) fold test for the Late Cretaceous sediments, lavas and Middle Eocene sites. (**c**) ChRM directions of pebble samples taken for a conglomerate test with statistical parameters. (**d**) Typical Zijderveld diagrams of conglomerate samples from site KS 37, showing the existence of a soft component with a MDF of 8 mT and a medium temperature component with unblocking temperature of 450°C.

test, demonstrating that these data pass the conglomerate test at the 99% confidence level. This indicates that the ChRM of the cobbles was acquired prior to the formation of the conglomerate. In addition, specimens from the matrix of the same formation yield a well-grouped mean direction of $D = 338.3°$, $I = 47.7°$ ($k = 73.2$, $\alpha_{95} = 5.7°$) and $D = 329.1°$, $I = 43.2°$ $k = 73.2$, $\alpha_{95} = 5.7°$) before and after tilt correction, respectively, which is consistent with directions from other sites in group C-1 (Table 1).

The mean direction of nine Middle Eocene sites (except sites KS 8 and 17) passes the McElhinny (1964) fold test, with a $k$-ratio of 3.6 at the 99% confidence level during progressive unfolding (critical values at 95% = 2.33; 99% = 3.37), indicating that the ChRM of the sites is primary. In the McFadden (1990) fold test, the precision parameter reached a maximum at 90% of unfolding (Fig. 9b). These data pass the reversal test with a Class C (McFadden & McElhinny 1990). The normal-polarity sites ($N = 4$) give a clockwise rotation angle (resultant vector of length $R = 3.9$ and $k = 69.2$), while the reversed sites ($N = 5$) give a resultant $R = 4.9$ and $k = 52.6$ in stratigraphic coordinates. The observed angular distance $g = 13.5°$ is smaller than the critical angle $g_c = 13.9°$, showing that the secondary overprints have been removed.

## AMS measurements

The AMS parameters for sites in Upper Cretaceous strata were evaluated using the statistical analysis of Jelínek (Jelínek 1977, 1978). However, some sites were not included because the AMS of their samples was too weak to be measured (Groups C-1, C-2). Parameters describing the magnitude of anisotropy and the shape of the AMS ellipsoid (Nagata 1961; Jelínek 1981; Hrouda 1982) are given in Table 2. The stereonet projections of the AMS principal axes after bedding correction are given in Figure 10a. The anisotropy degree is highly variable with $P$ ranging from 1.030 (KS 39) to 2.97 (KS 49) (Table 2). The high anisotropy degree of these sites should be interpreted as a result of tectonic strain experienced by the rocks of different lithology and in different tectonic situation (Kligfield *et al.* 1981; Hrouda 1982; Hirt *et al.* 1993; Tarling & Hrouda 1993; Borradaile & Henry 1997; Parés & van der Pluijm 2003). The ellipsoid shape is predominantly oblate but often close to neutral (Fig. 10b). The $\kappa_{min}$ axes are normal to the bedding plane at sites PT 26 and KS 49, which are neutral in shape. The magnetic lineation is defined at the $\kappa_{max}$ axes. The direction of the magnetic lineation is predominantly NE–SW except at sites KS 28, 37 and 51 where it is more along an east–west direction.

**Table 2.** *AMS data from Late Cretaceous sites*

| Site | AZ | PL | $\kappa_{max}$ $\alpha_{95}$ | $n$ | $\kappa_{mean}$ | VAR | $L$ | $F$ | $P$ |
|---|---|---|---|---|---|---|---|---|---|
| *GRP3* | | | | | | | | | |
| KS28 | 105.0 | 28.6 | 17.4 | 5 | $2.11 \times 10^{-4}$ | 35.4 | 1.019 | 1.039 | 1.082 |
| KS30 | 30.9 | 6.6 | 19.2 | 7 | $5.79 \times 10^{-3}$ | $-30.6$ | 1.152 | 1.504 | 1.716 |
| *GRP4* | | | | | | | | | |
| KS37 | 76.4 | 33.1 | 14.4 | 7 | $2.81 \times 10^{-4}$ | $-11.5$ | 1.035 | 1.039 | 1.075 |
| KS39 | 51.7 | 0.0 | 14.3 | 5 | $7.42 \times 10^{-4}$ | $-35.7$ | 1.010 | 1.024 | 1.030 |
| *GRP5* | | | | | | | | | |
| KS43 | 38.5 | 7.8 | 18.5 | 10 | $5.00 \times 10^{-5}$ | $-82.1$ | 1.025 | 1.049 | 1.075 |
| KS46 | 23.5 | 2.4 | 14.3 | 6 | $5.10 \times 10^{-6}$ | $-55.7$ | 1.187 | 1.503 | 1.837 |
| *GRP6* | | | | | | | | | |
| PT22 | 188.9 | 5.5 | 25.0 | 14 | $2.77 \times 10^{-3}$ | 163.3 | 1.015 | 1.031 | 1.048 |
| *GRP7* | | | | | | | | | |
| PT24 | 227.9 | 38.5 | 12.5 | 5 | $1.15 \times 10^{-3}$ | 156.4 | 1.056 | 1.143 | 1.207 |
| KS49 | 222.0 | 33.2 | 6.2 | 7 | $1.35 \times 10^{-3}$ | 173.3 | 1.727 | 1.725 | 2.97 |
| *GRP8* | | | | | | | | | |
| KS12 | 40.0 | 29.8 | 1.0 | 5 | $2.36 \times 10^{-3}$ | $-17.4$ | 1.043 | 1.287 | 1.392 |
| PT26 | 57.1 | 15.3 | 11.9 | 5 | $2.26 \times 10^{-5}$ | $-15.9$ | 1.063 | 1.075 | 1.143 |
| KS51 | 92.4 | 6.4 | 19.6 | 16 | $2.16 \times 10^{-3}$ | $-14.6$ | 1.017 | 1.019 | 1.036 |

*Abbreviations*: AZ: azimuth of the magnetic lineation; PL: plunge of the magnetic lineation; $\alpha_{95}$: 95% percent confidence circle of the $\kappa_{max}$ principal directions; $n$: number of samples; $\kappa_{mean}$: mean susceptibility; VAR: palaeomagnetic vertical-axis rotations; lineation $L = \kappa_{max}/\kappa_{int}$; foliation $F = \kappa_{int}/\kappa_{min}$; degree of anisotropy $P = \kappa_{max}/\kappa_{min}$.

## Discussion

### Palaeomagnetic rotations

Small and statistically insignificant clockwise rotations of $R \pm \Delta R = 3.7 \pm 7.8$ and $12.9 \pm 7.0$ are derived for the Late Cretaceous groups C-3 and C-8, respectively, with mostly normal polarity in the centre and southeast of the investigated area in places where the thrust directions are horizontal (Table 3; Fig. 11a). Extremely large clockwise rotations of $R \pm \Delta R = 141.4° \pm 8.6°$ and $162.6° \pm 8.0°$ are observed in the Late Cretaceous groups C-6 and C-7, respectively, on the eastern limb of the İzmir–Ankara–Erzincan ophiolitic belt to the east of the Kırşehir Block. Sites from group C-4 in the central northern belt show $R \pm \Delta R = -38.5° \pm 7.7$. Large anticlockwise rotations of $R \pm \Delta R = -79.3° \pm 7.3°$ and $R \pm \Delta R = -57.0° \pm 8.0°$ are seen on the western margin of the belt for C-5 and C-2, respectively. Group C-1 on the western flattened limb of the suture has been rotated by $R \pm \Delta R = -46.8° \pm 7.5°$ (Table 3; Fig. 11a). The rotation within the individual regions is consistent with the structure of the suture belt.

Further to the west, along the southwest flank of the IAES belt around Ankara, Paleocene rocks display a mean anticlockwise rotation of $R \pm \Delta R = -35 \pm 4.8°$. The Middle Eocene group mean directions (E1–E3) show estimated rotations between $R \pm \Delta R = -34.3 \pm 5.8$ and $-0.7 \pm 6.5°$, whereas further northeast in the suture belt palaeomagnetic declinations (groups G1–G11, reported in Çinku *et al.* 2011) of Middle Eocene rocks are rotated relative to the direction of the thrusts and the curvature of the North Anatolian Ophiolite Belt in the region (Table 3; Fig. 11b).

### Correlation between vertical axis rotation estimates and AMS results

Several methods, such as analysis of palaeostress, palaeoflows and magnetic anisotropy, especially AMS and palaeomagnetism, can be used to estimate the magnitude and sense of tectonic rotations. Sedimentary rocks that have experienced shortening generally exhibit $\kappa_{max}$ axes that are perpendicular to the shortening direction (e.g. Lowrie & Hirt 1986; Aubourg *et al.* 1995; Soto *et al.* 2009; Mochales *et al.* 2010; Pueyo-Anchuela *et al.* 2012; García-Lasanta *et al.* 2013). If the deformed zone has experienced vertical axis rotations, the original orientation of the magnetic fabric can be estimated from deviating palaeomagnetic mean declinations. The azimuth of the magnetic lineation ($\kappa_{max}$) can be used together with the estimation of vertical axis rotation to determine whether magnetic fabric acquisition took place before or after rotation.

For this purpose we compared the palaeomagnetic declinations in Upper Cretaceous rocks with the azimuth of magnetic lineations (Fig. 12).

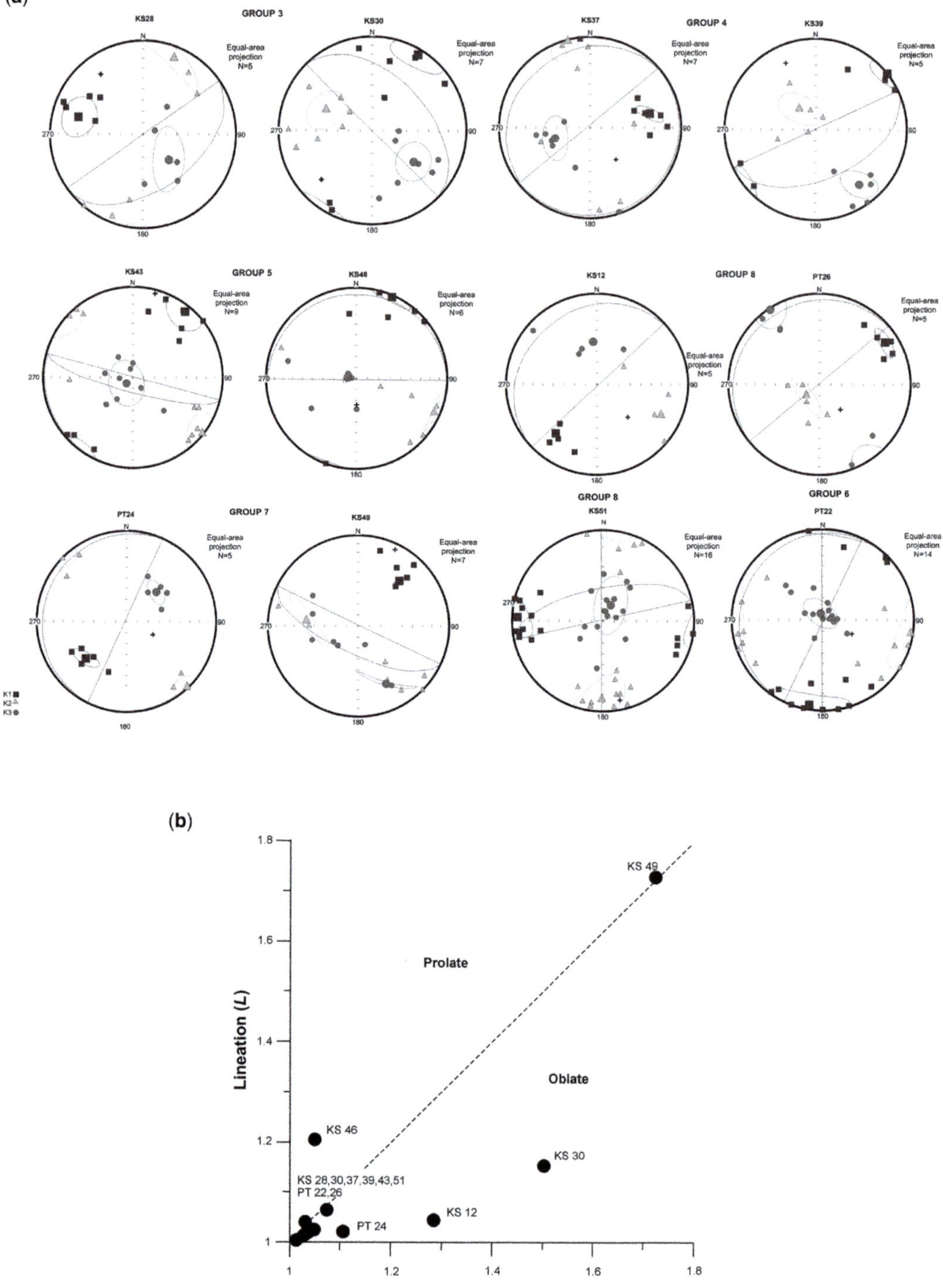



**Fig. 10.** (a) Stereographic projection of the principal susceptibility axes and site means with 95% circles of confidence in stratigraphic coordinates (great circle denotes the bedding plane). (b) Site means of *AMS lineation* v. *foliation.*

**Table 3.** *Group mean directions (D, I) and expected tectonic rotations (R) with respect to the European reference poles*

| Group | Site $\lambda$, $\phi$ (°N, °E) | D/I | $\alpha_{95}$ | $\lambda_{obs}$, $\phi_{obs}$ (°N, °E) | $\alpha_{95}$ | $R \pm \Delta R$ |
|---|---|---|---|---|---|---|
| C-1A | 37.98/32.46 | 316.6/34.9 | 6.9 | 47.6/286.0 | 6.0 | $-46.8 \pm 7.5$ |
| C-2A | 40.51/33.41 | 307.1/37.9 | 7.1 | 40.5/296.1 | 6.4 | $-57 \pm 8.0$ |
| C-3A | 40.78/33.78 | 7.9/38.0 | 6.6 | 69.4/192.4 | 6.0 | $-3.7 \pm 7.8$ |
| C-4A | 40.83/34.33 | 325.6/43.0 | 5.8 | 57.4/286.0 | 5.7 | $-38.5 \pm 7.7$ |
| C-5A | 39.83/33.30 | 285.0/36.3 | 6.0 | 24.0/310.2 | 5.3 | $-79.3 \pm 7.3$ |
| C-6A | 40.41/34.61 | 145.9/48.1 | 6.9 | $-13.6/64.9$ | 7.2 | $141.4 \pm 8.6$ |
| C-7A | 40.00/34.53 | 167.1/47.6 | 6.0 | $-20.3/46.6$ | 6.3 | $162.6 \pm 8.0$ |
| C-8A | 39.83/34.58 | 351.4/34.8 | 5.0 | 68.0/236.2 | 4.4 | $-12.9 \pm 7.0$ |
| KS3-5A | 39.8/32.45 | 329.2/38.5 | 4.2 | 58.2/277 | 3.8 | $-35 \pm 4.8$ |
| E-1A | 39.46/32.51 | 8.9/27.3 | 7.3 | 63.8/192.7 | 5.9 | $-0.7 \pm 6.5$ |
| E-2A | 39.78/35.66 | 155.6/$-31.7$ | 5.2 | 59.5/266.8 | 4.2 | $-34.3 \pm 5.8$ |
| E-3A | 39.88/33.41 | 340.0/25.0 | 6.9 | 58/252.3 | 5.4 | $-29.0 \pm 6.3$ |

Data calculated by Besse & Courtillot (2002) for the Late Cretaceous with $\lambda_{ref}$, $\phi_{ref} = 81.3°$ N, $188.6°$ E, $\alpha_{95} = 7.0°$, for the Paleocene with $\lambda_{ref}$, $\phi_{ref} = 80.5°$ N, $277°$ E, $\alpha_{95} = 3.8°$ and for the Middle Eocene with $\lambda_{ref}$, $\phi_{ref} = 81.1°$ N, $150.4°$ E, $\alpha_{95} = 5.3°$. The difference between the observed poles ($\lambda_{obs}$, $\phi_{obs}$) and the reference poles (computed using the pole-space method of Beck 1980), defines the amount of vertical axis rotation ($R$) and poleward transport ($\Gamma$) using Enkin's PMGSC (version 4.2) software (unpublished, 2004). The 95% confidence limits $\Delta R$ and $\Delta\Gamma$ were determined after Demarest (1983).

A linear correlation with $R = 0.95$ is found with $y = 1.16x - 90.26$ (Table 2), showing that the rotation values obtained from both palaeomagnetic data and the azimuth of magnetic lineations are in good agreement, giving evidence that the fabrics were acquired before oroclinal bending. The correlation also indicates that rocks that have not been rotated (i.e. VAR = 0 or equation in Figure 11 = 0) provide a measure for the general orientation of the magnetic lineation azimuth with a value of $77°/257°$, close to a supposedly east–west-aligned original belt.

### Palaeolatitude of the study area

Because of the large dispersion of observed declinations of individual site mean directions of lavas and sedimentary rocks (Table 1; Fig. 7), we have applied the inclination-only test of Enkin & Watson (1996) to obtain a reliable palaeolatitude for the IAES region during the Late Cretaceous. Eight out of ten volcanic site means were used to compute the mean inclination (sites PT 24 and KS 50 were omitted because of anomalous inclination values). The mean inclination before tilt correction is $I = 33.2°$ ($\alpha_{95} = +39.1 - 17.8°$, $k = 6.7$) and $I = 46.7°$ ($\alpha_{95} = \pm 7.8°$, $k = 25.6$) after tilt correction. A mean inclination for 22 site directions from sedimentary rocks was calculated (sites KS 45 and 51 were omitted because of their anomalously low initial inclinations). The mean inclination before and

after tilt correction is $I = 52.6°$ ($\alpha_{95} = +16.2°/-9.5°$, $k = 5.2$) and $I = 41.3°$ ($\alpha_{95} = \pm 5.5°$, $k = 18.9$), respectively. The corresponding palaeolatitude values are $\lambda = 27.9°$ N $+ 7.13°/-5.92°$ for data from the lavas and $\lambda = 23.7°$ N $+ 3.92°/-3.87°$ for data from the sedimentary rocks. These estimates are in excellent agreement with the Late Cretaceous palaeolatitudes obtained for the Pontides (Channell *et al.* 1996; Çinku 2004; Hisarlı 2011). The slightly higher value from the lavas is likely to be due to the smaller number of sedimentary sites, as discussed below (Table 1).

Paleocene rocks at the south-eastern limb of the belt yield a mean inclination of $I = 38.5°$ and an associated palaeolatitude of $\lambda = 21.7°$ N $+ 3.07°/-2.86°$. The Middle Eocene data (except KS 8 and 17) yield a reversed-polarity mean direction of $D = 152.9°$, $I = -47.2°$ ($k = 12.3$ and $\alpha_{95} = 15.3°$) and $D = 155.5°$, $I = -27.7°$ ($k = 44.4$ and $\alpha_{95} = 7.8°$) in geographic and stratigraphic coordinates, respectively. These Middle Eocene inclinations are shallower in comparison to the expected inclination in almost all sites. This inclination shallowing has been observed in many Middle Eocene results from Anatolia (Van der Voo 1968; Sarıbudak 1989; Kissel *et al.* 2003). Because the declination data are fairly well clustered, we use the Fisher (1953) analysis to define a mean inclination which yields a palaeolatitude of $\lambda = 14.7°$ N $+ 4.93°/-4.4°$, which is significantly more southwards displaced than expected.

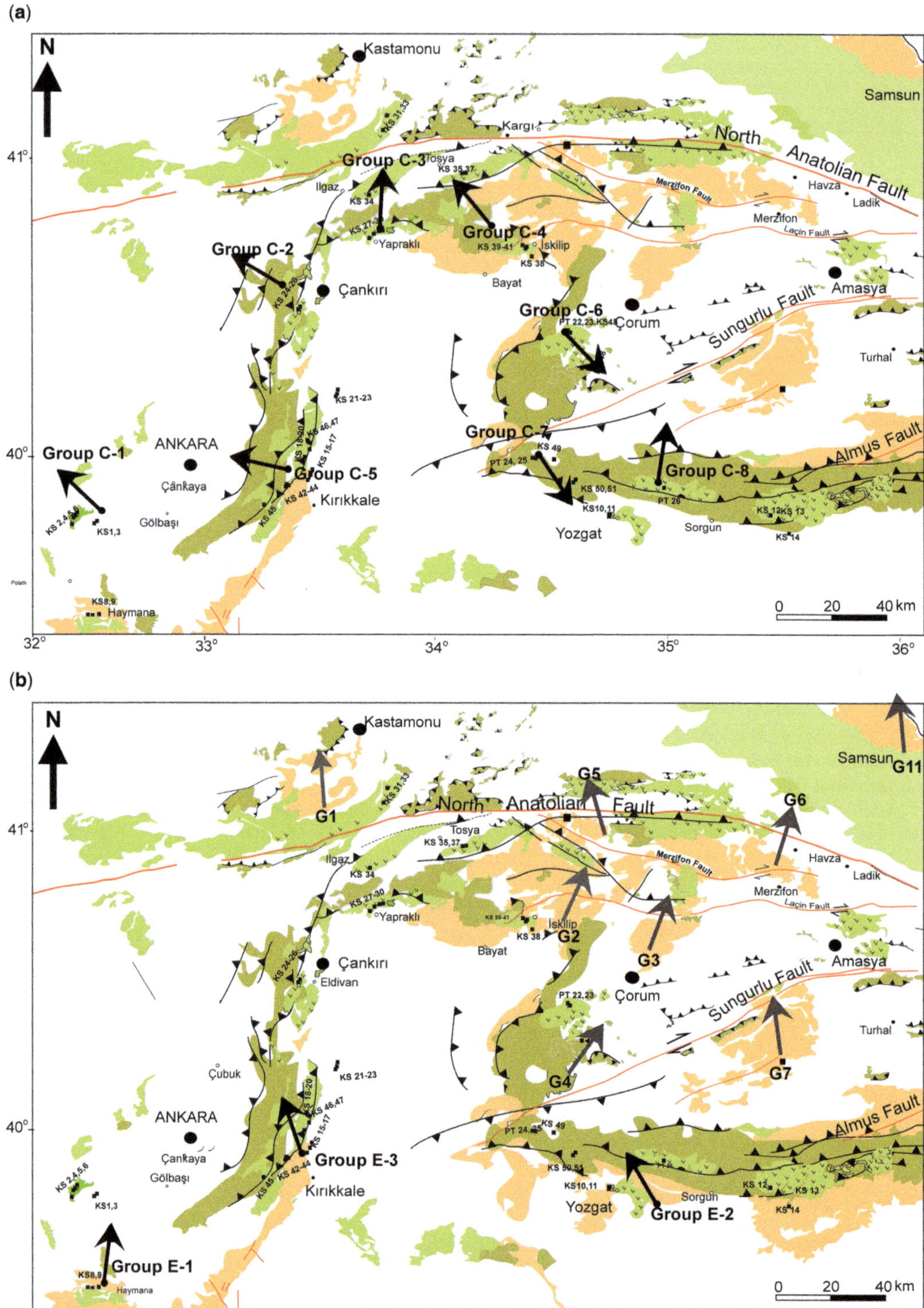



**Fig. 11.** Palaeomagnetic group mean declinations in the İzmir–Ankara–Erzincan Suture: (**a**) Late Cretaceous and (**b**) Middle Eocene, including grey-coloured vectors (G1–G11) added from Çinku *et al.* (2011).

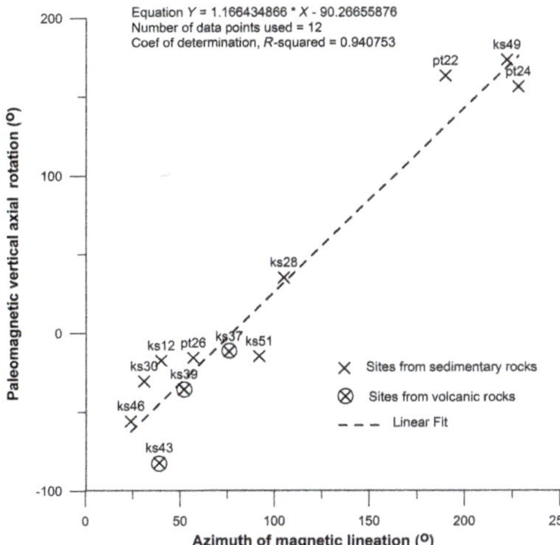



**Fig. 12.** Site mean palaeomagnetic declinations v. azimuths of AMS lineations.

The method of Tan & Kodama (2003) was applied in order to calculate the magnitude of flattening using the AMS data. Shallowing factors of 0.55 and 0.57 were obtained for Paleocene and Middle

Eocene rocks, respectively, which suggest that flattening occurred in both sequences of rocks. Applying the correction of Tan & Kodama (2003), we estimate mean inclinations of $I = 53°$ and $I = 43°$ for the Paleocene and Middle Eocene rocks, respectively. This yields a palaeolatitude of $\lambda = 33.6°$ N and $\lambda = 25.0°$ N for the Paleocene and Middle Eocene, respectively. However, the flattening in Middle Eocene rocks is still not removed when considering previous results from the surrounding area in lavas (Çinku *et al.* 2010), with inclination values lower than expected; the area may therefore have experienced significant shortening. Alternatively, the main blocks of Anatolia drifted southwards during the Middle Eocene.

## Kinematic model of the Northern Kırşehir Block during the Mesozoic

It is well documented that the northern Neotethys Ocean (İzmir–Ankara–Erzincan ocean) was active between the Pontides and the Kırşehir Block until the end of the Late Cretaceous (Görür *et al.* 1984; Tüysüz & Dellaloğlu 1992; Yılmaz *et al.* 1997; Nairn *et al.* 2012). The northwards subduction of this ocean led to widespread ophiolite emplacement and volcanic arcs. In a later phase, sedimentary

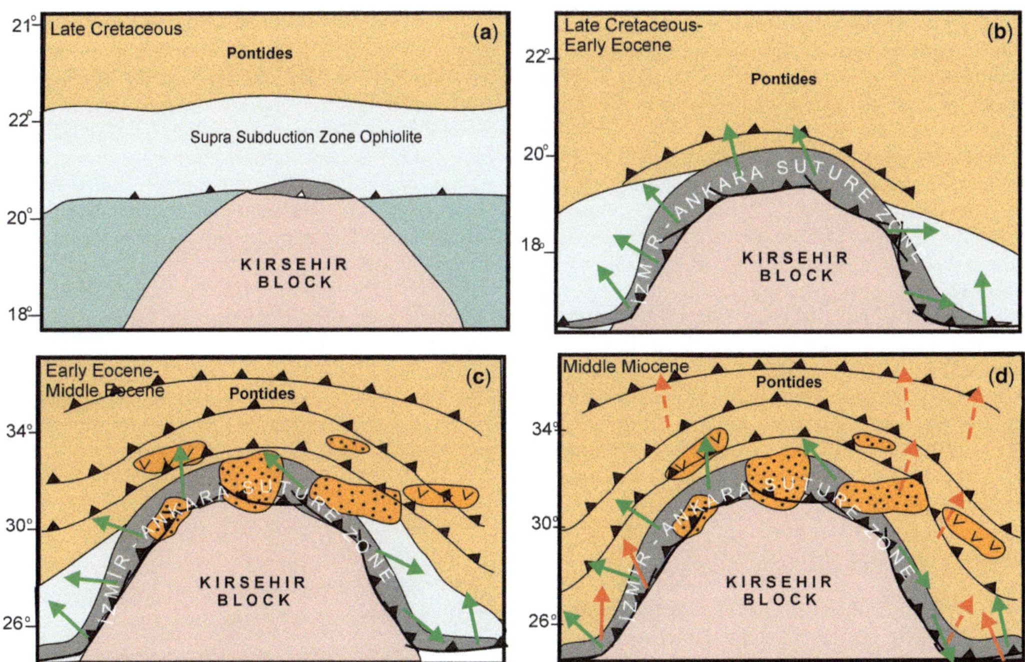



**Fig. 13.** Summary of the tectonic rotations around the northern Kırşehir Block according to the indentation model from (**a**) the Late Cretaceous through (**b**) the Early Eocene to (**c**) the Middle Eocene and (**d**) Middle Miocene. Green arrows refer to Late Cretaceous rotations while orange dashed and solid arrows indicate Middle Eocene rotations derived from our palaeomagnetic studies here and in Çinku *et al.* (2011).

cover rocks were emplaced mainly in the northern part of the Kırşehir Block. Palaeomagnetic evidence shows that these Late Cretaceous units, which were deformed by thrusts, are only slightly rotated in the northern part and the south-eastern rim of the Kırşehir Block, but underwent large rotations up to 180° along its western and eastern flanks.

The Late Cretaceous results from this study show large clockwise rotations up to >160° in the eastern limb and anticlockwise rotations of 75° in the western limb. In areas where the shape of the thrust traces is east–west, no significant rotation has been detected (Fig. 11a). Despite the large rotations found in the eastern and western limbs of the Kırşehir Block, Middle Eocene results from this work and a previous palaeomagnetic study (Çinku *et al.* 2010) clearly indicate clockwise rotation of *c.* 30° in the eastern corner of the Kırşehir Block. The degree of rotation decreases when moving further north. It has also been shown in the previous study that the magnitude of rotations in the Pontide block also decreases by a significant amount.

As can be seen in the geological map, the ophiolitic belt wrap them around the Kırşehir Block, forming a bending. The evolution of this belt could however be described by a tectonic model, presented in the following.

After the evolution of the Late Cretaceous ophiolitic sequences with related volcanic-arc rocks and sedimentary cover rocks (Fig. 13a), the Kırşehir Block started to indent these sequences and to wrap around the Kırşehir Block (Fig. 13b). The indentation led to large clockwise rotations in the eastern limb and anticlockwise rotations in the western limb; rotation in these limbs continued until the Middle Eocene (Fig. 13b, c). In areas where the shape of the thrust traces follows the more regional east–west-aligned trend, no significant rotation is detected (Fig. 13b, c). Net rotations for the Late Cretaceous formations have been obtained by subtracting the effect of rotation in the Middle Eocene and younger times from the amplitude of the Late Cretaceous rotations (Fig. 13c). In the Pontides, however, oroclinal bending continued after the Middle Eocene according to the palaeomagnetic results of Meijers *et al.* (2010).

Anticlockwise rotations up to 34° have been found in the Kırşehir Block according to data from Middle Eocene rocks reported by Kaymakçı *et al.* (2003a) and Çinku *et al.* (2011), and suggest that the movement of the Kırşehir Block and the collision with the Pontides was still ongoing until the Middle Miocene. This is based on palaeomagnetic results in the evaporites at site KS 22 on the Kırşehir Block, in which the palaeomagnetic directions are close to the present field (Fig. 13d). Although the

Middle Miocene results from the study of Lucifora *et al.* (2013) show fault-bounded block rotations in the Kırşehir Block, only one single site from this study and previous studies does not indicate significant rotation from the Middle Miocene to present (Platzman *et al.* 1998; Piper *et al.* 2002; Kaymakçı *et al.* 2003a). The tectonic movement, however, may still continue into modern times (Yavaşoğlu *et al.* 2011).

## Conclusions

The Upper Cretaceous and Middle Eocene rocks examined in this study yield palaeomagnetic directions that provide new insights into the collision-induced rotational deformation of the Kırşehir Block in northern Turkey during the closure of the İzmir–Ankara–Erzincan Suture. A comparison of the palaeomagnetic and AMS data shows that since the Late Cretaceous large clockwise rotations up to 163° occurred in the eastern limb of the belt and anticlockwise rotations ranging in magnitude between 50 and 80° in the western limb, both being related to bending after the closure of the İzmir–Ankara–Erzincan ocean. In contrast, small clockwise and anticlockwise rotations were defined in areas where the ophiolite thrust structure is quite elongated in an east–west direction. The large amount of rotations with contrasting sense may yield evidence about the indentation process between the Pontides and the Kırşehir Block.

Palaeomagnetic rotations derived in this study, as well as by Çinku *et al.* (2011), show that since the Middle Eocene anticlockwise rotations up to 30° occurred in the south-eastern flank of the Kırşehir Block, whereas outside the block both clockwise and anticlockwise rotations have been obtained, indicating that the deformation continued into the Middle Eocene. Most rotations, however, took place between the Late Cretaceous and the Middle Eocene.

This research was supported by the Scientific Research Projects Coordination Unit of Istanbul University, project number 6969 and 7272. Sincere thanks are extended to three anonymous reviewers and the editor of the special volume whose valuable comments contributed to the improvement of the manuscript.

## References

AKIMAN, O., ERLER, A., GÖNCÜOĞLU, M. C., GÜLEÇ, N., GEVEN, A., TÜRELI, T. K. & KADIOĞLU, Y. K. 1993. Geochemical characteristics of granitoids along the western margin of the Central Anatolian Crystalline Complex and their tectonic implications. *Geological Journal*, **28**, 371–382.

AUBOURG, C., ROCHETTE, P. & BERGMÜLLER, F. 1995. Composite magnetic fabric in weakly deformed black shales. *Physics of the Earth and Planetary Interiors*, **87**, 267–278.

BECK, M. E., Jr, 1980. Paleomagnetic record of plate-margin tectonic processes along the western edge of North America. *Journal of Geophysical Research*, **85**, 7115–7131, http://doi.org/10.1029/JB085iB12p07115

BESSE, J. & COURTILLOT, V. 2002. Apparent and true polar wander and the geometry of the geomagnetic field over the last 200 Myr. *Journal of Geophysical Research*, **107**, 2300, http://doi.org/10.1029/2000JB000050

BINGÖL, E., AKYÜREK, B. & KORKMAZER, B. 1975. The geology of the Biga Peninsula and some features of the Karakaya Formation. Cumhuriyetin 50. *In: Yılı Yerbilimleri Kongresi Tebliğleri, Maden Tetkik ve Arama Enstitüsü.* (MTA) Publications, 70–77 (in Turkish with English abstract).

BIRGILI, S., YOLDAŞ, R. & NALAN, U. G. 1974. *The Geology of Cankırı-Corum Basin and a Preliminary Report on Its Petroleum Possibilities.* Turkish Petroleum Company Report No. 1216 (unpublished).

BORRADAILE, G. J. & HENRY, B. 1997. Tectonic applications of magnetic susceptibility and its anisotropy. *Earth-Science Reviews*, **42**, 49–93.

BOZTUĞ, D. 2000. S–I–A-type intrusive associations: geodynamic significance of synchronism between metamorphism and magmatism in Central Anatolia, Turkey. *In*: BOZKURT, E., WINCHESTER, J., PIPER, J. A. D. (eds) *Tectonics and Magmatism in Turkey and the Surrounding Area.* Geological Society, London, Special Publications, **173**, 407–424.

BOZTUĞ, D. & JONCKHEERE, R. C. 2007. Apatite fission-track data from central-Anatolian granitoids (Turkey): constraints on Neo-Tethyan closure. *Tectonics*, **26**, TC3011.

CHADIMA, M. & JELÍNEK, V. 2008. Anisoft 4.2. Anisotropy data browser. *In*: HVOŽDARA, M. (ed.) *Paleo, Rock and Environmental Magnetism 11th Castle Meeting*, Abstract, ISSN 1335-2806, 22–28 June 2008, Geophysical Institute of the Slovak Academy of Sciences, Bojnice Castle, Slovac Republic. Contribution to Geophysics and Geodesy 2008, Special Issue **38**, 41.

CHANNELL, J. E. T., TÜYSÜZ, O., BEKTAS, O. & ŞENGÖR, A. M. C. 1996. Jurassic-Cretaceous paleomagnetism and paleogeography of the Pontides (Turkey). *Tectonics*, **15**, 201–212.

ÇINKU, C. M. 2004. Paleomagnetic evidence of the Western Black Sea Region. PhD thesis, İstanbul University, College of Science, Geophysical Engineering, İstanbul.

ÇINKU, C. M. & ORBAY, N. 2010. The origin of Neogene tectonic rotations in the Galatean volcanic massif, central Anatolia. *International Journal of Earth Sciences*, **99**, 413–426.

ÇINKU, C. M., USTAOMER, T., HIRT, A. M., HISARLI, Z. M., HELLER, F. & ORBAY, N. 2010. Southward migration of arc magmatism during latest Cretaceous associated with slab-steepening, East Pontides, N Turkey: new paleomagnetic data from the Amasya region. *Physics of the Earth and Planetary Interiors*, **182**, 18–29.

ÇINKU, C. M., MÜMTAZ HISARLI, Z., HELLER, F., ORBAY, N. & USTAÖMER, T. 2011. Middle Eocene paleomagnetic data from the eastern Sakarya Zone and the Central Pontides: implications for the tectonic evolution of north central Anatolia. *Tectonics*, **30**, TC1008, http://doi.org/10.1029/2010TC002705

CLARK, M. S. & ROBERTSON, A. H. F. 2002. The role of the Early Tertiary Ulukışla Basin, southern Turkey in suturing of the Mesozoic Tethys ocean. *Journal of the Geological Society, London*, **159**, 673–690.

CLARK, M. S. & ROBERTSON, A. H. F. 2005. Uppermost Cretaceous-Lower Tertiary Ulukışla Basin, south-central Turkey: sedimentary evolution of part of a unified basin complex within an evolving Neotethyan suture zone. *Sedimentary Geology*, **173**, 15–51.

DAY, R., FULLER, M. & SMITH, V. A. 1977. Hysteresis properties of titanomagnetites, grain size and composition dependence. *Physics of the Earth and Planetary Interiors*, **13**, 260–267.

DEMAREST, H. H., Jr, 1983. Error analysis for the determination of tectonic rotation from paleomagnetic data. *Journal of Geophysical Research*, **88**, 4321–4328, http://doi.org/10.1029/JB088iB05p04321

DILEK, Y. & THY, P. 2006. Age and petrogenesis of palgiogranite intrusions in the Ankara mélange, central Turkey. *Island Arc*, **15**, 44–57.

DUNLOP, D. J. 2002. Theory and application of the Day plot (Mrs/Ms v. Hcr/Hc) 1. Theoretical curves and tests using titanomagnetite data. *Journal of Geophysical Research*, **107**, 2056, EPM 4-1-4-22.

DUNLOP, D. J. & ÖZDEMIR, Ö. 1997. *Rock Magnetism: Fundamentals and Frontiers.* Cambridge University Press, New York, London and Cambridge.

ENKIN, R. J. & WATSON, G. S. 1996. Statistical analysis of paleomagnetic inclination data. *Geophysics Journal International*, **126**, 495–504.

FISHER, R. A. 1953. Dispersion on a sphere. *Proceedings of the Royal Society of London, Series A*, **217**, 295–305.

GARCÍA-LASANTA, C., OLIVA-URCIA, B., ROMÁN-BERDIEL, T., CASAS, A. M. & PÉREZ-LORENTE, F. 2013. Development of magnetic fabric in sedimentary rocks: insights from early compactional structures (ECS). *Geophysical Journal International*, **194**, 182–199.

GÖNCÜOĞLU, M. C., TOPRAK, G. M. V., KUŞCU, I., ERLER, A. & OLGUN, E. 1991. *Geology of the Western Part of the Central Anatolian Massif, Part 1: Southern part: Ankara, Turkey.* METU-TPC Project Report, 140 p. (in Turkish, unpublished).

GÖNCÜOĞLU, M. C., YALINIZ, K., KUŞCU, I., KÖKSAL, S. & DIRIK, K. 1993. *The Geology of the Central Area of the Central Anatolian Massif, Part 3: The Geological Evolution of the Central Kızılırmak Tertiary Basin.* Turkish Petroleum Coorporation. Report No. 3313 (unpublished).

GÖRÜR, N., OKTAY, F. Y., SEYMEN, İ. & ŞENGÖR, A. M. C. 1984. Paleotectonic evolution of the Tuzgölü basin complex, Central Anatolia, sedimentary record of a Neo-Tethyan closure. *In*: DIXON, J. E. & ROBERTSON, A. H. F. (eds) *The Geological Evolution of the Eastern Mediterranean.* Geological Society, London, Special Publications, **17**, 455–466.

GÖRÜR, N., TUYSUZ, O. & ŞENGOR, A. M. C. 1998. Tectonic evolution of the Central Anatolian Basins. *International Geology Review*, **40**, 831–850.

GÜRSOY, H., PIPER, J. D. A., TATAR, O. & MESCI, L. 1998. Paleomagnetic study of the Karaman and Karapinar volcanic complexes, central Turkey: neotectonic rotation in the south-central sector of the Anatolian block. *Tectonophysics*, **299**, 191–211.

HIRT, A., LOWRIE, W., CLENDENEN, W. S. & KLIGFIELD, R. 1993. Correlation of strain and the anisotropy of magnetic susceptibility in the Onaping Formation: evidence for a near-circular origin of the Sudbury basin. *Tectonophysics*, **225**, 231–254.

HISARLI, Z. M. 2011. New paleomagnetic constraints on the late Cretaceous and early Cenozoic tectonic history of the Eastern Pontides. *Journal of Geodynamics*, **52**, 114–128.

HROUDA, F. 1982. Magnetic anisotropy of rocks and its application in geology and geophysics. *Surveys in Geophysics*, **5**, 37–82.

İLBEYLI, N., PEARCE, J. A., THIRWALL, M. F. & MITCHELL, J. G. 2004. Petrogenesis of collision related plutonics in central Anatolia, Turkey. *Lithos*, **72**, 163–182.

İŞSEVEN, T. & TÜYSÜZ, O. 2006. Paleomagnetically defined rotations of fault-bounded continental blocks in the North Anatolian Shear Zone, North Central Anatolia. *Journal of Asian Earth Sciences*, **28**, 469–479.

JELÍNEK, V. 1977. *The Statistical Theory of Measuring Anisotropy of Magnetic Susceptibility of Rocks and its Application*. Geophysika, Brno.

JELÍNEK, V. 1978. Statistical processing of anisotropy of magnetic susceptibility measures on groups of specimens. *Studia Geophysica et Geodaetica*, **22**, 50–62.

JELÍNEK, V. 1981. Characterisation of the magnetic fabrics of rocks. *Tectonophysics*, **79**, 63–67.

KADIOĞLU, Y. K., DILEK, Y., GÜLEÇ, N. & FOLAND, K. A. 2003. Tectonomagmatic evolution of bimodal plutons in the Central Anatolian Crystalline Complex, Turkey. *Journal of Geology*, **111**, 671–690.

KADIOĞLU, Y. K., DILEK, Y. & FOLAND, K. A. 2006. Slab break-off and syncollisional origin of the Late Cretaceous magmatism in the Central Anatolian crystalline complex, Turkey. *In*: DILEK, Y. & PAVLIDES, S. (eds) *Postcollisional Tectonics and Magmatism in the Mediterranean Region and Asia*. Geological Society of America, Boulder, CO, Special Papers, **409**, 381–415.

KAYMAKÇI, N., WHITE, S. H. & VAN DIJK, P. M. 2000. Paleostress inversion in a multiphase deformed area: kinematic ve structural evolution of the Çankırı Basin (central Turkey), Part 1. *In*: BOZKURT, E., WINCHESTER, J. A. & PIPER, J. A. D. (eds) *Tectonics and Magmatism in Turkey and the Surrounding Area*. Geological Society, London, Special Publications, **173**, 445–473.

KAYMAKÇI, N., ÖZÇELIK, Y., WHITE, S. H. & VAN DIJK, P. M. 2001. Neogene tectonic development of the Çankırı basin (central Anatolia, Türkiye). *Bulletin of the Turkish Association of Petroleum Geology*, **13**, 27–56.

KAYMAKÇI, N., DUERMEIJER, C. E., LANGEREIS, C., WHITE, S. H. & VAN DIJK, P. M. 2003a. Palaeomagnetic evolution of the Çankırı basin (Central Anatolia, Turkey): implications for oroclinal bending due to indentation. *Geological Magazine*, **140**, 343–355.

KAYMAKÇI, N., WHITE, S. H. & VEIJK, P. M. 2003b. Kinematic and structural development of the Çankırı Basin (Central Anatolia, Turkey): a paleostress inversion study. *Tectonophysics*, **364**, 85–113.

KAYMAKÇI, N., ALDANMAZ, E., LANGEREIS, C., SPELL, T. L., GURER, O. F. & ZANETTI, K. A. 2007. Late Miocene transcurrent tectonics in NW Turkey: evidence from palaeomagnetism and 40Ar–39Ar dating of alkaline volcanic rocks. *Geological Magazine*, **144**, 379–392.

KAYMAKÇI, N., ÖZÇELIK, Y., WHITE, S. H. & VAN DIJK, P. M. 2009. Tectono-stratigraphy of the Çankırı Basin: Late Cretaceous to early Miocene evolution of the Neotethyan Suture Zone in Turkey. *In*: VAN HINSBERGEN, D. J., EDWARDS, M. A. & GOVERS, R. (eds) *Collision and Collapse at the Africa–Arabia–Eurasia Subduction Zone*. Geological Society, London, Special Publications, **311**, 67–106.

KIRSCHVINK, L. 1980. The least-squares line and plane and the analysis of palaeomagnetic data. *Geophysical Journal of the Royal Astronomical Society*, **62**, 699–718.

KISSEL, C., LAJ, C., POISSON, A. & GÖRÜR, N. 2003. Paleomagnetic reconstruction of the Cenozoic evolution of the Eastern Mediterranean. *Tectonophysics*, **362**, 199–217.

KLIGFIELD, R., OWENS, W. H. & LOWRIE, W. 1981. Magnetic susceptibility anisotropy, strain, and progressive deformation in Permian sediments from the Maritime Alps (France). *EPSL*, **55**, 181–189.

KOÇYIĞIT, A. 1991. An example of an accretionary forearc basin from northern central Anatolia and its implications for the history of subduction of Neo-Tethys in Turkey. *Bulletin of the Geological Society of America*, **103**, 22–36.

LOWRIE, W. 1990. Identification of ferromagnetic minerals in a rock by coercivity and unblocking temperature properties. *Geophysics Research Letters*, **17**, 159–162.

LOWRIE, W. & HIRT, A. M. 1986. Paleomagnetism in arcuate mountain belts. *In*: WEZEL, F. C. (ed.) *The Origin of Arcs*. Elsevier, Amsterdam, 141–158.

LUCIFORA, S., CIFELLI, F., ROJAY, F. B. & MATTEI, M. 2013. Paleomagnetic rotations in the Late Miocene sequence from the Çankırı Basin (Central Anatolia, Turkey): the role of strike–slip tectonics. *Turkish Journal of Earth Sciences*, **22**, 778–792.

MCELHINNY, M. W. 1964. Statistical significance of the fold test in palaeomagnetism, Geophys. *Journal of the Royal Astronomical Society*, **8**, 338–340, http://doi.org/10.1111/j.1365-246X.1964.tb06300.x

MCFADDEN, P. L. 1990. The fold test as an analytical tool. *Geophysics Journal International*, **135**, 329–338.

MCFADDEN, P. L. & MCELHINNY, M. W. 1990. Classification of the reversal test in paleomagnetism. *Geophysics Journal International*, **103**, 725–729.

MEIJERS, M. J. M., KAYMAKÇI, N., VAN HINSBERGEN, D. J. J., LANGEREIS, C. G., STEPHENSON, R. A. & HIPPOLYTE, J.-C. 2010. Late Cretaceous to Paleocene oroclinal bending in the Central Pontides (Turkey). *Tectonics*, **29**, TC4016, http://doi.org/10.1029/2009TC002620

MOCHALES, T., PUEYO, E. L., CASAS, A. M., BARNOLAS, A. & OLIVA-URCIA, B. 2010. Anisotropic magnetic susceptibility record of the kinematics of the Boltaña Anticline (Southern Pyrenees). *Geological Journal*, **45**, 562–581, http://doi.org/10.1002/gj.1207

MORRIS, A. & ROBERTSON, A. H. F. 1993. Miocene remagnetization of carbonate platform and Antalya Complex units within the Isparta Angle, SW Turkey. *Tectonophysics*, **220**, 243–266.

NAGATA, T. 1961. *Rock Magnetism*. Maruzen Ltd., Tokyo.

NAIRN, S., ROBERTSON, A. H. F., ÜNLÜGENÇ, U. C., TAŞLI, K. & İNAN, N. 2012. Tectonostratigraphic evolution of the Upper Cretaceous-Cenozoic central Anatolian basins: an integrated study of diachronous ocean basin closure and continental collision. *In*: ROBERTSON, A. H. F., PARLAK, O. & ÜNLÜGENÇ, U. C. (eds) *Geological Development of Anatolia and the Easternmost Mediterranean Region*. Geological Society, London, Special Publications, **372**, 343–383. First published online October 8, 2012, http://doi.org/10.1144/SP372.9

NORMAN, T. N. 1984. The role of the Ankara Mélange in the development of Anatolia. *In*: ROBERTSON, A. H. F. & DIXON, J. E. (eds) *The Geological Evolution of the Eastern Mediterranean*. Geological Society, London, Special Publications, **17**, 441–447.

OKAY, A. I. 1989. Tectonic units and sutures in the Pontides, northern Turkey. *In*: SENGÖR, A. M. C. (ed.) *Tectonic Evolution of the Tethyan Region, NATO Advance ASI Series*. Kluwer Academic Publications, 109–116.

OKAY, A. I. & GÖNCÜOĞLU, M. C. 2004. The Karakaya Complex: a review of data and concepts. *Turkish Journal of Earth Sciences*, **13**, 77–95.

OKAY, A. I., ŞENGÖR, A. M. C. & GÖRÜR, N. 1994. Kinematic history of the opening of the Black Sea and its effects on the surrounding regions. *Geology*, **22**, 267–270.

OKAY, A. I. & TÜYSÜZ, O. 1999. Tethyan sutures of northern Turkey. *In*: DURAND, B., JOLIVET, L., HORVÁT-HAND, F. & SÉRANNE, M. (eds) *The Mediterranean Basins: Tertiary extension within the Alpine orogen*. Geological Society, London, Special Publications, **156**, 475–515.

OKAY, A. I., TÜYSÜZ, O., SATIR, M., ÖZKAN-ALTINER, S., ALTINER, D., SHERLOCK, S. & EREN, R. H. 2006. Cretaceous and Triassic subduction-accretion, highpressure–low-temperature metamorphism, and continental growth in the Central Pontides, Turkey. *Geological Society of America Bulletin*, **118**, 1247–1269.

PARÉS, J. M. & VAN DER PLUIJM, B. A. 2003. Magnetic fabrics and strain in pencil structures of the Knobs Formation, Valley and Ridge Province, US Appalachians. *Journal of Structural Geology*, **25**, 1349–1358.

PETERS, M. K. 2010. Metamorphic evolution of the Hırkadağ and Kırşehir massifs, Central Turkey, based on geothermobarometry of HT metapelitic rocks. MSc thesis, Faculty of Geosciences, Department of Structural Geology and Tectonics, Utrecht University.

PICKETT, E. A. & ROBERTSON, A. H. F. 1996. Formation of the Late Palaeozoic–Early Mesozoic Karakaya Complex and related ophiolites in NW Turkey by Palaeotethyan subduction–accretion. *Journal of the Geological Society, London*, **153**, 995–1009.

PIPER, J. D. A., MOORE, J. M., TATAR, O., GÜRSOY, H. & PARK, R. G. 1996. Palaeomagnetic study of crustal deformation across an intracontinental transform: the North Anatolian Fault zone in northern Turkey. *In*: MORRIS, A. & TARLING, D. H. (eds) *Paleomagnetism of the Eastern Mediterranean Regions*. Geological Society, London, Special Publications, **105**, 299–310.

PIPER, J. D. A., GÜRSOY, H. & TATAR, O. 2002. Palaeomagnetism and magnetic properties of the Cappadocian ignimbrite succession, central Turkey and Neogene tectonics of the Anatolian collage. *Journal of Volcanology and Geothermal Research*, **117**, 237–262.

PIPER, J. D. A., GURSOY, H., TATAR, O., BECK, M. E., RAO, A., KOCBULUT, F. & MESCI, B. L. 2010. Distributed neotectonic deformation in the Anatolides of Turkey: a palaeomagnetic analysis. *Tectonophysics*, **488**, 31–50.

PLATZMAN, E. S., PLATT, J. P., TAPIRDAMAZ, M. C., SANVER, M. & RUNDLE, C. C. 1994. Why are there no clockwise rotations along the North Anatolian Fault Zone? *Journal of Geophysical Research*, **99**, 21705–21715.

PLATZMAN, E. S., TAPIRDAMAZ, C. & SANVER, M. 1998. Neogene anticlockwise rotation of central Anatolia (Turkey): preliminary paleomagnetic and geochronological results. *Tectonophysics*, **299**, 175–189.

POISSON, A., GUEZOU, J. C., ÖZTÜRK, A., INAN, S., TEMIZ, H., KAVAK, K. & ÖZDEN, S. 1996. Tectonic setting and evolution of the Sivas Basin, Central Anatolia, Turkey. *International Geology Review*, **38**, 838–853.

PUEYO-ANCHUELA, Ó., PUEYO, E. L., POCOVÍ JUAN, A. & GIL IMAZ, A. 2012. Vertical axis rotations in fold and thrust belts: comparison of AMS and paleomagnetic data in the Western External Sierras (Southern Pyrenees). *Tectonophysics*, **532–535**, 119–133.

RICE, S. P., ROBERTSON, A. H. F. & USTAOMER, T. 2006. Late Cretaceous–Early Cenozoic tectonic evolution of the Eurasian active margin in the central and eastern Pontides, Northern Turkey. *In*: ROBERTSON, A. H. F. & MOUNTRAKIS, D. (eds) *Tectonic Development of the Eastern Mediterranean Region*. Geological Society, London, Special Publications, **260**, 413–445.

RICE, S. P., ROBERTSON, A. H. F., USTAOMER, T., INAN, N. & TASLI, K. 2009. Late Cretaceous–Early Eocene tectonic development of the Tethyan suture zone in the Erzincan area, Eastern Pontides, Turkey. *Geological Magazine*, **146**, 567–590.

ROBERTSON, A. H. F. 2002. Overview of the genesis and emplacement of Mesozoic ophiolites in the Eastern Mediterranean Tethyan region. *Lithos*, **65**, 1–67.

ROBERTSON, A. H. F. & DIXON, J. E. 1984. Introduction: aspects of the geological evolution of the Eastern Mediterranean. *In*: DIXON, J. E. & ROBERTSON, A. H. F. (eds) *The Geological Evolution of the Eastern Mediterranean*. Geological Society, London, Special Publications, **17**, 1–74.

ROBERTSON, A. H. F., PARLAK, O. & USTAÖMER, T. 2009. Melange genesis and ophiolite emplacement related to subduction of the northern margin of the Tauride–Anatolide continent, central and western Turkey. *In*: VAN HINSBERGEN, D. J. J., EDWARDS, M. A. & GOVERS, R. (eds) *Collision and Collapse at the Africa–Arabia–Eurasia Subduction Zone*. Geological Society, London, Special Publications, **311**, 9–66.

ROJAY, B. 1995. Post-Triassic evolution of the Central Pontides: evidence from Amasya region, Northern Anatolia. *Geologica Romana*, **31**, 329–350.

ROJAY, B. & SÜZEN, M. L. 1997. Tectonostratigraphic evolution of the Cretaceous dynamic basins on accretionary ophiolitic melange prism, SW of Ankara region. *Turkish Petroleum Geology Bulletin*, **9**, 1–12.

SARIBUDAK, M. 1989. New results and a paleomagnetic overview of the Pontides in northern Turkey. *Geophysics Journal International*, **99**, 521–531.

ŞENGÖR, A. M. C. & YILMAZ, Y. 1981. Tethyan evolution of Turkey: a plate tectonic approach. *Tectonophysics*, **75**, 181–241.

ŞENGÖR, A. M. C., YILMAZ, Y. & KETIN, I. 1982. Remnants of a pre-late Jurassic ocean in northern Turkey – Fragments of Permian-Triassic Paleo-Tethys. *Geological Society of America Bulletin*, **93**, 932–936.

SOTO, R., LARRASOAÑA, J. C., ARLEGUI, L. E., BEAMUD, E., OLIVA-URCIA, B. & SIMÓ, J. L. 2009. Reliability of magnetic fabric of weakly deformed mudrocks as a palaeostress indicator in compressive settings. *Journal of Structural Geology*, **31**, 512–522.

TAN, X. & KODAMA, K. P. 2003. An analytical solution for correcting palaeomagnetic inclination error. *Geophysics Journal International*, **152**, 228–236.

TARLING, D. H. & HROUDA, F. 1993. *The Magnetic Anisotropy of Rocks*. Chapman and Hall, London.

TÜYSÜZ, O. & DELLALOĞLU, A. A. 1992. Tectonic units and geologic evolution of the Çankırı–Çorum Basin. *9th Turkish Petroleum Congress Proceedings*, Ankara, 1–17.

TÜYSÜZ, O., DELLALOĞLU, A. A. & TERZIOĞLU, N. 1995. A magmatic belt within the Neo-Tethyan suture zone and its role in the tectonic evolution of northern Turkey. *Tectonophysics*, **243**, 173–191.

USTAÖMER, T. & ROBERTSON, A. H. F. 1997. Tectonic sedimentary evolution of the North Tethyan margin in the Central Pontides of Northern Turkey. *In*: ROBINSON, A. G. (ed.) *Regional and Petroleum Geology of the Black Sea and Surrounding Region*. AAPG, Memoirs, Tulsa, **68**, 255–290.

VAN DER VOO, R. 1968. Palaeomagnetism and the Alpine tectonics of Eurasia, Part 4, Jurassic, Cretaceous and Eocene pole positions from NE Turkey. *Tectonophysics*, **6**, 251–269.

WATSON, G. S. 1956. A test for randomness. *Monthly Notices of the Royal Astronomical Society*, **7**, 153–159.

WHITNEY, D. L. & HAMILTON, M. 2004. Timing of high-grade metamorphism in central Turkey and the assembly of Anatolia. *Journal of the Geological Society of London*, **161**, 823–828.

YAVAŞOĞLU, H., TARI, E., TÜYSÜZ, O., ÇAKIR, Z. & ERGINTAV, S. 2011. Determining and modeling tectonic movements along the central part of the North Anatolian Fault (Turkey) using geodetic measurements. *Journal of Geodynamics*, **51**, 339–343.

YILMAZ, Y., YIĞITBAŞ, E. & GENÇ, Ş. C. 1993. Ophiolitic and metamorphic assemblages of southeast Anatolia and their significance in the geological evolution of the orogenic belt. *Tectonics*, **12**, 1280–1297.

YILMAZ, Y., TÜYSÜZ, O., YIĞITBAŞ, E., GENÇ, Ş. C., & ŞENGÖR, A. M. C. 1997. Geology and tectonic evolution of the Pontides. *In*: ROBINSON, A. G. (ed.) *Regional and Petroleum Geology of the Black Sea and Surrounding Region*. AAPG, Memoirs, Tulsa, **68**, 183–226.

YOLDAŞ, R. 1982. The geology between Tosya (Kastamonu) and Bayat (Corum) area. PhD thesis, University of Istanbul, College of Science.

ZIJDERVELD, J. D. A. 1967. AC Demagnetization of rocks: analysis of results. *In*: RUNCORN, S. K., CREER, K. M. & COLLINSON, D. W. (eds) *Methods in Paleomagnetism*. Elsevier, Amsterdam, 254–286.

# Timing of magnetization and vertical-axis rotations of the Cotiella massif (Late Cretaceous, South Central Pyrenees)


MIGUEL GARCÉS[1]*, JESÚS GARCÍA-SENZ[2], JOSEP ANTÓN MUÑOZ[3],
BERTA LÓPEZ-MIR[4] & ELISABET BEAMUD[3,5]

[1]*Institut Geomodels, Departament d'Estratigrafia Paleontologia i Geociències Marines, Facultat de Geologia, Universitat de Barcelona, 08028 Barcelona, Spain*

[2]*Instituto Geológico y Minero de España, La Calera 1, 28760 Tres Cantos, Madrid, Spain*

[3]*Institut Geomodels, Departament de Geodinàmica i Geofísica, Facultat de Geologia, Universitat de Barcelona, 08028 Barcelona, Spain*

[4]*CASP, West Building, 181A Huntingdon Road, Cambridge CB3 0DH, UK*

[5]*Laboratori de Paleomagnetisme CCiTUB-CSIC, Institut de Ciències de La Terra 'Jaume Almera', 08028 Barcelona, Spain*

**Corresponding author (e-mail: mgarces@ub.edu)*



**Abstract:** The Cotiella thrust sheet in the South Central Pyrenees is a well-studied example of an extensional basin that formed at the northern Iberian rift margin during the Late Cretaceous, and was inverted in the Late Cretaceous and Cenozoic on top of the South Pyrenean Central Unit. A palaeomagnetic study of the thick sequence of shallow marine syn-extensional carbonates filling the Cotiella Basin reveals the presence of a widespread secondary magnetization of both normal and reversed polarities. Neither *in situ* nor bedding tilt correction provides a satisfactory grouping of the palaeomagnetic directions, thus suggesting syn-deformational magnetization. Best clustering of palaeomagnetic vectors is obtained after unfolding into a post-extensional pre-compressional stage, coherent with an acquisition age between Santonian and Paleocene. Strong magnetic enhancement and very distinct magnetic properties with depth suggest a link with burial diagenesis of organic matter, precipitation of iron sulphides and subsequent oxidation to SD fine-grained magnetite. The secondary magnetizations of the Cotiella record significant vertical axis rotations ranging from 14° to 52° clockwise, coherent with the scenario of the Cotiella thrust sheet transported piggy-back and bent during the emplacement of the lower Gavarnie–Sierras Exteriores thrust sheet.


The importance of remagnetizations in orogenic belts has been increasingly recognized in palaeomagnetic studies since the observation of the widespread late Palaeozoic remagnetization in the Appalaches (McCabe *et al.* 1983). Remagnetizations were first detected as post-folding signatures, thus post-dating major deformational events. Early studies in the Appalaches tended to explain the widespread and pervasive late Palaeozoic remagnetization as an orogen-scale single event. This interpretation was supported by the fact that the remagnetization yielded always reversed polarities (acquired during the Kiaman superchron), suggesting a relative isochronous acquisition. An orogen-scale general model based on migrating orogenic fluids (Oliver 1986) was proposed to link the remagnetization process with mountain building. However, later studies have challenged the post-folding character and the relative synchronicity

of remagnetizations in orogenic belts. A well-documented case example is the Cantabrian Arc, showing a multi-phase remagnetization (Weil *et al.* 2000) with different remagnetization mechanisms occurring at different locations and ages. Studies on younger settings such as the alpine collisional belt have shown that, despite remagnetization appearing to be intimately linked with orogenesis, the variety of remagnetization mechanisms may be rather complex. In the Iberian Peninsula a widespread early Cretaceous remagnetization was found in the Mesozoic basins of the Iberian Chain (Juárez *et al.* 1998; Villalaín 2003). Its age correlates well with the low-grade Albian–Cenomanian metamorphism in the Cameros Basin, suggesting a thermoviscous origin associated with the rifting phase of the northern margin of the Iberian Plate. In the Pyrenean Chain a variety of Mesozoic to Cenozoic sedimentary formations have been proven to carry


*From*: PUEYO, E. L., CIFELLI, F., SUSSMAN, A. J. & OLIVA-URCIA, B. (eds) 2016. *Palaeomagnetism in Fold and Thrust Belts: New Perspectives*. Geological Society, London, Special Publications, **425**, 213–232.
First published online October 23, 2015, http://doi.org/10.1144/SP425.11

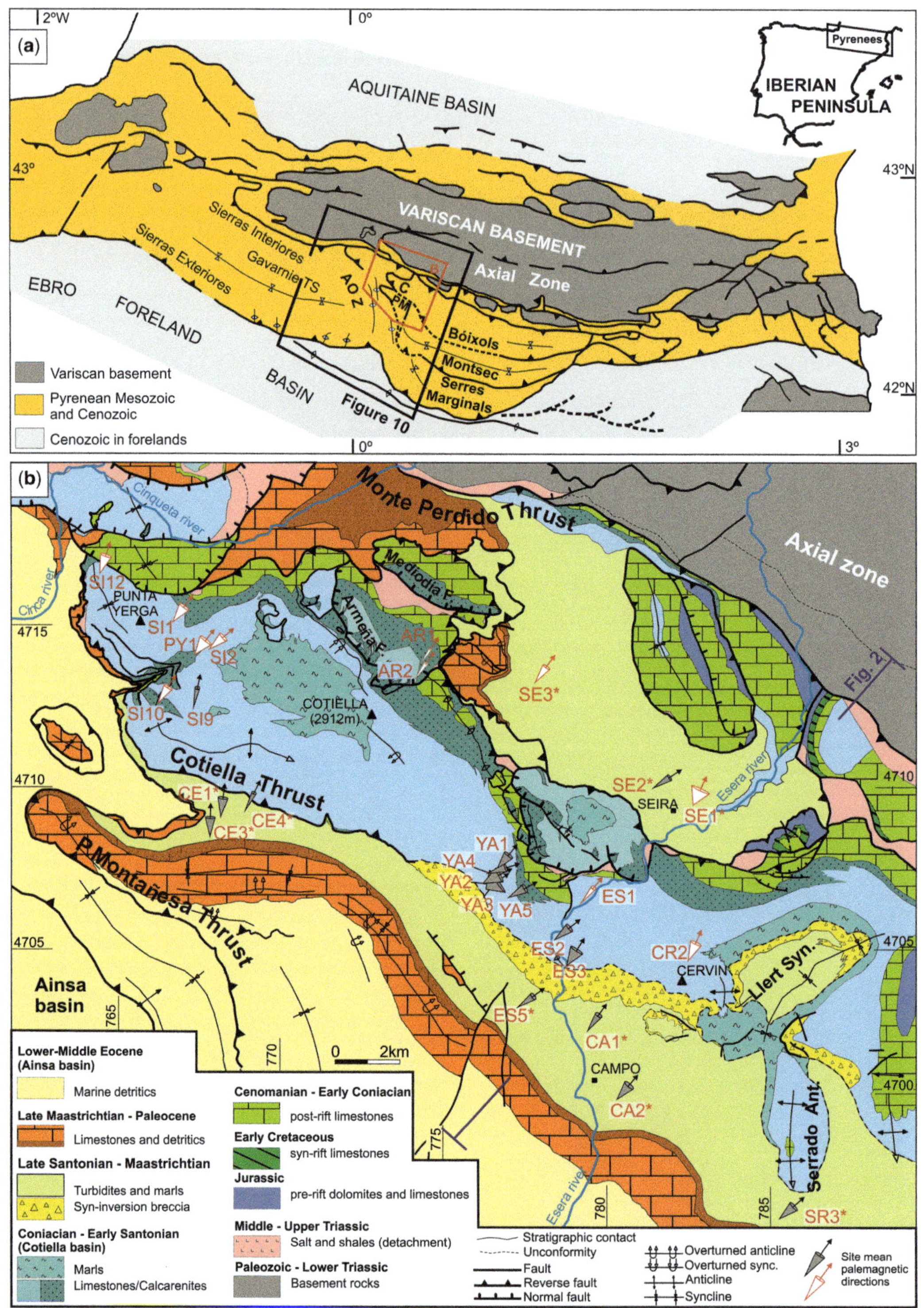





**Fig. 1.** Setting of the Cotiella massif in the frame of the south Pyrenean fold and thrust belt. (**a**) Structural map of the Pyrenees. (**b**) Detailed geological map of the Cotiella massif and surrounding areas (López-Mir 2013) with location

secondary magnetizations of different inferred ages and geographic distributions. In the western Pyrenees, a syn-folding secondary magnetization of the Permo-triassic red beds is interpreted as a Cretaceous thermoviscous component (Larrasoaña *et al.* 2003), similarly to what is observed in the Iberian chain. Contrastingly, in the Organyà Basin (eastern Pyrenees), remagnetization of early Cretaceous limestones is interpreted as resulting from burial diagenesis (Dinarès-Turell & García-Senz 2000; Gong *et al.* 2008). In addition to these early remagnetization signatures, various syn- to post-folding remagnetizations have been described affecting the Late Cretaceous sediments of the Pyrenean Internal Sierras (Izquierdo-Llavall *et al.* 2015) and the Eocene sediments of the Jaca Basin in the South-Central Pyrenees (Oliva-Urcia *et al.* 2008). In this latter case, a new remagnetization mechanism is suggested, which includes rotation of magnetic grains associated with processes of pressure-solution during cleavage formation.

Despite its often puzzling nature, the study of remagnetizations provides valuable information on key geological processes related to mountain building and associated foreland basins (Elmore *et al.* 2012). The nature and timing of diagenetic processes can often be constrained by palaeomagnetic data. In this study we describe a remagnetization event affecting Late Cretaceous limestones of the Cotiella massif in the South-Central Pyrenees and discuss both the timing and acquisition mechanisms.

## Setting of the Cotiella massif and tectonostratigraphic units

The southern fold-and-thrust belt of the Pyrenean orogen (Fig. 1) is an imbricate of cover thrust sheets of Mesozoic and Cenozoic sedimentary rocks transported over the autochthonous sediments of the Ebro foreland basin. The main sequence of thrust emplacement occurred from north to south in a piggy-back mode: the uppermost thrust-sheet, the Cotiella–Bóixols, developed during the Late Cretaceous, the Peña Montañesa–Montsec thrust sheet formed during the Paleocene–late Ypresian and the most external Sierras Exteriores thrust-sheet was emplaced during the Lutetian to Miocene (Séguret 1972; Garrido-Megías 1973; Muñoz 2002). The trailing edge of the Cotiella and Peña Montañesa–Montsec thrust sheets is folded at the basement duplex culmination of the Axial zone

(Fig. 2). In the footwall of the Cotiella and Peña Montañesa–Montsec thrust sheets the Ainsa piggy-back basin evolved on top of the Gavarnie–Sierras Exteriores thrust sheet. The structure of the Ainsa Basin is characterized by kilometre-scale north–south-trending folds that grew synchronously to the sedimentary filling.

The oldest rocks in the study area constitute the Variscan basement included in the Gavarnie thrust sheet. These units are unconformably covered by the Permian and Bunter sandstones and the Upper Triassic Keuper shales and evaporites, which acted as a detachment layer of the Late Cretaceous to Eocene cover units.

Pre-rift Jurassic and syn-rift Early Cretaceous sediments are poorly represented in the area because of their marginal position with respect to the main extensional basins and also because of the uplift episode in the middle–late Albian during the rift–postrift transition. As a result, the cover sedimentary sequence mostly consists of the following units:

(1) Late Albian–early Santonian sandstones, shelf limestones, marls and calcarenites. They have a typical thickness of few hundreds of metres that gradually decreases towards the south, shaping a tilted block. In the Cotiella extensional basin, however, they form an expanded sequence of hemipelagic limestones (Aguasalenz Fm and calcarenites (Maciños) that reach as much as 6 km in the hanging walls of the Cotiella, Armeña and Mediodía extensional faults (Figs 1b & 2; García-Senz 2002; McClay *et al.* 2004) and associated Keuper diapirs (López-Mir 2013; Lopez-Mir *et al.* 2014, 2015). The fanning of beds in the hanging wall of these faults and diapirs evidences the syn-extensional character of these units.

(2) Late Santonian–Maastrichtian turbidites (Mascarell Fm and Campo breccia Mb) and marls (Campo and Barbaruens Fms), which represent the filling of the syn-inversion trough linked to the reactivation of the extensional listric faults to form the Cotiella–Bóixols thrust sheet.

(3) Maastrichtian to Paleocene shelf limestones and dolomites that are transgressive southeastwards onto the Garumnian red beds and lacustrine limestones that completely fill the syn-inversion trough. They represent a period of tectonic quiescence preceding the acceleration of thrust activity during the Eocene.

---

**Fig. 1.** (*Continued*) of sampled sites. Palaeomagnetic directions represent either normal-polarity components (black) or the antipodal of reverse polarity components (red). Primary magnetizations (in tectonic corrected coordinates) are indicated with an asterisk. The remaining sites yielded secondary magnetizations (in syn- or post-folding coordinates). See Tables 1, 2, 3 for details of the tectonic restitution applied to each site.

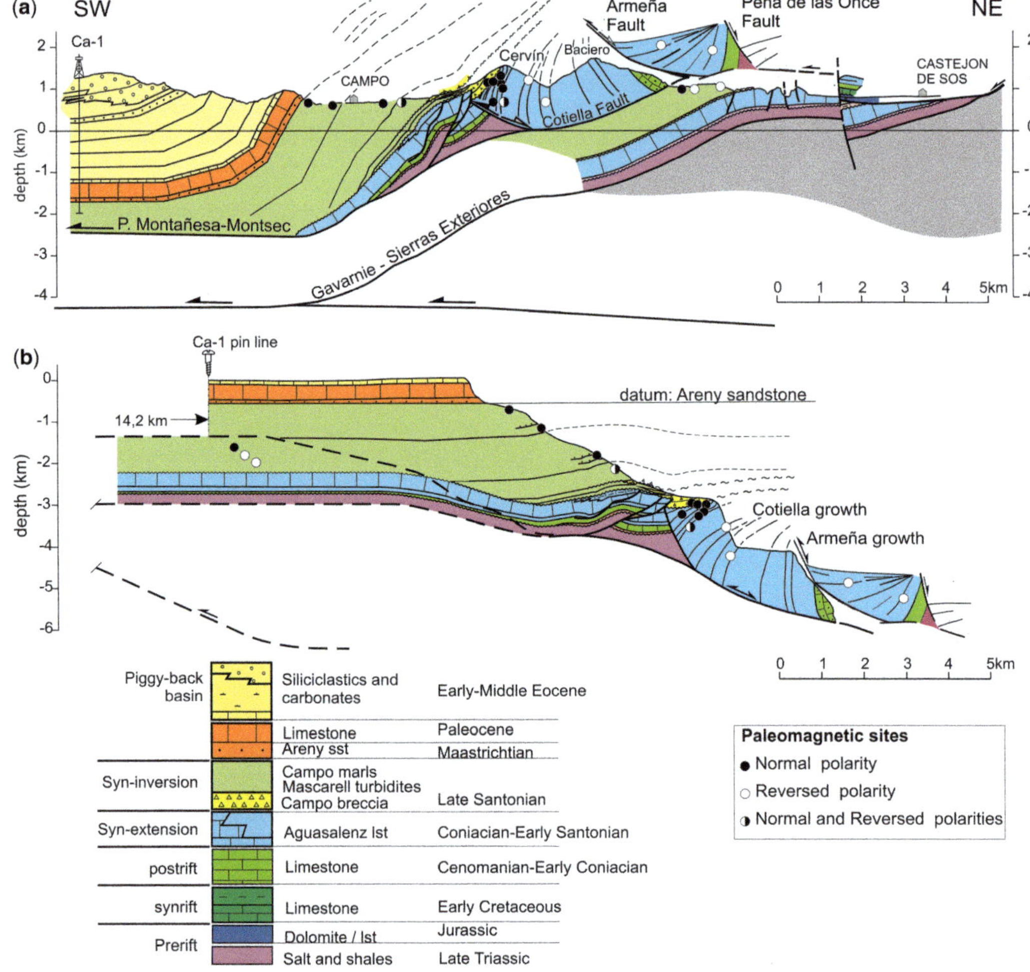



**Fig. 2.** (**a**) Geological cross-section along the Ésera River valley with projection of the sampled palaeomagnetic sites. (**b**) A latest Cretaceous cross-section restoration (modified from García-Senz 2002). See location in Fig 1.

(4) Early–Middle Eocene marls and sandstones that correspond to the progradation of the deltaic to slope-turbiditic complex of the Ainsa Basin. They fill the new foredeep developed synchronous to the southwards displacement of the Peña Montañesa–Montsec thrust sheet.

## Palaeomagnetic data

Earlier palaeomagnetic studies (Dinarès-Turell 1992; Pueyo-Morer 2000; Pueyo *et al.* 2002; Oliva-Urcia & Pueyo 2007; Mochales *et al.* 2012; Muñoz *et al.* 2013; Izquierdo-Llavall *et al.* 2015) have provided abundant information on vertical-axis block rotations associated with the emplacement of the Peña Montañesa–Montsec and the

Gavarnie–Sierras Exteriores thrust sheets. This paper aims to fill the gap in the Pyrenean palaeomagnetic database by studying the upper thrust sheets of Cotiella–Bóixols in the surrounding area of the Cotiella peak, between the Ésera and Cinqueta rivers (Fig. 1).

Seventeen sites were sampled in the Cotiella–Bóixols thrust sheet (13 in the Aguasalenz limestones, three in the Maciños calcarenites and one in the Campo breccia). Complementarily, and in order to check for differential rotations, 10 more sites were obtained in the underlying units of Gavarnie and Peña Montañesa–Montsec thrust sheets, yielding a total of 27 sites (Table 1).

The bulk magnetic susceptibility and anisotropy of magnetic susceptibility were measured in all samples by means of a Kappabridge KLY-2 (Geofyzika,

**Table 1.** *Palaeomagnetic data obtained in this study grouped into structural units (Gavarnie, Peña Montañesa–Montsec and Cotiella thrust sheet) and sectors. dec, declination; inc, inclination; N/R, Normal/Reverse polarity*

| Thrust sheet | Sector | Site | Code | UTM (30 T) X | UTM (30 T) Y | Age | Rock unit | $S_o$ Dip direction | $S_o$ Plunge | $n$ | N/R | ChRM A/m | In situ Dec | In situ Inc | In situ $k$ | In situ $\alpha_{95}$ | Tilt corrected Dec | Tilt corrected Inc | Tilt corrected $k$ | Tilt corrected $\alpha_{95}$ |
|---|---|---|---|---|---|---|---|---|---|---|---|---|---|---|---|---|---|---|---|---|
| Gavarnie | **Seira** | | | | | | | | | | | | | | | | | | | |
| | | *Seira 1* | SE1 | 782400 | 4704602 | Camp–Maast | Barbaruens marls | 225 | 16 | 10 | R | $6.05 \times 10^{-5}$ | 205 | −41 | 6 | 23 | 198 | −56 | | |
| | | *Seira 2* | SE2 | 781288 | 4709932 | Camp–Maast | Barbaruens marls | 231 | 20 | 8 | N | $4.23 \times 10^{-4}$ | 057 | 37 | 51 | 8 | 061 | 56 | | |
| | | *Seira 3* | SE3 | 777601 | 4712929 | Camp–Maast | Barbaruens marls | 230 | 10 | 8 | R | $1.74 \times 10^{-4}$ | 216 | −49 | 53 | 8 | 213 | −57 | | |
| | | **Mean** | | | | | | | | **3** | **N** | | **040** | **43** | **34** | **21** | **037** | **57** | **44.4** | **18.7** |
| Peña Montañesa–Montsec | **Campo** | | | | | | | | | | | | | | | | | | | |
| | | *Serrado 3* | SR3 | 785861 | 4696542 | Camp–Maast | Campo marls | 186 | 30 | 8 | N | $1.84 \times 10^{-4}$ | 036 | 23 | 27 | 11 | 049 | 48 | | |
| | | *Campo 2* | CA2 | 780223 | 4700340 | Camp–Maast | Campo marls | 197 | 45 | 9 | N | $1.39 \times 10^{-4}$ | 032 | 06 | 22 | 11 | 041 | 50 | | |
| | | *Campo 1* | CA1 | 779319 | 4702555 | Camp–Maast | Campo marls | 108 | 47 | 13 | N | $8.86 \times 10^{-5}$ | 031 | −09 | 24 | 9 | 035 | 37 | | |
| | | *Esera 5* | ES5 | 777367 | 4703281 | Camp–Maast | Mascarell turbidites | 232 | 56 | 8 | N(1R) | $1.31 \times 10^{-4}$ | 053 | −03 | 78 | 6 | 054 | 53 | | |
| | | **Mean** | | | | | | | | **4** | **N** | | **038** | **05** | **22** | **20** | **045** | **50** | **213.1** | **6.3** |
| | **Ceresa** | | | | | | | | | | | | | | | | | | | |
| | | *Ceresa 1* | CE1 | 768366 | 4708924 | U. Sant–Camp | Mascarell turbidites | 205 | 73 | 6 | N | $1.54 \times 10^{-4}$ | 016 | −11 | 91 | 7 | 007 | 61 | | |
| | | *Ceresa 3* | CE3 | 767696 | 4708601 | U. Sant–Camp | Mascarell turbidites | 191 | 25 | 8 | N | $3.14 \times 10^{-4}$ | 359 | 15 | 28 | 11 | 359 | 44 | | |
| | | *Ceresa 4* | CE4 | 769163 | 4708971 | U. Sant–Camp | Mascarell turbidites | 153 | 37 | 9 | N | $6.52 \times 10^{-4}$ | 009 | 20 | 112 | 5 | 027 | 47 | | |
| | | **Mean** | | | | | | | | **3** | **N** | | **008** | **08** | **19** | **29** | **011** | **51** | **37.8** | **20.3** |
| Cotiella–Boixols | **Punta Yerga–Santa Isabel** | | | | | | | | | | | | | | | | | | | |
| | | *Santa Isabel 1* | SI1 | 766586 | 4714907 | Coniac–Santon | Aguas Salenz Lst. | 309 | 06 | 6 | R | $5.36 \times 10^{-5}$ | 217 | −23 | 19 | 16 | 214 | −23 | | |
| | | *Santa Isabel 2* | SI2 | 767714 | 4714055 | Coniac–Santon | Aguas Salenz Lst. | 198 | 29 | 9 | R | $8.67 \times 10^{-5}$ | 230 | −42 | 23 | 11 | 250 | −69 | | |
| | | *Santa Isabel 9* | SI9 | 767159 | 4712341 | Coniac–Santon | Aguas Salenz Lst. | 209 | 34 | 10 | N | $3.40 \times 10^{-4}$ | 012 | 49 | 112 | 5 | 324 | 78 | | |
| | | *Santa Isabel 10* | SI10 | 766018 | 4712529 | Coniac–Santon | Maciños Cotiella | 244 | 69 | 5 | R | $5.88 \times 10^{-4}$ | 213 | −19 | 37 | 13 | 152 | −61 | | |
| | | *Santa Isabel 12* | SI12 | 764640 | 4716754 | Coniac–Santon | Maciños Cotiella | 204 | 80 | 6 | R | $6.39 \times 10^{-5}$ | 199 | −40 | 21 | 14 | 032 | −59 | | |
| | | *Punta Yerga 1* | PY1 | 767484 | 4714433 | Coniac–Santon | Maciños Cotiella | 207 | 55 | 4 | R | $1.15 \times 10^{-3}$ | 218 | −52 | 49 | 13 | 004 | −72 | | |
| | | **Mean** | | | | | | | | **6** | **N** | | **031** | **38** | **22** | **15** | **015** | **81** | **5.4** | **31.8** |
| | **Armeña** | | | | | | | | | | | | | | | | | | | |
| | | *Armeña 1* | AR1 | 774722 | 4714105 | Coniac–Santon | Aguas Salenz Lst. | 221 | 55 | 13 | R | $1.53 \times 10^{-3}$ | 214 | −42 | 190 | 3 | 079 | −82 | | |
| | | *Armeña 2* | AR2 | 774066 | 4713223 | Coniac–Santon | Aguas Salenz Lst. | 220 | 27 | 11 | R | $9.62 \times 10^{-4}$ | 215 | −49 | 172 | 4 | 207 | −75 | | |
| | | **Mean** | | | | | | | | **2** | **N** | | **035** | **45** | **259** | **16** | **351** | **84** | **30** | **47.3** |
| | **Yali – Esera** | | | | | | | | | | | | | | | | | | | |
| | | *Yali 1* | YA1 | 776194 | 4707428 | Upp. Santonian | Campo Breccia | 067 | 75* | 7 | N | $1.03 \times 10^{-4}$ | 079 | −52 | 42 | 9 | 079 | 52 | | |
| | | *Yali 2* | YA2 | 776184 | 4706980 | Coniac–Santon | Aguas Salenz Lst. | 090 | 76* | 8 | N(1R) | $7.70 \times 10^{-5}$ | 091 | −39 | 21 | 13 | 084 | 66 | | |
| | | *Yali 3* | YA3 | 776201 | 4706879 | Coniac–Santon | Aguas Salenz Lst. | 083 | 70* | 4 | N | $3.06 \times 10^{-5}$ | 084 | −39 | 19 | 22 | 087 | 78 | | |
| | | *Yali 4* | YA4 | 776381 | 4707184 | Coniac–Santon | Aguas Salenz Lst. | 102 | 69* | 6 | N | $6.24 \times 10^{-5}$ | 096 | −36 | 98 | 7 | 084 | 73 | | |
| | | *Yali 5* | YA5 | 776870 | 4706674 | Coniac–Santon | Aguas Salenz Lst. | 095 | 57* | 6 | N | $5.73 \times 10^{-5}$ | 100 | −47 | 50 | 10 | 102 | 73 | | |
| | | *Esera 3* | ES3 | 778718 | 4704477 | Coniac–Santon | Aguas Salenz Lst. | 209 | 89 | 10 | N | $8.24 \times 10^{-5}$ | 036 | 17 | 29 | 9 | 186 | 66 | | |
| | | *Esera 2* | ES2 | 778389 | 4705300 | Coniac–Santon | Aguas Salenz Lst. | 041 | 60* | 10 | N | $1.19 \times 10^{-4}$ | 053 | −01 | 22 | 11 | 176 | 55 | | |
| | | *Esera 1* | ES1 | 779364 | 4706407 | Coniac–Santon | Aguas Salenz Lst. | 028 | 74* | 9 | R | $4.13 \times 10^{-3}$ | 221 | −10 | 180 | 4 | 007 | −63 | | |
| | | *Cervín 2* | CR2 | 782220 | 4708842 | Coniac–Santon | Aguas Salenz Lst. | 039 | 87* | 8 | R | $9.24 \times 10^{-5}$ | 213 | −05 | 27 | 11 | 082 | −80 | | |
| | | **Mean** | | | | | | | | **9** | **N** | | **064** | **−23** | **5** | **26** | **129** | **76** | **14** | **14** |

Sector mean directions are calculated in geographic and tilt-corrected coordinates.

ChRM, Characteristic remanent magnetization.

*Overturned bedding.

$n$, number of samples.

$k$, fisherian precision parameter of the ChRM mean direction.

$\alpha_{95}$, 95% confidence angle about the mean.

Brno) following the standard 15 measurements routine (Jelinek 1981). One or two samples per core were stepwise demagnetized and the natural remanent magnetization (NRM) measured in a 2G superconducting rock magnetometer at the palaeomagnetic laboratory of the University of Barcelona. A minimum of eight samples per site were thermally demagnetized and, at least two samples per site were alternating field (AF) demagnetized and then subjected to stepwise Isothermal Remanent Magnetization (IRM) acquisition and three-axis IRM thermal demagnetization (Lowrie 1990) to constrain the mineralogy of magnetic carriers. Grain size distribution of magnetic mineralogy was constrained by means of observation of magnetic hysteresis loops on a limited number of samples using a MicroMag Vibrating Sample Magnetometer hosted at the

Paleomagnetic Laboratory Fort Hoffddijk (Utrecht University).

## Magnetic properties and mineralogy

The average magnetic susceptibility (MS) of the Campo and Barbaruens marls (Fig. 3a) was higher than that of the Aguasalenz limestones, in agreement with the higher content of clay-minerals in the marls relative to limestones. However, something intriguing was the distribution of MS data from the Aguasalenz limestones into two distinct groups. Despite its lithological homogeneity, a group of sites yielded low MS values, ranging from 10 to $20 \times 10^{-6}$ SI, while a second group clustered around notably high MS values ranging

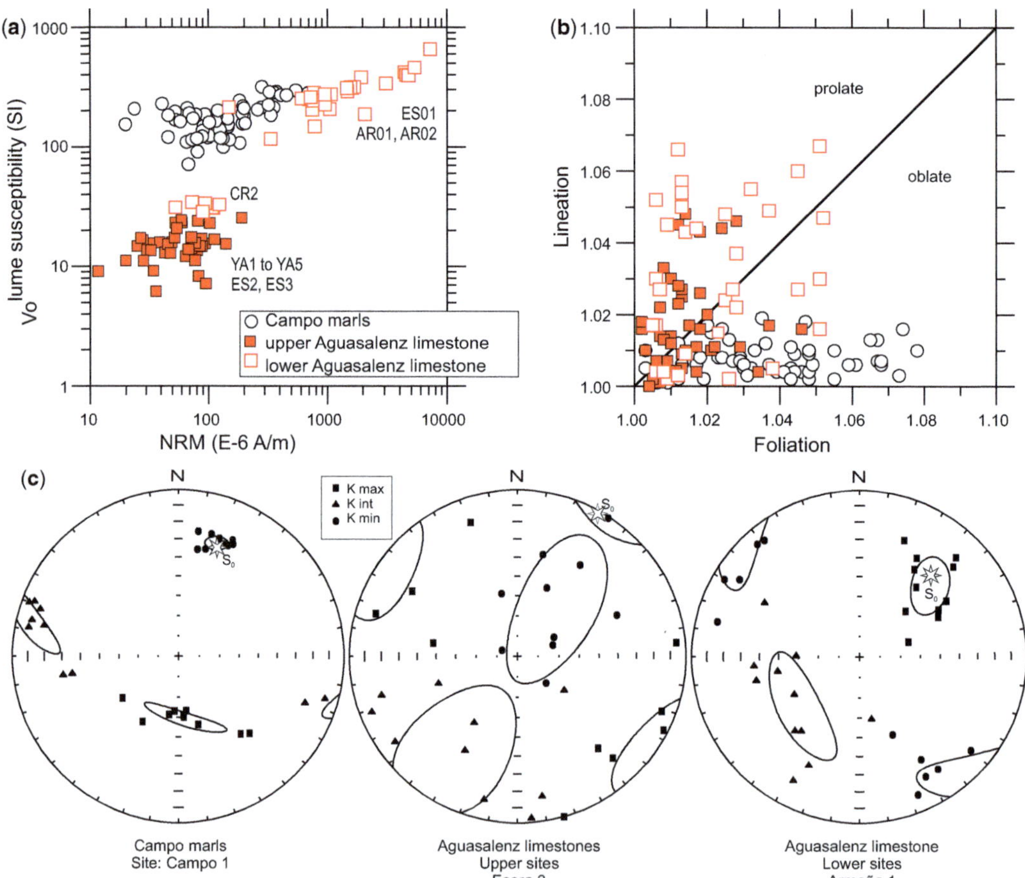



**Fig. 3.** Magnetic properties of representative samples from the Campo, Yali-Ésera and Armeña sectors. (**a**) Volume magnetic susceptibility plotted against NRM shows distribution of Aguasalenz limestone samples into two distinct groups (see text). (**b**) AMS parameters of Lineation and Foliation. (**c**) Stereoplots of the AMS ellipsoid axes. Oblate normal depositional fabrics in the Campo marls (left) to triaxial and prolate inverse fabrics in the Aguasalenz lst (right). $S_o$: bedding poles.

from 100 to $700 \times 10^{-6}$ SI. An apparent stratigraphic correspondence exists with these two types of magnetic behaviour, with much higher MS values being observed in samples from the lower units within the Aguasalenz Fm.

Anisotropy of magnetic susceptibility (AMS) data from the marl samples yielded characteristic oblate ellipsoids with lineation ($L$) lower than 1.020 and foliation ($F$) ranging from 1.000 to 1.080 (Fig. 3b). Minimum anisotropy axes (K3) aligned parallel to bedding poles and K1 and K2 distributed over the bedding plane. The fact that K1 does not match the orientation of folding axes indicates that AMS reflects a normal depositional fabric (Fig.

3c). AMS data of the Aguasalenz Limestone yielded triaxial to prolate ellipsoids with a wide range of anisotropy degree. The most remarkable feature was that samples with high MS, those taken from the lower part of the Aguasalenz limestone, yielded prolate inverse fabrics, with $L$ up to 1.080 and K1 perpendicular to the bedding plane (Fig. 3c). A discussion on the origin of the inverse fabrics is provided at the end of the chapter.

IRM experiments of representative samples provided some clues to the magnetic mineralogy (Fig. 4). IRM acquisition curves showed saturation at fields below 0.3 T in all samples supporting the presence of magnetically soft minerals in all rock

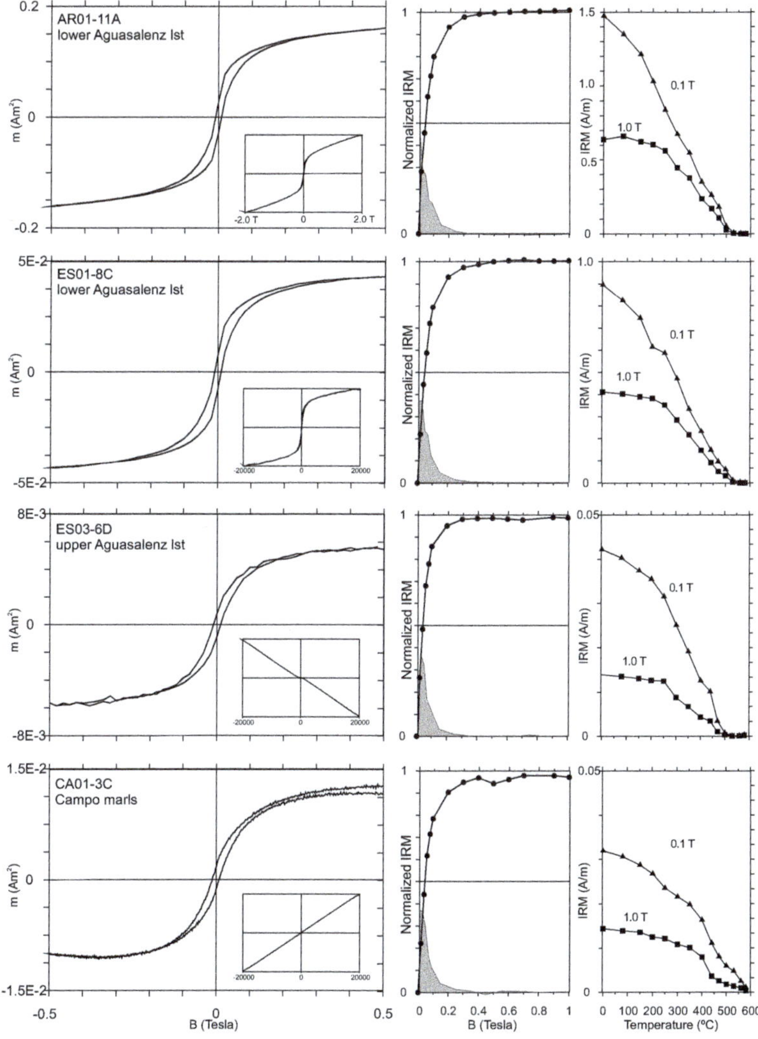


**Fig. 4.** Left: hysteresis loops (ferromagnetic component) of samples from the Campo marls and the Aguasalenz limestone. Inset plots represent full hysteresis loops in order to show dia/paramagnetic slopes. Right: IRM acquisition/thermal demagnetization plots of same samples. Grey-shaded areas in the acquisition plots represent coercivity spectra.

220                                    M. GARCÉS *ET AL.*

formations. Intensity of the saturation IRM showed a remarkable contrast between samples from the lower part of the Aguasalenz, with values as high as 1.5 A/m, and the rest of lithologies (Fig. 4), indicating a greater concentration of magnetic minerals at depth. Three-axis thermal demagnetization of the IRM revealed some differences among lithologies. IRM unblocking temperatures of the Campo marls were higher (570°C) than the Aguasalenz limestones (450–500°C), but in both cases magnetite could be inferred as the magnetic remanence carrier.

Hysteresis loops further revealed differences in magnetic properties between lithologies (Fig. 4) and, more interestingly, between the two groups of Aguasalenz limestone samples. Samples from the upper Aguasalenz Fm yielded characteristic diamagnetic slopes attributable to the predominance of calcite as the matrix mineral of limestones. Samples from the lower Aguasalenz on the other hand yielded paramagnetic slopes. In the absence of significant clay fraction in these rocks, a possibility was that the paramagnetic component was due to iron-bearing carbonates. This interpretation is supported by X-ray difraction (XRD) experiments which showed the presence of small but detectable amounts of ankerite or ferrous dolomite. A

magnesium-to-iron substitution could explain the paramagnetic anisotropy dominating the *c*-axis fabric of carbonates (Rochette 1988), as well as the inverse magnetic fabrics described above for this specific group of samples (Fig. 4).

Wasp-waisted hysteresis loops were observed in samples from the lower Aguasalenz Fm after removal of the paramagnetic component (Fig. 4), which suggests the presence of a mixture of superparamagnetic (SP) and single-domain (SD) magnetite grains. Samples from the upper Aguasalenz and Campo marls yielded hysteresis loops with no indication of superparamagnetic contribution. The ratios of hysteresis parameters in a Day plot (Day *et al.* 1977) yielded values within the Pseudo-Single Domain (PSD) region for the majority of samples (Fig. 5), but a clear tendency in samples from the lower Aguasalenz Fm to higher Mrs/Ms and Hcr/Hc ratios, approaching the trend for SP + SD mixtures (Dunlop 2002).

First-order reversal curve (FORC) diagrams (Roberts *et al.* 2000) were obtained for representative samples of the diverse magnetic behaviours described above in order to obtain further insights into grain size and distribution of magnetic particles within the rock samples (Fig. 5). Contours

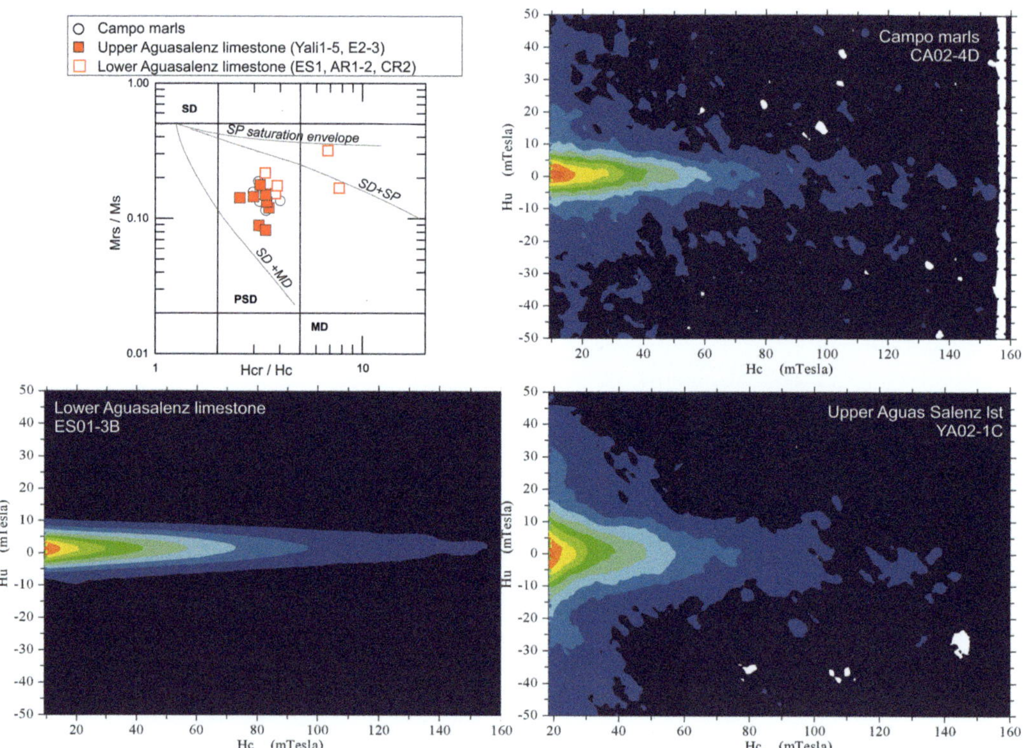

**Fig. 5.** Day plot and FORC diagrams of samples from the Campo marls and the Aguasalenz limestones.

centred about the origin of the diagram and aligned with the $H_u$ axis reveal a distribution of non-interacting and low coercivity magnetite particles in all cases. Divergence of contours away from the origin of the FORC diagram suggests the presence of Multi-Domain (MD) grains in samples from Campo marls and the upper Aguasalenz limestones. This is in contrast with non-diverging contours in the sample from the lower Aguasalenz, for which an assemblage dominated by SD particles is suggested. Contours centred at the origin of the FORC diagram, which are indicative for the presence SP particles, could not be differentiated because of insufficient resolution of the FORC data, which led to truncation of the FORC distribution in the lowest coercivity range.

Overall, the magnetic properties described above indicate that magnetite is the principal remanence carrier in all rock units. Low-field (0.1 T) IRM maximum unblocking temperatures close to 580°C in marl samples are coherent with this interpretation. Relatively low unblocking temperatures (<500°C) in the Aguasalenz limestones possibly indicate that magnetocrystalline rather than shape anisotropy dominates magnetic stability in these samples. This is a property which has been described in cases of authigenic growth of fine-grained equidimensional magnetite grains, where the low crystalline anisotropy of cubic magnetite leads to anomalously low unblocking temperatures (Jackson 1990). A striking dominance of fine-grained (SD + SP) magnetite grains is also inferred from hysteresis and FORC diagrams of samples from the lower Aguasalenz Fm. The intensity of magnetization of these same samples, up to 100 times higher than others, also indicates that a sort of magnetization enhancement occurred in a post-depositional stage.

Magnetization enhancement in the lower Aguasalenz Fm was not limited to the ferrimagnetic fraction. The presence of a new paramagnetic component is suggested by the positive slope of the hysteresis loops, which contrasts with the common diamagnetic component of carbonates of the upper Aguasalenz Fm. XRD further indicates the occurrence of a Fe-carbonate such as ankerite. Thus, there is a correspondence between the development of inverse AMS fabrics in the lower Aguasalenz Fm and the presence of both SD magnetite and ankerite, both minerals being potential carriers of inverse magnetic fabrics. However, magnetic properties do not suggest the predominance of uniaxial SD-magnetite particles, those most typically associated with the development of inverse magnetic fabrics. On the contrary, the low NRM unblocking temperature better agrees with predominance of low-anisotropy equidimensional magnetite, which carries a normal magnetic fabric. Therefore we suggest that ankerite, rather than magnetite, is responsible

for the inverse AMS fabric of the lower Aguasalenz limestones. The typical orientation of K1 perpendicular to bedding (Fig. 3c) is common to other studies (Rochette 1988; Winkler et al. 1996) that show that the c-axis preferred orientation of paramagnetic carbonates results in a maximum susceptibility aligned with the flattening direction.

## Palaeomagnetic components and timing of magnetization

Palaeomagnetic components were calculated by principal component analysis after visual inspection of stepwise demagnetization diagrams (Figs 6 & 7). Fisherian site mean directions were calculated for an average of eight samples/site (Table 1). Sites were grouped in different sectors according to their geographic coordinates and structural setting. Fold tests were carried out at the sector level (Fig. 8) in order to assess the relative age of magnetization. For purpose of comparison, sector mean directions in Table 1 are all shifted into normal polarity.

Marls from the Campo and Barbaruens Fms yielded remanence intensities in the range of 0.1 mA/m. Moderate thermal demagnetization up to 300°C allowed removal of a randomly oriented low-temperature component. At higher temperature a stable characteristic remanent magnetization (ChRM) pointing towards the origin could be determined. Normal- and reverse-polarity ChRM directions were observed at the sample level (Fig. 6) and between sites (Table 1). This ChRM component showed maximum unblocking temperatures of about 475°C and low coercivity (<100 mT). Unstable spurious magnetizations at temperatures above 400°C were relatively frequent, masking the clear observation of NRM unblocking temperatures. Two out of 10 sites yielded reversed polarity and eight sites normal polarity. Among these, one site (ES05) yielded both normal and reversed polarities. Fold tests were performed in three sectors (Fig. 8) with inconclusive (not statiscally significant) results in Ceresa and Seira because of the gentle bedding tilt changes between sites. Nonetheless, in the Campo sector fold test was positive with a 99% confidence level (McElhinny 1964), with clustering of directions after 100% unfolding. The mean directions from the Campo sector in the Peña Montañesa–Montsec thrust sheet, and the Seira sector in Gavarnie thrust sheet, both at the footwall of the Cotiella thrust, widely overlap in tilt corrected coordinates (Table 1) with an overall northeastwards direction, while the mean vector in the Ceresa sector was northwards directed.

Limestones from the Aguasalenz Fm yielded NRM intensities which ranged between 0.1 and 10 mA/m, the highest intensity being observed in

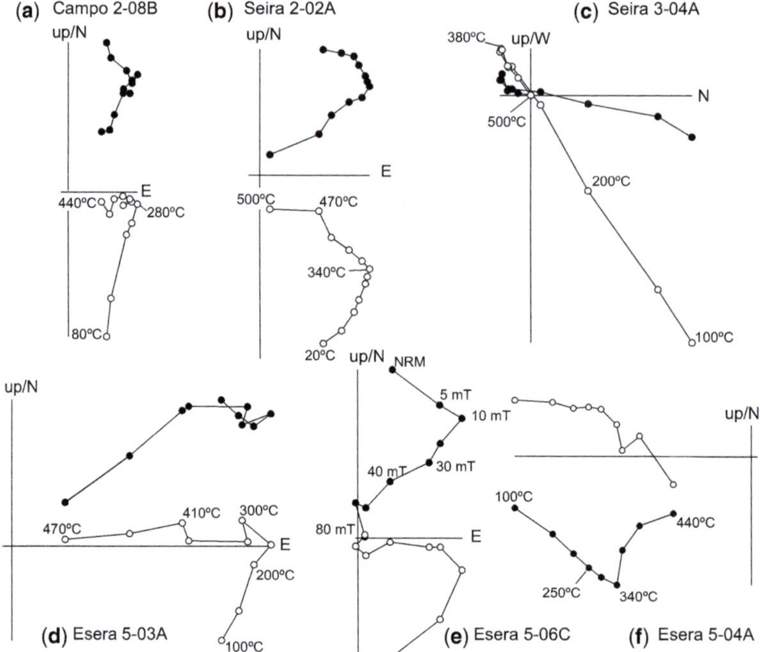


**Fig. 6.** NRM demagnetization Zijderveld plots (geographical coordinates) of samples from the Late Cretaceous marls and turbidites of the Montsec–Peña Montañesa (**a, d, e, f**) and Gavarnie (**b, c**) thrust sheets.

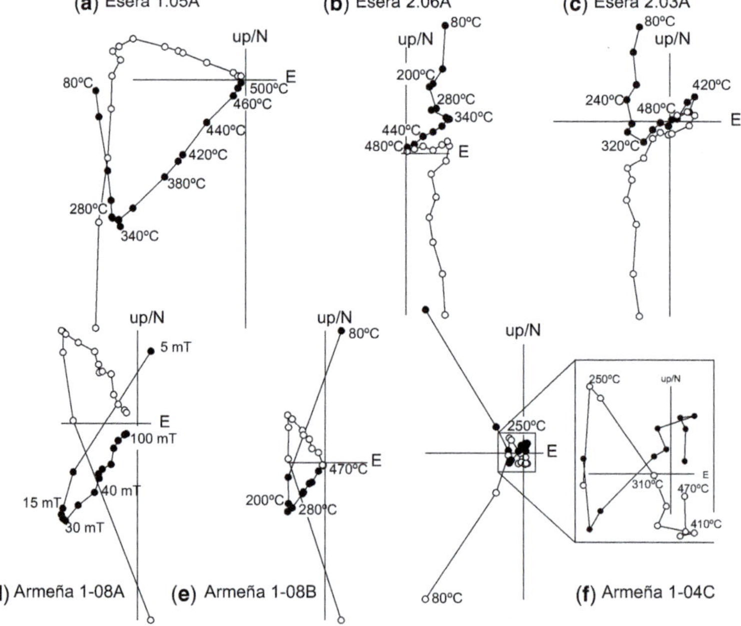


**Fig. 7.** NRM demagnetization Zijderveld plots (geographical coordinates) of samples from the Aguasalenz limestones of the Cotiella–Bóixols (**a, b, c**) and Armeña (**d, e, f**) thrust sheets.

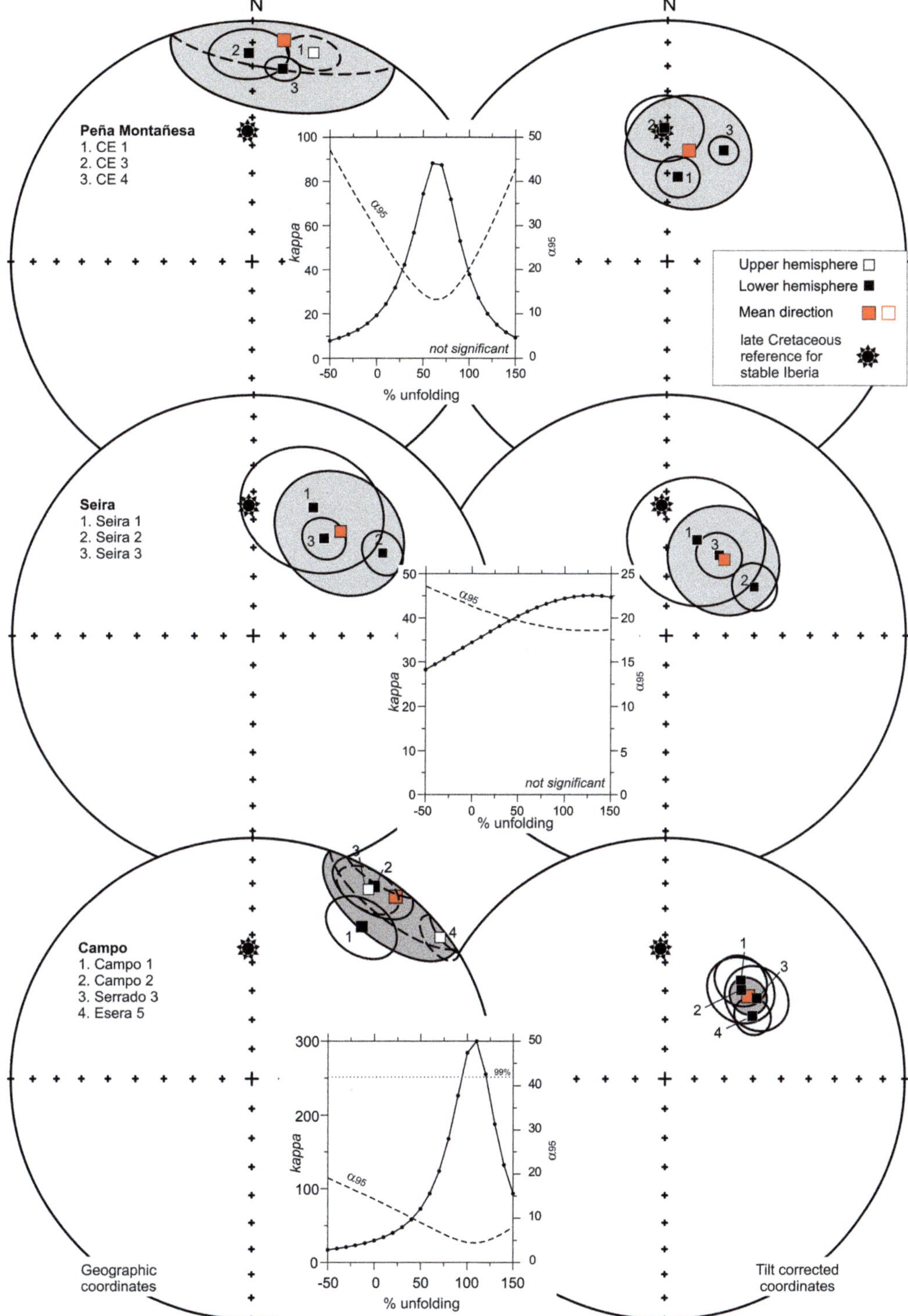



**Fig. 8.** Mean palaeomagnetic directions of sites from the Montsec–Peña Montañesa and Gavarnie thrust sheets before and after tilt correction. Star represents Late Cretaceous palaeomagnetic reference.

the lower stratigraphic intervals of the formation. Samples typically carried a northward and down-dipping component in geographic coordinates, interpreted as recent viscous overprint, which was removed after heating to 250°C or by applying moderate AF steps (<15 mT). A stable ChRM displaying a linear trend towards the origin was revealed after heating to 500°C and yielded both normal and reversed polarities (Fig. 7). Occasionally a third component that was antiparallel to the main ChRM component could be observed at intermediate temperature ranges in two samples (Fig. 7c, f). The distribution of normal- and reverse-polarity sites was found to match the stratigraphic relative position (depth in the restored cross-section of Fig. 2b): normal-polarity sites were found in the upper (shallower), and reversed polarities in lower (deeper) levels within the Aguasalenz Fm. Noticeably, site ES02, the lowest among the group of normally magnetized sites, had some samples with complex magnetization yielding both normal- and reversed-polarity components (Fig. 7c).

In the sector of Santa Isabel–Punta Yerga the Aguasalenz limestones form a tight south-verging syncline and anticline pair that parallels the Cotiella frontal thrust. Palaeomagnetic directions in this sector yielded a negative fold test (Fig. 9), indicating that magnetization was acquired after the main phase of folding. The mean palaeomagnetic vector in geographic coordinates yielded a northeastwards direction similar to the pre-folding magnetization of the younger Campo and Barbaruens Fms.

In the sector of Armeña, the oldest strata of the Aguasalenz Fm form a roll-over structure with fanning of beds against the Armeña fault (Fig. 2). The only two sites sampled in this unit yielded a northeastwards direction (after shifting to normal polarity) before tilt correction (Table 1). Progressive untitlting of beds gave best clustering of direction at 25% unfolding, indicating again that magnetization post-dates tilting.

In the sector of Yali–Ésera, extensional tectonics, which caused progressive fanning of beds against the Cotiella thrust fault, is overprinted by compressional folding as recorded in the beds of overlying Campo breccias. In this context palaeomagnetic directions did not yield consistent results either in geographic or in tilt-corrected coordinates (Fig. 9), thus indicating that rocks were magnetized at an intermediate stage. The relative age of magnetization was assessed by stepwise unfolding of beds, even though this could not be achieved by applying simple bedding tilt correction since rock units were affected by a relatively complex multi-phase folding. First of all, the fanning of beds was caused by syn-extension rotation along shallow listric faults. Secondly, fault inversion caused

compressional folding along a plunging axis oblique to the strike of the beds. In the sector between the Yali and Ésera valleys, beds of the Campo breccia describe a SE-plunging fold axis (N 150–30). Thus, tectonic correction of palaeomagnetic directions of the Ésera–Yali sector followed a four-fold procedure (Table 2): first, vectors and strata were rotated to untilt the Campo breccias fold axis; second, the Campo breccia unconformity was unfolded, bringing the underlying Aguasalenz Fm to its pre-compressional position; third, the growth fold axis was untilted; and fourth, the syn-extensional tilting was corrected.

The results of this stepwise folding exercise indicated the best grouping of palaeomagnetic vectors after 100% unfolding the Campo breccia unconformity (Fig. 9). Subsequent unfolding of the Aguasalenz growth strata increased the scattering of data. It can be inferred that the magnetization of the Aguasalenz Fm was acquired in a post-depositional stage, after extensional tilting of beds but before the folding associated with the inversion of the Cotiella extensional basin. The main folding of these beds possibly occurred during reactivation of the Cotiella faults associated with the emplacement of the Peña Montañesa–Montsec thrust sheet in the Paleocene. Therefore we concluded that the age of magnetization of the Aguasalenz limestones could be constrained between the Santonian and the Paleocene.

## Causes of the remagnetization in the Cotiella Basin

A positive fold test of the late Santonian to Maastrichtian marls and turbidites supported an early pre-folding magnetization in the studied sectors of the Peña Montañesa–Montsec and Gavarnie thrust sheets. In contrast, negative fold tests indicate that magnetization of the Aguasalenz limestones and Maciños calcarenites of the Cotiella is of secondary origin, acquired after tilting of beds during extensional rollover. An incremental fold test carried out in the Yali–Ésera sector also indicates that this secondary magnetization pre-dates compressional folding associated with the Late Cretaceous–Eocene tectonic inversion.

The occurrence of two polarities in the remagnetized samples of the Aguasalenz might suggest that the remanence acquisition was a long-living process related to burial diagenesis that elapsed over several geomagnetic chrons. Yet, on the basis of their magnetic properties, the Aguasalenz limestones can be divided into two very distinct groups. A first group of sites (AR01, AR02, ES01) represents the lower part of the rock formation, characterized by a high-intensity reversed-polarity magnetization. Its low

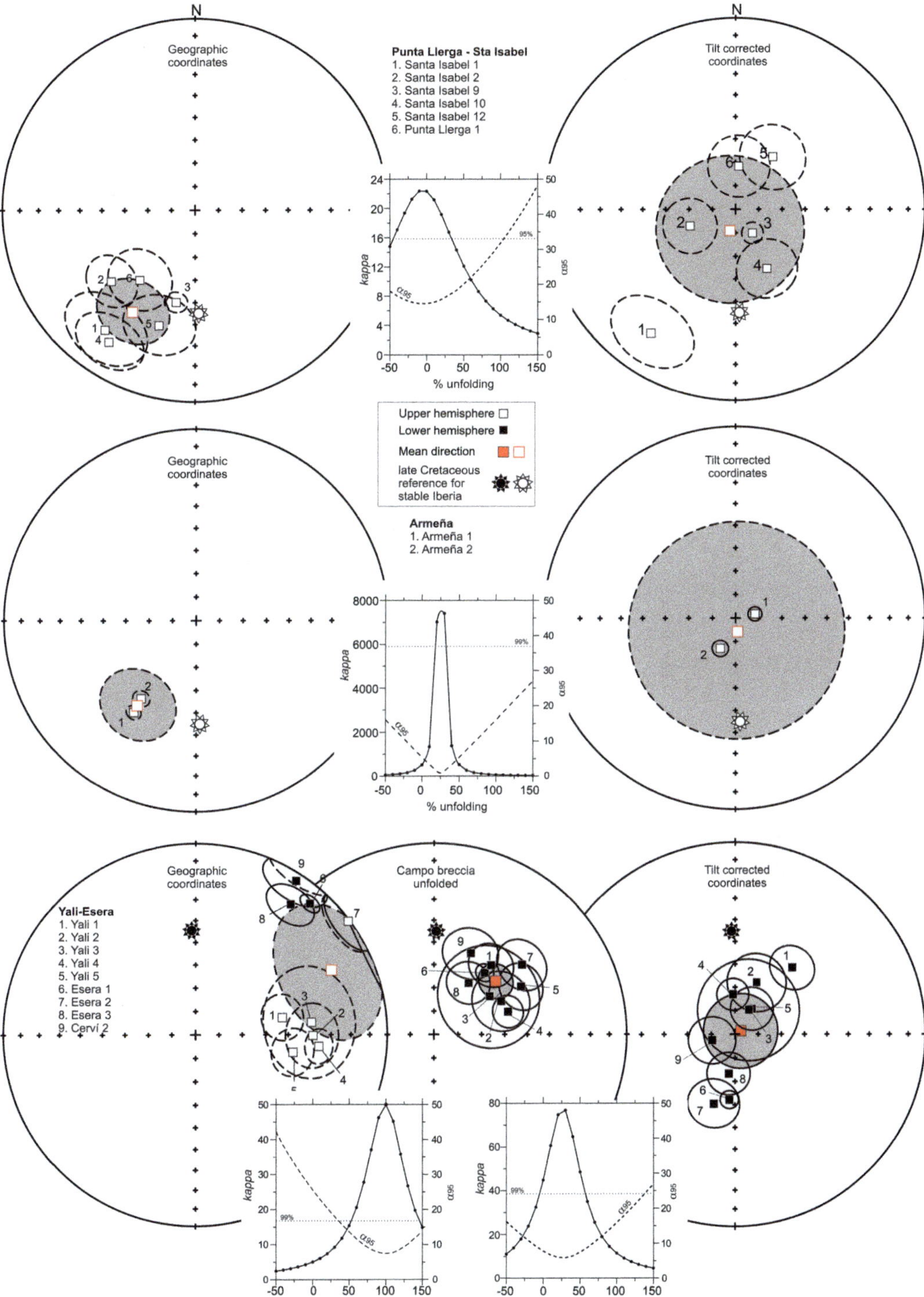



**Fig. 9.** Mean palaeomagnetic directions and fold tests of sites from the Cotiella–Bóixols and Armeña thrust sheets. Stepwise unfolding of sites from the Yali–Ésera sector reveals best clustering of directions at the stage of unfolding of the Brecha de Campo.

**Table 2.** *Stepwise unfolding of the Ésera-Yali site mean palaeomagnetic directions*

| | | | | | | | Unfolding Campo breccia unconformity | | | | | | | | | | | | Unfolding extensional growth | | | | | | | |
|---|---|---|---|---|---|---|---|---|---|---|---|---|---|---|---|---|---|---|---|---|---|---|---|---|---|---|---|
| | | | | | | | Fold axis (N150-30) untilted | | | | | | 100% unfolding | | | | | | Fold axis untilted | | | | 100% unfold | | | |
| | | *In situ* coordinates | | | | | $S_o$ | | Campo breccia | | ChRM | | $S_o$ | | ChRM | | | | $S_o$ | | ChRM | | ChRM | | | |
| Site | $N$ | N/R | Dec | Inc | $k$ | $\alpha_{95}$ | Dip direction | Dip | Dip direction | Dip | Dec | Inc | Dip direction | Dip | Dec | Inc | $k$ | $\alpha_{95}$ | Dip direction | Dip | Dec | Inc | Dec | Inc | $k$ | $\alpha_{95}$ |
| *Yali 1* | 7 | N | 079 | −52 | 42 | 9 | 059 | 74* | 059 | 74* | 039 | −51 | 196 | 00 | 039 | 51 | | | 098 | 24 | 007 | 59 | 041 | 52 | | |
| *Yali 2* | 8 | N | 091 | −39 | 21 | 13 | 081 | 64* | 059 | 74* | 062 | −48 | 299 | 23 | 063 | 58 | | | 024 | 08 | 022 | 74 | 023 | 66 | | |
| *Yali 3* | 4 | N | 084 | −39 | 19 | 22 | 067 | 56* | 059 | 74* | 056 | −45 | 259 | 19 | 055 | 61 | | | 144 | 09 | 003 | 72 | 030 | 78 | | |
| *Yali 4* | 6 | N | 096 | −36 | 98 | 7 | 089 | 53* | 059 | 74* | 070 | −48 | 285 | 34 | 073 | 57 | | | 300 | 11 | 037 | 76 | 357 | 73 | | |
| *Yali 5* | 6 | N | 100 | −47 | 54 | 9 | 074 | 45* | 059 | 74* | 062 | −59 | 259 | 32 | 061 | 47 | | | 221 | 12 | 037 | 65 | 034 | 77 | | |
| *Esera 3* | 10 | N | 036 | 17 | 29 | 9 | 213 | 75 | 241 | 39 | 048 | 27 | 199 | 42 | 033 | 63 | | | 172 | 43 | 341 | 63 | 189 | 73 | | |
| *Esera 2* | 10 | N | 053 | −01 | 22 | 11 | 030 | 72* | 241 | 39 | 054 | 03 | 210 | 74 | 051 | 41 | | | 202 | 67 | 026 | 54 | 197 | 59 | | |
| *Esera 1* | 9 | R | 221 | −10 | 180 | 4 | 205 | 90 | 241 | 39 | 228 | −18 | 198 | 59 | 219 | −56 | | | 183 | 58 | 181 | −60 | 005 | −62 | | |
| *Cervi 2* | 8 | R | 213 | −05 | 27 | 11 | 220 | 83 | 241 | 39 | 218 | −17 | 212 | 47 | 204 | −52 | | | 188 | 42 | 173 | −51 | 075 | −80 | | |
| **All sites** | **9** | | **064** | **−23** | **5** | **26** | | | | | | | | | **052** | **−22** | | | | | **049** | **55** | **50** | **7** | | | **061** | **87** | **12** | **16** |

First, data were rotated to unfold the Campo breccia unconformity. Second, the extensional growth strata of the Aguasalenz limestone was unfolded. Best statistical grouping of directions is achieved after unfolding the Campo breccia.
*Overturned bedding dip.

coercivity and and relatively low unblocking temperatures (<500°C) are characteristic of small magnetite grains lacking shape anisotropy. Waspwaisted hysteresis loops and hysteresis parameters are indicative of mixtures of fine-grained (SP + SD) particles; all of these properties are typically found in many chemical remagnetizations in carbonates (Jackson 1990; Elmore *et al.* 2012). The positive slope of the hysteresis loops reveals the existence of a paramagnetic fraction, identified in XRD data as ankerite, which accounts for the prolate inverse AMS fabric.

A second group of sites represents the upper part of the formation, characterized by lower MS and NRM intensities (Fig. 3), ChRM directions of both normal and reversed polarity (Table 1), low magnetic anisotropy and hysteresis loops with diamagnetic slopes, and lower Mrs/Ms and Hcr/Hc ratios within the PSD field.

The contrasting magnetic properties between the lower and upper strata of the Aguasalenz Fm point to two different magnetization mechanisms occurring at different depths. First, a delayed acquisition could operate at shallow levels in which postdepositional resetting or an early diagenetic growth of magnetite gave rise to an early but post-tilt magnetization. Second, a process of magnetic enhancement occurred below a depth threshold. According to the restored cross-section (Fig. 2), a maximum depth of 4–5 km could be reached in the deepest parts of the Cotiella Basin at the end of the Cretaceous. In this context, authigenic massive growth of fine-grained magnetite could account for the acquisition of a new chemical remanent magnetization of high intensity.

XRD analyses show that Aguasalenz samples consist fundamentally of calcite, quars and dolomite, with some amounts of ankerite and pyrite as the principal Fe bearing minerals. The presence of pyrite points to a reducing diagenetic environment, most likely associated with oxidation of organic matter, and the presence of ankerite indicates that Fe was available. Considering the geological context of the Cotiella Basin, that is, very fast sediment accumulation in minibasins developed on the hanging wall of the synsedimentary listric faults and extensional diapir fall, it is likely that large amounts of organic matter were buried during deposition of Aguasalenz Fm. In this scenario it is well known that bacteria-mediated sulphate reduction causes dissolution of iron oxides and the iron released into pore fluids is rapidly fixed as Fe-sulphide (pyrite) and Fe-carbonates (ankerite and ferroan dolomite; Machel 2001). Under these conditions, detrital (and biogenic) magnetite grains carrying a primary remanence could be almost completely removed. At greater depths, temperature gradient favoured conditions for inorganic

precipitation of fine-grained magnetite, inducing a higher-intensity, reversed-polarity remagnetization. The boundary between normal and reversed remagnetizations would represent an ancient upper limit for the thermogenic redox reactions, causing precipitation of the late secondary magnetite. The high content of organic matter in the Aguasalenz limestones, together with evidences of hydrocarbon seepage in the region, indicates that thermal maturation of organic rich sediments could be the process associated with the precipitation of magnetite in a deep reducing environment (Machel 1995; Cioppa & Symons 2000; Manning & Elmore 2015). Laboratory experiments reproducing the late diagenetic geochemical conditions associated with burial of organic-rich carbonates have been successful in producing magnetite from dissolution–reprecipitation of pyrite (Brothers *et al.* 1996). Conditions for hydrocarbon maturation were feasible in the Cotiella Basin because the burial depth of the Aguasalenz limestones increased up to 5 km by Paleocene times as a result of accumulation of syninversion sediments (Fig. 2).

A regional-scale remagnetization associated with burial of carbonate sediments under elevated geothermal gradients has been documented in a number of northern Iberia rift basins (Juárez *et al.* 1998; Dinarès-Turell & García-Senz 2000; Soto *et al.* 2008). For some of these, a thermo-viscous resetting remagnetization mechanism was suggested in the light of evidence of low-grade metamorphism (Villalaín 2003). In most cases fold tests allowed the magnetization to be constrained between a postextensional and a pre-compressional stage. In all cases remagnetization was of normal polarity, in agreement with a preferred age of acquisition during the early Cretaceous Normal Superchron.

Despite its much younger age, the Cotiella remagnetization shares with the above discussed examples a deep basinal context that favoured an increase in the burial temperature as a crucial factor for remagnetization. In contrast to the Early Cretaceous Organyà Basin, accummulation of a thick sedimentary sequence in Cotiella resulted not from fast syn-rift crustal subsidence, but from localized gravitational extension and salt withdrawal in a post-rift stage.

Acquisition most probably took place by the time of tectonic inversion of the Cotiella Basin, and remagnetization was completed well before the formation of the Monte Perdido antiformal stack and initiation of orogenic relief in the Axial Zone. Therefore, potential influence of topographically and tectonically driven expulsion of orogenic fluids is unlikely. Its age, pre-dating major compressional phases, and the absence of significant cleavage in the Aguasalenz Fm, rules out a tectonic strain-related remagnetization mechanism, as suggested in the Eocene sediments of the adjacent Ainsa Basin (Oliva-Urcia *et al.* 2008).

## Vertical axis rotations

Palaeomagnetic vectors corrected for time of acquisition coordinates (tectonic corrected in Peña Montañesa–Montsec and Gavarnie trust sheets; *in situ* coordinates in Punta Yerga–Sta Isabel and Armeña; partially unfolded in Yali–Ésera) are listed in Table 3. These palaeomagnetic directions were used to calculate vertical axis rotations by reference to a Late Cretaceous pole (longitude/latitude: 190.1/75.1; $k$, 296; $\alpha_{95}$, 3.9), which was

**Table 3.** *Palaeomagnetic directions of the studied sites corrected for their time of acquisition coordinates (TC, tectonic coordinates; IS, in situ, or geographic coordinates; CBU, Campo breccia unfolded), and vertical axis rotation (VAR) relative to the Late Cretaceous reference for Stable Iberia (dec/inc: 357.2/46.5). VAR error $= \alpha_{95}/cos(inc)$*

| Thrust-sheet | Site | Code | Correction | $n$ | N/R | Dec | Inc | $\alpha_{95}$ | Vertical axis rotation |
|---|---|---|---|---|---|---|---|---|---|
| *Gavarnie* | **Seira** | | | | | | | | |
| | Seira 1 | SE1 | TC | 10 | R | 205 | −41 | 23 | **27.8 ± 30.9** |
| | Seira 2 | SE2 | TC | 8 | N | 057 | 37 | 8 | **60.1 ± 12.5** |
| | Seira 3 | SE3 | TC | 8 | R | 216 | −49 | 8 | **38.8 ± 14.0** |
| | | | **Mean** | **3** | **N** | **040** | **43** | **21** | **42.7 ± 41.9** |
| *Peña Montañesa–Montsec* | **Campo** | | | | | | | | |
| | Serrado 3 | SR3 | TC | 8 | N | 049 | 48 | 11 | **51.6 ± 18.0** |
| | Campo 2 | CA2 | TC | 9 | N | 041 | 50 | 11 | **43.4 ± 19.0** |
| | Campo 1 | CA1 | TC | 13 | N | 035 | 37 | 9 | **37.5 ± 13.3** |
| | Esera 5 | ES5 | TC | 8 | N(1R) | 054 | 53 | 6 | **57.2 ± 13.0** |
| | | | **Mean** | **4** | **N** | **045** | **50** | **6.3** | **48.0 ± 12.5** |
| | **Peña Montañesa** | | | | | | | | |
| | Ceresa 1 | CE1 | TC | 6 | N | 007 | 61 | 7 | **9.9 ± 16.6** |
| | Ceresa 3 | CE3 | TC | 8 | N | 359 | 44 | 11 | **1.6 ± 16.7** |
| | Ceresa 4 | CE4 | TC | 9 | N | 027 | 47 | 5 | **29.8 ± 10.6** |
| | | | **Mean** | **3** | **N** | **011** | **51** | **20.3** | **13.9 ± 33.4** |
| *Cotiella–Boixols* | **Punta Yerga–Santa Isabel** | | | | | | | | |
| | Santa Isabel 1 | SI1 | IS | 6 | R | 217 | −23 | 16 | **39.6 ± 19.0** |
| | Santa Isabel 2 | SI2 | IS | 9 | R | 230 | −42 | 11 | **52.3 ± 16.6** |
| | Santa Isabel 9 | SI9 | IS | 10 | N | 012 | 49 | 5 | **14.4 ± 10.5** |
| | Santa Isabel 10 | SI10 | IS | 5 | R | 213 | −19 | 13 | **35.9 ± 15.6** |
| | Santa Isabel 12 | SI12 | IS | 6 | R | 199 | −40 | 14 | **21.6 ± 19.5** |
| | Punta Yerga 1 | PY1 | IS | 4 | R | 218 | −52 | 13 | **41.1 ± 22.8** |
| | | | **Mean** | **6** | **N** | **031** | **38** | **15** | **34.2 ± 19.9** |
| | **Esera–Yali** | | | | | | | | |
| | Yali 1 | YA1 | CBU | 7 | N | 039 | 51 | 9 | **42.0 ± 16.9** |
| | Yali 2 | YA2 | CBU | 8 | N(1R) | 063 | 58 | 13 | **66.2 ± 24.9** |
| | Yali 3 | YA3 | CBU | 4 | N | 055 | 61 | 22 | **58.0 ± 45.8** |
| | Yali 4 | YA4 | CBU | 6 | N | 073 | 57 | 7 | **75.5 ± 14.7** |
| | Yali 5 | YA5 | CBU | 6 | N | 061 | 47 | 10 | **63.7 ± 16.1** |
| | Esera 3 | ES3 | CBU | 10 | N | 033 | 63 | 9 | **36.2 ± 21.8** |
| | Esera 2 | ES2 | CBU | 10 | N | 051 | 41 | 11 | **54.1 ± 16.1** |
| | Esera 1 | ES1 | CBU | 9 | R | 219 | −56 | 4 | **41.8 ± 10.3** |
| | Cervín 2 | CR2 | CBU | 8 | R | 204 | −52 | 11 | **26.8 ± 19.1** |
| | | | **Mean** | **9** | **N** | **049** | **55** | **7** | **51.6 ± 15.2** |
| | **Armeña** | | | | | | | | |
| | Armeña 1 | AR1 | IS | 13 | R | 214 | −42 | 3 | **36.9 ± 8.7** |
| | Armeña 2 | AR2 | IS | 11 | R | 215 | −49 | 4 | **37.8 ± 9.4** |
| | | | **Mean** | **2** | **R** | **215** | **−45** | **16** | **37.3 ± 23.5** |

taken from earlier studies on igneous rocks from the stable Iberian plate (van der Voo 1969; van der Voo & Zijderveld 1971; Storetvedt *et al.* 1990; Neres *et al.* 2012). Mean sector rotations ranged from 14° to 52° clockwise, and their distribution on map view (Figs 1 & 10) reveals congruence with data from earlier studies in the Eocene sediments of the Gavarnie thrust sheet in the Ainsa Basin (Muñoz *et al.* 2013). It must be noticed that, since the palaeomagnetic signal in the Cotiella unit is

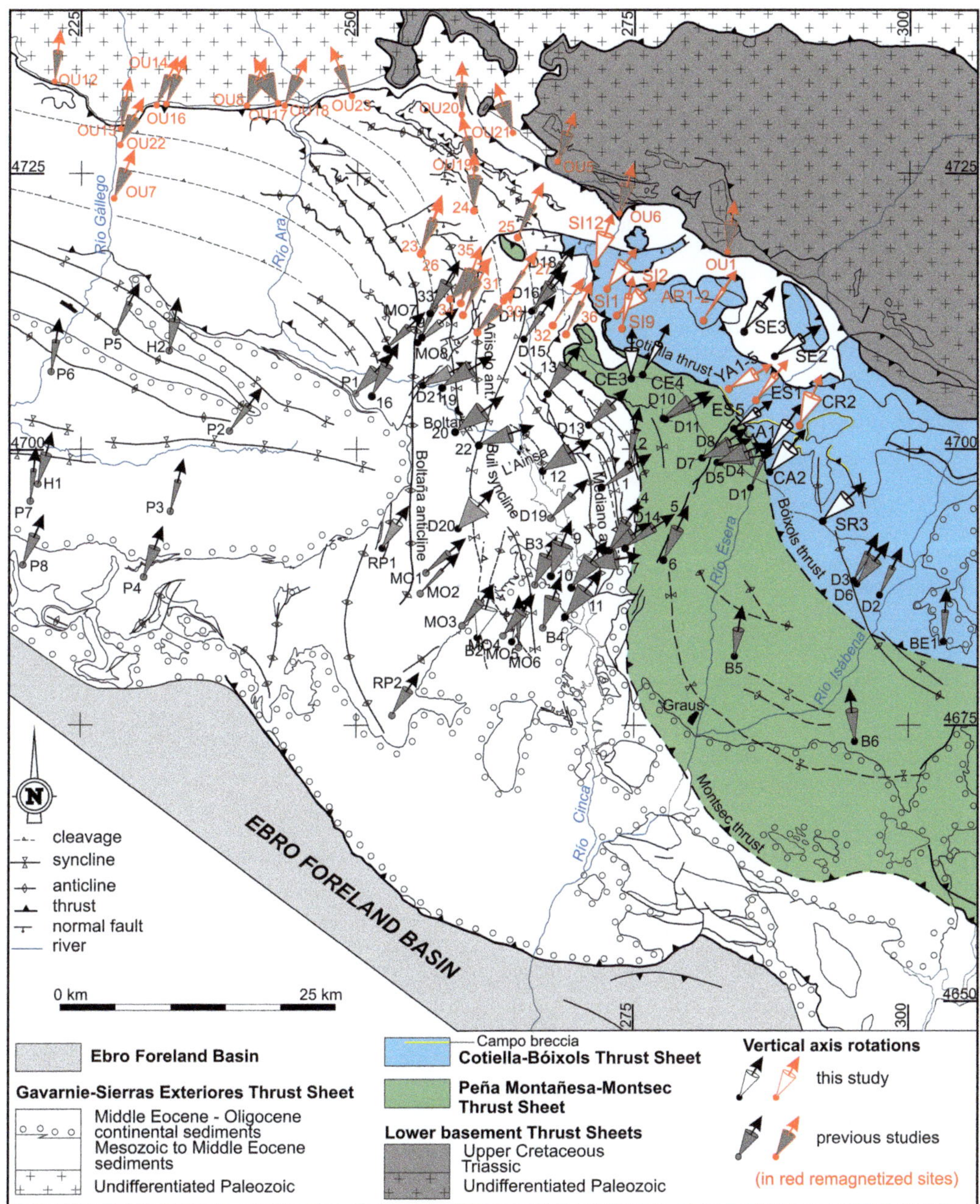



**Fig. 10.** A structural map of the Ainsa Basin and surrounding areas with location of palaeomagnetic data from earlier studies (adapted from Muñoz *et al.* (2013) and the new data of the present study).

not primary, earlier block rotations associated with the extensional stage, if existing, are not recorded. Thus, calculated vertical axis rotations only account for the compressional stage. The results presented here do not reveal significant rotational gradients between the Cotiella–Bóixols, the Peña Montañesa–Montsec and Gavarnie thrust sheets. Hence the rotations recorded in the Cotiella and Peña Montañesa–Montsec could be exclusively associated with the emplacement of the lower Gavarnie–Sierras Exteriores thrust sheet. Integration of palaeomagnetic, structural and stratigraphic data demonstrated that rotations of the Gavarnie–Sierras Exteriores thrust-sheet occurred synchronously to fold growth and the sedimentation of the Lutetian to Priabonian stratigraphic sequences of the Ainsa Basin. The proposed kinematic model involves the formation of a fold-and-thrust belt that bent progressively as a result of differential shortening during the emplacement of the Gavarnie thrust sheet, localized by the distribution of detachment levels at the base of the Mesozoic and Palaeogene successions (Muñoz *et al.* 2013). In this scenario, the previously emplaced upper thrust sheets of Cotiella and Peña Montañesa–Montsec were transported piggy-back and bent to form a thrust salient that parallels the structural trends in the Ainsa Basin (Fig. 10).

## Conclusions

Palaeomagnetic data from the Aguasalenz limestones of the Cotiella thrust-sheet reveal a complex secondary magnetization that is best understood as the result of two separate remagnetization mechanisms operating at different burial depths. First, a shallow, early post-depositional, possibly bacterial-mediated remagnetization was responsible for a dual-polarity remanence carried by magnetite. Second, a later magnetization enhancement occurred at greater depth, where higher burial temperatures favoured precipitation of large quantities of very-fine-grained magnetite, possibly related to thermal maturation of organic matter. Fold tests indicate that both remagnetization processes gave rise to a palaeomagnetic signature that post-dates extensional tilting and pre-dates compressional folding, basin inversion and thrusting of the Cotiella–Bóixols thrust sheet.

The pattern of vertical axis rotations in the Cotiella–Bóixols and Peña Montañesa–Montsec thrust sheet matches earlier observations from the Ainsa Basin and neighbouring regions (Fig. 10). It further supports the vastly documented important clockwise rotations associated with differential shortening during the emplacement of the Gavarnie–Sierras Exteriores thrust sheet.

This is a contribution of the Research Group of *Geodinàmica i Análisi de Conques* 2014SGR467, *Secretaria d'Universitat i Recerca del Departament d'Economia i Coneixement de la Generalitat de Catalunya*, and the Spanish project CGL2014-55900-P. Thanks to the Paleomagnetic Laboratory of Barcelona (CCiTUB-CSIC) and the Paleomagnetic Laboratory Fort Hoffddijk (Utrecht University). Antonio M. Casas and Fabio Speranza are kindly acknowledged for their helpful reviews.

## References

BROTHERS, L. A., ENGEL, M. H. & ELMORE, R. D. 1996. The late diagenetic conversion of pyrite to magnetite by organically complexed ferric iron. *Chemical Geology*, **130**, 1–14.

CIOPPA, M. T. & SYMONS, D. T. A. 2000. Timing of hydrocarbon generation and migration: paleomagnetic and rock magnetic analysis of the Devonian Duvernay Formation. *Journal of Geochemical Exploration*, **69–70**, 387–390.

DAY, R., FULLER, M. & SCHMIDT, V. A. 1977. Hysteresis properties of titanomagnetites: grain-size and compositional dependence. *Physics of the Earth and Planetary Interiors*, **13**, 260–267.

DINARÈS-TURELL, J. 1992. *Paleomagnetisme a Les Unitats Sudpirenenques Superiors: Implicacions Estructurals*. PhD thesis. Universitat de Barcelona.

DINARÈS-TURELL, J. & GARCÍA-SENZ, J. 2000. Remagnetization of Lower Cretaceous limestones from the southern Pyrenees and relation to the Iberian plate geodynamic evolution. *Journal of Geophysical Research*, ,**105**, 19 405–19 418.

DUNLOP, D. J. 2002. Theory and application of the Day plot (M rs/M s v. Hcr/H c). 2. Application to data for rocks, sediments , and soils. *Journal of Geophysical Research*, **107**, 2057.

ELMORE, R. D., MUXWORTHY, A. R. & ALDANA, M. 2012. Remagnetization and chemical alteration of sedimentary rocks. Geological Society, London, Special Publications, **371**, 1–21. First published on November 7, 2012, http://doi.org/10.1144/SP371.15

GARCÍA-SENZ, J. 2002. *Cuencas Extensivas Del Cretácico Inferior En Los Pirineos Centrales: Formación y Subsecuente Inversión*. PhD thesis. Universitat de Barcelona.

GARRIDO-MEGÍAS, A. 1973. *Estudio Geológico y Relación Entre Tectónica y Sedimentación Del Secundario y Terciario de La Vertiente Meridional Pirenaica En Su Zona Central (provincias de Huesca y Lérida)*. PhD thesis, Universidad de Granada.

GONG, Z., DEKKERS, M. J., DINARÈS-TURELL, J. & MULLENDER, T. a. T. 2008. Remagnetization mechanism of Lower Cretaceous rocks from the Organyà Basin (Pyrenees, Spain). *Studia Geophysica et Geodaetica*, **52**, 187–210, http://doi.org/10.1007/s11200-008-0013-3

IZQUIERDO-LLAVALL, E., SAINZ, A. C., OLIVA-URCIA, B., BURMESTER, R., PUEYO, E. L. & HOUSEN, B. 2015. Multi-episodic remagnetization related to deformation in the Pyrenean Internal Sierras. *Geophysical Journal International*, **201**, 891–914, http://doi.org/10.1093/gji/ggv042

JACKSON, M. 1990. Diagenetic sources of stable remanence in remagnetized paleozoic cratonic carbonates: a rock magnetic study. *Journal of Geophysical Research: Solid Earth*, **95**, 2753–2761, http://doi.org/10.1029/JB095iB03p02753

JELINEK, V. 1981. Characterization of the magnetic fabric of rocks. *Tectonophysics*, **79**, T63–T67, http://doi.org/10.1016/0040-1951(81)90110-4

JUÁREZ, M. T., LOWRIE, W., OSETE, M. L. & MELÉNDEZ, G. 1998. Evidence of widespread cretaceous remagnetisation in the Iberian Range and its relation with the rotation if Iberia. *Earth and Planetary Science Letters*, **160**, 729–743.

LARRASOAÑA, J. C., PARÉS, J. M., DEL VALLE, J. & MILLÁN, H. 2003. Triassic paleomagnetism from the western Pyrenees revisited: implications for the Iberian- Eurasia Mesozoic plate boundary. *Tectonophysics*, 161–182.

LÓPEZ-MIR, B. 2013. *Extensional Salt Tectonics in the Cotiella Post-Rift Basin (south-Central Pyrenees): 3D Structure and Evolution*. PhD thesis, Universitat de Barcelona.

LOPEZ-MIR, B., MUÑOZ, J.-A. & GARCÍA-SENZ, J. 2014. Restoration of basins driven by extension and salt tectonics: example from the Cotiella Basin in the central Pyrenees. *Journal of Structural Geology*, **69A**, 147–162, http://doi.org/10.1016/j.jsg.2014.09.022

LOPEZ-MIR, B., MUÑOZ, J.-A. & GARCÍA-SENZ, J. 2015. Extensional salt tectonics in the partially inverted Cotiella post-rift basin (south-central Pyrenees): structure and evolution. *International Journal of Earth Sciences*, **104**, 419–434, http://doi.org/10.1007/s00531-014-1091-9

LOWRIE, W. 1990. Identification of ferromagnetic minerals in a rock by coercivity and un- blocking temperature properties. *Geophysical Research Letters*, **17**, 159–162.

MACHEL, H. G. 1995. Magnetic mineral assemblages and magnetic contrasts in diagenetic environments – with implications for studies of paleomagnetism, hydrocarbon migration and exploration. *In*: TURNER, P. & TURNER, A. (eds) *Palaeomagnetic Applications in Hydrocarbon Exploration and Production*. Geological Society, London, Special Publications, **98**, 9–29.

MACHEL, H. G. 2001. Bacterial and thermochemical sulfate reduction in diagenetic settings — old and new insights. *Sedimentary Geology*, **140**, 143–175.

MANNING, E. B. & ELMORE, R. D. 2015. An integrated paleomagnetic, rock magnetic, and geochemical study of the Marcellus shale in the Valley and Ridge province in Pennsylvania and West Virginia. *Journal of Geophysical Research: Solid Earth*, **120**, 705–724, http://doi.org/10.1002/2014JB011418

MCCABE, C., VAN DER VOO, R., PEACOR, D. R., SCOTESE, C. R. & FREEMAN, R. 1983. Diagenetic magnetite carries ancient yet secondary remanence in some Paleozoic sedimentary carbonates. *Geology*, **11**, 221–223.

MCCLAY, K., MUÑOZ, J.-A. & GARCÍA-SENZ, J. 2004. Extensional salt tectonics in a contractional orogen: a newly identified tectonic event in the Spanish Pyrenees. *Geology*, **32**, 737, http://doi.org/10.1130/G20565.1

MCELHINNY, M. W. 1964. Statistical Significance of the Fold Test in Palaeomagnetism. *Geophysical Journal International*, **8**, 338–340.

MOCHALES, T., PUEYO-MORER, E., CASAS-SAINZ, A. M. & BARNOLAS, A. 2012. Rotational velocity for oblique structures (Boltaña anticline, Southern Pyrenees). *Journal of Structural Geology*, **35**, 2–16.

MUÑOZ, J. A. 2002. The Pyrenees. *In*: GIBBONS, W. & MORENO, T. (eds) *The Geology of Spain*. Geological Society of London, London, 370–385.

MUÑOZ, J.-A., BEAMUD, E., FERNÁNDEZ, O., ARBUÉS, P., DINARÈS-TURELL, J. & POBLET, J. 2013. The Ainsa Fold and Thrust Oblique Zone of the Central Pyrenees: kinematics of a curved contractional system from paleomagnetic and structural data. *Tectonics*, **32**, 1142–1175.

NERES, M., FONT, E., MIRANDA, J. M., CAMPS, P., TERRINHA, P. & MIRO, J. 2012. Reconciling Cretaceous paleomagnetic and marine magnetic data for Iberia: new Iberian paleomagnetic poles. *Journal of Geophysical Research: Solid Earth*, **117**, 1–21, http://doi.org/10.1029/2011JB009067

OLIVA-URCIA, B. & PUEYO, E. L. 2007. Rotational basement kinematics deduced from remagnetized cover rocks (Internal Sierras, southwestern Pyrenees). *Tectonics*, **26**, http://doi.org/10.1029/2006TC001955

OLIVA-URCIA, B., PUEYO, E. L. & LARRASOAÑA, J. C. 2008. Magnetic reorientation induced by pressure solution: a potential mechanism for orogenic-scale remagnetizations. *Earth and Planetary Science Letters*, **265**, 525–534, http://doi.org/10.1016/j.epsl.2007.10.032

OLIVER, J. 1986. Fluids expelled tectonically from orogenic belts: their role in hydrocarbon migration and other geologic phenomena. *Geology*, **14**, 99–102.

PUEYO, E., MILLÁN, H. & POCOVÍ, A. 2002. Rotation velocity of a thrust: a paleomagnetic study in the External Sierras (Southern Pyrenees). *Sedimentary Geology*, **146**, 191–208.

PUEYO-MORER, E. 2000. *Rotaciones Paleomagnéticas En Sistemas de Pliegues Y Cabalgamientos. Tipos, Causas, Significado Y Aplicaciones. (Ejemplos de Las Sierras Exteriores Y Cuenca de Jaca, Pirineo Aragonés)*. PhD thesis, Universidad de Zaragoza.

ROBERTS, A. P., PIKE, C. R. & VEROSUB, K. L. 2000. First-order reversal curve diagrams: a new tool for characterizing the magnetic properties of natural samples. *Journal of Geophysical Research: Solid Earth*, **105**, 28 461–28 475, http://doi.org/10.1029/2000JB900326

ROCHETTE, P. 1988. Inverse magnetic fabric in carbonate-bearing rocks. *Earth and Planetary Science Letters*, **90**, 229–237.

SÉGURET, M. 1972. *Étude Tectonique Des Nappes et Séries Décollées de La Partie Centrale Du Versant Sud Des Pyrénées – Caractère Synsédimentaire, Rôle de La Compression et de La Gravité*. Publications de l'Universite des Sciences et Techniques du Languedoc (USTELA), Montpellier, France.

SOTO, R., VILLALAÍN, J. J. & CASAS-SAINZ, A. M. 2008. Remagnetizations as a tool to analyze the tectonic history of inverted sedimentary basins: a case study from the Basque-Cantabrian basin (north Spain). *Tectonics*, **27**, http://doi.org/10.1029/2007TC002208

STORETVEDT, K. M., MITCHELL, J. G., ABRANCHES, M. C. & OFTE-DAHL, S. 1990. A new kinematic model for Iberia; further palaeomagnetic results and isotopic age evidence. *Physics of the Earth and Planetary Interiors*, **62**, 109–125.

VAN DER VOO, R. 1969. Paleomagnetic evidence for the rotation of the Iberian Península. *Tectonophysics*, **7**, 5–56.

VAN DER VOO, R. & ZIJDERVELD, J. D. A. 1971. Renewed paleomagnetic study of the Lisbon Volcanics and implications for the rotation of the Iberian Peninsula. *Journal of Geophysical Research*, **76**, 3913–3921.

VILLALAÍN, J. 2003. Evidence of a Cretaceous remagnetization in the Cameros Basin (North Spain): implications for basin geometry. *Tectonophysics*, **377**, 101–117, http://doi.org/10.1016/j.tecto.2003.08.024

WEIL, A. B., VOO, R., VAN DER, PLUIJM, B. A., VAN DER, & PARÉS, J. M. 2000. The formation of an orocline by multiphase deformation: a paleomagnetic investigation of the Cantabria-Asturias Arc (northern Spain ). *Journal of Structural Geology*, **22**, 735–756.

WINKLER, A., FLORINDO, F., SAGNOTTI, L. & SARTI, G. 1996. Inverse to normal magnetic fabric transition in an Upper Miocene Marly Sequence from Tuscany, Italy. *Geophysical Research Letters*, **23**, 909–912, http://doi.org/10.1029/96GL00745

# Reconstruction of inverted sedimentary basins from syn-tectonic remagnetizations. A methodological proposal

JUAN J. VILLALAÍN[1]*, ANTONIO M. CASAS-SAINZ[2] & RUTH SOTO[3]

[1]*Departamento Física, Escuela Politécnica Superior, Universidad de Burgos, Avda Cantabria S/N 09006, Burgos, Spain*

[2]*Departamento Ciencias de la Tierra, Universidad de Zaragoza, C/Pedro Cerbuna, 12, 50009, Zaragoza, Spain*

[3]*Instituto Geológico y Minero de España. C/ Manuel Lasala 44, 9B. 50006, Zaragoza, Spain*

**Corresponding author (e-mail: villa@ubu.es)*

**Abstract:** Syn-tectonic remagnetizations related to burial processes that occurred during their extensional stage are frequently recognized in inverted sedimentary basins. The incremental fold test is the main analytical tool used to detect these syn-tectonic remagnetizations. However this test gives spurious results when asymmetrical folding occurs (i.e. both limbs were not tilted simultaneously at the same rate). Asymmetrical folding is very common in sedimentary basins, especially at their margins, during the basinal stage and subsequent tectonic inversion. To correctly analyse syn-tectonic remagnetizations in these scenarios, we propose a method to restore palaeo-magnetic vectors, which allows determining the tilting of beds at the remagnetization stage and therefore gives some hints on the geometry of sedimentary basins and/or folds at the overprint acquisition time. In this paper we consider the analysis of syn-tectonic remagnetization directions for basin reconstruction in two end-member basin models (i.e. syncline vs roll-over geometry). Finally, we compare the results obtained with several northern Iberian examples (Cameros basin in the Iberian Chain, Cabuérniga and Polientes Basin in the Pyrenées), formed during the Mesozoic as extensional basins and subsequently inverted during the Cenozoic compression. From these examples we propose generalizations for the application of syn-tectonic remagnetizations to constrain the geometry of sedimentary basins.

Palaeomagnetic methods have since long been applied to the study of sedimentary basins. Most of these studies were devoted to determine rotations around vertical axes of the sedimentary bodies during or after basin formation (i.e. thrust systems, subduction or collision zones, strike-slip fault zones; e.g. McCaig & McClelland 1992). In palaeomagnetic analysis, the primary palaeomagnetic orientation is usually the parameter used to constrain geological models. However, secondary magnetizations can also be used for these purposes in spite of (a) the difficulties of dating remagnetization events and (b) the uncertainty in applying the valid tectonic correction when dealing with syn-tectonic remagnetizations. Several characteristics of palaeomagnetic vectors corresponding to remagnetizations allow this component to be used as a fundamental feature for constraining basin models: (a) the possibility of comparing the orientation of vectors between different levels (with similar lithology) along a stratigraphic log; (b) the angular relationships between palaeomagnetic vectors and bedding that can be clearly defined in each site; and (c) post-depositional variations of the magnetic vector owing to compaction that are usually minimized with respect to primary palaeomagnetic vectors and can be controlled testing different rock types, provided that the age of magnetization relative to deposition, diagenesis, folding and exhumation can be reasonably bracketed.

Many remagnetizations described in the literature have been characterized as 'syn-tectonic'. The incremental fold test is the traditional technique used to define whether magnetization has been recorded in an intermediate phase of deformation, and to determine the direction of remagnetization at the acquisition time (e.g. Kodama 1988; Villalaín *et al.* 1994, 1996; Cederquist *et al.* 2006). However, the use of the incremental fold test implies assumptions that are not always fulfilled (Villalaín *et al.* 2003), as we describe in the next section. Methods based on the use of secondary palaeomagnetic orientations provide an alternative tool to restore bedding and reconstruct the geometry of folds and/or sedimentary basins at the moment of acquisition of the remagnetization (e.g. Villalaín *et al.* 2003; Henry *et al.* 2004).

*From*: PUEYO, E. L., CIFELLI, F., SUSSMAN, A. J. & OLIVA-URCIA, B. (eds) 2016. *Palaeomagnetism in Fold and Thrust Belts: New Perspectives*. Geological Society, London, Special Publications, **425**, 233–246. First published online October 2, 2015, http://doi.org/10.1144/SP425.10

Starting from geometrical models, we analyse in this work how the orientation of palaeomagnetic vectors evolves as sedimentary basins develop and are subsequently inverted. The main contribution of this method is the filtering of the compressional deformation (tilting related to folds and thrusts) linked to the inversion period of basin evolution, allowing a view of the extensional geometry to be generated. Two end-members of inverted basin models (half-graben with a roll-over monocline and half-graben syncline basin) are analysed in this paper. These two extreme models are the result of normal faulting and extension acting on different upper crustal profiles, the existence of a shallow, low-strength detachment level being one of the main factors conditioning basin geometry (Soto *et al.* 2007*a*). Finally, geometrical basin models are compared with the reconstructions obtained using the restoration methodology based upon remagnetization in several Mesozoic sedimentary basins located in northern Iberia. These sedimentary basins (Cameros Basin in the Iberian Chain and Cabuérniga and Polientes basins in the Basque–Cantabrian Pyrenées; Villalaín *et al.* 2003; Soto *et al.* 2008, 2011; Casas *et al.* 2009) (a) display excellent outcrop conditions that clearly constrain the present-day geometry and allow definition of complete across-strike transects, (b) have a very good preserved remagnetization in their syn-rift rocks and (c) underwent nearly coaxial deformation during their extensional and compressional stages. From these examples we propose generalizations for the application of palaeomagnetic remagnetization vectors to constrain the geometry of sedimentary basins and properly apply tectonic corrections.

## Syn-folding remagnetizations and the reliability of the incremental fold test solution

The fold test is the most widely used technique of the palaeomagnetic stability tests for determining the timing of magnetization relative to deformation (Graham 1949). This simple test indicates whether magnetization is pre- or post-folding (i.e. if palaeomagnetic directions are clustered or dispersed after application of bedding correction). However, in many cases, magnetizations neither 'pass' nor 'fail' the fold test (it is neither pre- nor post-folding) and the palaeomagnetic directions of both limbs do not group with the required statistic significance neither after nor before bedding correction. These results can be explained if one of the fold test hypotheses is not met, for example if the components are not properly isolated (e.g. overlapping of directions), or can arise from incorrect tectonic corrections (e.g. not detected fold axis inclination), or

deformation at the site level (e.g. simple shear). However, in many cases, failures of the fold test are the consequence of 'syn-tectonic' remagnetizations acquired in an intermediate stage of the deformational history of the studied units.

The incremental fold test is the most widespread technique used to analyse syn-tectonic remagnetizations (McClelland-Brown 1983; Scotese & Van der Voo 1983; Kent & Opdyke 1985). It consists in rotating the magnetic directions around the strike of beds at incremental unfolding steps. A statistical parameter (precision parameter $k$ of Fisher 1953; $f$ of McFadden & Jones 1981; etc.) measures the grouping of directions as a function of unfolding (Fig. 1c). When the maximum grouping (minimum value of parameter $f$ in Fig. 1c) is obtained with the required statistical significance at an intermediate value of unfolding, a magnetization contemporaneous with folding (also called 'syn-folding' or 'syn-tectonic') is usually interpreted (e.g. Van der Voo 1993). The mean direction of the best grouping configuration can also be used for determining the palaeomagnetic direction at the remagnetization acquisition time (e.g. Villalaín *et al.* 1994; Cederquist *et al.* 2006). Nevertheless, a similar 'syn-folding' result can be obtained if magnetization is acquired between two different tectonic events, a feasible process in inverted sedimentary basins. In these cases, remagnetizations are acquired before the tectonic inversion of the basin but after syn-sedimentary deformation associated with the basinal stage (Villalaín *et al.* 2003). This situation is relatively common because many palaeomagnetic studies have demonstrated that remagnetizations occur in sedimentary basins owing to different mechanisms during burial (Katz *et al.* 1998; Villalaín *et al.* 2003; Aubourg *et al.* 2012; Torres-López *et al.* 2014).

When analysing 'syn-folding' remagnetizations in inverted sedimentary basins, basin margins are key areas to study how magnetic vectors evolve with deformation. These areas are particularly interesting because of localization of tectonic structures both during extensional deformation and subsequent tectonic inversion. However, because of the polyphased nature of deformation in these areas, it is often observed that the two limbs of folds associated with the basin margin did not develop at the same time (or during the same process), one of the hypotheses assumed in the incremental fold test. In order to highlight the errors of applying the incremental fold test to asymmetric development of folds, a simulation has been done (Fig. 1), considering the margin of a half-graben syncline basin remagnetized during the extensional stage by a magnetic field with direction $D = 0°/I = 45°$. The statistical significance of this fold test has been determined by the McFadden & Jones (1981)

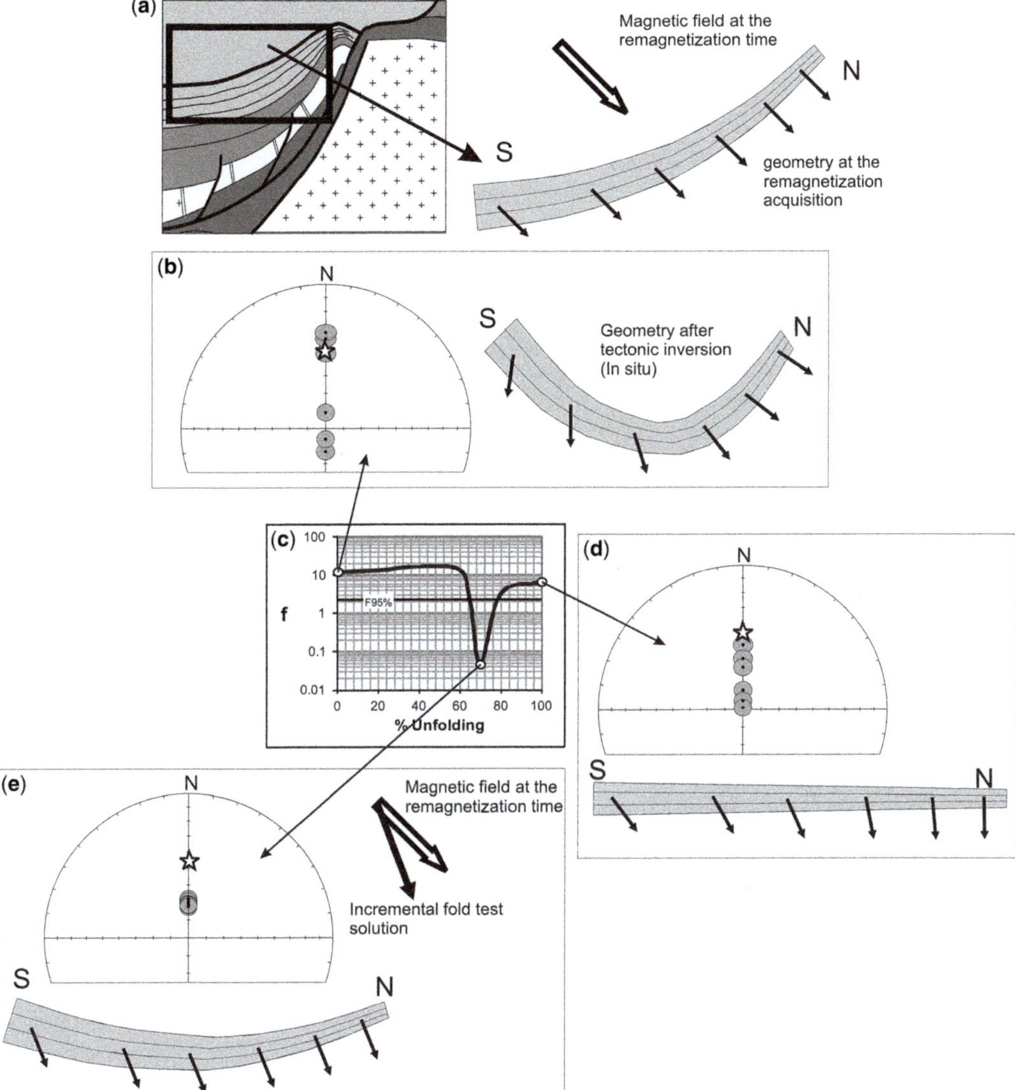



**Fig. 1.** Simulation of an incremental fold test applied in a margin of a half-graben syncline basin remagnetized before the tectonic inversion, as example of non-symmetric folding evolution. (**a**) Original basin geometry before tectonic inversion. The units show an increasing dip towards the border of the basin (from $8°$ to $44°$) and a flexure of $36°$. The units are remagnetized by an original palaeofield direction $D = 0°/I = 45°$. (**b**) Geometry after the tectonic inversion. The remagnetization directions subsequently deflected by the compression are also depicted. (**c**) Incremental fold test: parameter $f$ (McFadden & Jones 1981) as a function of unfolding between 0% (*in situ* configuration of b) and 100% (complete bedding correction of d). (**d**) Geometry and equal area projection for a complete bedding correction (i.e. restitution of beds to the horizontal or 100% unfolding). (**e**) Geometry corresponding to the best fit configuration (minimum $f$), that is, incremental fold test solution. The obtained remagnetization direction ($D = 0°/I = 71°$) is compared with the original palaeofield direction, showing a deflection of $26°$. Stars in the equal area projections are the magnetic field at the remagnetization acquisition time.

method considering six site-mean directions. At the remagnetization acquisition time, units show a basinwards dip, progressively increasing towards the basin border (from 8 to $44°$, Fig. 1a). After tectonic inversion, *in situ* palaeomagnetic directions are scattered along a north–south girdle (actually a small circle coinciding in this case with the great circle; Fig. 1b). If we perform an incremental fold test taking into account these directions, they do not converge either before or after bedding correction, indicating a syn-tectonic acquisition (Fig. 1c). The parameter $f$ (McFadden & Jones 1981) shows the minimum value (i.e. tighter clustering) at 70% of unfolding. This solution shows a statistical value for significant grouping at the 95% confidence level ($f < F_{95\%}$, where $F_{95\%}$ is the critical value of $f$ at the 95% confidence level). However, if we reconstruct the geometry of the basin margin at 70% of unfolding (in a symmetrical way), indicated as the optimal solution by the incremental fold test, both the obtained mean palaeomagnetic direction ($D = 0°/I = 71°$) and the resulting geometry (Fig. 1a) do not match with the real ones (Fig. 1a). The palaeomagnetic direction inferred from this solution would show a significant deflection of $26°$ with respect to the original magnetic field and a symmetrical geometry of the units at the magnetization acquisition time different from the real one (Fig. 1e). Therefore, although the incremental fold test gives very good statistical significance, the remagnetization direction solution and the geometry of the basin at the acquisition time are incorrect. Several examples of incremental fold tests showing this effect can be seen in Villalaín *et al.* (2003). These errors arise because the incremental fold test assumes symmetrical development of deformation and it must be cautiously used when dealing with asymmetric development of folding, as in the example described (Villalaín *et al.* 1994; Delaunay *et al.* 2002).

## Method to restore syn-tectonic remagnetizations

Because symmetrical evolution of folds is not granted, the incremental fold test does not give acceptable solutions in remagnetized inverted basins. We propose an alternative method that considers a diachronous, non-symmetric deformational history and takes into account that remagnetizations were acquired at different stages of tilting in the different limbs. The only hypothesis that must be satisfied in this method is that the different phases of folding are broadly coaxial, a condition often met because compressional folding usually follows previous anisotropies at the basin scale. If that is the case, the *in situ* mean palaeomagnetic direction defines a small circle (SC) in a complete rotation about the strike of bedding (i.e. the apical angle of its cone is the angle between the palaeomagnetic vector and the strike of beds; Fig. 2a). This SC gives all possible original directions of the magnetization and all possible values of tectonic correction can be considered (Waldhör *et al.* 2001).

Since the angle of tectonic correction for each palaeomagnetic site is unknown, we need another constraint to correctly restore beds. This could possibly be given by the palaeofield at the remagnetization time. However, since the age of remagnetization is in general also unknown, the apparent polar wander path (APWP) of the plate cannot be used to determine the expected direction.

The Small Circle Intersection (SCI) method is a reliable technique for obtaining the expected palaeomagnetic vector at the time of acquisition of syn-tectonic remagnetizations. This method was proposed by Shipunov (1997) and modified by Waldhör & Appel (2006). It assumes that during tilting the direction of magnetization rotates around a horizontal axis parallel to the bedding strike in each individual palaeomagnetic site (i.e. along a SC; see Fig. 2a). The intersection of the SCs from different sites represents the single possible common magnetization direction at the remagnetization acquisition time (Shipunov 1997). The statistical solution for the intersection is obtained by minimizing the parameter $A = \Sigma|\alpha_n|$, where $\alpha_n$ is the angular distance between a particular direction and each SC (Fig. 2b; Waldhör & Appel 2006). Figure 2 illustrates the application of the SCI method to the Polientes Basin (Basque–Cantabrian Pyrenées; Soto *et al.* 2011). Figure 2c shows the projection of the set of SCs, and Figure 2d depicts the contours of parameter $A$ (Waldhör & Appel 2006). This stereoplot allows observation of how much the solution is constrained from the contours shape. The star (minimum value of $A$) represents the palaeofield direction at the remagnetization acquisition time.

Once the optimum expected direction is determined from the SCI method, the orientation of each palaeomagnetic vector and, consequently, the bedding corresponding to each site at the moment of remagnetization, can be easily calculated. Instead of completely restoring bedding to the horizontal, beds are restored to the orientation that best approaches the palaeomagnetic vector of the site to the expected direction previously calculated. The orientation of bedding following this process is considered, according to the method proposed, the palaeodip at the moment of remagnetization. This palaeodip can be horizontal or steeper or shallower than the post-inversion dip (Fig. 1).

The amount of unfolding needed in each palaeomagnetic site can be different for each limb of the same fold. This solution is termed 'asymmetric solution' by Villalaín *et al.* (1994) and Villalaín *et al.* (2003) and is obtained by rotating the magnetic directions of each site around the strike of bedding until the palaeomagnetic mean vector reaches the expected direction (or the nearest possible orientation). The sketch of Figure 3 shows this procedure. The plane and vector in blue represent the bedding

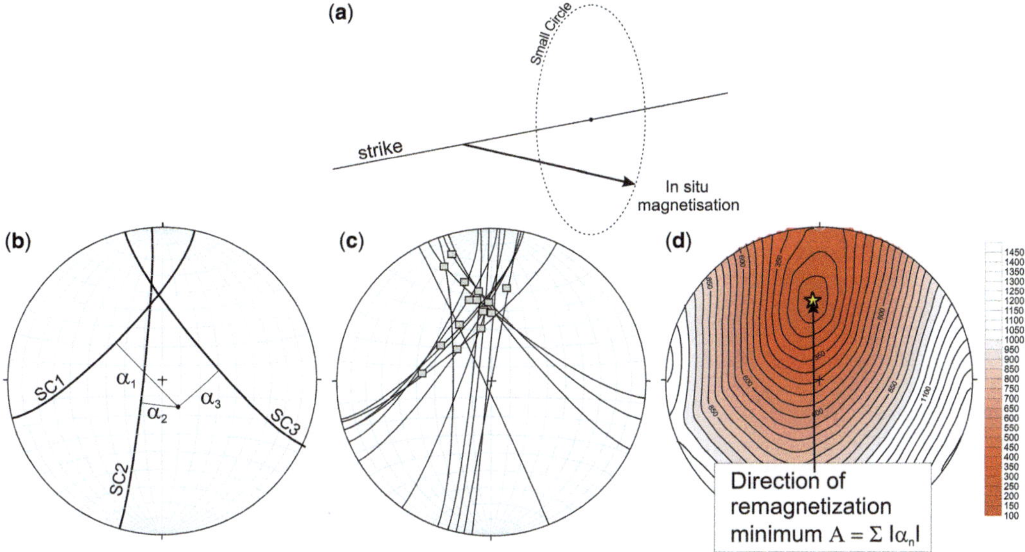



**Fig. 2.** An example of application of the SCI method (Shipunov 1997; Waldhör & Appel 2006). (**a**) Sketch explaining the generation of small circles (SCs). (**b**) Equal area projection showing examples of SC and angular distances $\alpha_n$ between a particular direction and each SC in order to calculate the parameter $A = \Sigma |\alpha_n|$. (**c**) Equal area projection with a set of SCs obtained in the Polientes Basin (Soto *et al.* 2011). Squares are the *in situ* remagnetization site mean directions. (**d**) Equal area projection showing contours of equal value $A$ corresponding to the SC set of Polientes Basin. The star (minimum value of $A$) is the solution of the SCI and represents the palaeofield direction at the remagnetization acquisition time.

and the *in situ* remagnetization, respectively. If both are rotated around the strike as a rigid body, when the remagnetization direction reaches the reference direction obtained from SCI method (best fit remagnetization direction, in red) the plane (in red) shows

the bedding orientation at the remagnetization acquisition time (Fig. 3a). The equal area projection of Figure 3b also shows these geometrical elements. The angle of rotation between the *in situ* magnetization and the best fit direction gives the unfolding

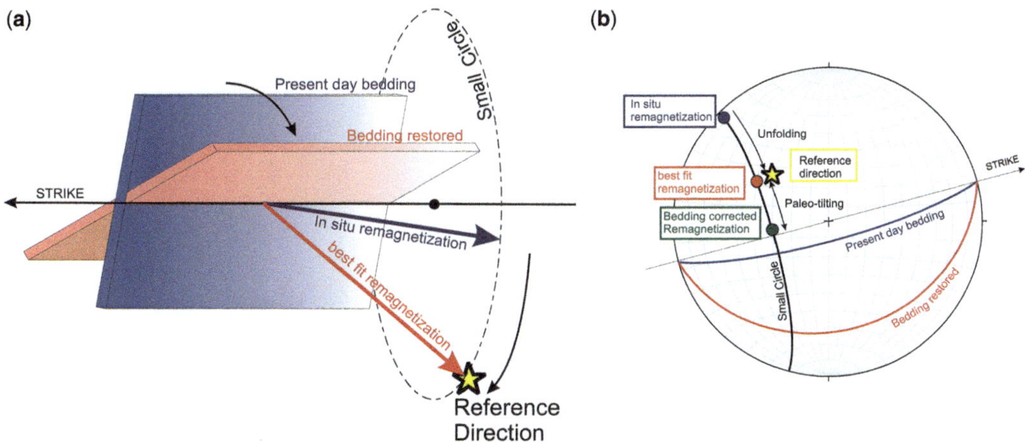



**Fig. 3.** Sketch explaining the method for restoring the bedding planes at the remagnetization acquisition time. Present-day bedding and the *in situ* remagnetization are represented in blue in both the figure (**a**) and the equal area projection (**b**). If both are rotated around the strike, when the remagnetization direction reaches the reference direction (yellow star), the plane (in red) gives the bedding orientation at the remagnetization acquisition time.

restitution. The angle between the best fit direction and the palaeomagnetic direction after the complete bedding correction (green) gives the palaeotilting of the beds at the remagnetization acquisition time. This technique allows the deformation that occurred after the remagnetization to be filtered.

The above-mentioned hypothesis of the method implies some limitations that we explain below. The method assumes that relative vertical axis rotations did not occur because only restitutions around horizontal axes are considered. On the other hand, the remagnetization direction cannot be parallel to the strike of beds, lest the SC described by the palaeomagnetic vector be too small for obtaining sufficient resolution. This last condition is always achieved outside equatorial palaeolatitudes.

Accuracy in the restoration of magnetization and bedding requires that the axes of folding phases separated by the remagnetization event must form low angles (certain degree of coaxiality). If the two phases of folding are far from coaxiality, the *in situ* strikes of beds are different from the axis of rotation of the last folding phase to be restored. In this case the considered SCs are not appropriate and consequently the restored palaeomagnetic direction and the calculated palaeobedding can differ strongly from the true values at the remagnetization time. However, the explained coaxiality condition can represent a limitation for the determination of the optimum expected direction of the remagnetization from the SCI method. If the two phases are strongly coaxial, the variability of bedding strikes between sites becomes moderate and the SCI method gives an SC intersection area (minimum value of A parameter) that is very elongated

along the SC orientation (a banana-like shape). This effect generates uncertainty in the calculation of the palaeofield direction by the SCI method. In these cases the age of remagnetization can be determined by combining the uncertainty area (equal-*A* contours) with the APWP. Figure 4 shows an example of determination of the palaeofield direction at the remagnetization time by the SCI method from High Atlas data (Torres-López *et al.* 2014). The minimum value of *A* is obtained at $D = 336.4°$; $I = 29.2°$; $\alpha_{95} = 4.1°$. However the high degree of coaxiality in this area generates elongated equal-*A* contours, extending the uncertainty of the solution in a NNW direction (Fig. 4b). The comparison between the expected palaeomagnetic directions at the High Atlas obtained from the global apparent polar wander path (GAPWP) in Africa coordinates with the minimum values of equal-*A* contours gives the most probable direction corresponding to the 100 Ma pole position. This age for the High Atlas remagnetization has been confirmed by other palaeomagnetic results (e.g. Moussaid *et al.* 2015).

Regarding the condition of no relative vertical axis rotations, this requirement can be avoided if only the inclination of the reference direction is considered, However this modification of the method affects very significantly the accuracy of the results. Waldhör *et al.* (2001) propose SC reconstruction by rotating the site mean direction along the remanence SC to the intersection point with the SC of constant inclination. If the direction obtained has a declination significantly different from the reference, a vertical axis rotation can be determined. However this technique has a drawback − the uncertainty in the

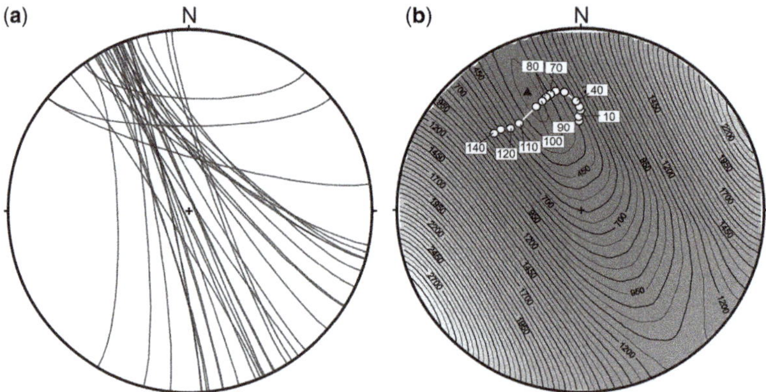



**Fig. 4.** (**a**) Small circles corresponding to the Cretaceous remagnetization directions obtained in the Moroccan High Atlas (Torres-López *et al.* 2014). (**b**) Equal area projection showing contours of equal value *A*. Black triangle shows the solution obtained from the SCI method (Waldhör & Appel 2006), that is, the minimum value of *A* ($A = \Sigma|\alpha_M|$). The white dots show the expected palaeomagnetic directions at the High Atlas from the GAPWP in Africa coordinates (Torsvik *et al.* 2012). Modified from Torres-López *et al.* (2014).

solution of the restored direction and the amount of restoration are in many cases very large. In general there are two points of intersection between the SC and the constant inclination line. On the other hand if the error $\alpha_{95}$ of the site mean direction is considered, in many cases, most of the whole sector between both intersection points must be considered as a feasible result. This great uncertainty enables the detection of only very large vertical axis rotations.

As we summarize in the next sections, this method has been successfully used to provide palaeogeometric reconstruction in different areas (Villalaín *et al.* 2003; Henry *et al.* 2004; Soto *et al.* 2008, 2011; Casas *et al.* 2009).

## Syn-folding remagnetizations in sedimentary basin models

In a previous section, we simulated syn-folding remagnetization directions evolving at the margin of an inverted half-graben syncline basin in order to analyse the reliability of the incremental fold test. In this section and in parallel we determine from forward modelling the theoretical orientation of syn-folding remagnetizations considering the two end-members of basin models: syncline basin geometry (normal drag fold) and the roll-over, half-graben geometry (reverse drag fold; Fig. 5).

The two end-member basin models proposed show either constant dip direction, with variable

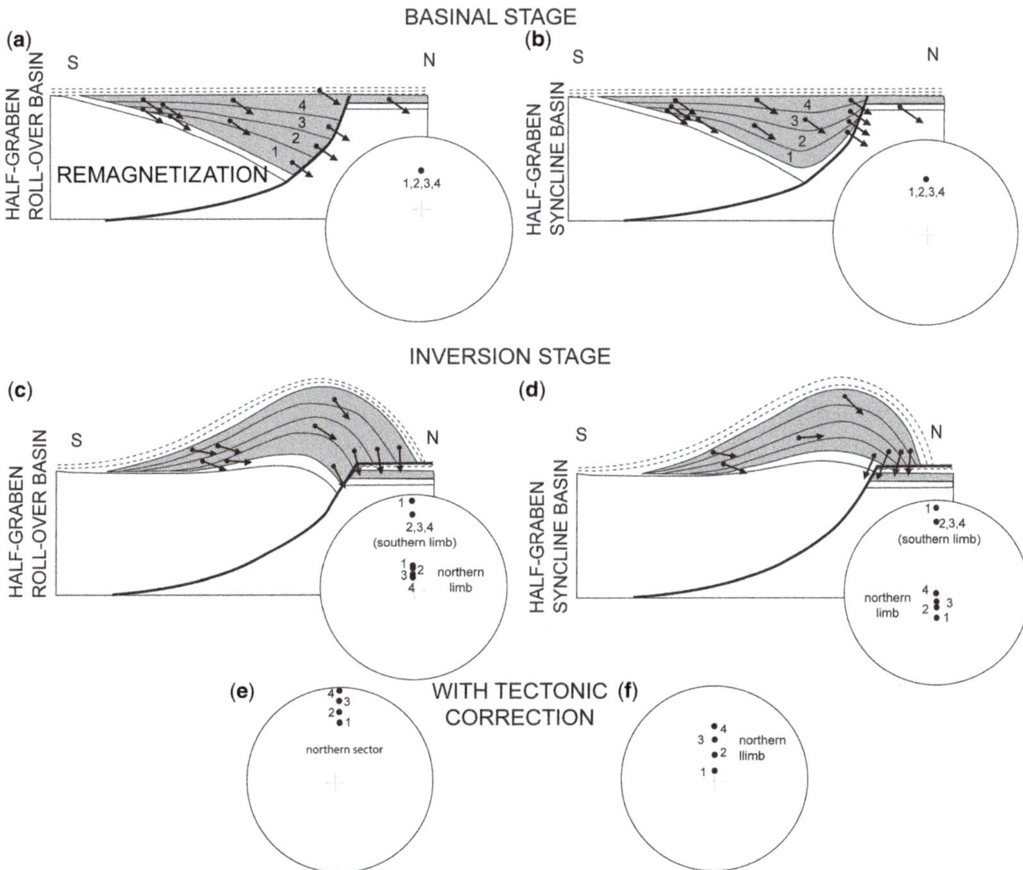


**Fig. 5.** Different basin models showing the position of remagnetization vectors before and after tectonic correction. These models were built considering a basin orientation with the main faults perpendicular to the trend of the magnetic north. Substantial variations in the directions of *in situ* remagnetization are expected to occur with different basin orientations. Numbers (1, 2, 3, 4) correspond to successive stratigraphic units deposited during extension. Sketches show the orientation of vectors during the pre-inversion, post-remagnetization stage for half-graben rollover (**a**) and syncline (**b**) basin geometry. (**c**) Orientation of paleomagnetic vectors after the inversion for the half-graben with roll-over basin. (**d**) Idem for the syncline basin geometry. Stereoplots for palaeomagnetic vectors after tectonic correction in the case of inverted basins with half-graben, roll-over geometry (**e**) and syncline geometry (**f**). See text for explanation.

dip values along the stratigraphic pile (half-graben with a roll-over monocline, Fig. 5a), and conversely, centripetal dips at both limbs of the syncline in the case of the half-graben syncline basin (Fig. 5b). In the first case, bed thickness increases towards the main fault, whereas in the second case the maximum thickness is at a distance from the fault. If remagnetization occurred in the pre-inversion stage, the inclination of palaeomagnetic vectors is the same regardless of the dip of beds in the basinal stage. The palaeomagnetic vectors show different angular relationships to bedding according to their position within the basin, especially in the case of the syncline basin, where palaeomagnetic vectors can be perpendicular to bedding in the older units (Fig. 1a, b).

The classical model of inversion geometry of half-graben basins shows a wide hanging-wall ramp resulting from the uplift and fault-bend folding during compression of the syn-rift sequence deposited in the hanging-wall of the normal fault (Williams *et al.* 1989). This model is valid when there is no syn-tectonic sedimentation at the front of the inverted normal fault, and should be modified when sedimentation takes place in the foreland of the main thrust. Nevertheless, the main results obtained can be applicable to examples of this kind, modifying the dip of beds at the main thrust front. After basin inversion, the magnetic vectors corresponding to different remagnetized units will distribute along an SC (coincident with a great circle in the case of the figure because the trend of the palaeomagnetic vector is perpendicular to the bedding strike). If tectonic correction is applied to remagnetized units (Fig. 5e, f) without further considerations, it can lead to non-fitting fold tests and overestimation (or underestimation) of palaeomagnetic inclinations (Fig. 5e, f). Nevertheless, the distribution pattern of palaeomagnetic vectors gives some clues on the original basin geometry, with shallower inclinations (for the presented case) in relation to the expected direction in the case of the roll-over basin geometry and generally steeper inclinations in the syncline basin model.

## Comparison with examples from the Iberian plate

Reconstruction of the extensional geometry of inverted sedimentary basins using remagnetizations has already been successfully done in several basins in northern Iberia and the High Atlas (Morocco) (Villalaín *et al.* 2003; Soto *et al.* 2008, 2011; Casas *et al.* 2009; Torres-López *et al.* 2014). Sedimentary basins of northern Iberia display a roughly 090–100° N trend and Mesozoic remagnetizations

obtained using the SCI method with a declination ranging between 316° in the Organyà Basin (Gong *et al.* 2009) and 359° in the Cameros Basin (Casas *et al.* 2009).

## *The Cameros Basin: two end-members basin models*

The Cameros massif forms the northwesternmost part of the Iberian Cordillera, the main intraplate mountain range located within the Iberian peninsula. The Iberian Cordillera resulted from the inversion of several Mesozoic extensional basins (Salas & Casas 1993) formed between the Triassic and the Late Cretaceous. The Cameros Basin formed during the Late Jurassic–Early Cretaceous as an extensional basin, with its main, southward-dipping faults located near its northern margin (Tischer 1966; Guiraud & Séguret 1984; Guimerà *et al.* 1995; Casas-Sainz & Gil-Imaz 1998). During the Cenozoic, the basin was inverted, overthusting toward the north the syn-tectonic continental deposits of the Rioja Trough (Muñoz-Jiménez & Casas-Sainz 1997), allowing the syn-extensional series to be exposed at present in the hanging-wall of the main thrust. The syn-extensional deposits of the Cameros Basin are arranged in five lithostratigraphic units: the Tera, Oncala, Urbión, Enciso and Oliván groups (Tischer 1966), reaching a total cumulate thickness of about 8000 m (Fig. 6b, c). These five units were deposited in continental environments (fluvial and lacustrine), between two well-dated marine episodes (Kimmeridgian and Albian in age) in the Cameros Basin. A heating event, associated with thermal metamorphism forming chloritoid crystals in the deepest areas of the basin (about 5000 m deep) exposed at present, has been dated at 100 Ma (Albian), coinciding with the end of the rifting stage (Guiraud & Séguret 1984; Goldberg *et al.* 1988).

The inversion, Cenozoic geometry of the Cameros massif shows a 100 km-long continuous basement-involved thrust with a wide hanging-wall flat that can be followed down to 4 km depth (Casas-Sainz 1993), overlying a footwall ramp with subhorizontal Cenozoic strata. The main structures identified within the Cameros massif are the northern Cameros syncline (Fig. 6b), that follows approximately the shape of its northern border, and the Oncala anticline, with a WNW–ESE trend, in the central part of the massif. It is probable that these structures were partly outlined during the extensional stage in the Cameros Basin and tightened during the inversion stage. Phenomena of hinge migration in the Tertiary inversion stage are witnessed by geometrical relationships between folds and slaty cleavage (Casas-Sainz & Gil-Imaz 1998).

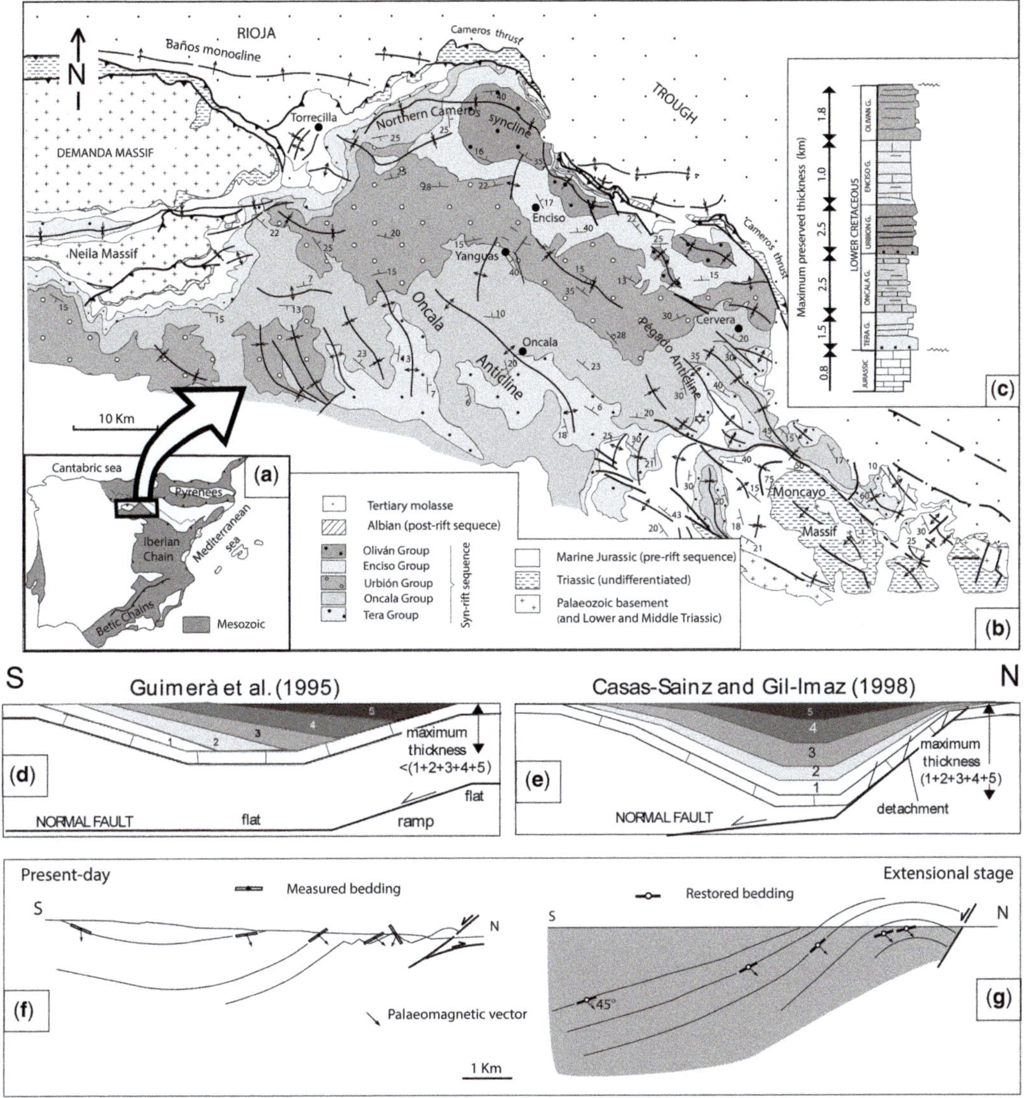



**Fig. 6.** Remagnetization analysis applied to the Cameros Basin. (**a**) Sketch showing the location of the Cameros Basin within the frame of Mesozoic basins in the Iberian Peninsula. (**b**) Geological map of the Cameros Basin (modified from Casas-Sainz & Gil-Imaz 1998). (**c**) Stratigraphic log showing the maximum thickness of units in the basin centre. (**d**) Extensional basin model proposed by Guimerà *et al.* (1995). Modified from Mata *et al.* (2001). (**e**) Basin model according to Casas-Sainz & Gil-Imaz (1998). Modified after Mata *et al.* (2001). (**f**) Cross-section of the Northern Cameros Syncline showing orientation of bedding and palaeomagnetic (*in situ*) vectors, modified after Villalaín *et al.* (2003). (**g**) Restored palaeomagnetic vectors of the time of acquisition of remagnetization, and the corresponding restored bedding. Modified from Villalaín *et al.* (2003). Bedding dip angle and palaeomagnetic inclination have been projected onto the vertical plane of the cross-section (i.e. apparent bedding dip angle and apparent palaeomagnetic inclination).

The abnormally high thickness of Upper Jurassic-Lower Cretaceous units in the Mesozoic Cameros Basin, within the context of the Iberian intraplate ranges, has led to two hypotheses being established concerning its geometry and evolution (Guimerà *et al.* 1995; Casas-Sainz & Gil-Imaz 1998). The main differences between the two basin models are:

(1)  The geometry of syn-rift units in the northern basin border – according to Guimerà *et al.*

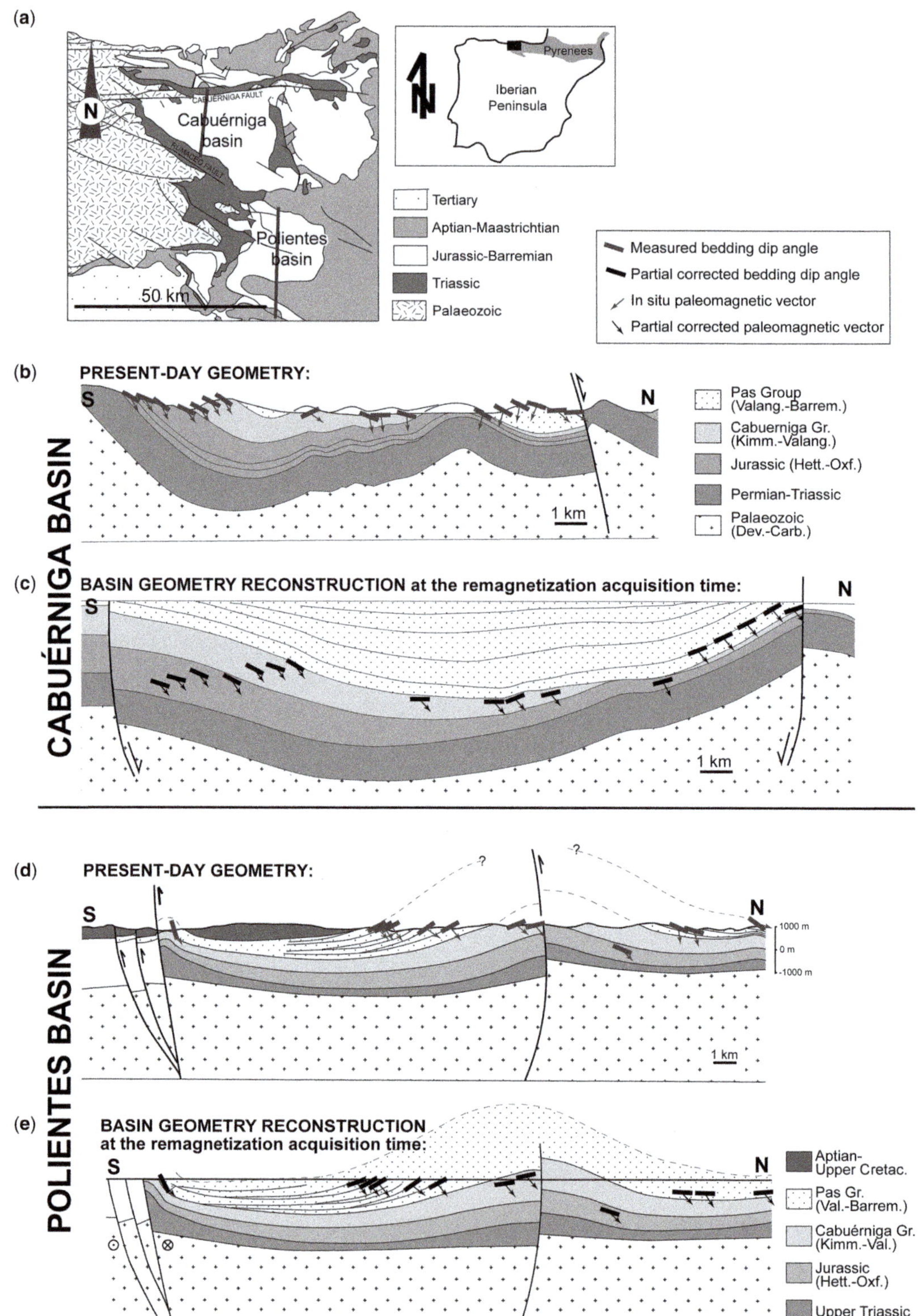

(a)
N
CABUERNIGA FAULT
Cabuérniga basin
RUMACEO FAULT
Polientes basin
50 km
Pyrenees
Iberian Peninsula
N
Tertiary
Aptian-Maastrichtian
Jurassic-Barremian
Triassic
Palaeozoic
Measured bedding dip angle
Partial corrected bedding dip angle
In situ paleomagnetic vector
Partial corrected paleomagnetic vector

(b)
PRESENT-DAY GEOMETRY:
S
N
CABUÉRNIGA BASIN
1 km
Pas Group (Valang.-Barrem.)
Cabuerniga Gr. (Kimm.-Valang.)
Jurassic (Hett.-Oxf.)
Permian-Triassic
Palaeozoic (Dev.-Carb.)

(c)
BASIN GEOMETRY RECONSTRUCTION at the remagnetization acquisition time:
S
N
1 km

(d)
PRESENT-DAY GEOMETRY:
?
?
S
N
POLIENTES BASIN
1000 m
0 m
-1000 m
1 km

(e)
BASIN GEOMETRY RECONSTRUCTION at the remagnetization acquisition time:
S
N
1 km
Aptian-Upper Cretac.
Pas Gr. (Val.-Barrem.)
Cabuérniga Gr. (Kimm.-Val.)
Jurassic (Hett.-Oxf.)
Upper Triassic
Palaeozoic

(1995), beds deposited with a syn-sedimentary dip towards the north, with a half-graben-like basin geometry, and each extensional units lies directly onto the pre-rift marine Jurassic sequence. The total thickness of sediments in the basin centre was below 5 km and sedimentary units were juxtaposed, and not superimposed in a vertical section (Fig. 6d).

(2) In the basin model proposed by Casas-Sainz & Gil-Imaz (1998) the syn-rift sedimentary units dip originally basinward near its northern border, defining a synclinal basin geometry. The younger units lap onto the previous units, and not on the pre-rift sequence. Therefore, in the basin centre the total thickness of the basin sediments would approximate the sum of the maximum thickness of the individual sequences (Fig. 6e).

The main faults limiting the northern basin margin show different geometry in the two models: Guimerà *et al.* (1995) propose a buried normal fault with two horizontal flats separated by a shallowly dipping ramp. The southward movement of the hanging-wall would create the accommodation space for the basin, without cutting across the upper part of the basement and the pre-rift sequence. In the model proposed by Casas-Sainz & Gil-Imaz (1998), the basin is created during the movement of a basement normal fault with listric geometry. The Mesozoic cover is detached from the basement by the Upper Triassic regional detachment level and the pre-rift sequence is stretched by means of minor faults. Large-scale drag associated with the main faults would contribute to the formation of a synclinal-like geometry in the northern basin border. Studies based on thermochronology (Mata *et al.* 2001; see also Casas *et al.* 2012) have shown that the thickness of syn-rift deposits could approximate the maximum depth reached by the pre-rift sequence in the basin centre, which seems to corroborate the syncline basin model. However no definitive data have been given to date to confirm either one of the two basin models. The use of remagnetization directions can allow an attempt to determine the original dips of beds and which of the basin models fits better with the original (syn- or inmediately post-sedimentary) dips obtained.

Palaeomagnetic sites analysed in previous works (Villalaín *et al.* 2003; Casas *et al.* 2009) are located along north–south transects (Fig. 6f shows one of them) and show a pervasive remagnetization recorded in both red beds and limestones. Fold and conglomerate tests (Villalaín *et al.* 2003, 2012) are consistent with a remagnetization age between the basinal and the inversion stages. The aplication of the proposed method to the remagnetization directions (Fig. 6g) indicates basinward dips for units located near the northern basin border, and also in the southern half of the basin, thus corroborating the syncline basin geometry pre-dating the Cenozoic compression. However, to accurately define the position of the syn-sedimentary syncline and the hinge migration processes between the basin formation and inversion, new palaeomagnetic data are necessary.

## The Cabuérniga and Polientes basins

The Cabuérniga Basin is one of the main depositional areas close to the northern Iberian plate margin during the Mesozoic rifting stage (Fig. 7). Syn-rift deposits are also continental in origin (Cabuérniga and Pas groups, Lower Cretaceous in age), although their thickness is smaller than in the Cameros Basin. The main extension direction during the rifting stage, as inferred from anisotropy of magnetic susceptibility and mesostructural studies, is north–south to NE–SW (Soto *et al.* 2007*b*). The present-day structure of the basin is a gentle east–west syncline with an anticline in its central part (Fig. 7b). The northern basin border is an east–west normal fault partly folded during the Cenozoic inversion, whereas its southern border is defined by the outcrops of the pre-rift sequence, probably uplifted in relation with an extensional NW–SE fault. Compressional cleavage is present in the central part of the basin (Oliva-Urcia *et al.* 2013), although its origin in terms of age is not completely solved.

Widespread Cretaceous remagnetization in the Cabuérniga Basin is demonstrated by a conglomerate test in the syn-rift sequence and a fold test in compressional (Cenozoic) folds (Soto *et al.* 2008). In the Cabuérniga Basin the overall geometry after restoring palaeomagnetic vectors to the remagnetization orientation is also a syncline (Fig. 7c), with some complications in the basin centre, which probably nucleated the intra-basinal anticline during the Cenozoic inversion, which enhanced some of the

---

**Fig. 7.** Reconstruction of the Cabuérniga and Polientes basin (Basque–Cantabrian Pyrenées) using syn-folding remagnetizations. (**a**) Geological sketch of the area and location of cross-sections. (**b** and **e**) Present-day cross-sections of the Cabuérniga and Polientes basins, respectively. (**c** and **d**) North–south cross-section at the remagnetization acquisition time from syn-folding remagnetizations and partial bedding restoration. Bedding dip angle and palaeomagnetic inclination have been projected onto the vertical plane of the cross-section (i.e. apparent bedding dip angle and apparent palaeomagnetic inclination).

features outlined during extension. The expected direction for remagnetization obtained from the SCI is slightly different from the one obtained in the Cameros Basin, probably resulting from a weak diachronism of remagnesitation in northern Iberia (Gong *et al.* 2009).

The Polientes Basin is another of the three basins (Pas, Cabuérniga and Polientes) constituting the main depocentral areas during the Early Cretaceous in North Iberia. The syn-rift continental sequence is also represented by the Cabuérniga (lacustrine) and Pas (fluvial) groups, limited to the south by an important, east–west-striking normal fault (Fig. 7d). Conversely, the northern basin border cannot be identified as the syn-rift deposits are below the Albian–Cenomanian unconformity.

Regarding palaeomagnetic data, the Polientes Basin is especially interesting because, in spite of giving a similar remagnetization age, according to palaeomagnetic criteria and the application of the SCI method (Soto *et al.* 2011), there is a disagreement between the distribution of thickness (well constrained by seismic reflection profiles) within the basin and the palaeomagnetic vectors (and palaeodips) obtained for the remagnetization stage. An intra-Cretaceous unconformity, determined both from field data and from seismic reflection profiles, is the clue for defining an early inversion, pre-remagnetization deformation stage, linked to strike-slip events according to Soto *et al.* (2011).

## Interpretation. Patterns of palaeomagnetic vector distribution

The palaeomagnetic results previously obtained in the three studied basins can be interpreted in terms of a secondary magnetization combined with different previous dips, many of the latter being basinward-directed. In the three cases a simple tectonic correction gives unrealistic values (too high inclinations) for the original palaeomagnetic orientation in Cretaceous (and also in Cenozoic) times. Therefore, the simplest explanation is that the sampled units had original basinward dips when remagnetization occurred (Villalaín *et al.* 2003; Soto *et al.* 2008, 2011; Casas *et al.* 2009). Variations of the direction of remagnetization vectors in units located at different heights along the stratigraphic pile can be interpreted as the result of different original dips when remagnetization was fixed (see Fig. 5).

What consequences can be obtained from geometrical modelling of palaeomagnetic directions and palaeomagnetic field results in the analysed cases? The models shown indicate that the differences in orientation along a stratigraphic sequence (with data obtained in similar rock types and sharing remagnetization features) are useful for determining the possibility of original dips, and also the existence of remagnetizations predating the inversion stage, provided that data can be tested in a fold of proven inversion stage (see Villalaín *et al.* 2003). The distribution of palaeomagnetic vectors, before and after tectonic correction, allows determination of the original basin geometry. In the case of the Cameros Basin, the results obtained from palaeomagnetic analysis are consistent with a synclinal basin geometry, with units remagnetized at the end of the rifting stage, probably coinciding with the heating event involving the whole basin at 100 Ma. The Cabuérniga and Polientes basins show a remagnetization orientation consistent with a similar age, but different tectonic regimes, in spite of being located in nearby areas. In the case of the Polientes Basin, the remagnetization post-dates an early, Cretaceous inversion (that can be also recognized from sedimentary features in seismic profiles; Soto *et al.* 2011), thus indicating that not only basinal geometries, but also other kinds of processes, can be inferred from the geometrical analysis of palaeomagnetic vectors linked to remagnetization.

## Conclusions

The results obtained in this work show that tectonic corrections of syn-tectonic remagnetizations must be carefully treated when dealing with basinal strata. Under- or overcorrection can lead to over- or under-estimations of the original magnetic vector.

The classic incremental fold test can give erroneous results when the hypothesis of a symmetric evolution of folding is not fulfilled. This has been proved in the case of remagnetization in inverted sedimentary basins. In this paper a method to restore syn-tectonic remagnetizations is proposed. The method that combines the SCI method with the asymmetric tectonic correction gives a clue to understand the post-sedimentary orientation of palaeomagnetic vectors. In this way, palaeomagnetic remagnetization vectors can be used to determine differential sedimentary dips along a stratigraphic series, and therefore to define the geometry of sedimentary basins.

This method has been applied in the Cameros, Cabuérniga and Polientes basins, which were remagnetized during Cretaceous times, probably associated with the thermal metamorphism at 100 Ma in this sector of the Iberian peninsula. In the Cameros Basin, the palaeotilting obtained after appropriate restoration of palaeomagnetic vectors indicates that during the remagnetization stage the sedimentary units had basinward dips, which is consistent with a synclinal basin geometry, with a southward-dipping northern limb owing to large-scale drag associated with the movement of the basement normal fault limiting the basin northwards. In the Cabuérniga Basin, palaeomagnetic results also indicate an overall syncline geometry, whereas in the Polientes Basin remagnetization predates an early inversion probably related to transpression along the northern Iberian plate margin.

This work was done within the frame of projects CGL2009-10840 and CGL2012-38481 of the Dirección General de Investigación Científica y Técnica, Ministerio de Economía y Competitividad of the Spanish Government and European Regional Development Fund. We thank Emilio Pueyo, Rob Van der Voo and Samantha Nemkin-Wilson for their valuable comments, which improved the paper. We also thank Sara Torres for her help with figures.

# References

AUBOURG, C., POZZI, J.-P. & KARS, M. 2012. Burial, claystones remagnetizations and some consequences for magnetistratigraphy. *In*: ELMORE, R. D., MUXWORTHY, A. R., ALDANA, M. M. & MENA, M. (eds) *Remagnetization and Chemical Alteration of Sedimentary Rocks*. Geological Society, London, Special Publications, **371**, 181–188.

CASAS, A., VILLALAÍN, J. J., SOTO, R., GIL-IMAZ, A., DEL RÍO, P. & FERNÁNDEZ, G. 2009. Multidisciplinary approach to an extensional syncline model for the Mesozoic Cameros Basin (N Spain). *Tectonophysics*, **470**, 3–20.

CASAS, A. M., DEL RÍO, P., MATA, P., VILLALAÍN, J. & BARBERO, L. 2012. Comment on González-Acebrón ET AL. Criteria for the recognition of localization and timing of multiple events of hydrothermal alteration in sandstones illustrated by petrographic, fluid inclusion, and isotopic analysis of the Tera Group. Northern Spain International Journal of Earth Sciences (2011), **100**, 1811–1826. *International Journal of Earth Sciences*, **101**, 1811–1826.

CASAS-SAINZ, A. M. 1993. Oblique tectonic inversion and basement thrusting in the Cameros Massif (Northern Spain). *Geodinamica Acta*, **6**, 202–216.

CASAS-SAINZ, A. M. & GIL-IMAZ, A. 1998. Extensional subsidence, contractional folding and thrust inversion of the Eastern Cameros Massif, northern Spain. *Geologische Rundschau*, **86**, 802–818.

CEDERQUIST, D. P., VAN DER VOO, R. & VAN DER PLUIJM, B. A. 2006. Syn-folding remagnetization of Cambro-Ordovician carbonates from the Pennsylvania Salient post-dates oroclinal rotation. *Tectonophysics*, **422**, 41–54.

DELAUNAY, S., SMITH, B. & AUBOURG, C. 2002. Asymmetrical fold test in the case of overfolding: two examples from the Makran accretionary prism (Southern Iran). *Physics and Chemistry of the Earth*, **27**, 1195–1203.

FISHER, R. A. 1953. Dispersion on a sphere. *Proceedings of the Royal Society, London, Series A*, **217**, 295–305.

GOLDBERG, J. M., GUIRAUD, M., MALUSKI, H. & SÉGURET, M. 1988. Caractères pétrologiques et âge du métamorphisme en contexte distensif du bassin sur décrochement de Soria (Crétacé inférieur, Nord Espagne). *Comptes Rendues de l'Académie des Sciences Paris*, **307**, 521–527.

GONG, Z., VAN HINSBERGEN, D. J. J. & DEKKERS, M. J. 2009. Diachronous pervasive remagnetization in northern Iberian basins during Cretaceous rotation and extension. *Earth and Planetary Science Letters*, **284**, 292–301.

GRAHAM, J. W. 1949. The stability and significance of magnetism in sedimentary rocks. *Journal of Geophysics Research*, **54**, 131–167.

GUIMERÀ, J., ALONSO, A. & MAS, J. R. 1995. Inversion of an extensional-ramp basin by a newly formed thrust: the Cameros basin (NSpain). *In*: BUCHANAN, J. G. & BUCHANAN, P. G. (eds) *Basin Inversion*. Geological Society, London, Special Publications, **88**, 433–453.

GUIRAUD, M. & SÉGURET, M. 1984. Releasing solitary overstep model for the Late Jurassic–Early Cretaceous (Wealdien) Soria strike-slip basin (North Spain) *In*: BIDDLE, K. T. & CRISTHIE-BLICK, N. (eds) *Strike-Slip Deformation, Basin Formation and Sedimentation*. SEPM, Special Publications, Tulsa, OK, **37**, 159–175.

HENRY, B., ROUVIER, H. & LE GOFF, M. 2004. Using syntectonic remagnetizations for fold geometry and vertical axis rotation: the example of the Cévennes border (France). *Geophysics Journal International*, **157**, 1061–1070.

KATZ, B., ELMORE, R. D., COGOINI, M. & FERRY, S. 1998. Widespread chemical remagnetization: orogenic fluids or burial diagenesis of clays? *Geology*, **26**, 603–606.

KENT, D. V. & OPDYKE, N. D. 1985. Multicomponent magnetizations from the Mississippian Mauch Chunk Formation of the Central Appalachians and their tectonic implications. *Journal of Geophysical Research*, **90**, 5371–5383.

KODAMA, K. P. 1988. Remanence rotation due to rock strain during folding and the step-wise application of the fold test. *Journal of Geophysical Research*, **93**, 3357–3371.

MATA, M. P., CASAS, A. M., CANALS, A., GIL, A. & POCOVI, A. 2001. Thermal history during Mesozoic extension and Tertiary uplift in the Cameros Basin, northern Spain. *Basin Research*, **13**, 1–22.

MCCAIG, A. M. & MCCLELLAND, E. 1992. Palaeomagnetic techniques applied to thrust belts. *In*: MCCLAY, K. R. (ed.) *Thrust Tectonics*. Chapman & Hall, London, 209–216.

MCCLELLAND-BROWN, E. 1983. Paleomagnetic studies of fold development and propagation in the Pembrokeshire Old Red Sandstone. *Tectonophysics*, **98**, 131–149.

MCFADDEN, P. L. & JONES, D. L. 1981. The fold test in palaeomagnetism. *Geophysics Journal of the Royal Astronomy Society*, **67**, 53–58.

MOUSSAID, B., VILLALAÍN, J. J., CASAS-SAINZ, A., EL OUARDI, H., OLIVA-URCIA, B., SOTO, R., ROMÁN-BERDIEL, T. & TORRES-LÓPEZ, S. 2015. Primary vs. secondary curved fold axes: deciphering the origin of the Aït Attab syncline (Moroccan High Atlas) using paleomagnetic data. *Journal of Structural Geology*, **70**, 65–77.

Muñoz-Jiménez, A. & Casas-Sainz, A. M. 1997. The Rioja Trough: tecto-sedimentary evolution of a foreland symmetric basin. *Basin Research*, **9**, 65–85.

Oliva-Urcia, B., Román-Berdiel, T., Casas, A. M., Bógalo, M. F., Osácar, M. C. & García-Lasanta, C. 2013. Transition from extensional to compressional magnetic fabrics in the Cretaceous Cabuérniga basin (North Spain). *Journal of Structural Geology*, **46**, 220–234.

Salas, R. & Casas, A. 1993. Mesozoic extensional tectonics, stratigraphy and crustal evolution during the Alpine cycle of the eastern Iberian basin. *Tectonophysics*, **228**, 33–35.

Scotese, C. R. & Van der Voo, R. 1983. Paleomagnetic dating of Alleghenian folding (abstract). *EOS Transactions of the American Geophysics Union*, **64**, 218.

Shipunov, S. V. 1997. Synfolding magnetization: detection, testing and geological applications. *Geophysical Journal International*, **130**, 405–410.

Soto, R., Casas-Sainz, A. M. & Del Río, P. 2007a. Geometry of half-grabens containing a mid-level viscous décollement. *Basin Research*, **19**, 437–450.

Soto, R., Casas-Sainz, A. M., Villalaín, J. J. & Oliva-Urcia, B. 2007b. Mesozoic extension in the Basque–Cantabrian basin (N Spain): contributions from AMS and brittle mesostructures. *Tectonophysics*, **445**, 373–394.

Soto, R., Villalaín, J. J. & Casas-Sainz, A. M. 2008. Remagnetizations as a tool to analyze the tectonic history of inverted sedimentary basins: a case study from the Basque–Cantabrian basin (north Spain). *Tectonics*, **27**, TC1017, http://doi.org/10.1029/2007 TC002208

Soto, R., Casas-Sainz, A. M. & Villalaín, J. J. 2011. Widespread Cretaceous inversion event in northern Spain; evidence from subsurface and paleomagnetic data. *Journal of the Geological Society, London*, **168**, 899–912.

Torres-López, S., Villalaín, J. J., Casas, A. M., El Ouardi, H., Moussaid, B. & Ruiz-Martínez, V. C. 2014. Widespread Cretaceous secondary magnetization in the High Atlas (Morocco). A common origin for the Cretaceous remagnetizations in the western Tethys? *Journal of the Geological Society, London*, **171**, 673–687.

Torsvik, T. H., Van der Voo, R. *et al.* 2012. Phanerozoic polar wander, palaeogeography and dynamics. *Earth-Science Reviews*, **114**, 325–368.

Tischer, G. 1966. El delta Weáldico de las montañas Ibéricas Occidentales y sus enlaces tectónicos. *Notas y Comunicaciones de Instituto Geológico y Minero de España*, **81**, 53–78.

Van der Voo, R. 1993. *Paleomagnetism of the Atlantic, Tethys and Iapetus Oceans*. Cambridge University Press, Cambridge.

Villalaín, J. J., Osete, M. L., Vegas, R., García-Dueñas, V. & Heller, F. 1994. Widespread Neogene remagnetization in Jurassic limestones of the south Iberian paleomargin (Western Betics, Gibraltar Arc). *Physics of the Earth and Planetary Interiors*, **85**, 15–33.

Villalaín, J. J., Osete, M. L., Vegas, R., Garcia-Duenas, V. & Heller, F. 1996. The Neogene remagnetization in the western Betics: a brief comment on the reliability of palaeomagnetic directions. *In*: Morris, A. & Tarling, D.-H. (eds) *Palaeomagnetism, Tectonics of the Mediterranean Region*. Geological Society, London, Special Publications, **105**, 33–41.

Villalaín, J. J., Fernández-Gonzalez, G., Casas, A. M. & Gil-Imaz, A. 2003. Evidence of a Cretaceous remagnetization in the Cameros Basin (North Spain): implications for basin geometry. *Tectonophysics*, **377**, 101–117.

Villalaín, J. J., Casas-Sainz, A., Soto, R. & Torres-López, S. 2012. The widespread Cretaceous remagnetizations in Mesozoic intraplate basins from Iberia and North Africa. Contribution to tectonic studies. *Contributions to Geophysics and Geodesy*, **42**, 121–122.

Waldhör, M. & Appel, E. 2006. Intersections of remanence small circles: new tools to improve data processing and interpretation in palaeomagnetism. *Geophysical Journal International*, **166**, 33–45.

Waldhör, M., Appel, E., Frisch, W. & Patzelt, A. 2001. Palaeomagnetic investigation in the Pamirs and its tectonic implications. *Journal of Asian Earth Sciences*, **19**, 429–451.

Williams, G. D., Powell, C. M. & Cooper, M. A. 1989. Geometry and kinematics of inversion tectonics. *In*: Cooper, M. A. & Williams, G. D. (eds) *Inversion Tectonics*. Geological Society, London, Special Publications, **44**, 3–15.

# Parametric unfolding of flexural folds using palaeomagnetic vectors

MARIA JOSÉ RAMÓN[1]*, EMILIO L. PUEYO[1], GUILLAUME CAUMON[2] &
JOSÉ LUIS BRIZ[3]

[1]*Instituto Geológico y Minero de España (IGME), Unidad de Zaragoza, Spain*

[2]*Gocad Research Group – ASGA, Institut National Polytechnique de Lorraine, France*

[3]*Depto. de Informática e Ingeniería de Sistemas e Instituto de Investigación
en Ingeniería de Aragón (I3A), Universidad de Zaragoza, Spain*

**Corresponding author (e-mail: mj.ramon.ortiga@gmail.com)*


**Abstract:** Providing that a primary and reliable record of the magnetic field and its reference in a deformed area exists, the incorporation of palaeomagnetic constraints in restoration methods reduces uncertainty of rotation because such constraints can be applied both before and after deformation. In this paper, we utilize palaeomagnetic data to improve an unfolding algorithm based on the parameterization of the surface using isometric constraints. This method is more robust than others based on piecewise restoration of a triangulated surface, which are dependent on the meshing and, especially, on the pin-element. A disadvantage of this approach is that parametric restoration is sensitive to the initial solution, which hampers results for complex non-coaxial or non-cylindrical structures. We show that the use of palaeomagnetism as the initial gradient of one of the parameters improves the results of the method. We use analogue models to test the method because the expected restoration result can be stated, since the initial surface is known. We study the restoration sensitivity to surface meshing and the initial palaeomagnetic orientation. All in all, the use of palaeomagnetic vectors in the studied analogue models achieves the best restoration results. The implementation of palaeomagnetic vectors is crucial to obtain reliable 3D restorations of complex structures.

Reconstruction and restoration methods of geological structures aim to provide a plausible image of the subsurface structure through time, back to the undeformed state. These methods are based on geometric/mechanic laws, and are designed to tackle areas with scarce and heterogeneous data (e.g. Caumon *et al.* 2009). Restoration algorithms are an important tool to validate 3D geological reconstructions of the subsurface. The application of geometric rules to retro deform with geological meaning is useful to validate the initial reconstruction of the folded structure and the deformation processes assumed.

Many restoration methods exist and are based on suites of assumptions regarding deformation processes, geomechanics and initial states. Methods include cross-section balancing (2D) (Dahlstrom 1969); map-view restoration (2D) (Audibert 1991; Rouby *et al.* 1993); surface restoration (so-called '2.5D') (Gratier *et al.* 1991; Rouby *et al.* 2000; Ramón *et al.* 2012); or real volume restoration (3D) (Muron 2005; Maerten & Maerten 2006; Durand-Riard *et al.* 2010, 2013; González *et al.* 2012). One of two main assumptions is usually made (Moretti 2008): the flexural slip method and the simple shear method. The first assumption can

be valid for fold and thrust belts where internal deformation occurs mostly by layer-parallel shortening and where bed length and bed thickness remained constant during the deformation process. The second assumption requires preservation of distances in the shear direction. Depending on the mathematical approach considered, we present geometric and geomechanical methods. Geometric methods are based on triangulation (Gratier *et al.* 1991; Gratier & Guiller 1993) or parameterization (Massot 2002) of the surface. The main drawback of the geometric approach is that it neglects the mechanical properties of materials. The geomechanical method uses finite element calculus (Muron 2005), but has the drawback that it is too dependent on the boundary conditions.

Most existing restoration methods do not succeed for complex structures. This is the case for example for non-cylindrical geometries, non-coaxial superposition of deformations and/or areas undergoing vertical-axis rotations. The latter use 2D references (bedding surface) and therefore cannot resolve 'out of plan' motions, a classic drawback already identified in the cross-section balancing techniques (Dahlstrom 1969; Elliott 1976). In this sense palaeomagnetism can shed some light

*From*: PUEYO, E. L., CIFELLI, F., SUSSMAN, A. J. & OLIVA-URCIA, B. (eds) 2016. *Palaeomagnetism in Fold and Thrust Belts: New Perspectives*. Geological Society, London, Special Publications, **425**, 247–258.
First published online August 12, 2015, http://doi.org/10.1144/SP425.6

on the problem, since it is likely the only vector indicator that can be determined for time intervals both before and after the deformation. As a result of this capability, palaeomagnetism can be used as an additional and powerful constraint to improve restoration methods and reduce the uncertainty of the results. The use of palaeomagnetism in restoration was suggested in the early 1990s (McCaig & McClelland 1992), and was later used to double check the rotation estimates derived from map-view restorations (Rouby *et al.* 1996; Bourgeois *et al.* 1997). In this 2D approach, the restoration result is transformed into the displacement, internal deformation and rotation fields. Remarkably, rotations are always underestimated when compared to the observed rotations, as demonstrated in Afar and in the Tajik depression. Palaeomagnetism was later extended to map view reconstructions (2D) and applied to numerically restore cases with rigid-body rotations constrained by palaeomagnetic data (Arriagada *et al.* 2008). In 2D approaches, palaeomagnetism has also been used to numerically correct shortening estimates in balanced cross-sections, as well as to relate the shortening differences to the rotation magnitudes (Pueyo *et al.* 2004; Soto *et al.* 2006; Sussman *et al.* 2012).

More recently, Ramón *et al.* (2012) incorporated palaeomagnetism into a surface restoration method (2.5D). The latter approach is inspired in the UNFOLD method by Gratier & Guiller (1993), a geometric unfolding algorithm valid for surfaces folded under flexural conditions (global preservation of lengths and angles). Results suggest that palaeomagnetism provides a good constraint to anchor the surface and characterize deformation patterns, and to reduce the uncertainty of the restoration output. On the downside, the fact that it is based on the triangulation and piecewise restoration of the surface makes the proposal sensitive to the density and method of triangulation and, especially, the pin-element location (Ramón *et al.* in press).

In this paper we propose an alternative approach to reach more stable and reliable solutions from the geometric restoration of a single folded horizon. We incorporate palaeomagnetic information to an unfolding algorithm based on the parameterization of the surface (Massot 2002), a method also valid for surfaces folded under flexural conditions. Some analogue models, where deformation patterns are known, are used to test the quality of the restoration.

## Method

In this section, we explain how to use primary palaeomagnetic vectors in the parametric method. We then introduce analogue models of complex structures and use them to check the accuracy of the restoration.

## *Unfolding: parametric approach*

Any surface $S$ ($\aleph^3$) can be projected onto a map $C$ ($\aleph^2$) using a parametric representation (Fig. 1a). This so-called surface parameterization problem has been extensively studied in the computer graphics community. We summarize the main ideas below and refer the reader to Botsch *et al.* (2010) and Floater & Hormann (2005) for further discussion, and Mallet (2002), Moretti *et al.* (2005) and Moretti (2008) for more details about the application of the free-boundary parameterization to surface restoration. Each point in $C$ defined by the coordinates $\mathbf{u} = (u, v)$ has a unique associated image in $S$ $\mathbf{x}(\mathbf{u})$ with coordinates $(x, y, z)$:

$$\mathbf{u} = \begin{bmatrix} u \\ v \end{bmatrix} \in C \xrightleftharpoons[\mathbf{u}(\mathbf{x})]{\mathbf{x}(\mathbf{u})} \mathbf{x} = \begin{bmatrix} x \\ y \\ z \end{bmatrix} \in S \quad (1)$$

For a single surface there are many possible parametric representations. However, for any parameterization it is possible to define the metric tensor $G(u, v)$ which defines the intrinsic properties of the surface.

The main assumption for most restoration algorithms is the horizontality of the initial surface. The restoration problem is therefore equivalent to looking for a particular parameterization $\mathbf{u}(\mathbf{x})$. Moreover, we focus on surfaces folded under flexural conditions that lead to *globally developable* surfaces (Gaussian curvature equal to zero *almost* everywhere). This assumption implies the preservation of lengths and angles in the folded surface.

The palaeogeographic coordinate system $(u, v)$ can be freely and arbitrarily chosen and is assumed to be rectilinear and orthonormal for simplicity. The metric tensor $G$ associated with the local parameterization must be close to the unit tensor (Equation (2)) by the principle of minimum deformation:

$$\varepsilon = \frac{1}{2}[G(u, v) - I] \cong 0 \rightarrow G(u, v) \cong I = \begin{bmatrix} 1 & 0 \\ 0 & 1 \end{bmatrix}$$
$$(2)$$

From a given 3D surface, the restoration problem can be addressed by computing two coordinates $(u, v)$ on the surface so that:

$$\nabla u \cdot \nabla v = 0$$
$$\|\nabla u\| = \|\nabla v\| = 1 \quad (3)$$

The first constraint means that the gradients are orthogonal, and the second that they have the same norm (Fig. 1). These two conformality

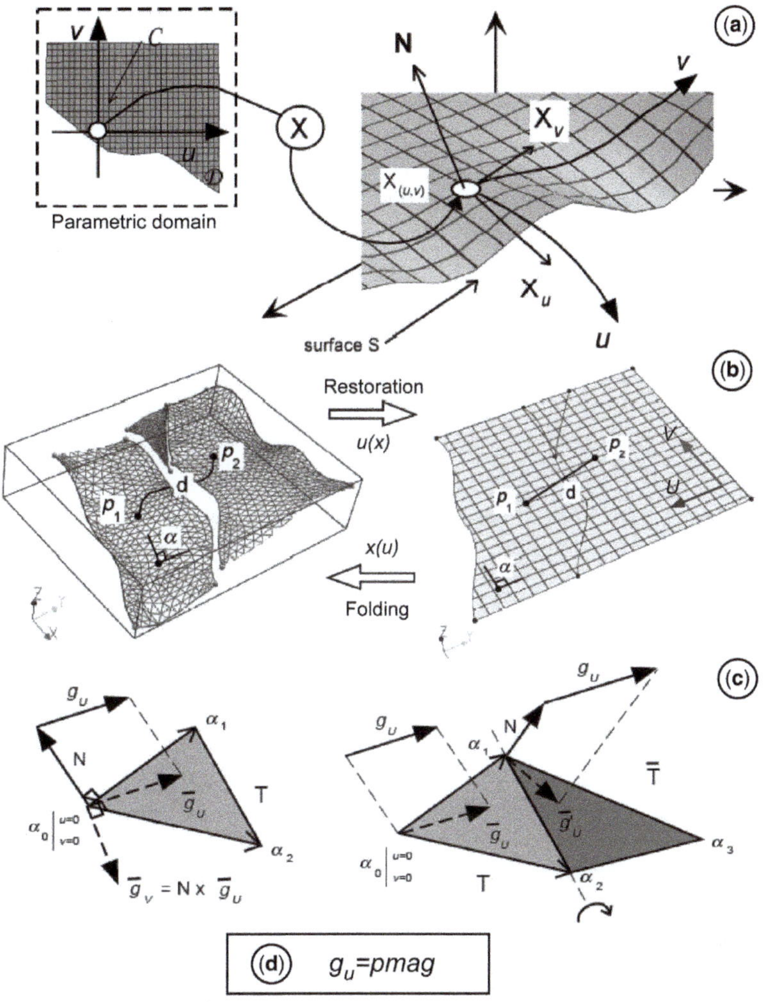



**Fig. 1.** (**a**) A 3D surface can be mapped onto a 2D plane by computing a curvilinear coordinate system $u(x, y, z)$ which is chosen to be rectilinear and orthonormal. (**b, c**) Algorithm used to compute the initial parameterization. (**d**) Parameterized surface.

constraints can be discretized on the surface in order to compute the parameterization (e.g. Lévy & Mallet 1998). In this work we started with a parameterization method developed by Massot (2002), which begins from a pinpoint on the surface where two directions, $\nabla u_0$ and $\nabla v_0$, are chosen in order to be mutually orthogonal, and orthogonal to the local normal vector $\mathbf{n}_0$ (Fig. 1b). The method starts by propagating this local frame to all the nodes of the surface. The frame is rotated along an axis $l$ given by the cross product of the normal at the pinpoint, $\mathbf{n}_0$, and the normal at the current point, $\mathbf{n}_1$. (Fig. 1b, c). Last, the $u$ and $v$ coordinates are integrated numerically from the pinpoint coordinate using a least-squares method in order to honour

the input gradients at all locations (Fig. 1d); see details in Massot (2002) and Mallet (2002).

Although the parametric method often provides good results in moderately deformed areas, it remains quite sensitive to the initial solution due to the method of propagating the local frame to the whole surface. In this paper, we present an improvement to this restoration approach by incorporating palaeomagnetic data, and show significant improvements to the robustness of the parametric approach. Palaeomagnetism defines the initial solution by using palaeomagnetic values defined at all points of the surface as the initial gradient values $\nabla u$. Since the gradient is a vector contained in the surface, the initial palaeomagnetic vector is rotated to have null inclination and is therefore embedded in the surface.

## Palaeomagnetic data

Before using palaeomagnetism as an additional constraint in the restoration method, we must check its validity. Palaeomagnetic vectors in the deformed state need to be a proven record of the primary magnetic field (existing at the time of rock formation). Local datasets must therefore honour some reliability criteria (such as defined in Van der Voo 1990; Pueyo 2010; Pueyo *et al.* in press) such as: proper isolation of magnetization components (overlapping or inclinations problems are absent); suitable statistical parameters ($n$, confidence angle $\alpha_{95}$, $R$, $k$) and stability tests (fold, reversal); and the proven absence of internal deformation of the rock volume containing the data. Similarly, the palaeomagnetic reference (recorded in the undeformed stage) also has to be estimated, and must fulfil identical quality criteria. The palaeomagnetic reference should be deduced from the undeformed foreland whenever possible, and should be contrasted to the expected direction derived from the correspondent apparent polar wander path in that plate. Note that these two datasets might not always be available, limiting the application of this constraint in 3D restoration.

In addition to the reliability and quality of the data, the studied structure must be characterized by a dense and evenly distributed net of palaeomagnetic sites. In addition to a reliable interpolation algorithm, it would be necessary to spread the data from the surface into the underground all along the studied geometry, not only the part which is cropping out (Ramón 2013). In this paper, we apply the method to a theoretical model where palaeomagnetism is perfectly known at all points of the surface, and future studies should tackle all these problems.

## Analogue models

We evaluate the quality of the palaeomagnetic-parametric method utilizing analogue models as in Ramón *et al.* (2012). We simulate complex structures folded under flexural conditions at the laboratory scale using pseudo-rigid (developable) and pseudo-elastic (globally developable) materials, plastic nets and EVA (ethylene vinyl acetate) plates, respectively. All surfaces have a local orthogonal reference system, and palaeomagnetism is defined using the orientation of one of those lineations. Note that, in contrast to natural cases, we can theoretically establish a primary record of the magnetic field. The digitalization or reconstruction of the upper surface of these analogue models is performed using photogrammetry, although X-ray computed tomography (CT) scanners can also be used for a 3D reconstruction (Ramón *et al.* 2013). The importance of these models is that they provide the position of all points on the surface in both the initial (horizontal) and the folded states. Consequently, we can compare the obtained restored surface with the initial (horizontal) surface.

Analogue models used in this work were inspired by a set of complex geometries from the External Sierras front at the south Pyrenean thrust system (Fig. 2a). Trending WNW–ESE and 100 km long, the External Sierras were developed during Late Lutetian–Early Miocene times (Puigdefábregas 1975; Millán *et al.* 1995, 2000; Arenas *et al.* 2001), and caused the separation of the Jaca piggyback basin in the north from the main part of the Ebro foreland basin in the south (e.g. Ori & Friend 1984; Millán *et al.* 1995; Teixell & García Sansegundo 1995; Teixell 1996). The deformations of the Middle–Late Triassic–Early Miocene sediments were heavily influenced by the weak rheology of the Triassic evaporite deposits that served as a regional detachment horizon. The External Sierras display remarkable interference patterns between transverse (north–south to NW–SE) structures and the north–south-trending Pyrenean folds and thrusts (e.g. Fachar, Bentué, Pico del Aguila, Balzes anticlines). The kinematic evolution can be synthesized by two diachronous, many times non-coaxial and distinctive episodes of deformation (Pueyo *et al.* 1997; Millán *et al.* 2000) which were derived in a complex fold and thrust belt comprising conical folds, superposed folding, oblique thrust ramps and folds, etc. Intrinsic to this kinematics evolution, significant magnitudes of vertical axis rotations have been detected by primary palaeomagnetic records all along the front (Pueyo *et al.* 2002, 2003, 2004; Oliva-Urcia & Pueyo 2007; Mochales *et al.* 2012; Rodríguez-Pintó *et al.* 2012, 2013).

We have selected three significant structures from this suite of complex geometries (Fig. 2b). The first structure models the San Marzal pericline, which is the termination of the Santo Domingo anticline, the westernmost structure of the External Sierras, that is, the Pyrenean sole thrust. San Marzal is a large conical fold involving more than 60° of differential rotation among its flanks (Pueyo 2000; Oliva-Urcia *et al.* 2012; Sánchez *et al.* 2012). The model is a conical fold made with one EVA plate of 30 × 40 × 0.5 cm. The second selected structure is a curved fold similar to the Balzes anticline. Variable clockwise rotations between 50° and 25°, detected in the northern and southern sectors of this structure, allow the primary curvature of the structure to be reconstructed. The model is made with two EVA plates of dimension 58 × 38 × 0.5 cm. The third structure is an interference fold

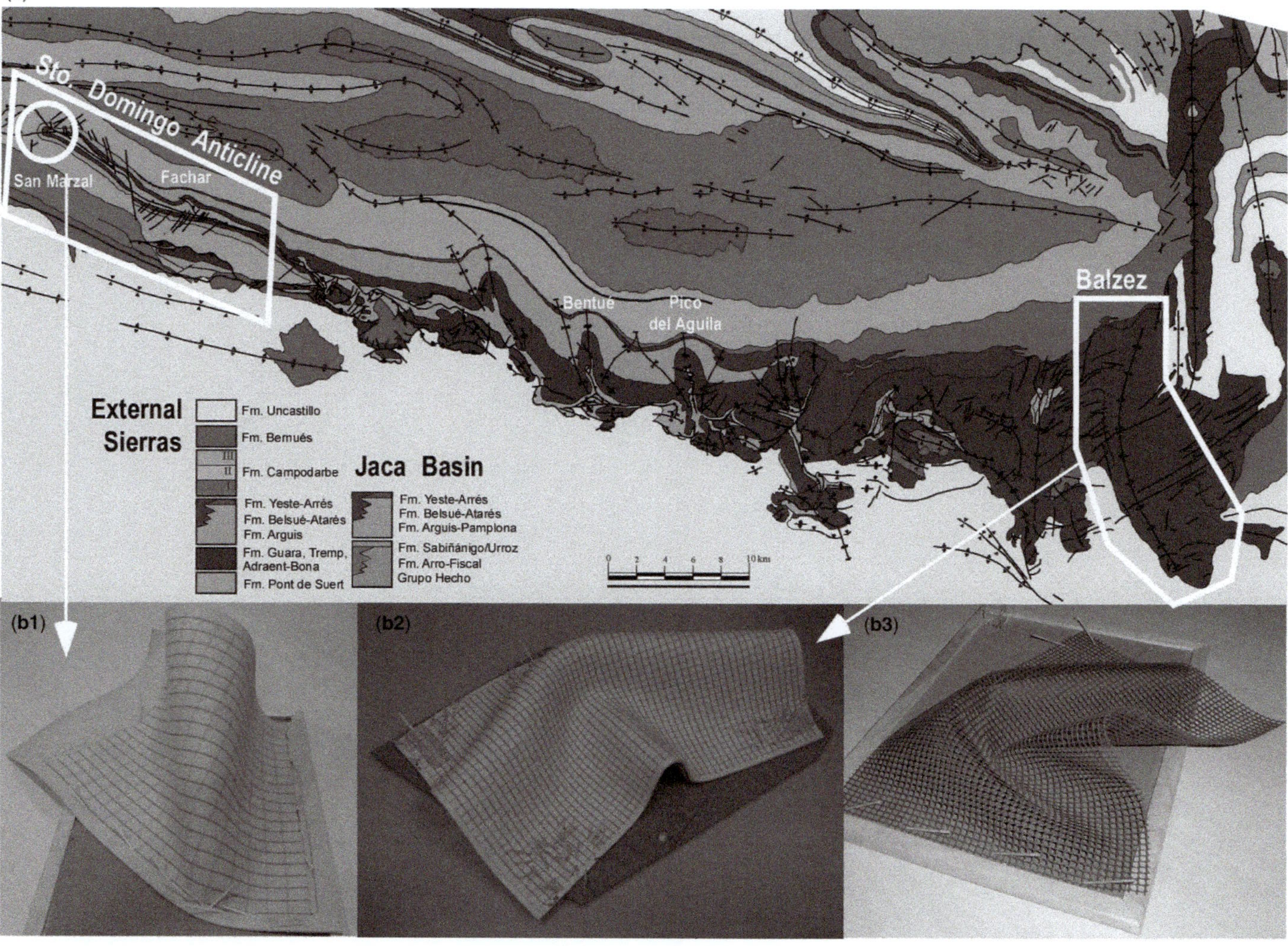


**Fig. 2.** (**a**) Geological background of structures that have inspired the analogue models. (**b**) Analog models: (1) San Marzal pericline (conical fold); (2) Balzes anticline (curved fold); and (3) interference fold.

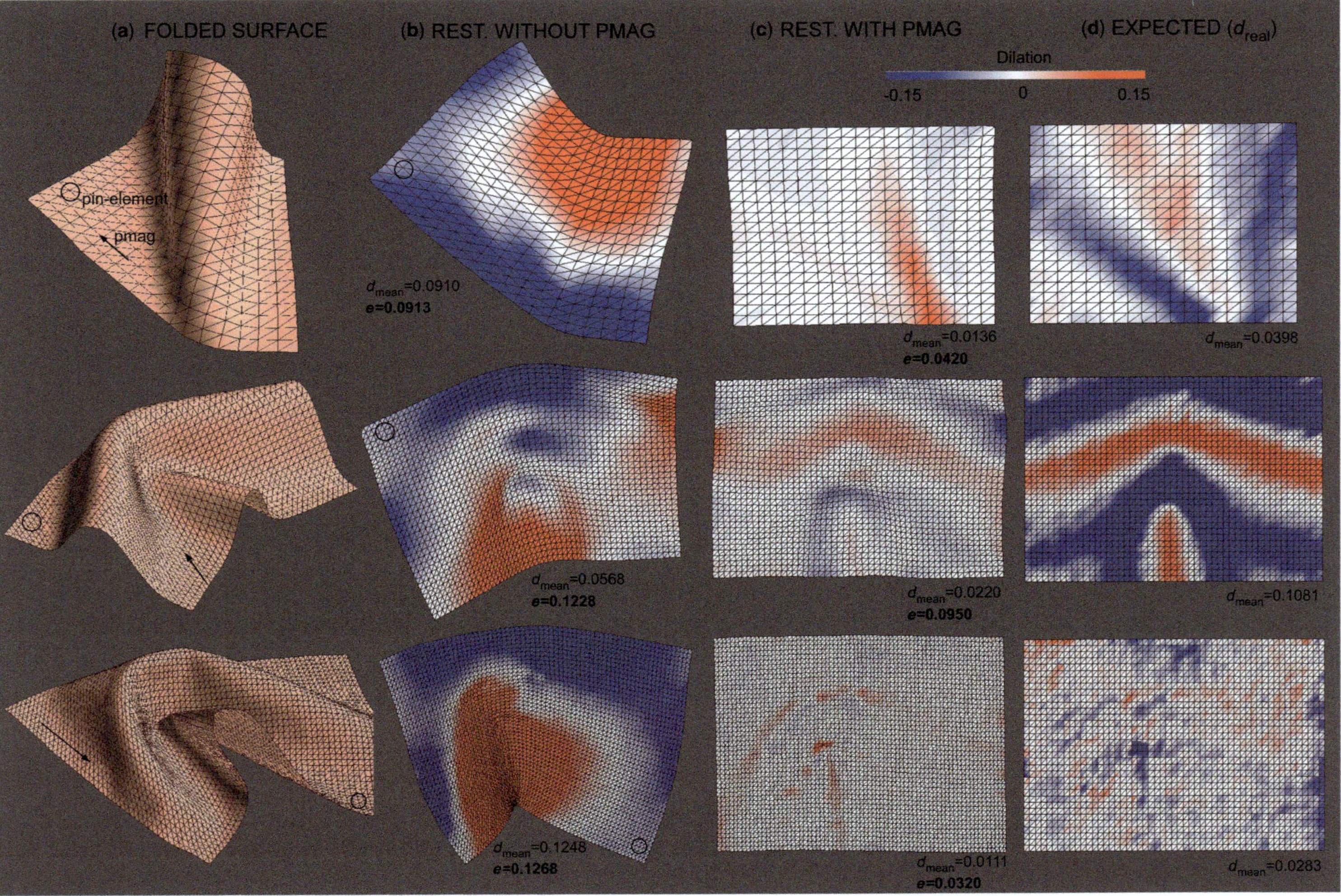



**Fig. 3.** Restoration of three analogue models with and without palaeomagnetic data. (**a**) Folded surface with the pin-element and pin-vector orientation displayed. (**b**) Restored surface without palaeomagnetism. (**c**) Restored surface with palaeomagnetism. Restored surfaces are plotted with dilation values. (**d**) Initial horizontal surface with real dilation ($d_{real}$): expected result. $e$, the mean error between real and calculated dilation ($e = \mathrm{mean}(|d_{real} - d|)$) and $d_{mean}$ is the mean dilation ($d_{mean} = \mathrm{mean}(|d|)$).

which magnifies the deformation of the Balzes and Boltaña structures. The model is made with a plastic net of $34.5 \times 53.25$ cm.

The great advantage of these analogues is that we accurately know their deformed and undeformed stages. We can therefore evaluate the restoration method on the analogues and then eventually apply the method to natural cases, while being aware of its limits (Ramón *et al.* 2013).

## Results and discussion

In this section, we first compare the restoration results with and without the palaeomagnetic information. Secondly, we evaluate the effect of the mesh density and sampling strategy of the surface. Finally, we analyse the influence of the initial palaeomagnetic orientation.

## *Restoration with and without palaeomagnetic data*

We apply the parametric-restoration method to the upper surfaces derived from the three analogue models with and without the palaeomagnetic information as a constraint for the initial $u$ gradient. In this section we use palaeomagnetic data defined at all points of the surface. Results for the restoration without palaeomagnetism clearly show that $U$ and $V$ are not orthonormal and that we do not reach the initial rectangular-shape surface (Fig. 3b). The analogue folds presented here are all non-coaxial and non-cylindrical and imply out-of-plane movements in terms of 2D balancing techniques (Hossack 1979). All these factors make the restoration difficult, and the final result is too dependent on the initial solution. For this reason, using

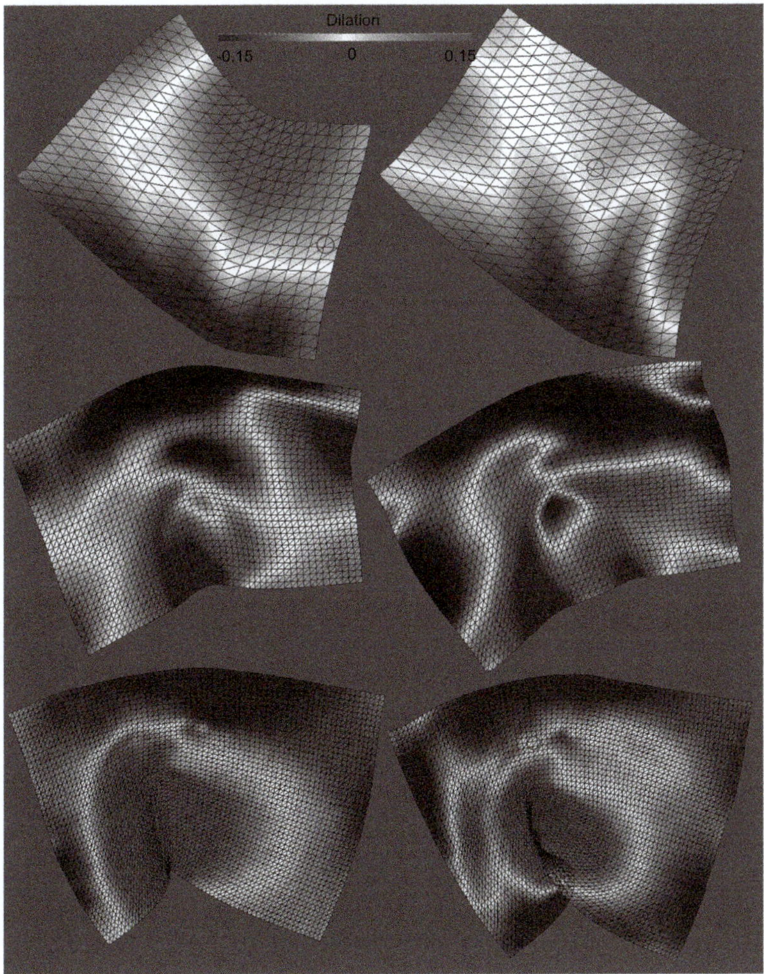


**Fig. 4.** Restoration sensitivity to the pin-element location without using palaeomagnetism.

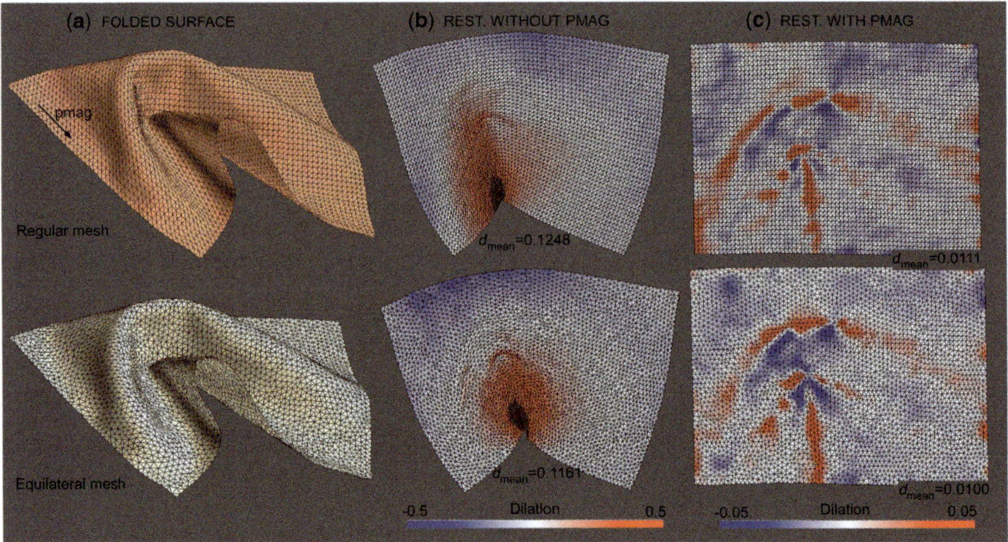



**Fig. 5.** Mesh type sensitivity. (**a**) Folded surface defined with several triangulations. First row: regular mesh; second row: equilateral mesh. (**b**) Restored surface without using palaeomagnetism. (**c**) Restored surface with palaeomagnetism.

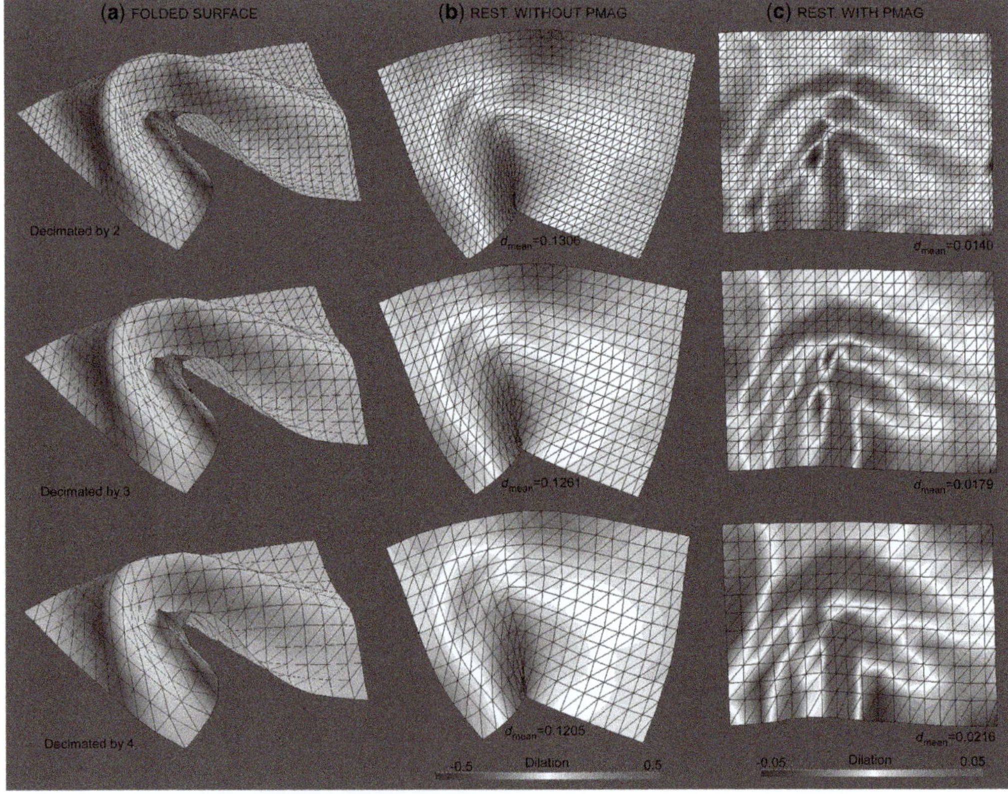



**Fig. 6.** Mesh density sensitivity. (**a**) Folded surface defined with the regular grid decimated by 2, 3 and 4. (**b**) Restored surface without using palaeomagnetism. (**c**) Restored surface with palaeomagnetism.

palaeomagnetism is crucial to define this initial solution. The shape of restored surfaces using palaeomagnetism almost perfectly matches the expected surfaces and orthogonality is preserved in the restored state. We visualize the dilation in the restored state as a measure of area variation between the folded and the restored surface, $d =$ (area$_{\text{fold}}$ − area$_{\text{rest}}$)/area$_{\text{rest}}$ (Fig. 3c). We contrast this variable with the real dilation, calculated between the folded surface and the initial horizontal surface, $d_{\text{real}} =$ (area$_{\text{fold}}$ − area$_{\text{ini}}$)/area$_{\text{ini}}$ (Fig. 3d). The distribution of maximum dilation values roughly corresponds to the expected result. In order to quantify the difference, we measure the mean error between real (expected from the model) and calculated dilation $e = \text{mean}(|d_{\text{real}} − d|)$, and observe that it diminishes when using palaeomagnetism in the restoration process. Moreover, the mean dilation $d_{\text{mean}} = \text{mean}(|d|)$ indicates that the degree of deformation of the surface is lower in the restoration with palaeomagnetism.

The restoration method assumes isometric parameterization based on developable surfaces folded under flexural conditions. We have used it to restore globally developable surfaces with small deformation in strongly and/or complexly folded areas (where the expected $d_{\text{mean}}$ is small but not null). The unfolding algorithm minimizes deformation showing dilation patterns similar to the real patterns. Nevertheless, no restoration method can ever predict the real deformation with total accuracy.

Piece-way restoration methods (Gratier *et al.* 1991) are very sensitive to the location of the pin-elements (Ramón *et al.* 2013; in press) and depend heavily on the location of the starting point of the restoration iteration. The pin-elements are conceptually similar to the pinpoint used in cross-section balancing. We have therefore attempted the restoration starting from different initial points (pin-element or seed). The restoration without palaeomagnetism is very sensitive to the pin-element location which conditions the initial solution (as already suggested by Lallier *et al.* 2008), and we found as many solutions as starting pin-element locations (Fig. 4). In contrast, restoration with palaeomagnetism is completely independent of the pin-element because the initial solution is determined by the orientation of the palaeomagnetic data; all the solutions are therefore identical.

## Mesh sensitivity

In discrete computer-based models some algorithms may lead to artefacts when surface sampling changes. We have therefore checked the robustness of the method towards variations in the density and the structure of the mesh. For this analysis we have selected the interference fold (third row in Fig. 3) because it is the most complex structure among the three examples and has the highest mesh density. We use two different triangulations to evaluate the mesh-type effect: a regular grid (right-angled triangles defined by the reference mesh) and an

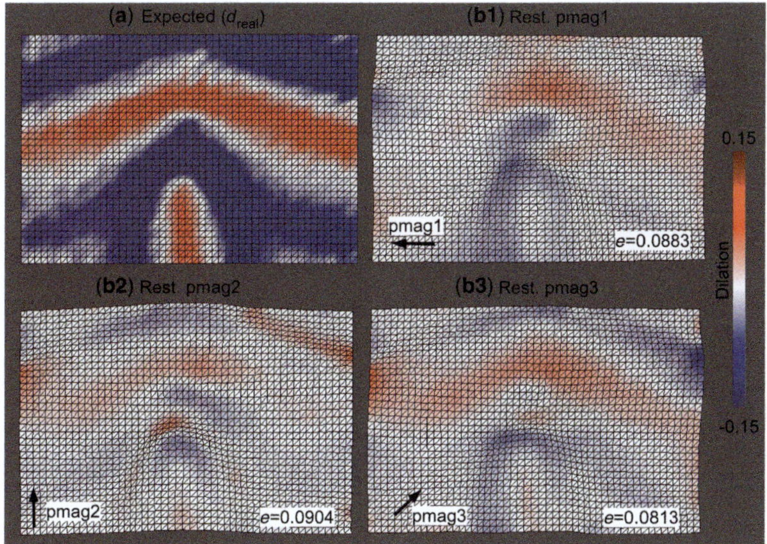



**Fig. 7.** Palaeomagnetic initial orientation sensitivity. (**a**) Expected result, real dilation. (**b**) Parametric restoration with dilation patterns using three orthogonal palaeomagnetic datasets (pmag1, pmag2 and pmag3). $e$, is the mean error between real and calculated dilation.

equilateral grid, obtained by adaptive Delaunay (1934) sampling, performed in gOcad™ using the 'beautify for equilaterality' command. To evaluate the density of the mesh we further divide the initial regular mesh by 2, 3 and 4 while preserving the boundary.

Figure 5 shows the comparison of restoration results obtained for the regular mesh (upper tier) and the equilateral grid (lower tier). The restoration without palaeomagnetic data (Fig. 5b) yields poor and rather different results for both triangulations. On the contrary, when palaeomagnetic data are utilized, surfaces and dilation patterns are accurate and equivalent, irrespective of the meshing.

Figure 6 displays the results obtained with different mesh densities. The divided regular meshes lead to an equivalent result, although with a smoother definition of the dilation pattern inversely to the density. Aside from this consideration, palaeomagnetism guarantees the reliability of the results and the difference between these and the restoration results when palaeomagnetic vectors are neglected is significant.

*Palaeomagnetic initial orientation sensitivity*

The pin-vector-orientation ($u$ gradient) determined by the primary palaeomagnetic vector may condition the final result in the parametric restoration, as happens when using piecewise restoration (Pmag3DRest, Ramón *et al.* 2012). In a natural case we cannot modify this parameter, but we need to know if the orientation of initial palaeomagnetic data conditions the result. In this section we compare the restoration results for the second model (Balzes curved fold) because is the most deformed among the three case studies. We use three palaeomagnetic references, defined by the three sides of the triangles pmag1, pmag2 and pmag3, based on the real network of points that define the original grid. All the restored surfaces have equivalent shapes, and we compare the dilation patterns with the real dilation (Fig. 7). In order to quantify the difference we measure the mean error between real and calculated dilation.

The parametric restoration with pmag3 (oblique to most structural trends) produces a dilation pattern more similar to the real pattern, with the same extension (red colour) and compression (blue) areas corresponding to anticline and syncline hinges, respectively. This observation is coherent with the mean dilation error measured ($e\_pmag1 = 0.0883$, $e\_pmag2 = 0.0904$ and $e\_pmag3 = 0.0813$). Surfaces restored with orthogonal pin-orientation-vectors display similar restored surfaces but not necessarily identical deformation patterns, reminding us that restoration algorithms do not locate the deformation with total accuracy because of limits in the method itself.

## Conclusions

The incorporation of palaeomagnetism as an additional constraint in the restoration methods reduces the uncertainty in the result and can help to validate the reconstruction of complex structures. The previous approximation (Ramón *et al.* 2012) based on the piecewise restoration of a triangulated surface presented some undesired dependence on the mesh and on the pin-element (although much less dependence than the same method without palaeomagnetism). If we look for alternatives to methods that rely on surface triangulation, parametric restoration is a good candidate. However, the original method is too dependent on the initial solution and it is only valid for smoothly folded surfaces. It is unfeasible for complex surfaces such as the examples we examine here based on laboratory-scale analogue models. We have demonstrated that parametric restoration can be much improved by considering palaeomagnetic data as a constraint in the initial solution. Moreover, palaeomagnetism avoids the dependency of the initial solution on the pin-element. A problem still remains regarding the pin-vector-orientation, which slightly influences the deformation pattern of the restored surface. In any case, the restoration results obtained using palaeomagnetic vectors are much closer to the expected results (known from analogue models) than standard parametric approaches. However, the main drawback of this method is the need to define the palaeomagnetic vector at all points. In this paper we use analogue models (vectors are known everywhere); future research should therefore tackle an interpolation algorithm to extrapolate individual sites to the whole surface, in order that it can be used in natural cases.

The research was funded by grants from the projects Pmag3Drest (CGL-2006-2289-BTE, CGL2009-14214) of the Spanish Ministry of Science and 3DR3 (PI165/09) of the Government of Aragon. In addition, MJRO receives funding (PTA2007-0282) from the Spanish Ministry of Science. We thank Paradigm for the Gocad software and API and the affiliates of the Gocad Research 14 Consortium for the restoration plugging used in this study.

## References

ARENAS, C., MILLÁN, H., PARDO, G. & POCOVÍ, A. 2001. Ebro Basin continental sedimentation associated with late compresional pyrenean tectonics (NE Iberia): controls on margin fans and alluvial systems. *Basin Research*, **13**, 65–89.

Arriagada, C., Roperch, P., Mpodozis, C. & Cobbold, P. 2008. Paleogene building of the Bolivian Orocline: tectonic restoration of the central Andes in 2-D map view. *Tectonics*, **27**, TC601, http://doi.org/10.1029/2008TC002269

Audibert, M. 1991. *Déformation Discontinue et Rotations de Blocs, Méthodes Numériques de Restauration*. Memoires et documents du Centre Armoricain d'Etude Structurale des Socles. Rennes University, France, **40**.

Botsch, M., Kobbelt, L., Pauly, M., Alliez, P. & Lévy, B. 2010. *Polygon Mesh Processing (Chapter 5)*. A. K. Peters/CRC Press, USA.

Bourgeois, O., Cobbold, P. R., Rouby, D., Thomas, J. C. & Shein, V. 1997. Least squares restoration of Tertiary thrust sheets in map view, Tajik Depression, Central Asia. *Journal of Geophysical Research, B, Solid Earth and Planets*, **102**, 27 553–27 573.

Caumon, G., Collon-Drouaillet, P., Le Carlier De Veslud, C., Viseur, S. & Sausse, J. 2009. Surface-based 3D modeling of geological structures. *Mathematical Geosciences*, **41**, 927–945.

Dahlstrom, C. D. 1969. Balanced cross sections. *Canadian Journal Earth Sciences*, **6**, 743–757.

Delaunay, B. 1934. Sur la sphère vide. *Bulletin of the Academy of Sciences of the USSR, Classe des Sciences Mathématiques et Naturelle*, **7**, 793–800.

Durand-Riard, P., Caumon, G. & Muron, P. 2010. Balanced restoration of geological volumes with relaxing meshing constraints. *Computer and Geosciences*, **36**, 441–452.

Durand-Riard, P., Shaw, J., Plesch, A. & Lufadeju, G. 2013. Enabling 3D geomechanical restoration of strike- and oblique-slip faults using geological constraints, with applications to the deepwater. *Niger Delta Journal of Structural Geology*, **48**, 33–44.

Elliott, D. 1976. The energy balance and deformation mechanisms of thrust sheets. *Philosophical Transactions of the Royal Society*, **A283**, 289–312.

Floater, M. S. & Hormann, K. 2005. Surface parameterization: a tutorial and survey. *In*: Dodgson, N. A., Floater, M. S. & Sabin, M. A. (eds) *Advances in Multiresolution for Geometric Modelling, Mathematics and Visualization*. Springer, Berlin Heidelberg, 157–186.

González, D., Pinto, L., Peña, M. & Arriagada, C. 2012. 3D deformation in strike-slip systems: analogue modelling and numerical restoration. *Andean Geology*, **39**, 295–316.

Gratier, J. P. & Guiller, B. 1993. Compatibility constraints on folded and faulted strata and calculation of total displacement using computational restoration (UNFOLD program). *Journal of Structural Geology*, **15**, 391–402.

Gratier, J. P., Guillier, B., Delorme, A. & Odonne, F. 1991. Restoration and balance of a folded and faulted surface by best-fitting of finite elements; principle and applications. *Journal of Structural Geology*, **13**, 111–115.

Hossack, J. R. 1979. The use of balanced cross-sections in the calculation of orogenic contraction: a review. *Journal of the Geological Society, London*, **136**, 705–711.

Lallier, F., Durand-Riard, P., Titeux, M. O. & Caumon, G. 2008. *Map restoration: latest advances. 28th Gocad Meeting Proceedings, Nancy, France.*

Lévy, B. & Mallet, J. L. 1998. Non-distorted texture mapping for sheared triangulated meshes. *In*: Cunningham, S. *et al.* (eds) *Proceedings of the 25th Annual Conference on Computer Graphics and Interactive Techniques*. ACM, 343–352.

Maerten, L. & Maerten, F. 2006. Chronologic modelling of faulted and fractured reservoirs using geomechanically based restoration: technique and industry applications. *AAPG Bulletin*, **90**, 1201–1226.

Mallet, J. L. 2002. *Geomodeling: Applied Geostatistics*. Oxford University Press.

Massot, J. 2002. *Implémentation de méthodes de restauration équilibrée 3D*. PhD thesis, Institut National Polytechnique de Lorraine.

McCaig, A. M. & McClelland, E. 1992. Palaeomagnetic techniques applied to thrust belts. *In*: McClay, K. R. (ed.) *Thrust Tectonics*. Chapman-Hall, London, 209–216.

Millán, H., Pocoví, A. & Casas, A. 1995. El frente de cabalgamiento surpirenaico en el extremo occidental de las Sierras Exteriores: sistemas imbricados y pliegues de despegue. *Revista de la Sociedad Geológica de España*, **8**, 73–90.

Millán, H., Pueyo, E. L. *et al.* 2000. Actividad tectónica registrada en los depósitos terciarios del frente meridional del Pirineo central. *Revista de la Sociedad Geológica de España*, **13**, 279–300.

Mochales, T., Casas, A. M., Pueyo, E. L. & Barnolas, A. 2012. Rotational velocity for oblique structures (Boltaña anticline, southern Pyrenees). *Journal of Structural Geology*, **35**, 2–16, http://doi.org/10.1016/j.jsg.2011.11.009

Moretti, I. 2008. Working in complex areas: new restoration workflow based on quality control, 2D and 3D restorations. *Marine and Petroleum Geology*, **25**, 205–218.

Moretti, I., Lepage, F. & Guiton, M. 2005. KINE3D: a new 3D restoration method based on a mixed approach linking geometry and geomechanics. *Oils & Gas Science and Technology (Rev IFP)*, **60**, 1–12.

Muron, P. 2005. *Méthodes numériques 3-D de restauration des structures géologiques faillées*. PhD École Nationale Supérieure de Géologie, Institut National Polytechnique de Lorraine.

Oliva-Urcia, B. & Pueyo, E. L. 2007. Gradient of shortening and vertical-axis rotations in the southwestern Pyrenees (Spain). *Revista de la Sociedad Geológica de España*, **20**, 105–118.

Oliva-Urcia, B., Casas, A. M., Pueyo, E. L. & Pocoví, A. 2012. Structural and paleomagnetic evidence for non-rotational kinematics in the western termination of the External Sierras (southwestern central Pyrenees). *Geologica Acta*, **10**, 1–22.

Ori, G. G. & Friend, P. F. 1984. Sedimentary basins formed and carried piggyback on active thrust sheets. *Geology*, **12**, 475–478.

Pueyo, E. L. 2000. *Rotaciones paleomagnéticas en sistemas de pliegues y cabalgamientos. Tipos, causas, significado y aplicaciones (ejemplos del Pirineo Aragonés)*. PhD thesis, Universidad de Zaragoza, Spain.

Pueyo, E. L. 2010. Evaluating the paleomagnetic reliability in fold and thrust belt studies. *Trabajos de Geología*, **30**, 145–154.

Pueyo, E. L., Millán, H. & Pocoví, A. 1997. Cinemática rotacional del emplazamiento del cabalgamiento basal surpirenaico en las Sierras Exteriores Aragonesas: Datos magnetotectónicos. *Acta Geológica Hispánica*, **32**, 119–138.

Pueyo, E. L., Millán, H. & Pocoví, A. 2002. Rotation velocity of a thrust: a paleomagnetic study in the External Sierras (Southern Pyrenees). *Sedimentary Geology*, **146**, 191–208.

Pueyo, E. L., Parés, J. M., Millán, H. & Pocoví, A. 2003. Conical folds and apparent rotations in paleomagnetism (A case study in the Pyrenees). *Tectonophysics*, **362**, 345–366.

Pueyo, E. L., Pocoví, A., Millán, H. & Sussman,. 2004. Map-view models for correcting and calculating shortening estimates in rotated thrust fronts using paleomagnetic data. *In:* Weil, A. & Sussman, A. (eds) *Special Publication on Orogenic Curvature: Integrating Paleomagnetic and Structural Analyses*. Geological Society of America, **383**, 57–71.

Pueyo, E. L., Oliva, B., Cifelli, F. & Sussman, A. J. In press. Palaeomagnetic analysis of fold and thrust belts: some reliability criteria. *In*: Pueyo, E. L., Cifelli, F., Sussman, A. J. & Oliva-Urcia, B. (eds) *Palaeomagnetism in Fold and Thrust Belts: New Perspectives*. Geological Society, London, Special Publications, **425**.

Puigdefábregas, C. 1975. La sedimentación molásica en la cuenca de Jaca. *Pirineos*, **104**, 1–188.

Ramón, M. J. 2013. *Flexural Unfolding of Complex Geometries in Fold and Thrust Belts Using Paleomagnetic Vectors*. Unpublished PhD, University of Zaragoza, Spain, http://zaguan.unizar.es/record/11750

Ramón, M. J., Pueyo, E. L., Briz, J. L., Pocoví, A. & Ciria, J. C. 2012. Flexural unfolding using paleomagnetic vectors. *Journal of Structural Geology*, **35**, 28–39.

Ramón, M. J., Pueyo, E. L., Rodríguez-Pintó, A., Ros, L. H., Pocoví, A., Briz, J. L. & Ciria, J. C. 2013. A computed tomography approach to understanding 3D deformation patterns in complex flexural folds. *Tectonophysics*, **593**, 57–72.

Ramón, M. J., Briz, J. L., Fernández, O. & Pueyo, E. L. in press. Horizon Restoration by Best Fitting of Finite Elements and Rotation Constraints: Sensitivity to the Mesh Geometry and Pin-Element Location. *Mathematical Geosciences*, http://doi.org/10.1007/s11004-015-9602-1

Rodríguez-Pintó, A., Pueyo, E. L., Serra-Kiel, J., Samsó, J. M., Barnolas, A. & Pocoví, A. 2012. Lutetian chronostratigraphic calibration based on magnetostratigraphy and shallow benthic zones biostratigraphy at the Isuela section (Southern Pyrenees). *Palaeogeography, Palaeoclimatology, Palaeoecology*, **333–334**, 107–120, http://doi.org/10.1016/j.palaeo.2012.03.012

Rodríguez-Pintó, A., Pueyo, E. L., Pocoví, A., Ramón, M. J. & Oliva-Urcia, B. 2013. Structural control on overlapped paleomagnetic vectors: a case study in the Balzes anticline (Southern Pyrenees). *Physics of the Earth and Planetary Interiors*, **215**, 43–57, http://doi.org/10.1016/j.pepi.2012.10.005

Rouby, D., Cobbold, P. R., Szatmari, P., Demercian, S., Coelho, D. & Rici, J. A. 1993. Restoration in plan view of faulted Upper Cretaceous and Oligocene horizons and its bearing on the history of salt tectonics in the Campos Basin (Brazil). *Tectonophysics*, **228**, 435–445.

Rouby, D., Souriot, T., Brun, J. P. & Cobbold, P. R. 1996. Displacements, strains, and rotations within the Afar depression (Djibouti) from restoration in map view. *Tectonics*, **15**, 952–965.

Rouby, D., Suppe, J. & Xiao, H. 2000. 3D restoration of complexly faulted and folded surfaces using multiple unfolding mechanisms. *American Association of Petroleum Geology*, **84**, 805–829.

Sánchez, E., Pueyo, E. L., Bausa, J., Beamud, B., Ramón, M. J., Oliva-Urcia, B. & Pocoví, A. 2012. Geometría no-coaxial del anticlinal de Sto. Domingo (Pirineo Occidental) deducida de la fábrica magnética (ASM). *Geotemas*, **13**, 1188–1191.

Soto, R., Casas Sainz, A. M. & Pueyo, E. L. 2006. Along strike variation of orogenic wedges associated with vertical axis rotations. *Journal of Geophysical Research: Solid Earth*, **111**, 1978–2012.

Sussman, A. J., Pueyo, E. L., Chase, C. G., Mitra, G. & Weil, A. B. 2012. The impact of vertical-axis rotations on shortening estimates. *Lithosphere*, **4**, 383–394.

Teixell, A. 1996. The Ansó transect of the Southern Pyrenees: basement and cover thrust geometries. *Journal of the Geological Society, London*, **153**, 301–310.

Teixell, A. & García Sansegundo, J. 1995. Estructura del sector central de la Cuenca de Jaca (Pirineos meridionales). *Revista de la Sociedad Geológica de España*, **8**, 215–228.

Van der Voo, R. 1990. The reliability of paleomagnetic data. *Tectonophysics*, **184**, 1–9.

# Palaeomagnetism in fold and thrust belts: use with caution

E. L. PUEYO[1]*, A. J. SUSSMAN[2], B. OLIVA-URCIA[3] & F. CIFELLI[4]

[1]*Instituto Geológico y Minero de España, Unidad de Zaragoza
c/Manuel Lasala 44, 9°, 50006 Zaragoza, Spain*

[2]*Earth And Environmental Sciences, MS-D452, Los Alamos National Laboratory,
Los Alamos, New Mexico 87545, USA*

[3]*Dpto. Geología y Geoquímica Facultad de Ciencias,
Universidad Autónoma de Madrid, Madrid, Spain*

[4]*Dipartimento Scienze, Università degli Studi di Roma TRE,
Largo San Leonardo Murialdo 1, 00146 Roma, Italy*

**Corresponding author (e-mail: unaim@igme.es)*

**Abstract:** The application of palaeomagnetism in fold and thrust belts is a unique way to obtain kinematic information regarding the evolution of these systems. However, since many potential problems can affect the reliability of palaeomagnetic datasets and their interpretations, such data should be used with caution. In this paper, we thoroughly review the sources of error from palaeomagnetism with a particular focus on deciphering vertical-axis rotations and the assumptions behind the method. Recent investigations have demonstrated that the age of the magnetization and syn-folding results from the fold test must also be carefully examined: factors such as internal deformation, deficient isolation of components (i.e. overlapping) or incorrect restoration procedures may produce apparent syn-folding results. In fact, the restoration procedure used to return the palaeomagnetic signal to the palaeogeographic coordinate system may itself inhibit accurate estimations of vertical-axis rotations when complex deformation histories induce different, non-coaxial, deformation axes. We recommend the auxiliary use of the inclination v. dip diagram as an efficient tool for identifying many errors. Finally, to determine accurate vertical axis rotations, the reference direction should honour standard reliability criteria and would ideally be measured within the undeformed foreland of the thrust system. In this paper, we review five decades of palaeomagnetic research in fold and thrust belts by concentrating on maximizing standard reliability criteria procedures to reduce uncertainty and increase confidence when applying palaeomagnetic data to unravel the tectonic evolution of fold and thrust belts.

Palaeomagnetism is the study of Earth's ancient magnetic fields as recorded in the stratigraphic archive. Given that the magnetic field is a global reference system, a palaeomagnetic signal can be used to reconstruct relative motions of geological elements on a range of spatial scales. On a global scale, palaeomagnetism can track the position of tectonic plates as they move in time (e.g. Opdyke 1995; Irving 2005). At the scale of orogenic systems, palaeomagnetism can be used to quantify processes such as oroclinal bending, plate indentation or escape tectonics (Eldredge *et al.* 1985; Van der Voo 2004; Weil & Sussman 2004 and many others). At smaller scales, such as thrust sheets, palaeomagnetism can be applied to detect rotations associated with structural deformation (i.e. differential shortening). Rotations about a horizontal axis (i.e. tilt) are responsible for folding in the cross-section plane and are usually straightforward to characterize using stratigraphic horizons (bedding surfaces). Moreover, vertical axis rotations (VARs) are more difficult to detect and quantify. For instance, to fully define VARs in a thrust sheet, it is necessary to compare palaeomagnetic vectors in both the hanging wall and the footwall of a thrust. VARs comprise an especially valuable dataset since other kinematic indicators typically do not allow for the determination of rotations about a vertical axis.

The conceptual application of palaeomagnetic techniques to detect relative motions in fold and thrust belts (FTBs) was first described by Norris & Black (1961), who proposed the use of palaeomagnetism to decipher the origin of along-strike changes associated with the Lewis Thrust sheet in western Montana. Other early authors (see reviews by Hospers & Van Andel 1969; Tarling 1969) applied

*From*: PUEYO, E. L., CIFELLI, F., SUSSMAN, A. J. & OLIVA-URCIA, B. (eds) 2016. *Palaeomagnetism in Fold and Thrust Belts: New Perspectives*. Geological Society, London, Special Publications, **425**, 259–276.
First published online July 07, 2016, http://doi.org/10.1144/SP425.14

this idea and the technique has since grown in its utilization (see Van der Voo & Channell 1980; Kissel & Laj 1989; Sussman & Weil 2004).

The causes of VARs at the thrust fault scale are commonly associated with differential displacement along strike (McCaig & McClelland 1992; Allerton 1998; Pueyo *et al.* 2004; Soto *et al.* 2006; Sussman *et al.* 2012). Locally, non-cylindrical and non-coaxial folding (Sellés-Martínez 1988; Allerton 1994; Pueyo *et al.* 2003*a*), superposed folding (Hirt *et al.* 1992; Weil *et al.* 2000; Mochales *et al.* 2016) and plunging folds (Stewart 1995; Pueyo *et al.* 2002*a*) may also play a key role. FTBs may also be affected by significant secondary VARs during later orogenic processes including (Fig. 1) piggy-back movements during thrust stacking (Oliva-Urcia & Pueyo 2007*a*, *b*), indentation (Achache *et al.* 1983; Thomas *et al.* 1994; Collombet *et al.* 2002), buttressing (Grubbs & Van der Voo 1976; Eldredge & Van der Voo 1988), oroclinal

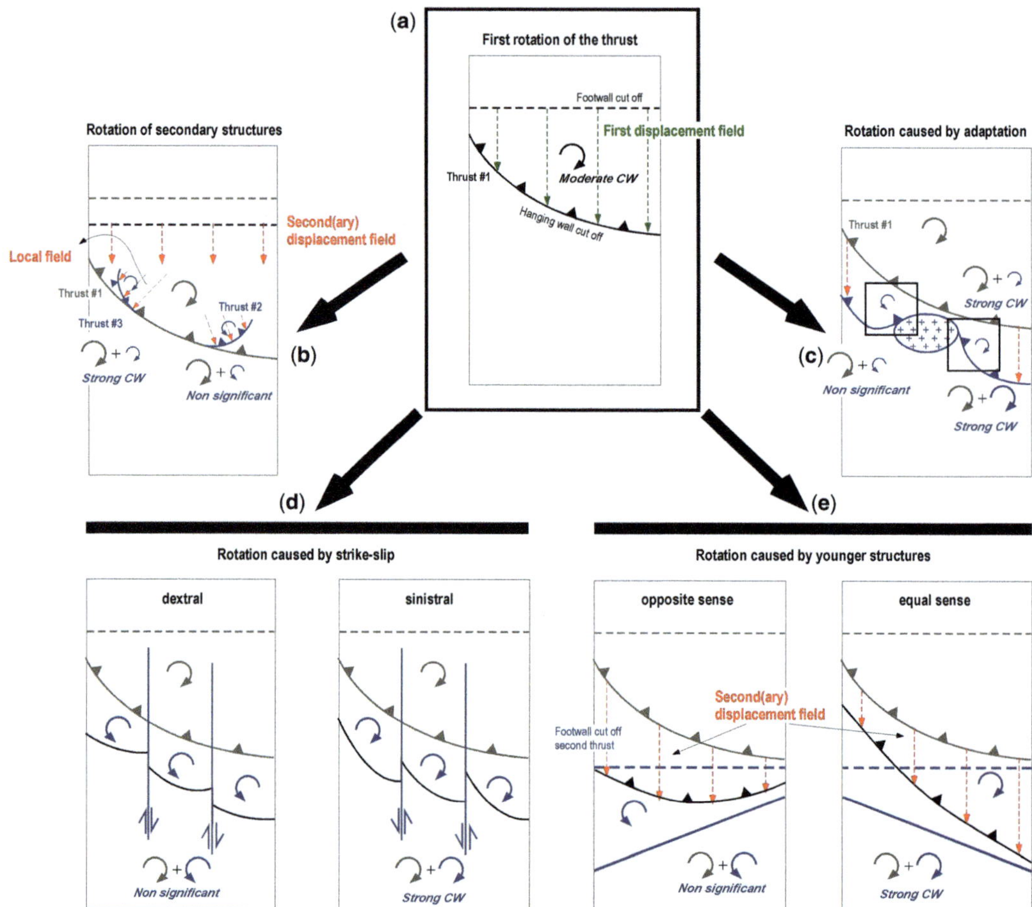


**Fig. 1.** Superimposed VARs in FTBs. (**a**) A given segment of a FTB contains a lateral shortening gradient with an associated VAR (as attested by the displacement field in green). (**b**) However, later deformation processes, shown in blue, may affect the primary VAR record, shown in grey (second displacement field is shown in red). For example, younger or coeval smaller structures could accommodate additional rotations in both senses (local displacement field induces the local rotation). (**c**) The progress of deformation may also incorporate previous anisotropies such that the FTB will undergo additional rotations (here the second displacement field could be homogeneous). (**d**) A similar process can occur with strike-slip motion: here, the sense of fault movement conditions the VAR. (**e**) FTB evolution can be characterized by sequences of different thrusts, with hanging-wall (piggyback) sequences being most typical. In these cases, every thrust may have accommodated different VARs. All of these processes are scale-independent and may take place at a number and range of scales. The complete VAR affecting a segment of an FTB will be the cumulative addition of all these processes and must be unravelled taking into account the structural and tectonic history.

bending (Eldredge *et al.* 1985; Weil & Sussman 2004) and strike-slip rearrangements (Ron *et al.* 1984; Nur *et al.* 1986).

Palaeomagnetic and structural investigations should be coordinated as a best practice method in deciphering and delineating the many physical processes associated with deformation. Unravelling the complex spatial–temporal deformation processes responsible for along-strike changes associated with thrust movement is an important research topic in structural geology (Hindle & Burkhard 1999; Strayer & Suppe 2002; Wilkerson *et al.* 2002; Hnat *et al.* 2008; Adam *et al.* 2013; Muñoz *et al.* 2013). Palaeomagnetic techniques play a key role in providing geometric and kinematic information in orogenic belts to comprehend deformation patterns in four dimensions.

VARs have been reported in most orogenic belts investigated for such data, for example Himalaya and Tibet (Bazhenov *et al.* 1999; Dupont-Nivet *et al.* 2002; Schill *et al.* 2002; Crouzet *et al.* 2003), Zagros (Smith *et al.* 2005; Aubourg *et al.* 2008), Taurides and Aegean (Kissel *et al.* 1993; Duermeijer *et al.* 2000; van Hinsbergen *et al.* 2007; **Çinku *et al.* 2015**), Carpathian (Márton *et al.* 2007, **2015**; Dupont-Nivet *et al.* 2005), Alps (Collombet *et al.* 2002; Sonnette *et al.* 2014; Cardello *et al.* 2015), Pyrenees (Sussman *et al.* 2004; Oliva-Urcia & Pueyo 2007*a*; Oliva-Urcia *et al.* 2010, 2012*a, b*; Muñoz *et al.* 2013; Izquierdo-Llavall *et al.* 2015) Cantabrian (Weil *et al.* 2001; Weil 2006), Betics, Rif and Atlas (Platt *et al.* 2003; Mattei *et al.* 2006; Moussaid *et al.* 2015), Calabrian Arc (Channell *et al.* 1990; Speranza *et al.* 1999; Cifelli *et al.* 2007, 2008*a, b*), Apennines (Speranza *et al.* 1997; Muttoni *et al.* 1998; Satolli *et al.* 2005; Caricchi *et al.* 2014), North American Cordillera (Beck 1980; Eldredge & Van der Voo 1988; Conder *et al.* 2003; Sears & Hendrix 2004; Harlan *et al.* 2008), Appalachians (Bayona *et al.* 2003; Ong *et al.* 2007; Hnat *et al.* 2008), Andes (Roperch & Carlier 1992; Mac-Fadden *et al.* 1995; Prezzi *et al.* 2004; Richards *et al.* 2004; Rousse *et al.* 2005; Arriagada *et al.* 2006; Rapalini 2007; **Japas *et al.* 2015**; Rapalini *et al.* 2015) and New Zealand (Nicol *et al.* 2007).

Palaeomagnetic analyses of FTBs usually aim to obtain geometric and/or kinematic information. To date, most palaeomagnetic studies of FTBs have focused on determining rotation magnitudes at distributed locations within thrust sheets, observing along-strike variations of VARs (Otofuji *et al.* 1985; Butler *et al.* 1995; Hnat *et al.* 2008) or differential block rotations (Nur *et al.* 1986; Mattei *et al.* 1995; Thöny *et al.* 2006; Pueyo *et al.* 2007). Detailed applications at smaller scales such as sigmoidal or curved folds are now more frequent (Smith *et al.* 2005; Rouvier *et al.* 2012; Rodríguez-Pintó *et al.* 2016), but applications for superposed folding

(Bonhommet *et al.* 1981; Weil 2006; Mochales *et al.* 2016), non-cylindrical folding and non-coaxial superposed folding geometries (Zotkevich 1972; Sellés-Martínez 1988; Stewart 1995; Pueyo *et al.* 2003*a, b*), as well as fold closures (Stewart & Jackson 1995), have not been widely published. On the other hand, kinematic information can be deduced when syn-orogenic sedimentary sequences are dated using magnetostratigraphic records. With such datasets, thrust timing, displacement velocities and other parameters can be determined (Sempere *et al.* 1990; Burbank *et al.* 1992; Powers *et al.* 1998; **Oliva-Urcia *et al.* 2015**). In combination with fission tracks, accurate exhumation velocities (Beamud *et al.* 2011 among many references) can be constrained. However, understanding the kinematics that accompany rotational processes, such as rotational velocities, is still enigmatic. Despite the importance of determining rotational velocity for understanding the four-dimensional nature of deformation, there are very few such datasets for either orogens (Duermeijer *et al.* 2000; Mattei *et al.* 2004) or individual thrusts (Pueyo *et al.* 2002*b*; Mochales *et al.* 2012; Rodríguez-Pintó *et al.* 2016). Rotational velocity can be obtained only if syn-tectonic sedimentation and rotation were simultaneous; however, this results in dispersion of the palaeomagnetic declination, requiring coordinated structural analyses to understand the deformation process and pattern. Finally, rotations are governed by a pivot point, that is, the physical rotation axis about which a part of the hanging wall is displaced. Determining the location and evolution of the pivot point is key to understanding the kinematics of FTBs (Bates 1989; McCaig & McClelland 1992; Allerton 1994, 1998; Pueyo *et al.* 2004; Sussman *et al.* 2012); we recommend more research on this topic.

Palaeomagnetic analyses of FTBs require integration with structural and tectonic data to achieve reliable, quantitatively constrained interpretations. For instance, some researchers have already shown the importance of VARs for palinspastic reconstruction of FTBs (Bourgeois *et al.* 1997; Arriagada *et al.* 2006, 2008; Muñoz *et al.* 2013; Ramón 2013). Palaeomagnetic rotations have also been used to correct errors caused in shortening estimates in balanced sections (Pueyo *et al.* 2004; Oliva-Urcia & Pueyo 2007*b*; Sussman *et al.* 2012), to validate rotations deduced by anisotropy of magnetic susceptibility (Pueyo-Anchuela *et al.* 2012), or as an additional constraint in 3D restoration (Ramón *et al.* 2012, 2015*a, b*). Some recent studies have tackled the qualitative reconstruction of FTBs derived from inverted extensional basins (partial restoration in 2D) using remagnetization components that represent snapshots of the deformation history (Villalaín *et al.* 2003, **2015**; Soto *et al.* 2008). These studies show the great potential for numerically combining palaeomagnetism and structural datasets at the FTB scale.

Despite the wide range of applications for understanding the evolution of FTBs, palaeomagnetic datasets should be used with caution. This paper reviews issues associated with the field and laboratory procedures associated with collecting, measuring, processing and interpreting palaeomagnetic data, as well as the application of palaeomagnetic techniques to FTBs. We also suggest best practice approaches to help minimize the problems and to increase the reliability of the palaeomagnetic data. While several of the issues reported here are common and/or well known and have been presented in the classic paper by Van der Voo (1990), we believe an updated review will help reinvigorate attention to these topics.

## Determination of vertical-axis rotations: common problems and solutions

The successful application of palaeomagnetism is based on three assumptions (e.g. Butler 1992): (A) over long periods of time, the Earth's magnetic field behaves like a geocentric axial dipole (GAD hypothesis); (B) the ferromagnetic minerals (s.l.) contained in the rocks efficiently record the Earth's magnetic field during rock formation and can be isolated and measured in the laboratory; and (C) the palaeomagnetic signal recorded in the rocks remain stable over geological time. Here we review common scenarios that may void the three main assumptions (above) as well as other secondary assumptions (grouped as D) that, if not met, make the palaeomagnetic results invalidate the method. Scenarios may include issues related to global reference, structural position, restoration methods, statistics and deformation events, etc. Whenever possible, we offer solutions to mitigate such complications.

### A.1. The signal does not represent the geocentric axial dipole field

For sedimentary sequences, sampling within a stratigraphic thickness equivalent to *c.* 10 ky timeframe might be sufficient to represent the GAD, as models for secular variations of the magnetic field during the last eight millennia predict fulfillment of this assumption (Pavón-Carrasco *et al.* 2010). A population of 10–15 palaeomagnetic samples should yield the distinctive fisherian (Fisher 1953) dispersion around the mean value and confidence parameters within expected reliability boundaries (Van der Voo 1990). An additional problem may occur if the magnetic field displays a significant non-dipolar component, as has been proposed for

certain periods of Earth history (Van der Voo & Torsvik 2001). If this hypothesis is true (a matter of debate: Tauxe & Kent 2004; Meert 2009), then the rotations deduced from palaeomagnetic data in FTB in those periods may be questioned.

### B.1. There is not a primary palaeomagnetic signal or the signal is weak

A primary component is defined as the record of the Earth magnetic field at the time of rock formation. In some scenarios, the Earth's magnetic field was not efficiently recorded or the palaeomagnetic signal is of poor quality. This can occur when there is an insufficient amount of ferromagnetic minerals and/or those minerals are not stable. In this instance, the blocking mechanism is ineffective (e.g. some detrital rocks) and/or subsequent processes (e.g. dolomitization, shock magnetization, burning, exhumation) erased or disturbed the original palaeomagnetic record and no new stable or nonsense magnetization is produced. Unfortunately, negative results and/or the attempts to alleviate them are seldom published and thus similar problems have not been fully understood for mitigation in future studies. Perturbation of a primary magnetization by remagnetization is discussed in C1.

### B.2. The palaeomagnetic signal cannot be fully isolated

Isolation of two or more components of remanence can be complicated when magnetizations of different ages are simultaneously demagnetized in the laboratory, since available techniques (alternating fields or thermal) cannot always successfully distinguish among them (overlapping of unblocking spectra). For instance, palaeomagnetic data from tilted strata affected by a secondary overprint that overlapped the original signal may result in unreliable palaeomagnetic directions once bedding is restored (Fig. 2). This can manifest as large errors in declination and inclination, and inexact palaeomagnetic age determinations as estimated from fold test results (Rodríguez-Pintó *et al.* 2011, 2013). The simultaneous demagnetization of two overlapping components yields a demagnetization circle, represented as a great circle on an equal area projection. The intersection of girdles signifies the less scattered component (Khramov 1958; Halls 1976). This technique can be very useful in determining the primary component in a scenario with overlapping components, (Halls 1978; Bailey & Halls 1984), but only when the primary component is the less scattered one. The demagnetization circle method searches for the grouping of great circle intersections; since the number of intersections of $n$ planes is an exponential function ($[n^2 - n]/2$),

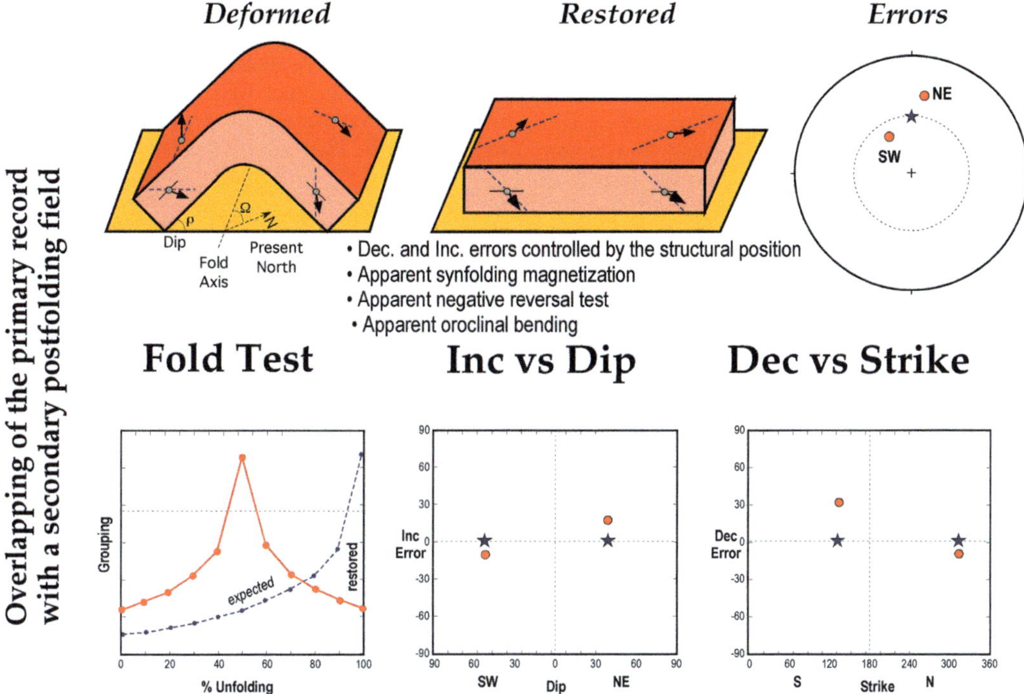


**Fig. 2.** Errors caused by overlapping of palaeomagnetic components. A primary (pre-folding) record could not be distinguished from a post-folding overprint. Ideally, the discontinuous line (and blue star in the stereonet) would have been obtained, but the overlapped vector is strongly conditioned by the structural location of the sampling sites along the fold geometry. The overlapped vectors may display both declination and inclination errors (compared with the expected result) and will produce anomalous results in the fold test (apparent syn-folding), in a plot of inclination v. dip of the bed, and in the oroclinal test (declination v. strike). Although not shown, the polarity of the primary and secondary components also plays an important role.

the application of the Fisher statistics to this population cannot be compared to standard populations of vectors derived from direct estimation. The problem of achieving a statically comparable result using the combination of remagnetization circles and direct observations was tackled by McFadden & McElhinny (1988).

### C.1. The primary palaeomagnetic signal has not been stable during the geological time due to remagnetization

Sources of recently acquired secondary magnetizations include lightning, insolation, burning, blasting, viscous overprinting of the present-day field and/or sampling-induced magnetizations. However, ancient secondary magnetizations (remagnetizations; Elmore *et al.* 2012) reveal physicochemical changes in the rock volume related to significant geological processes. Remagnetizations were described in earlier palaeomagnetic studies (Creer

1962) and a historical review was compiled by Van der Voo & Torsvik (2012). In ideal cases, the information that secondary magnetizations provide can be very useful as snapshots of deformation processes and can help temporally constrain tectonic events (Cullen *et al.* 2012; Çinku *et al.* 2013; Kirscher *et al.* 2013; Izquierdo-Llavall *et al.* 2015). In addition, secondary magnetizations can allow for the dating of hydrothermal events or orogenic fluid migration (Evans *et al.* 2000; Elmore *et al.* 2001; Ribeiro *et al.* 2013) and can assist in constraining the time of burial diagenesis (Blumstein *et al.* 2004; Aubourg *et al.* 2012). However, a great disadvantage of using remagnetizations is that the palaeohorizontal reference frame is absent, thus limiting the potentiality of palaeomagnetism as a 3D reference indicator and reducing accuracy for determining VARs. Recently developed techniques for determining the ages of illitization (Nemkin *et al.* 2015) may allow for the accurate timing of the remagnetization and, thus, can help overcome associated problems.

 E. L. PUEYO *ET AL.*

## C.2. The palaeomagnetic signal has not been stable over geological time owing to reorientation of palaeomagnetic vectors

Even for scenarios in which the palaeomagnetic record is of good quality and has survived over geological time, other processes may disturb the palaeomagnetic vectors. If the third assumption of record stability over time is met, complete stability of the palaeomagnetic signal can still only be achieved if a rock volume behaves like a rigid solid. However, in some cases, palaeomagnetic vectors are reoriented. During lithostatic loading sediments may lose volume during burial and subsequently palaeomagentic vectors will undergo inclination shallowing (van Andel & Hospers 1966); published solutions can help account for inclination shallowing (Kodama 1997; Tauxe 2005; Bilardello & Kodama 2010). On the other hand, reorientation is significantly more complicated when a rock volume is subject to penetrative, internal tectonic deformation. Simple and pure shear strain are common during tectonic deformation (i.e. during flexural folding) and their possible influences on palaeomagnetic vectors has been described by many researchers (Perroud 1982; Facer 1983; Lowrie *et al.* 1986; Cogné 1987; van der Pluijm 1987; Kodama 1988; Stamatakos & Kodama 1991; Borradaile 1997; Oliva-Urcia *et al.* 2010, among others). Error correction caused by internal deformation is not straightforward owing to the difficulty in knowing the deformation tensor (magnitude and orientation of the strain axes). Thus, identification of this problem is critical (Fig. 3) and any dataset from an FTB with penetrative internal tectonic deformation has to be considered cautiously.

## D.1. The palaeomagnetic reference direction is unknown

Estimates of VARs require comparison with a coeval palaeomagnetic reference direction for the same tectonic element. As such, estimates may be inaccurate if the age of a remagnetization cannot

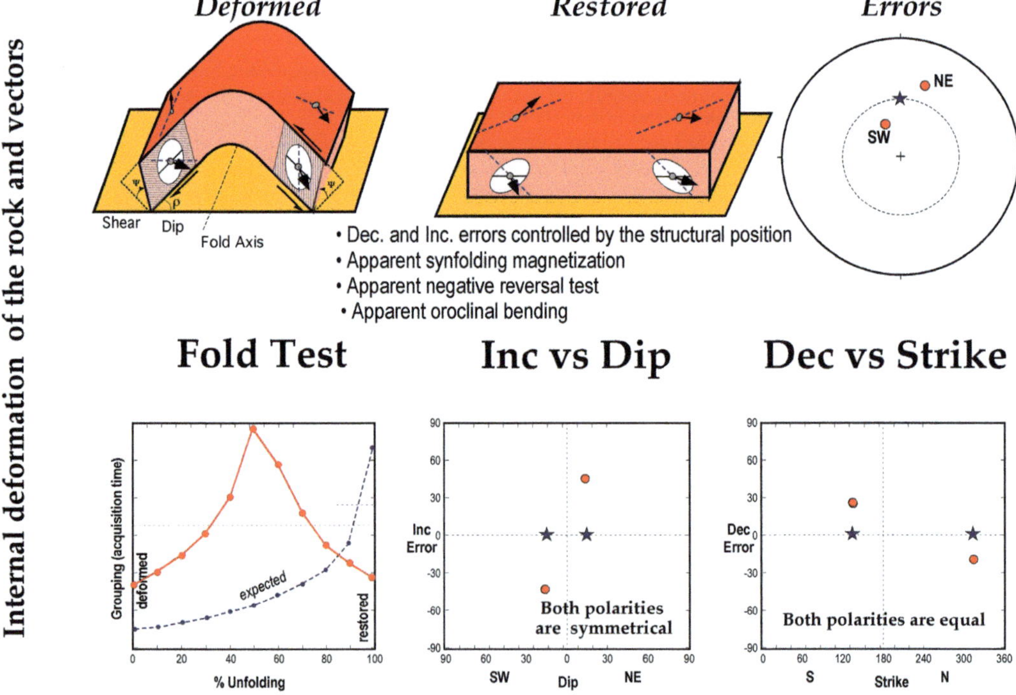



**Fig. 3.** Errors caused by internal deformation of rocks. In this case, flexural folding in the fold limbs produced simple shear and deformed the original palaeomagnetic vectors. Similar to the problem of overlapped directions, the final deformed vectors will depend upon the original vector, the fold axis orientation, and the magnitude of shear. Internal deformation will modify the palaeomagnetic declination and inclination as well as results from fold and reversal tests and will produce erroneous interpretations of palaeomagnetic datasets if it is not taken into account. This spurious effect may be critical in the detection and mitigation of these kinds of errors. In this case, the magnetic polarity did not exert any influence.

be associated with a specific (and well-dated) deformation event and must be deduced from apparent polar wander paths.

The absence of well-constrained palaeomagnetic references will decrease the reliability of the palaeomagnetic data to determine VARs. This occurs when either primary or secondary palaeomagnetic datasets are compromised, such as when references for the same tectonic element and equal age are not reliable or do not exist. In these scenarios, comparison between the hanging wall and footwall of the thrust system will only yield relative rotations between them.

## D.2. Lack of a palaeohorizontal reference (such as bedding planes)

Bedding planes usually represent a palaeohorizontal reference in sedimentary rocks and some volcanic rocks (i.e. ignimbrites, basaltic flows). In terms of primary magnetizations, the bedding plane is the only surface that allows for confident restoration of the palaeomagnetic data to the 'palaeogeographic reference system' or reference frame in which the VARs are estimated. In fact, the combination of a stratigraphic horizon and palaeomagnetic vectors is the only 3D marker that can be related both before and after the deformation. Therefore, in the absence of palaeohorizontal references, or with uncertain palaeohorizontal markers in some stratigraphic surfaces (i.e. in delta fans, cross-bedding, channels, some igneous bodies or in remagnetized rocks) the reliability of the palaeomagnetic data to provide an accurate VARs or palaeolatitudes is reduced.

## D.3. Cylindrical bedding correction to restore the palaeomagnetic data

Mountain building processes may rotate, translate and/or distort rock volumes several times and in several different ways. Non-coaxial deformation may produce conical and plunging folds, superposed folding, forced folds, fold closures and/or oblique thrust ramps along with any kind of superimposed deformation. The cylindrical bedding correction, for example, involves tilting the palaeomagnetic vector and the bedding plane by the bedding strike for an angle equal to the dip. This correction will return the vectors to palaeohorizontal but does not guarantee proper restoration to the palaeogeographic reference system (Fig. 4). In scenarios for which several deformation events have affected a given location, restoration should follow the reverse chronological order of the deformation events (restoration is not commutative; Ramsay 1960). Therefore, application of the bedding correction in complex parts of FTBs will generate an error in the declination component ('apparent rotation' by MacDonald 1980; Chan 1988; 'spurious rotation' by Pueyo et al. 2003b) as well producing errors in results of the fold test or in the strike v. declination (oroclinal) diagram. Quantifying these errors in different scenarios was an active structural geology research topic in the 1960s (Norman 1960; Ramsay 1960, 1961; Stauffer 1964; Cummins 1964, 1966), but has attracted little attention since (Zotkevich 1972; Scott 1984; Bazhenov 1988; Sellés-Martínez 1988; Setiabudidaya et al. 1994; Stewart 1995; Weinberger et al. 1995; Pueyo et al. 2003a, b). A full understanding of the deformation sequence (derived from a thorough structural analysis) is critical in order to perform the correct restoration sequence with the palaeomagnetic data and to determine accurate estimates of VARs.

## D.4. Resolution, accuracy and statistical significance of VARs in FTBs

In order to characterize VARs in parts of FTBs sharing similar structural trends, several sites must be investigated to obtain a reliable palaeomagnetic signal. In addition, detailed studies regarding the statistical significance of the palaeomagnetic data from a structural point of view (i.e. the number of sites needed to characterize a trend-domain, dip-domain in terms by Suppe 1985) still need to be standardized. Pastor-Galán et al. (2016) have recently simulated the relationship between the site standard deviation (confidence angle) and the number of sites needed to define the curvature of an orocline; 12–13 sites may be enough to characterize $45°$ of structural bend when the $\alpha_{95}$ of those sites is below $10°$, but the number of sites rises to 50 if $\alpha_{95} < 20°$. These results indicate that the characterization of segments of a FTB with variable trends would require at least five sites for every $15°$ of strike of the thrust.

## D.4. Multiple rotations

A common problem with interpreting palaeomagnetic data from FTBs is the superposition of different rotational processes (Fig. 1). For a scenario with a given thrust that initially underwent variable shortening along strike (i.e. rotations), secondary and/or passive younger rotational movements would superpose and must be first determined and then subtracted because VARs are additive and commutative and require complete structural and geometric understanding of the deformation processes and their timing. Unresolved multiple rotations may affect the oroclinal bedding diagram and will add noise to the estimation of the oroclinal slope (Yonkee & Weil 2010).

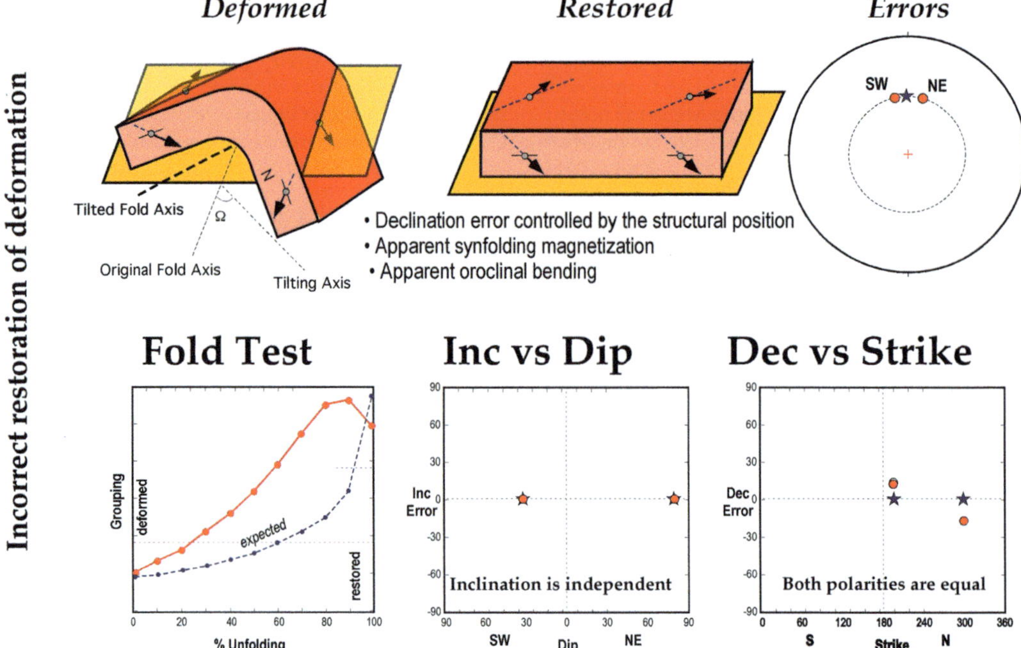



**Fig. 4.** Errors caused by application of the bedding correction in areas of superposed folding. The fold from previous examples maintains the primary palaeomagnetic signal but has undergone a secondary tilting. The application of the bedding correction will produce anomalies in the declination component and in the fold test, but not in the reversal test or in the inclination v. dip diagram. Here the magnetic polarity has no influence.

## Palaeomagnetic data and FTBs: key steps

In order to mitigate the difficulties and challenges presented in the previous section, we discuss techniques that cover all elements (i.e. data collection, processing and interpretation), of palaeomagnetic analysis. Proposed best practice steps and procedures, of which most are well accepted by the palaeomagnetic community, are specified below:

(1) *Sampling sites.* A sampling site is a location from which individual palaeomagnetic samples are collected. A site should be geo-referenced with precision and samples (10–15 for each site) should be collected from homogeneous rock types such that a consistent palaeomagnetic signal can be obtained. In addition, several meters of stratigraphic section (5–10 m or more, depending upon inferred sedimentation rates) should be sampled to average secular variation, thus fulfilling the GAD assumption. Some lithologies (i.e. condensed pelagic carbonates) may meet this requirement in a few centimetres. Ideally, the stratigraphic position and age of the sampling site should be determined and reported in detail, including the stratigraphic age and the magnetic polarity stratigraphy. In addition, the site should be structurally uniform with constant and well-characterized bedding planes (ideally all samples will contain bedding planes allowing for fisherian calculation of the mean plane; Fisher 1953). Both the regional and local structural settings must be properly characterized. On the other hand, drilling in multiple directions at the outcrop scale during the sampling can be key for detecting spurious components in the laboratory.

(2) *Network of rotations.* The distribution of the sites within a structural domain designed to estimate the rotation magnitudes needs to be carefully planned. Ideally, sites within individual folds, thrusts and other faults should be selected to meet the following criteria. (a) Sites should contain rocks with good-quality palaeomagnetic signals based on pilot campaigns. (b) Sufficient sites are chosen to allow for a robust rotation estimate (mean and error) of the structural trend. (c) Sites should be evenly distributed to account for along-strike changes in order to add certainty to the statistical significance of the results. (d) Sufficient sites are

chosen along the full arc of a fold or thrust width to permit achieving statistically significant fold tests. Individual folds (at any scale) with different dips within the fold surface must be sampled in every thrust unit to ensure a reliable result of the fold test. Thus, regional scale fold tests are undesirable because they may be biased by numerous sources of errors. (e) In syn-orogenic sediments, sites should be evenly distributed within the stratigraphic sequence of the hanging-wall, potentially allowing for characterization of the rotational velocity of the associated thrust.

(3) *Laboratory procedures.* In order to meet standards for a successful laboratory campaign, a number of published criteria should be met. (a) A sufficient number of demagnetized samples per site will assure a reliable site mean characterization; between 10 and 15 specimens usually yields confident results (Kirschvink 1980; Van der Voo 1990). (2) Detailed pilot alternating field (AF) or thermal (TH) demagnetizations, or combinations of those approaches, help to define the unblocking coercivity and temperature spectrums. (3) Characteristic directions and demagnetization planes must be defined with at least four or five demagnetization steps (Kirschvink 1980). (4) Rock magnetic carriers should be unambiguously identified. While there are many possibilities (low/ high temperature magnetization/susceptibility runs, hysteresis loops, FORC diagrams, etc.), thermal demagnetization of three-component isothermal remanent magnetization (IRM) (Lowrie 1990), although a qualitative approach, yields a useful definition of demagnetization strategy.

(4) *Sample level characterization.* The fitting of individual (specimen) demagnetization results should follow some published/accepted criteria, as follows. (a) Vector directions must be fitted by principal component analyses (PCA) after visual inspection of demagnetization diagrams and the maximum angular deviation (MAD) should be below 15° (Kirschvink 1980). (2) Demagnetization circles (Halls 1978), defined by two remanences, can be used to double check the PCA fitting of the directions and may help to define overlapping components of remanence. Other ancillary methods (or combinations of them), such as linearity spectrum analysis (Schmidt 1982), stacking routine (Scheepers & Zijderveld 1992) and virtual palaeomagnetic directions (Pueyo 2000; Ramón 2013), may be very useful to understand the palaeomagnetic behaviour from a global perspective and to obtain a first-order palaeomagnetic signal. Individual directions or planes

should be obtained using standard PCA analysis (see a partial compilation in https:// magwiki.wikispaces.com/Paleomagnetic+and+ Rock+Magnetic+Software). In addition, rock density should be always calculated from standard samples to help the processing of some rock magnetism experiments. Further, palaeomagnetists should consider (and publish) the petrophysical properties like density ($\rho$), magnetic susceptibility ($k$) or the natural remanent magnetization (NRM) of their samples because of their potential additional value in exploration geophysics.

(5) *Site level characterization. Next statistical level includes the characterization of the mean palaeomagnetic vector for a site.* (a) Fisher (1953) statistics, mean and confidence parameters ($\alpha_{95}$, $k$ and $R$) have been widely used by palaemagnetists for decades. As proposed by Van der Voo (1990), the $\alpha_{95}$ value, which may have structural implications in shortening estimates, should be lower than 10° (never higher than 15°). The precision parameter ($k$) should be larger than 20. (b) To double check the correct application of the Fisher (1953) distribution, the orientation tensor (Bingham 1974; Scheidegger 1965) could be calculated at the site scale. Ratios between eigenvectors (Tauxe 1998) could ascertain the suitability of the Fisher (1953) distribution and thus could detect possible sources of error (i.e. overlapping, internal deformation, improper restoration). (c) Mean bedding planes or other structural markers should be fitted using Fisher (1953) or Bingham (1974) statistics (depending upon the nature of the indicator) and should be reported in publications.

(6) *Restoration of site means.* The palaeomagnetic vectors must be corrected to the palaeogeographical reference system for the time when their magnetization was blocked in. This is accomplished as follows. (a) The geometry of folds and thrusts should be accurately defined. Bedding measurements within the area (exceeding the sampling sites) should be collected during fieldwork to characterize fold axes and thrust planes, as their geometry (i.e. conical br cylindric folds, and oblique, lateral or frontal thrusts) may be key to performing an appropriate restoration. (b) Restoration of the palaeomagnetic vectors must strictly follow the reverse order of the deformational sequence because deformation processes are non-commutative and may impart errors in the final estimation of VARs. (c) While a kinematic model to understand the structures is needed, it is also likely that palaeomagnetic analyses will help to improve such a model.

(7) *Combined palaeomagnetic and structural analyses.* A key factor during palaeomagnetic data processing is derived from the application of the fold test (Graham 1949), which assesses the relative age between folding and magnetization acquisition. (The fold test is a powerful and effective technique that could be extended to analysing other structural indicators.) Collecting a wide distribution of palaeomagnetic data obtained along fold surfaces is critical for minimizing many potential sources of errors in the application of the fold test. (a) All sampling sites should unequivocally belong to the same fold. (b) Results of many fold tests show that statistical approaches (McElhinny 1964; McFadden & Jones 1981; McFadden 1990, 1998; Bazhenov & Shipunov 1991; Tauxe & Watson 1994; Shipunov 1997; Weil & Van der Voo 2002; Enkin 2003) will bring similar results in case of postfolding and prefolding magnetizations. In these cases, progressive unfolding techniques or the small circle intersection method (SCI: Waldhör 1999; Waldhör & Appel 2006) will also yield similar results. (c) However, 'syn-folding' magnetizations derived from any method must be carefully evaluated because significant sources of errors will produce an apparent syn-folding result in the fold test (i.e. partially overlapped components, internally deformed vectors or incorrectly restored directions). In these circumstances, the use of auxiliary methods such as evaluation of trends in the declination v. strike diagram and/or the inclination v. dip plot (Figs 2, 3 & 4) will help distinguish between real syn-folding remagnetizations and apparent ones. (4) Real syn-folding magnetizations should be better defined by the SCI method because the progressive and proportional untilting technique assumes a kinematic model that may not represent how the structure developed (Cairanne *et al.* 2002; Delaunay *et al.* 2002). The results derived from the SCI approach may help to accurately reconstruct the geometry of the fold when remagnetization took place (Villalaín *et al.* 2003, 2015). (5) Palaeomagnetic declinations without inclination errors are usually represented as cones on geological maps (Mochales & Blenkinsop 2014) and are an informative way to visualize VARs and their relationship to the structure. However, the $\alpha_{95}$ must be converted to the A95 angle (Demarest 1983) to avoid a declination underestimation caused by latitude.

(8) *Palaeomagnetic reference and rotation confidence.* Accurate quantification of VAR magnitudes requires a high-quality reference field direction. To determine absolute rotation values, angles from a deformed area should be contrasted against a palaeomagnetic reference direction obtained in a close and undeformed part of the tectonic element. Reference data from the undisturbed foreland basin of the FTB will most often yield the best result. Relative VAR values in the hanging wall can be estimated from the data obtained from rocks in the footwall of the thrust.

## Conclusion: reliability of palaeomagnetic data in fold and thrust belts

Considering the complexities of obtaining reliable VARs from palaeomagnetic data in FTBs, it is important to follow the philosophy of the reliability criteria established by Van der Voo (1990) and Opdyke and Channell (1996) to evaluate the quality of palaeopoles and the quality of magnetostratigraphic sections, respectively. We present and update these quality criteria as a best practice approach for using palaeomagnetic data to characterize VARs in FTBs:

(1) The age(s) of the rock(s), timing of deformation (i.e. folding, thrusting and rotation) and timing of acquisition of magnetization must be known.

(2) A minimum of five sites (10 is desirable) per thrust unit (10–15 samples per site), with a site mean characterized by $\alpha_{95} \leq 10°$ (never $> 15°$) and $k > 20$ (never $< 10$). In cases with variable structural trends (along-strike changes), at least five sites should be targeted fir every $15°$ of trend domain.

(3) Detailed demagnetization procedures must isolate all magnetization components and allow for a trusted calculation of remanence directions with demagnetization circles fitted by PCA (Kirschvink 1980). More than four or five steps must be involved to fit vectors and planes (respectively) and a MAD $< 15°$ is desirable. The use of auxiliary methods (demagnetization circles, stacking routine, linearity spectrum, virtual directions, etc.) is recommended to best interpret the palaeomagnetic signal. Combined use of difference and resultant vectors should be carried out to facilitate the detection of instrument problems.

(4) Stability (field) tests and error-control techniques (i.e. conglomerate-, reversal- and fold tests and the small-circle intersection method) should be performed to support the timing of magnetization acquisition. Additionally, strike v. declination and dip v. inclination diagrams should be plotted to detect and to try to avoid errors, especially for syn-folding magnetizations.

(5) The geometry, kinematics and timing of all structures studied should be known to avoid restoration errors and apparent results in the fold and reversal tests and in the declination (rotation) deflection. This information is critical to restoring multiple rotations.

(6) When an inclination error is found, its origin (i.e. compaction, internal deformation and overlapping of directions) must be determined by means of geometric techniques. Identifying the source of the inclination error is critical to mitigate possible errors that may also affect declination.

(7) Rotations must be compared with a reliable palaeomagnetic reference direction obtained from the undeformed foreland (absolute VAR) or in the nearest footwall (relative VAR). Multiple rotations must be taken into account.

Research funding provided by the project DR3AM (CGL2014-54118-) from the Spanish National Research Program is acknowledged. We are indebted to John Geissman and Valerian Bachtadse for their constructive comments and to the editor Randell Stephenson, who helped us improve the original version of the paper.

## References

ACHACHE, J., COURTILLOT, V. & BESSE, J. 1983. Paleomagnetic constraints on the late Cretaceous and Cenozoic tectonics of southeastern Asia. *Earth and Planetary Science Letters*, **63**, 123–136.

ADAM, J., KLINKMÜLLER, M., SCHREURS, G. & WIENEKE, B. 2013. Quantitative 3D strain analysis in analogue experiments simulating tectonic deformation: integration of X-ray computed tomography and digital volume correlation techniques. *Journal of Structural Geology*, **55**, 127–149.

ALLERTON, S. 1994. Vertical-axis rotation associated with folding and thrusting: an example from the eastern Subbetic zone of southern Spain. *Geology*, **22**, 1039–1042.

ALLERTON, S. 1998. Geometry and kinematics of vertical-axis rotations in fold and thrust belts. *Tectonophysics*, **299**, 15–30.

ARRIAGADA, C., ROPERCH, P., MPODOZIS, C. & FERNANDEZ, R. 2006. Paleomagnetism and tectonics of the southern Atacama Desert (25–28 S), northern Chile. *Tectonics*, **25**, TC4001–TC4026.

ARRIAGADA, C., ROPERCH, P., MPODOZIS, C. & COBBOLD, P.R. 2008. Paleogene building of the Bolivian Orocline: tectonic restoration of the central Andes in 2-D map view. *Tectonics*, **27**, TC6014–TC6027.

AUBOURG, C., SMITH, B., BAKHTARI, H.R., GUYA, N. & ESHRAGHI, A. 2008. Tertiary block rotations in the Fars arc (Zagros, Iran). *Geophysical Journal International*, **173**, 659–673.

AUBOURG, C., POZZI, J.-P. & KARS, M. 2012. Burial, claystones remagnetization and some consequences for magnetostratigraphy. *In*: ELMORE, R.D., MUXWORTHY, A.R., ALDANA, M.M. & MENA, M. (eds) *Remagnetization and Chemical Alteration of Sedimentary Rocks*. Geological Society, London, Special Publications, **371**, 181–188, http://doi.org/10.1144/SP371.4

BAILEY, R.C. & HALLS, H.C. 1984. Estimate of confidence in paleomagnetic directions derived from mixed remagnetization circle and direct observational data. *Journal of Geophysics*, **54**, 174–182.

BATES, M.P. 1989. Palaeomagnetic evidence for rotations and deformation in the Nogueras Zone, central southern Pyrenees, Spain. *Journal of the Geological Society, London*, **146**, 459–476, http://doi.org/10.1144/gsjgs.146.3.0459

BAYONA, G., THOMAS, W.A. & VAN DER VOO, R. 2003. Kinematics of thrust sheets within transverse zones; a structural and paleomagnetic investigation in the Appalachian thrust belt of Georgia and Alabama. *Journal of Structural Geology*, **25**, 1193–1212.

BAZHENOV, M.L. 1988. Analysis of the resolution of the paleomagnetic method in solving tectonic problems. *Geotectonics*, **22**, 204–212.

BAZHENOV, M.L. & SHIPUNOV, S.V. 1991. Fold test in paleomagnetism: new approaches and reappraisal of data. *Earth and Planetary Science Letters*, **104**, 16–24.

BAZHENOV, M.L., BURTMAN, V.S. & DVOROVA, A.V. 1999. Permian paleomagnetism of the Tien Shan fold belt, Central Asia: post-collisional rotations and deformation. *Tectonophysics*, **312**, 303–329.

BEAMUD, E., MUÑOZ, J.A., FITZGERALD, P.G., BALDWIN, S.L., GARCÉS, M., CABRERA, L. & METCALF, J.R. 2011. Magnetostratigraphy and detrital apatite fission track thermochronology in syntectonic conglomerates: constraints on the exhumation of the South-Central Pyrenees. *Basin Research*, **23**, 309–331.

BECK, M.E. 1980. Paleomagnetic record of plate-margin tectonic processes along the western edge of North America. *Journal of Geophysical Research: Solid Earth (1978–2012)*, **85**, 7115–7131.

BILARDELLO, D. & KODAMA, K.P. 2010. A new inclination shallowing correction of the Mauch Chunk Formation of Pennsylvania, based on high-field AIR results: implications for the Carboniferous North American APW path and Pangea reconstructions. *Earth and Planetary Science Letters*, **299**, 218–227.

BINGHAM, C. 1974. An antipodally symmetric distribution on the sphere. *Annals of Statistics*, **2**, 1201–1225.

BLUMSTEIN, A.M., ELMORE, R.D., ENGEL, M.H., ELLIOT, C. & BASU, A. 2004. Paleomagnetic dating of burial diagenesis in Mississippian carbonates, Utah. *Journal of Geophysical Research: Solid Earth (1978–2012)*, **109**, B04101–B04116.

BONHOMMET, N., COBBOLD, P.R., PERROUD, H. & RICHARDSON, A. 1981. Paleomagnetism and cross-folding in a key area of the Asturian Arc (Spain). *Journal of Geophysical Research: Solid Earth (1978–2012)*, **86**, 1873–1887.

BORRADAILE, G.J. 1997. Deformation and paleomagnetism. *Surveys in Geophysics*, **18**, 405–435.

BOURGEOIS, O., COBBOLD, P.R., ROUBY, D., THOMAS, J.C. & SHEIN, V. 1997. Least squares restoration of Tertiary thrust sheets in map view, Tajik Depression, Central Asia. *Journal of Geophysical Research, B, Solid Earth and Planets*, **102**, 27553–27573.

BURBANK, D.W., VERGÉS, J., MUÑOZ, J.A. & BENTHAM, P. 1992. Coeval hindward-and forward-imbricating thrusting in the south-central Pyrenees, Spain: timing and rates of shortening and deposition. *Geological Society of America Bulletin*, **104**, 3–17.

BUTLER, R.F. 1992. *Paleomagnetism: Magnetic Domains to Geologic Terranes*. Blackwell, Boston, MA.

BUTLER, R.F., RICHARDS, D.R., SEMPERE, T. & MARSHALL, L.G. 1995. Paleomagnetic determinations of vertical-axis tectonic rotations from Late Cretaceous and Paleocene strata of Bolivia. *Geology*, **23**, 799–802.

CAIRANNE, C., AUBOURG, C. & POZZI, J.P. 2002. Synfolding remagnetization and the significance of the small circle test: examples from the Vocontian trough (SE France). *Physics and Chemistry of the Earth*, **27**, 1151–1159.

CARDELLO, G.L., ALMQVIST, B.S.G., HIRT, A.M. & MANCKTELOW, N.S. 2015. Determining the timing of formation of the Rawil Depression in the Helvetic Alps by paleomagnetic and structural methods. *In*: PUEYO, E.L., CIFELLI, F., SUSSMAN, A.J. & OLIVA-URCIA, B. (eds) *Palaeomagnetism in Fold and Thrust Belts: New Perspectives*. Geological Society, London, Special Publications, **425**. First published online July 22, 2015, updated August 18, 2015, http://doi.org/10.1144/SP425.4

CARICCHI, C., CIFELLI, F., SAGNOTTI, L., SANI, F., SPERANZA, F. & MATTEI, M. 2014. Paleomagnetic evidence for a post-Eocene 90° CCW rotation of internal Apennine units: a linkage with Corsica-Sardinia rotation? *Tectonics*, **33**, 374–392.

CHAN, L.S. 1988. Apparent tectonic rotations, declination anomaly equations and declination anomaly charts. *Journal of Geophysical Research*, **93**, 12151–12158.

CHANNELL, J.E.T., OLDOW, J.S., CATALANO, R. & D'ARGENIO, B. 1990. Paleomagnetically determined rotations in the western Sicilian fold and thrust belt. *Tectonics*, **9**, 641–660.

CIFELLI, F., MATTEI, M. & ROSSETTI, F. 2007. Tectonic evolution of arcuate mountain belts on top of a retreating subduction slab: the example of the Calabrian Arc. *Journal of Geophysical Research*, **112**, B09101–B09120.

CIFELLI, F., MATTEI, M. & PORRECA, M. 2008*a*. New paleomagnetic data from Oligocene-upper Miocene sediments in the Rif chain (northern Morocco): insights on the Neogene tectonic evolution of the Gibraltar Arc. *Journal of Geophysical Research*, B02104, http://doi.org/10.1029/2007JB005271

CIFELLI, F., MATTEI, M. & DELLA SETA, M. 2008*b*. Calabrian Arc oroclinal bending: the role of subduction. *Tectonics*, **27**, TC5001–TC5015, http://doi.org/10.1029/2008TC002272

ÇINKU, M.C., HISARLI, Z.M. *ET AL.* 2013. Evidence of Early Cretaceous remagnetization in the Crimean Peninsula: a palaeomagnetic study from Mesozoic rocks in the Crimean and Western Pontides, conjugate margins of the Western Black Sea. *Geophysical Journal International*, **195**, 821–843.

ÇINKU, M.C., HISARLI, M. *ET AL.* 2015. Evidence of Late Cretaceous oroclinal bending in north–central Anatolia: palaeomagnetic results from Mesozoic and Cenozoic rocks along the İzmir–Ankara–Erzincan Suture Zone. *In*: PUEYO, E.L., CIFELLI, F., SUSSMAN, A.J. & OLIVA-URCIA, B. (eds) *Palaeomagnetism in Fold and Thrust Belts: New Perspectives*. Geological Society, London, Special Publications, **425**. First published online August 3, 2015, updated August 14, 2015, http://doi.org/10.1144/SP425.2

COGNÉ, J.P. 1987. Paleomagnetic direction obtained by strain removal in the Pyrenean Permian redbeds at the 'Col du Somport' (France). *Earth and Planetary Science Letters*, **85**, 162–172.

COLLOMBET, M., THOMAS, J.C., CHAUVIN, A., TRICART, P., BOUILLIN, J.P. & GRATIER, J.P. 2002. Counterclockwise rotation of the Western Alps since the Oligocene; new insights from paleomagnetic data. *Tectonics*, **21**, 1401–1415.

CONDER, J., BUTLER, R.F., DeCELLES, P.G. & CONSTENIUS, K. 2003. Paleomagnetic determination of vertical-axis rotations within the Charleston–Nebo salient, Utah. *Geology*, **31**, 1113–1116.

CREER, K.M. 1962. A statistical enquiry into the partial remagnetization of folded Old Red Sandstone rocks. *Journal of Geophysical Research*, **67**, 1899–1906.

CROUZET, C., GAUTAM, P., SCHILL, E. & APPEL, E. 2003. Multicomponent magnetization in western Dolpo (Tethyan Himalaya, Nepal); tectonic implications. *Tectonophysics*, **377**, 179–196.

CULLEN, A.B., ZECHMEISTER, M.S., ELMORE, R.D. & PANNALAL, S.J. 2012. Paleomagnetism of the Crocker Formation, northwest Borneo: implications for late Cenozoic tectonics. *Geosphere*, **8**, 1146–1169.

CUMMINS, W.A. 1964. Current directions from folded strata. *Geological Magazine*, **101/2**, 169–173.

CUMMINS, W.A. 1966. Conical folding and sedimentary lineations. *Geological Magazine*, **103**, 197–203.

DELAUNAY, S., SMITH, B. & AUBOURG, C. 2002. Asymmetrical fold test in the case of overfolding: two examples from the Makran accretionary prism (Southern Iran). *Physics and Chemistry of the Earth*, **27**, 1195–1203.

DEMAREST, H.H. 1983. Error analysis for the determination of tectonic rotation from paleomagnetic data. *Journal of Geophysical Research: Solid Earth (1978–2012)*, **88**, 4321–4328.

DUERMEIJER, C.E., NYST, M., MEIJER, P.T., LANGEREIS, C.G. & SPAKMAN, W. 2000. Neogene evolution of the Aegean arc: paleomagnetic and geodetic evidence for a rapid and young rotation phase. *Earth and Planetary Science Letters*, **176**, 509–525.

DUPONT-NIVET, G., BUTLER, R.F., YIN, A. & CHEN, X. 2002. Paleomagnetism indicates no Neogene rotation of the Qaidam Basin in northern Tibet during Indo-Asian collision. *Geology*, **30**, 263–266.

DUPONT-NIVET, G., VASILIEV, I., LANGEREIS, C.G., KRIJGSMAN, W. & PANAIOTU, C. 2005. Neogene tectonic evolution of the Southern and Eastern Carpathians constrained by paleomagnetism. *Earth and Planetary Science Letters*, **236**, 374–387.

ELDREDGE, S. & VAN DER VOO, R. 1988. Paleomagnetic study of thrust sheet rotations in the Helena and Wyoming salients of the northern Rocky Mountains. *Geological Society of America Memoirs*, **171**, 319–332.

ELDREDGE, S., BACHTADSE, V. & VAN DER VOO, R. 1985. Paleomagnetism and the orocline hypothesis. *Tectonophysics*, **119**, 153–179.

ELMORE, R.D., KELLEY, J., EVANS, M. & LEWCHUK, M.T. 2001. Remagnetization and orogenic fluids: testing the hypothesis in the central Appalachians. *Geophysical Journal International*, **144**, 568–576.

ELMORE, R.D., MUXWORTHY, A.R. & ALDANA, M. 2012. Remagnetization and chemical alteration of sedimentary rocks. *In*: ELMORE, R.D., MUXWORTHY, A.R., ALDANA, M.M. & MENA, M. (eds) *Remagnetization and Chemical Alteration of Sedimentary Rocks*. Geological Society, London, Special Publications, **371**, 1–21, http://doi.org/10.1144/SP371.15

ENKIN, R.J. 2003. The direction-correction tilt test: an all-purpose tilt/fold test for paleomagnetic studies. *Earth and Planetary Science Letters*, **212**, 151–166.

EVANS, M.A., ELMORE, R.D. & LEWCHUK, M.T. 2000. Examining the relationship between remagnetization and orogenic fluids: central Appalachians. *Journal of Geochemical Exploration*, **69**, 139–142.

FACER, R.A. 1983. Folding, strain and Graham's fold test in palaeomagnetic investigations. *Geophysical Journal International*, **72**, 165–171.

FISHER, R.A. 1953. Dispersion on a sphere. *Proceedings of the Royal Society, London*, **A217**, 295–305.

GRAHAM, J.W. 1949. The stability and significance of magnetism in sedimentary rocks. *Journal of Geophysical Research*, **54**, 131–167.

GRUBBS, K.L. & VAN DER VOO, R. 1976. Structural deformation of the Idaho-Wyoming overthrust belt (U.S.A.), as determined by Triassic paleomagnetism. *Tectonophysics*, **33**, 321–336.

HALLS, H.C. 1976. A least-squares method to find a remanence direction from converging remagnetization circles. *Geophysical Journal International*, **45**, 297–304.

HALLS, H.C. 1978. The use of converging remagnetization circles in paleomagnetism. *Physics of the Earth and Planetary Interiors*, **16**, 1–11.

HARLAN, S.S., GEISSMAN, J.W., WHISNER, S.C. & SCHMIDT, C.J. 2008. Paleomagnetism and geochronology of sills of the Doherty Mountain area, southwestern Montana; implications for the timing of fold-and-thrust belt deformation and vertical-axis rotations along the southern margin of the Helena Salient. *Geological Society of America Bulletin*, **120**, 1091–1104.

HINDLE, D. & BURKHARD, M. 1999. Strain, displacement and rotation associated with the formation of curvature in fold belts; the example of the Jura arc. *Journal of Structural Geology*, **21**, 1089–1101.

HIRT, A.M., LOWRIE, W., JULIVERT, M. & ARBOLEYA, M.L. 1992. Paleomagnetic results in support of a model for the origin of the Asturian Arc. *Tectonophysics*, **213**, 321–339.

HNAT, J.S., VAN DER PLUIJM, B.A., VAN DER VOO, R. & THOMAS, W.A. 2008. Differential displacement and rotation in thrust fronts: a magnetic, calcite twinning and palinspastic study of the Jones Valley thrust, Alabama, US Appalachians. *Journal of Structural Geology*, **30**, 725–738.

HOSPERS, J. & VAN ANDEL, S.I. 1969. Palaeomagnetism and tectonics, a review. *Earth-Science Reviews*, **5**, 5–44.

IRVING, E. 2005. The role of latitude in mobilism debate. *Proceedings of the National Academy of Science, USA*, **102**, 1821–1828.

IZQUIERDO-LLAVALL, E., SAINZ, A.C., OLIVA-URCIA, B., BURMESTER, R., PUEYO, E.L. & HOUSEN, B. 2015. Multi-episodic remagnetization related to deformation in the Pyrenean Internal Sierras. *Geophysical Journal International*, **201**, 891–914.

JAPAS, M.S., RÉ, G.H., ORIOLO, S. & VILAS, J.F. 2015. Palaeomagnetic data from the Precordillera fold and thrust belt constraining Neogene foreland evolution of the Pampean flat-slab segment (Central Andes, Argentina). *In*: PUEYO, E.L., CIFELLI, F., SUSSMAN, A.J. & OLIVA-URCIA, B. (eds) *Palaeomagnetism in Fold and Thrust Belts: New Perspectives*. Geological Society, London, Special Publications, **425**. First published online August 18, 2015, http://doi.org/10.1144/SP425.9

KHRAMOV, A.N. 1958. *Paleomagnetism and Stratigraphic Correlation*. Gostoptechizdat, Leningrad.

KIRSCHER, U., ZWING, A., ALEXEIEV, D.V., ECHTLER, H.P. & BACHTADSE, V. 2013. Paleomagnetism of Paleozoic sedimentary rocks from the Karatau Range, Southern Kazakhstan: multiple remagnetization events correlate with phases of deformation. *Journal of Geophysical Research: Solid Earth*, **118**, 3871–3885.

KIRSCHVINK, J.L. 1980. The least-squares line and plane and the analysis of the paleomagnetic data. *Geophysical Journal of the Royal Astronomical Society*, **62**, 699–718.

KISSEL, C. & LAJ, C. 1989. *Paleomagnetic Rotations and Continental Deformation*. NATO ASI Science Series C, **254**. Kluwer Academic Publishers, Dordrecht.

KISSEL, C., AVERBUCH, O., DE LAMOTTE, D.F., MONOD, O. & ALLERTON, S. 1993. First paleomagnetic evidence for a post-Eocene clockwise rotation of the Western Taurides thrust belt east of the Isparta reentrant (Southwestern Turkey). *Earth and Planetary Science Letters*, **117**, 1–14.

KODAMA, K.P. 1988. Remanence rotation due to rock strain during folding and the stepwise application of the fold test. *Journal of Geophysical Research, B, Solid Earth and Planets*, **93**, 3357–3371.

KODAMA, K.P. 1997. A successful rock magnetic technique for correctng paleomagnetic inclination shallowing: case study of the nacimiento formation, New Mexico. *Journal of Geophysical Research: Solid Earth (1978–2012)*, **102**, 5193–5205.

LOWRIE, W. 1990. Identification of ferromagnetic minerals in a rock by coercitivity and un-blocking temperature properties. *Geophysics Research Letters*, **17**, 159–162.

LOWRIE, W., HIRT, A.M. & KLIGFIELD, R. 1986. Effects of tectonic deformation on the remanent magnetization of rocks. *Tectonics*, **5**, 713–722.

MACDONALD, W.D. 1980. Net tectonic rotation, apparent tectonic rotation and the structural tilt correction in paleomagnetism studies. *Journal of Geophysical Research*, **85**, 3659–3669.

MACFADDEN, B.J., ANAYA, F. & SWISHER, C.C. 1995. Neogene paleomagnetism and oroclinal bending of the central Andes of Bolivia. *Journal of Geophysical Research: Solid Earth*, **100**, 8153–8167.

MÁRTON, E., TISCHLER, M., CSONTOS, L., FUEGENSCHUH, B. & SCHMID, S.M. 2007. The contact zone between the ALCAPA and Tisza-Dacia megatectonic units of northern Romania in the light of new paleomagnetic data. *Swiss Journal of Geosciences*, **100**, 109–124.

MÁRTON, E., GRABOWSKI, J., TOKARSKI, A.K. & TÚNYI, I. 2015. Palaeomagnetic results from the fold and thrust belt of the Western Carpathians: an overview. *In*: PUEYO, E.L., CIFELLI, F., SUSSMAN, A.J. & OLIVA-URCIA, B. (eds) *Palaeomagnetism in Fold and Thrust Belts: New Perspectives*. Geological Society, London, Special Publications, **425**. First published online August 12, 2015, http://doi.org/10.1144/SP425.1

MATTEI, M., FUNICIELLO, R. & KISSEL, C. 1995. Paleomagnetic and structural evidence for Neogene block rotations in the Central Apennines, Italy. *Journal of Geophysical Research: Solid Earth (1978–2012)*, **100**, 17863–17883.

MATTEI, M., PETROCELLI, V., LACAVA, D. & SCHIATTARELLA, M. 2004. Geodynamic implications of Pleistocene ultrarapid vertical-axis rotations in the Southern Apennines, Italy. *Geology*, **32**, 789–792.

MATTEI, M., CIFELLI, F., ROJAS, I.M., CRESPO-BLANC, A., COMAS, M., FACCENNA, C. & PORRECA, M. 2006. Neogene tectonic evolution of the Gibraltar Arc; new paleomagnetic constrains from the Betic chain. *Earth and Planetary Science Letters*, **250**, 522–540.

McCAIG, A.M. & McCLELLAND, E. 1992. Palaeomagnetic techniques applied to thrust belts. *In*: McCLAY, K.R. (ed.) *Thrust Tectonics*. Chapman and Hall, London, 209–216.

McELHINNY, M.W. 1964. Statistical significance of the fold test in palaeomagnetism. *Geophysical Journal of the Royal Astronomical Society*, **8**, 338–340.

McFADDEN, P.L. 1990. A new fold test for palaeomagnetic studies. *Geophysical Journal International.*, **103**, 163–169.

McFADDEN, P.L. 1998. The fold test as an analytical tool. *Geophysical Journal International*, **135**, 329–338.

McFADDEN, P.L. & JONES, D.L. 1981. The fold test in palaeomagnetism. *Geophysical Journal of the Royal Astronomical Society*, **67**, 53–58.

McFADDEN, P.L. & McELHINNY, M.W. 1988. The combined analysis of remagnetization circles and direct observations in paleomagnetism. *EPSL*, **87**, 161–172.

MEERT, J.G. 2009. Palaeomagnetism: in GAD we trust. *Nature Geoscience*, **2**, 673–674.

MOCHALES, T. & BLENKINSOP, T.G. 2014. Representation of paleomagnetic data in virtual globes: a case study from the Pyrenees. *Computers & Geosciences*, **70**, 56–62.

MOCHALES, T., CASAS, A.M., PUEYO, E.L. & BARNOLAS, A. 2012. Rotational velocity for oblique structures (Boltaña anticline, Southern Pyrenees). *Journal of Structural Geology*, **35**, 2–16.

MOCHALES, T., PUEYO, E.L., CASAS, A.M. & BARNOLAS, A. 2016. Restoring paleomagnetic data in complex superposed folding settings: the Boltaña anticline (Southern Pyrenees). *Tectonophysics*, **671**, 281–298, http://doi.org/10.1016/j.tecto.2016.01.008

MOUSSAID, B., VILLALAÍN, J.J., CASAS-SAINZ, A., EL OUARDI, H., OLIVA-URCIA, B., SOTO, R. & TORRES-LÓPEZ, S. 2015. Primary v. secondary curved fold axes: deciphering the origin of the Aït Attab syncline (Moroccan High Atlas) using paleomagnetic data. *Journal of Structural Geology*, **70**, 65–77.

MUÑOZ, J.A., BEAMUD, E., FERNÁNDEZ, O., ARBUÉS, P., DINARÈS-TURELL, J. & POBLET, J. 2013. The Ainsa Fold and thrust oblique zone of the central Pyrenees: kinematics of a curved contractional system from paleomagnetic and structural data. *Tectonics*, **32**, 1142–1175.

MUTTONI, G., ARGNANI, A., KENT, D.V., ABRAHAMSEN, N. & CIBIN, U. 1998. Paleomagnetic evidence for Neogene tectonic rotations in the northern Apennines, Italy. *Earth and Planetary Science Letters*, **154**, 25–40.

NEMKIN, S.R., FITZ-DÍAZ, E., VAN DER PLUIJM, B. & VAN DER VOO, R. 2015. Dating synfolding remagnetization: approach and field application (central Sierra Madre Oriental, Mexico). *Geosphere*, **11**, 1617–1628.

NICOL, A., MAZENGARB, C., CHANIER, F., RAIT, G., URUSKI, C. & WALLACE, L. 2007. Tectonic evolution of the active Hikurangi subduction margin, New Zealand, since the Oligocene. *Tectonics*, **26**, 1–5.

NORMAN, T.N. 1960. Azimuths of primary linear structures in folded strata. *Geological Magazine*, **97**, 338–343.

NORRIS, D.K. & BLACK, R.F. 1961. Application of palaeomagnetism to thrust mechanics. *Nature (London)*, **192**, 933–935.

NUR, A., RON, H. & SCOTTI, O. 1986. Fault mechanics and the kinematics of block rotations. *Geology*, **14**, 746–749.

OLIVA-URCIA, B. & PUEYO, E.L. 2007a. Basement kinematic constrains from paleomagnetic data of remagnetized cover rocks (Internal Sierras, Southwestern Pyrenees). *Tectonics*, **26**, TC4014–TC4036.

OLIVA-URCIA, B. & PUEYO, E.L. 2007b. Gradient of shortening and vertical-axis rotations in the Southern Pyrenees (Spain), insights from a synthesis of paleomagnetic data. *Revista de la Sociedad Geológica de España*, **20**, 105–118.

OLIVA-URCIA, B., CASAS, A.M., PUEYO, E.L., ROMÁN-BERDIEL, T. & GEISSMAN, J.W. 2010. Paleomagnetic evidence for dextral strike-slip motion in the Pyrenees during the Alpine convergence (Mauléon basin, France). *Tectonophysics*, **494**, 165–179, http://doi.org/10.1016/j.tecto.2010.09.018

OLIVA-URCIA, B., CASAS, A.M., PUEYO, E.L. & POCOVÍ, A. 2012a. Structural and paleomagnetic evidence for non-rotational kinematics in the western termination of the External Sierras (southwestern central Pyrenees). *Geologica Acta*, **10**, 1–22.

OLIVA-URCIA, B., PUEYO, E.L., LARRASOAÑA, J.C., CASAS, A.M., ROMÁN-BERDIEL, T., VAN DER VOO, R. & SCHOLGER, R. 2012b. New and revisited paleomagnetic data from Permian-Triassic red beds: Two kinematic domains in the West-Central Pyrenees. *Tectonophysics*, **522–523**, 158–175, http://doi.org/10.1016/j.tecto.2011.11.023

OLIVA-URCIA, B., BEAMUD, E., GARCÉS, M., ARENAS, C., SOTO, R., PUEYO, E.L. & PARDO, G. 2015. New magnetostratigraphic dating of the Palaeogene syntectonic sediments of the west-central Pyrenees: tectonostratigraphic implications. *In*: PUEYO, E.L., CIFELLI, F., SUSSMAN, A.J. & OLIVA-URCIA, B. (eds) *Palaeomagnetism in Fold and Thrust Belts: New Perspectives*. Geological Society, London, Special Publications, **425**. First published online August 3, 2015, updated August 21, 2015, http://doi.org/10.1144/SP425.5

ONG, P.F., VAN DER PLUIJM, B.A. & VAN DER VOO, R. 2007. Early rotation and late folding in the Pennsylvania Salient (U.S. Appalachians); evidence from calcite-twinning analysis of Paleozoic carbonates. *Geological Society of America Bulletin*, **119**, 796–804.

OPDYKE, M.D. 1995. Paleomagnetism, polar wondering, and the rejuvenation of crustal mobility. *Journal of Geophysical Research*, **100**, B24361–24366.

OPDYKE, M.D. & CHANNELL, J.E. 1996. *Magnetic Stratigraphy*. Academic Press, San Diego.

OTOFUJI, Y.I., MATSUDA, T. & NOHDA, S. 1985. Paleomagnetic evidence for the Miocene counter-clockwise rotation of Northeast Japan – rifting process of the Japan Arc. *Earth and Planetary Science Letters*, **75**, 265–277.

PASTOR-GALÁN, D., MULCHRONE, K.F., KOYMANS, M.R., VAN HINSBERGEN, D.J.J. & LANGEREIS, C.G. 2016. Total Least Squares Orocline test: A robust method to quantify vertical axis rotation patterns in orogens, with examples from the Cantabrian and Aegean oroclines. *Lithosphere* (in press).

PAVÓN-CARRASCO, F.J., OSETE, M.L. & TORTA, J.M. 2010. Regional modeling of the geomagnetic field in Europe from 6000 to 1000 BC. *Geochemistry, Geophysics, Geosystems*, **11**, 1–20, http://doi.org/10.1029/2010GC003197

PERROUD, H. 1982. Relations paléomagnétisme et déformation: exemple de la région de Cabo de Penas (Espagne). *Comptes Rendus de l'Académie des Sciences Paris*, **294**, 45–48.

PLATT, J.P., ALLERTON, S., KIRKER, A., MANDEVILLE, C., MAYFIELD, A., PLATZMAN, E.S. & RIMI, A. 2003. The ultimate arc: differential displacement, oroclinal bending, and vertical axis rotation in the External Betic–Rif arc. *Tectonics*, **22**, 201–231.

POWERS, P.M., LILLIE, R.J. & YEATS, R.S. 1998. Structure and shortening of the Kangra and Dehra Dun reentrants, sub-Himalaya, India. *Geological Society of America Bulletin*, **110**, 1010–1027.

PREZZI, C., CAFFE, P.J. & SOMOZA, R. 2004. New paleomagnetic data from the northern Puna and western Cordillera Oriental, Argentina; a new insight on the timing of rotational deformation. *Journal of Geodynamics*, **38**, 93–115.

PUEYO, E.L. 2000. *Rotaciones paleomagnéticas en sistemas de pliegues y cabalgamientos. Tipos, causas, significado y aplicaciones (ejemplos del Pirineo Aragonés)*. Unpublished PhD thesis, Universidad de Zaragoza.

PUEYO, E.L., POCOVÍ, A. & PARÉS, J.M. 2002a. Flexural folding of linear (paleomagnetic) data by non-coaxial axes of deformation: restoration and errors. *Mitteilungen Naturwissenschaftlicher Verein für Steiermark*, **132**, 35–38.

PUEYO, E.L., MILLÁN, H. & POCOVÍ, A. 2002b. Rotation velocity of a thrust: a paleomagnetic study in the External Sierras (Southern Pyrenees). *Sedimentary Geology*, **146**, 191–208.

PUEYO, E.L., PARÉS, J.M., MILLÁN, H. & POCOVÍ, A. 2003a. Conical folds and apparent rotations in paleomagnetism (A case studied in the Southern Pyrenees). *Tectonophysics*, **362**, 345–366.

PUEYO, E.L., POCOVÍ, A., PARÉS, J.M., MILLÁN, H. & LARRASOAÑA, J.C. 2003b. Thrust ramp geometry and spurious rotations of paleomagnetic vectors. *Studia Geophysica Geodetica*, **47**, 331–357.

PUEYO, E.L., POCOVÍ, A., MILLÁN, H. & SUSSMAN, A. 2004. Map-view models for correcting and calculating shortening estimates in rotated thrust fronts using paleomagnetic data. *In*: WEIL, A. & SUSSMAN, A. (eds) *Orogenic Curvature: Integrating Paleomagnetic and Structural Analyses*. Geological Society of America, Special Papers, **383**, 57–71.

PUEYO, E.L., MAURITSCH, H.J., GAWLICK, H.J., SCHOLGER, R. & FRISCH, W. 2007. New evidence for block and thrust sheet rotations in the Central Northern Calcareous Alps deduced from two pervasive remagnetization events. *Tectonics*, **26**, TC5011–TC5036. http://doi.org/10.1029/2006TC0 01965

PUEYO-ANCHUELA, O., PUEYO, E.L., POCOVÍ, A. & GIL-IMAZ, A. 2012. Vertical axis rotations in fold and thrust belts: comparison of AMS and paleomagnetic data in the Western External Sierras (Southern Pyrenees). *Tectonophysics*, **532–535**, 119–133.

RAMÓN, M.J. 2013. *Flexural unfolding of complex geometries in fold and thrust belts using paleomagnetic vectors*. Unpublished PhD, University of Zaragoza, http://zaguan.unizar.es/record/11750

RAMÓN, M.J., PUEYO, E.L., BRIZ, J.L., POCOVÍ, A. & CIRIA, J.C. 2012. Flexural unfolding in 3D using paleomagnetic vectors. *Journal of Structural Geology*, **35**, 28–39, http://doi.org/10.1016/j.jsg.2011.11.015

RAMÓN, M.J., BRIZ, J.L., PUEYO, E.L. & FERNÁNDEZ, O. 2015a. Horizon restoration by best fitting of finite elements and rotation constraints: sensitivity to the mesh geometry and pin-element location. *Mathematical Geosciences*, **48**, 419–437.

RAMÓN, M.J., PUEYO, E.L., CAUMON, G. & BRIZ, J.L. 2015b. Parametric unfolding of flexural folds using palaeomagnetic vectors. *In*: PUEYO, E.L., CIFELLI, F., SUSSMAN, A.J. & OLIVA-URCIA, B. (eds) *Palaeomagnetism in Fold and Thrust Belts: New Perspectives*. Geological Society, London, Special Publications, **425**. First published online August 12, 2015, http://doi.org/10.1144/SP425.6

RAMSAY, J.G. 1960. The deformation of early linear structures in areas of repeated folding. *Journal of Geology*, **68**, 75–93.

RAMSAY, J.G. 1961. The effects of folding upon the orientation of sedimentation structures. *Journal of Geology*, **69**, 84–100.

RAPALINI, A.E. 2007. A paleomagnetic analysis of the patagonian orocline. *Geologica Acta*, **5**, 287–294.

RAPALINI, A.E., PERONI, J. ET AL. 2015. Palaeomagnetism of Mesozoic magmatic bodies of the Fuegian Cordillera: implications for the formation of the Patagonian Orocline. *In*: PUEYO, E.L., CIFELLI, F., SUSSMAN, A.J. & OLIVA-URCIA, B. (eds) *Palaeomagnetism in Fold and Thrust Belts: New Perspectives*. Geological Society, London, Special Publications, **425**. First published online July 22, 2015, http://doi.org/10.1144/SP425.3

RIBEIRO, P., SILVA, P.F., MOITA, P., KRATINOVA, Z., MARQUES, F.O. & HENRY, B. 2013. Palaeomagnetism in the Sines massif (SW Iberia) revisited: evidences for Late Cretaceous hydrothermal alteration and associated partial remagnetization. *Geophysical Journal International*, **195**, 176–191.

RICHARDS, D.R., BUTLER, R.F. & SEMPERE, T. 2004. Vertical-axis rotations determined from paleomagnetism of Mesozoic and Cenozoic strata of the Bolivian Andes. *Journal of Geophysical Research: Solid Earth*, **109**, B07104.

RODRÍGUEZ-PINTÓ, A., RAMÓN, M.J., OLIVA-URCIA, B., PUEYO, E.L. & POCOVÍ, A. 2011. Errors in paleomagnetism: structural control on overlapped vectors, mathematical models. *Physics of the Earth and Planetary Interiors*, **186**, 11–22.

RODRÍGUEZ-PINTÓ, A., PUEYO, E.L., BARNOLAS, A., POCOVÍ, A., RAMÓN, M.J. & OLIVA-URCIA, B. 2013. Overlapped paleomagnetic vectors and fold geometry: a case study in the Balzes anticline (Southern Pyrenees). *Physics of the Earth and Planetary Interiors*, **215**, 43–57. http://doi.org/10.1016/j.pepi.2012.10.005

RODRÍGUEZ-PINTÓ, A., PUEYO, E.L.M. *ET AL.* 2016. Rotational kinematics of a curved fold: a structural and paleomagnetic study in the Balzes anticline (Southern Pyrenees). *Tectonophysics*, **677–678**, 171–189.

RON, H., FREUND, R., GARFUNKEL, Z. & NUR, A. 1984. Block rotation by strike-slip faulting: structural and paleomagnetic evidence. *Journal of Geophysical Research: Solid Earth (1978–2012)*, **89**, 6256–6270.

ROPERCH, P. & CARLIER, G. 1992. Paleomagnetism of Mesozoic rocks from the central Andes of southern Peru: importance of rotations in the development of the Bolivian orocline. *Journal of Geophysical Research: Solid Earth (1978–2012)*, **97**, 17233–17249.

ROUVIER, H., HENRY, B. & LE GOFF, M. 2012. Mise en évidence par le paléomagnétisme de rotations régionales dans la virgation des Corbières (France). *Bulletin de la Société Geologique de France*, **183**, 409–424.

ROUSSE, S., GILDER, S., FORNARI, M. & SEMPERE, T. 2005. Insight into the Neogene tectonic history of the northern Bolivian Orocline from new paleomagnetic and geochronologic data. *Tectonics*, **24**, 1–23.

SATOLLI, S., SPERANZA, F. & CALAMITA, F. 2005. Paleomagnetism of the Gran Sasso range salient (central Apennines, Italy): pattern of orogenic rotations due to translation of a massive carbonate indenter. *Tectonics*, **24**, 1–22.

SCHEEPERS, P.J.J. & ZIJDERVELD, J.D.A. 1992. Stacking in Paleomagnetism: application to marine sediments with weak NRM. *Geophysical Research Letters*, **1914**, 1519–1522.

SCHEIDEGGER, A.M. 1965. On the statistics of the orientation of bedding planes, grain axes, and similar sedimentological data. *US Geological Survey Professional Papers*, **525C**, 164–167.

SCHILL, E., CROUZET, C., GAUTAM, P., SINGH, V.K. & APPEL, E. 2002. Where did rotational shortening occur in the Himalayas? Inferences from palaeomagnetic remagnetisations. *Earth and Planetary Science Letters*, **203**, 45–57.

SCHMIDT, P.W. 1982. Linearity spectrum analysis of multi-component magnetizations and its application to some igneous rocks from south-eastern Australia. *Geophysical Journal International*, **70**, 647–665.

SCOTT, G.D. 1984. Anomalous declinations from plunging structures. *EOS Transactions AGU*, **65**, 866.

SEARS, J.W. & HENDRIX, M.S. 2004. Lewis and Clark line and the rotational origin of the Alberta and Helena Salients, North American Cordillera. *In*: SUSSMAN, A.J. & WEIL, A.B. (eds) *Orogenic Curvature: Integrating Paleomagnetic and Structural Analyses*. Geological Society of America Special Papers, **383**, 173–186.

SELLÉS-MARTÍNEZ, J. 1988. Las correcciones estructural y tectónica en el tratamiento de los datos magnéticos. *Geofísica Internacional*, **27-3**, 379–393.

SEMPERE, T., HÉRAIL, G., OLLER, J. & BONHOMME, M.G. 1990. Late Oligocene-early Miocene major tectonic crisis and related basins in Bolivia. *Geology*, **18**, 946–949.

SETIABUDIDAYA, D.J., PIPER, D.A. & SHAW, J. 1994. Paleomagnetism of the (Early Devonian) Lower Old Red sandstones of South Wales: implications to Variscan overprinting and differential rotations. *Tectonophysics*, **231**, 257–280.

SHIPUNOV, S.V. 1997. Synfolding magnetization; detection, testing and geological applications. *Geophysical Journal International*, **130**, 405–410.

SMITH, B., AUBOURG, C., GUEZOU, J.C., NAZARI, H., MOLINARO, M., BRAUD, X. & GUYA, N. 2005. Kinematics of a sigmoidal fold and vertical axis rotation in the east of the Zagros–Makran syntaxis (southern Iran); paleomagnetic, magnetic fabric and microtectonic approaches. *Tectonophysics*, **411**, 89–109.

SONNETTE, L., HUMBERT, F., AUBOURG, C., GATTACCECA, J., LEE, J.C. & ANGELIER, J. 2014. Significant rotations related to cover–substratum decoupling: example of the Dôme de Barrôt (Southwestern Alps, France). *Tectonophysics*, **629**, 275–289.

SOTO, R., CASAS-SAINZ, A.M. & PUEYO, E.L. 2006. Along-strike variation of orogenic wedges associated with vertical axis rotations. *Journal of Geophysical Research (Solid Earth)*, **111**, B10402–B10423.

SOTO, R., VILLALAIN, J.J. & CASAS-SAINZ, A.M. 2008. Remagnetizations as a tool to analyze the tectonic history of inverted sedimentary basins: a case study from the Basque–Cantabrian basin (north Spain). *Tectonics*, **27**, TC1017–33.

SPERANZA, F., SAGNOTTI, L. & MATTEI, M. 1997. Tectonics of the Umbria–Marche–Romagna Arc (central northern Apennines, Italy): new paleomagnetic constraints. *Journal of Geophysical Research: Solid Earth (1978–2012)*, **102**, 3153–3166.

SPERANZA, F., MANISCALCO, R., MATTEI, M., DI STEFANO, A., BUTLER, R.W.H. & FUNICIELLO, R. 1999. Timing and magnitude of rotations in the frontal thrust systems of southwestern Sicily. *Tectonics*, **18**, 1178–1197.

STAMATAKOS, J. & KODAMA, K.P. 1991. Flexural flow folding and the paleomagnetic fold test; an example of strain reorientation of remanence in the Mauch Chunk Formation. *Tectonics*, **10**, 807–819.

STAUFFER, M.R. 1964. The geometry of conical folds. *New Zealand Journal of Geology and Geophysics*, **7**, 340–347.

STEWART, S.A. 1995. Paleomagnetic analysis of plunging fold structures: errors and a simple fold test. *Earth and Planetary Science Letters*, **130**, 57–67.

STEWART, S.A. & JACKSON, K.C. 1995. Palaeomagnetic analysis of fold closure growth and volumetrics. *In*: TURNER, P. & TURNER, A. (eds) *Palaeomagnetic Applications in Hydrocarbon Exploration and*

*Production*. Geological Society, London, Special Publications, **98**, 283–295, http://doi.org/10.1144/GSL.SP.1995.098.01.19

STRAYER, L.M. & SUPPE, J. 2002. Out-of-plane motion of a thrust sheet during along-strike propagation of a thrust ramp: a distinct-element approach. *Journal of Structural Geology*, **24**, 637–650.

SUPPE, J. 1985. *Principles of Structural Geology*. Prentice-Hall, Englewood Cliffs, NJ.

SUSSMAN, A.J. & WEIL, A.B. (eds) 2004. *Orogenic Curvature: Integrating Paleomagnetic and Structural Analyses*. Geological Society of America, Special Papers **383**.

SUSSMAN, A.J., BUTLER, R.F., DINARÈS-TURELL, J. & VERGÉS, J. 2004. Vertical-axis rotation of a foreland fold and implications for orogenic curvature: an example from the Southern Pyrenees, Spain. *Earth and Planetary Science Letters*, **218**, 435–449.

SUSSMAN, A.J., PUEYO, E.L., CHASE, C.G., MITRA, G. & WEIL, A.J. 2012. The impact of vertical-axis rotations on shortening estimates. *Lithosphere*, **4**, 383–394.

TARLING, D.H. 1969. The palaeomagnetic evidence of displacements within continents. *In*: KENT, E.P.E., SATTERTHWAITE, A.M.S. & SPENCER, A.M. (eds) *Time and Place in Orogeny*. Geological Society, London, Special Publications, **3**, 95–113, http://doi.org/10.1144/GSL.SP.1969.003.01.06

TAUXE, L. 1998. *Paleomagnetic Principles and Practice*. Springer Science & Business Media, **1**. Dordrecht, The Netherlands.

TAUXE, L. 2005. Inclination flattening and the geocentric axial dipole hipótesis. *Earth and Planetary Science Letters*, **233**, 247–261, 15.

TAUXE, L. & KENT, D.V. 2004. A simplified statistical model for the geomagnetic field and the detection of shallow bias in paleomagnetic inclinations: was the ancient magnetic field dipolar? *Timescales of the Paleomagnetic Field*. Geophysical Monograph Series, **145**, 101–115.

TAUXE, L. & WATSON, G.S. 1994. The fold test: an eigen analysis approach. *Earth and Planetary Science Letters*, **122**, 331–341.

THOMAS, J.C., CHAUVIN, A., GAPAIS, D., BAZHENOV, M.L., PERROUD, H., COBBOLD, P.R. & BURTMAN, V.S. 1994. Paleomagnetic evidence for Cenozoic block rotations in the Tadjik depression (Central Asia). *Journal of Geophysical Research: Solid Earth (1978–2012)*, **99**, B15141–15160 (Research, 112(B1), B01102).

THÖNY, W., ORTNER, H. & SCHOLGER, R. 2006. Paleomagnetic evidence for large en-bloc rotations in the Eastern Alps during Neogene orogeny. *Tectonophysics*, **414**, 169–189.

VAN ANDEL, S.I. & HOSPERS, J. 1966. Systematic errors in the palaeomagnetic inclination of sedimentary rocks. *Nature (London)*, **212**, 891–893.

VAN DER, & PLUIJM, B.A. 1987. Grain scale deformation and the fold test-evaluation of synfolding remagnetization. *Geophysical Research Letters*, **14**, 155–157.

VAN DER VOO, R. 1990. The reliability of paleomagnetic data. *Tectonophysics*, **184**, 1–9.

VAN DER VOO, R. 2004. Paleomagnetism, oroclines, and growth of the continental crust. *GSA Today*, **14**, 4–9.

VAN DER VOO, R. & CHANNELL, J.E.T. 1980. Paleomagnetism in orogenic belts. *Reviews of Geophysics and Space Physics*, **18**, 455–481.

VAN DER VOO, R. & TORSVIK, T.H. 2001. Evidence for late Paleozoic and Mesozoic non-dipole fields provides an explanation for the Pangea reconstruction problems. *Earth and Planetary Science Letters*, **187**, 71–81.

VAN DER VOO, R. & TORSVIK, T.H. 2012. The history of remagnetization of sedimentary rocks: deceptions, developments and discoveries. *In*: ELMORE, R.D., MUXWORTHY, A.R., ALDANA, M.M. & MENA, M. (eds) *Remagnetization and Chemical Alteration of Sedimentary Rocks*. Geological Society, London, Special Publications, **371**, 23–53, http://doi.org/10.1144/SP371.2

VAN HINSBERGEN, D.J.J., KRIJGSMAN, W., LANGEREIS, C.G., CORNEE, J.J., DUERMEIJER, C.E. & VAN VUGT, N. 2007. Discrete Plio-Pleistocene phases of tilting and counterclockwise rotation in the southeastern Aegean Arc (Rhodos, Greece); early Pliocene formation of the south Aegean left-lateral strike-slip system. *Journal of the Geological Society, London*, **164**, 1133–1144, http://doi.org/10.1144/0016-76492006-061

VILLALAÍN, J.J., FERNÁNDEZ, G., CASAS, A. & GIL IMAZ, A. 2003. Evidence of a Cretaceous remagnetization in the Cameros Basin (North Spain): implications for basin geometry. *Tectonophysics*, **377**, 101–117.

VILLALAÍN, J.J., CASAS-SAINZ, A.M. & SOTO, R. 2015. Reconstruction of inverted sedimentary basins from syn-tectonic remagnetizations. A methodological proposal. *In*: PUEYO, E.L., CIFELLI, F., SUSSMAN, A.J. & OLIVA-URCIA, B. (eds) *Palaeomagnetism in Fold and Thrust Belts: New Perspectives*. Geological Society, London, Special Publications, **425**. First published online October 2, 2015, http://doi.org/10.1144/SP425.10

WALDHÖR, M. 1999. The small-circle reconstruction in palaeomagnetism and its application to palaeomagnetic data from the Pamirs. Tuebinger Geowissenschaftliche Arbeiten. *Reihe A, Geologie, Palaeontologie, Stratigraphie*, **45**, 1–104.

WALDHÖR, M. & APPEL, E. 2006. Intersections of remanence small circles; new tools to improve data processing interpretation in palaeomagnetism. *Geophysical Journal International*, **166**, 33–45.

WEIL, A.B. 2006. Kinematics of orocline tightening in the core of an arc; paleomagnetic analysis of the Ponga Unit, Cantabrian Arc, northern Spain. *Tectonics*, **25**, 1–23.

WEIL, A.B. & SUSSMAN, A. 2004. Classification of curved ororgens based on the timing relationships between structural development and vertical-axis rotations. *In*: SUSSMAN, A.J. & WEIL, A.B. (eds) *Orogenic Curvature: Integrating Paleomagnetic and Structural Analyses*. Geological Society of America, Special Papers, **383**, 1–17.

WEIL, A.B. & VAN DER VOO, R. 2002. The evolution of the paleomagnetic fold test as applied to complex geologic situations, illustrated by a case study from northern Spain. *Physics and Chemistry of the Earth*, **27**, 1223–1235.

WEIL, A.B., VAN DER VOO, R., VAN DER PLUIJM, B.A. & PARÉS, J.M. 2000. The formation of an orocline by multiphase deformation: a paleomagnetic investigation of the Cantabria–Asturias Arc (northern Spain). *Journal of Structural Geology*, **22**, 735–756.

WEIL, A.B., VAN DER VOO, R. & VAN DER PLUIJM, B.A. 2001. Oroclinal bending and evidence against the Pangea megashear; the Cantabria-Asturias Arc (northern Spain). *Geology*, **29**, 991–994.

WEINBERGER, R., AGNON, A., RON, H. & GARFUNKEL, Z. 1995. Rotation about an inclined axis: three dimensional matrices for reconstructing paleomagnetic and structural data. *Journal of Structural Geology*, **17**, 777–782.

WILKERSON, M.S., APOTRIA, T. & FARID, T. 2002. Interpreting the geologic map expression of contractional fault-related fold terminations: lateral/oblique ramps versus displacement gradients. *Journal of Structural Geology*, **24**, 593–607.

YONKEE, A. & WEIL, A.B. 2010. Quantifying vertical axis rotation in curved orogens: correlating multiple data sets with a refined weighted least squares strike test. *Tectonics*, **29**, TC3012–TC3043.

ZOTKEVICH, I.A. 1972. Reduction of the natural remanent magnetization of a plunging fold to the ancient coordinate system in paleomagnetic studies. *Earth Physics*, **2**, 95–99.

# Index

Page numbers in *italics* refer to Figures. Page numbers in **bold** refer to Tables.